Multiphase Flow Analysis Using Population Balance Modeling

Multiphase Flow Analysis Using Population Balance Modeling

Bubbles, Drops and Particles

Guan Heng Yeoh
Chi Pok Cheung
Jiyuan Tu

ELSEVIER

AMSTERDAM • BOSTON • HEIDELBERG • LONDON
NEW YORK • OXFORD • PARIS • SAN DIEGO
SAN FRANCISCO • SINGAPORE • SYDNEY • TOKYO
Butterworth-Heinemann is an imprint of Elsevier

BH

Butterworth-Heinemann is an imprint of Elsevier
The Boulevard, Langford Lane, Kidlington, Oxford, OX5 1GB, UK
225 Wyman Street, Waltham, MA 02451, USA

First published 2014

Notices
Knowledge and best practice in this field are constantly changing. As new research and experience broaden our understanding, changes in research methods, professional practices, or medical treatment may become necessary.

Practitioners and researchers must always rely on their own experience and knowledge in evaluating and using any information, methods, compounds, or experiments described herein. In using such information or methods they should be mindful of their own safety and the safety of others, including parties for whom they have a professional responsibility.

To the fullest extent of the law, neither the Publisher nor the authors, contributors, or editors, assume any liability for any injury and/or damage to persons or property as a matter of products liability, negligence or otherwise, or from any use or operation of any methods, products, instructions, or ideas contained in the material herein.

British Library Cataloguing in Publication Data
A catalogue record for this book is available from the British Library

Library of Congress Cataloguing in Publication Data
A catalog record for this book is available from the Library of Congress

ISBN: 978-0-08-098229-8

For information on all Butterworth-Heinemann publications
visit our website at www.store.elsevier.com

Printed and bound by CPI Group (UK) Ltd, Croydon, CR0 4YY

Contents

Preface . . . . ***ix***

Foreword . . . . ***xi***

Acknowledgments . . . . ***xiii***

Introduction . . . . ***xv***

Chapter 1 Introduction ***1***

1.1. Classification and Application of Multiphase Flows. 1
1.2. Complexity of Multiphase Flows 2
1.3. Multiscale Characteristics of Multiphase Flows 5
1.4. Need of Population Balance Modeling for Multiphase Flows 12
1.5. Scope of this Book 13

Chapter 2 Computational Multiphase Fluid Dynamics Framework ***17***

2.1. Eulerian Formulation Based on Interpenetrating Media Framework 17
2.1.1. Mass Conservation. 19
2.1.2. Momentum Conservation 23
2.1.3. Energy Conservation 27
2.1.4. Physical Description of Interfacial Exchange Terms 34
2.1.5. Effective Conservation Equations 37
2.2. Lagrangian Description on Discrete Element Framework. 43
2.2.1. Equations of Motion 43
2.2.2. Fluid—Particle Interaction (Forces Related to Fluid Acting on Particle One-Way, Two-Way Coupling) 44
2.2.3. Particle—Particle Interaction (Four-Way Coupling Concept: Collisions and Turbulent Dispersion of Particles) 49
2.3. Differential, Generic and Integral Form of the Transport Equations for Multiphase Flow. 59
2.4. Boundary Conditions for Multiphase Flow 62
2.5. Summary. 67

Chapter 3 Population Balance Approach—A Generic Framework ***69***

3.1. What is a Population Balance Approach? 69
3.2. Basic Definitions 70

3.2.1. Coordinate System and Density Function 70
3.2.2. Particle State Vector. 71
3.2.3. Continuous Phase Vector 72
3.2.4. Rate of Change of Particle State Vector and Particle State Continuum 72
3.3. Fundamentals of Population Balance Equation 73
3.3.1. Basic Consideration. 73
3.3.2. Various Integrated Forms of Transport Equations 77
3.3.3. Breakage/Break up Processes 80
3.3.4. Aggregation/Coalescence Processes 82
3.3.5. Net Generation of Particles 84
3.4. Practical Considerations of Population Balance Framework 85
3.5. Comments on the Coupling Between Population Balance and Computational Multiphase Fluid Dynamics. 87
3.6. Summary. 89

Chapter 4 Mechanistic Models for Gas–Liquid/Liquid–Liquid Flows 91
4.1. Introduction. 91
4.2. Mechanisms and Kernels of Fluid Particle Coalescence 92
4.2.1. Collision Frequency due to Turbulent Fluctuation and Random Collision 94
4.2.2. Collision Frequency due to Wake Entrainment 98
4.2.3. Collision Frequency due to Other Mechanisms 103
4.2.4. Coalescence Efficiency due to Film Drainage Model. 105
4.2.5. Coalescence Efficiency due to Energy Model 112
4.2.6. Coalescence Efficiency due to Critical Approach Velocity Model. . . .113
4.3. Mechanisms and Kernels of Fluid Particle Break up 114
4.3.1. Break up due to Turbulent Shearing 115
4.3.2. Break up due to Viscous Shear Force 127
4.3.3. Break up due to Interfacial Instability and Shearing Off. 128
4.3.4. Comments on Daughter Particle Size Distribution 128
4.4. Mechanisms and Kernels of Fluid Particle Coalescence and Break up for One-Group, Two-Group and Multigroup for Mulation 133
4.5. Summary. 136

Chapter 5 Mechanistic Models for Gas–Particle Liquid–Particle Flows 137
5.1. Introduction. 137
5.2. Mechanisms and Kernel Models of Solid Particle Aggregation. 138
5.2.1. Aggregation due to Interparticle Collision. .139
5.3. Mechanisms and Kernel Models of Solid Particle Breakage. 144
5.3.1. Breakage due to Hydrodynamic Stresses. 145
5.3.2. Breakage due to Other Mechanisms 148

5.4. Discrete Element Method—Soft-Sphere Model 150
5.4.1. Particle–Particle Interaction without Adhesion 151
5.4.2. Particle–Particle Interaction due to Adhesion 159
5.5. Summary 167

Chapter 6 Solution Methods and Turbulence Modeling 169
6.1. Introduction 169
6.2. Solution Methods for Eulerian Models 170
6.3. Mesh Systems 172
6.4. Numerical Discretization 177
6.4.1. Finite Volume Method 177
6.4.2. Basic Approximation of the Diffusion Term 184
6.4.3. Basic Approximation of Advection Term 186
6.4.4. Basic Approximation of Time-Advancing Solutions 191
6.4.5. Algebraic Form of Discretized Equations 194
6.5. Numerical Solvers 199
6.5.1. Iterative Calculations for the Segregated Approach 199
6.5.2. Application of IPSA or IPSA-C for the Segregated Approach 203
6.5.3. Comments on Matrix Solvers 210
6.5.4. Coupled Equation System 217
6.6. Solution Methods for Population Balance Equation 218
6.6.1. Class Method 219
6.6.2. Standard Method of Moments 223
6.6.3. Numerical Quadrature 228
6.6.4. Other Population Balance Methods 233
6.7. Solution Methods for Lagrangian Models 234
6.7.1. Molecular Dynamics 235
6.7.2. Brownian Dynamics 238
6.7.3. Discrete Element Method 240
6.8. Turbulence Modeling for Multiphase Flows 244
6.8.1. Reynolds-Averaged Equations and Closure 244
6.8.2. Large Eddy Simulation 253
6.9. Summary 261

Chapter 7 Some Applications of Population Balance with Examples 263
7.1. Introduction 263
7.2. Population Balance Solutions to Gas–Liquid Flow 264
7.2.1. Background 264
7.2.2. Modeling Interfacial Momentum Transfer for Gas–Liquid Flow . . . 264
7.2.3. Worked Examples 271

7.3. Population Balance Solutions to Liquid–Liquid Flow 298
7.3.1. Background. 298
7.3.2. Multiblock Model for Heterogeneous Turbulent Flow Structure in a Stirred Tank. 299
7.3.3. Worked Example 303
7.4. Population Balance Solutions to Gas–Particle Flow 308
7.4.1. Background. 308
7.4.2. Modeling Gas–Particle Flow via Direct Quadrature Method of Moment Multifluid Model 310
7.4.3. Worked Example 312
7.5. Population Balance Solutions to Liquid–Particle Flow 317
7.5.1. Background. 317
7.5.2. Modeling Liquid–Particle Flow via Quadrature Method of Moment 319
7.5.3. Worked Example 321
7.6. Summary. 326

Chapter 8 Future of the Population Balance Approach 329
8.1. Introduction. 329
8.2. Emerging Areas on the Use of the Population Balance Approach. 329
8.2.1. Natural and Biological Systems 329
8.2.2. Bulk Attrition 332
8.2.3. Crystallization 334
8.2.4. Synthesis of Nanoparticles 336
8.3. Summary. 337

References 339
Index 353

Preface

Given the increasing power of computers and the accelerated pace of hardware development of digital computers, computer models are gaining significant traction and increasingly being employed to investigate the fluid dynamics, heat transfer, and mass transfer of gas bubbles, liquid drops, and solid particles affecting natural, biological, and industrial systems. Nevertheless, the prime difficulty experienced by multiphase modelers is the modeling of multiphase flows due to large-scale separation. The existence of large-scale flow structures encompassing different particles within a mixture can be readily observed at the device or macroscale. These structures are, nonetheless, influenced by the physical behavior of particles on the microscale. Significant interactions between particles may result in local structural changes due to agglomeration/coalescence and breakage/breakup processes of gas bubbles, liquid drops, and solid particles on the mesoscale. It is on the mesoscale where the ability to accurately determine the size distributions of these particles is the primary focus of this book. Interphase mass, momentum, and energy exchanges that appear in the transport equations of computational multiphase fluid dynamics are usually found to be strongly dependent on the bubble, drop, and particle size distribution, which in any case represent the cornerstone for a fully predictive thermal and hydrodynamics in any multiphase system. In order to aptly describe the changes in the size distribution, population balance modeling is increasingly being adopted in predicting the representative bubble, drop, or particle size. Thus, it is the consideration of population balance modeling in conjunction with computational multiphase fluid dynamics which makes this book rather unique. One main feature of population balance modeling is its capability to accommodate the essential kinetics of particle–particle interaction. Since it is mechanistically driven, the fusion between computational multiphase fluid dynamics and population balance modeling will revolutionize how different classifications of multiphase flows can be feasibly treated and solved.

Guan Heng Yeoh
Chi Pok Cheung
Jiyuan Tu

Foreword

I am delighted to write the Foreword for this book, which captures the recent advances in both analytical and computational techniques in describing accurately multiphase flow phenomena in engineering practice and complex flow systems comprising bubbles, drops, and particles.

Bubbles, drops, and particles are of fundamental importance in many natural physical processes and in a host of industrial activities. Meteorologists constantly study and survey the behavior of raindrops and of hailstones. Applied mathematicians and physicists are always concerned with the fundamental aspects of the interaction between fluids and discrete entities. Chemical engineers rely on bubbles or drops for critical operations of engineering systems associated with distillation, absorption, and spray drying while employing solid particles as means of catalysts or chemical reactants in such systems. Mechanical engineers require the knowledge of bubbles in electromachining and boiling. In all of these phenomena and processes, the relative motion between bubbles, drops, and particles on the one hand and surrounding fluid on the other hand has a major effect. The flow of these discrete entities co-flowing with the surrounding fluid can be treated through computational multiphase fluid dynamics simulations, of which the authors of this book are prominent experts in this field.

In complex multiphase flows, the treatment of the interactions between discrete entities is also of major importance. Consideration of the so-called *population balance equations* brings another dimension in the prediction of the size distribution of bubbles, drops, and particles in a fluid. The use of these population balance equations is considered, in addition to the computational multiphase fluid dynamics simulations. It should be noted that population balance equations are not entirely new. The famous Boltzmann equation was first introduced in chemical physics and is now more than a century old. It can be readily seen that the atomic and the molecular processes that are involved in solid crystal nucleation, growth, agglomeration, and breakage, for example, are entirely different from those that occur in the nucleation, coalescence, and breakup of bubbles or drops in a boiling flow, or even the growth and reproduction of a population of microbes or animal cells. Nevertheless, certain general concepts, such as those embedded in population balance equations, can be extended to deal with all of the aforementioned processes. It is this generic framework that the authors have gone to great lengths to promote in this book with the specific aim of drawing in readers, who will not only understand the basic concepts as well as specific techniques of population balance modeling but also acquire a holistic view of a situation that involves a collection of discrete entities with an evolving distribution of properties applicable to a wide range of multiphase flows.

In my view, the authors can be confident that there will be many grateful readers who will gain broader perspectives of computational multiphase fluid dynamics and of population balance modeling as result of their efforts. The challenge posed to the authors has been to amalgamate these two large frameworks into a single text; the authors can be proud that they have managed to achieve such a task. The description of the complex multiphase flows is broken down to easy-to-understand exposition beginning from first principles, and then moving toward a thorough description of different models targeted for specific classifications of multiphase flows. More importantly, the authors have provided useful best practice solution methods, which are accompanied by a practical approach in solving different classifications of multiphase flows. Whether readers view this text from either the computational multiphase fluid dynamics or the population balance framework, they will discover that it has been written in a manner to guide them into a clear understanding of the comprehensive treatment of the complicated nature of multiphase flows. This text is certainly an enthusiastic celebration of three basic elements—theory, modeling and practice—in handling a multitude of multiphase flows.

Goodarz Ahmadi
Clarkson University,
Potsdam, New York
13 May, 2013

Acknowledgments

Guan Heng Yeoh would like to acknowledge the untiring support and endless tolerance of his wife, Natalie, and his three daughters, Genevieve, Ellana, and Clarissa, for their enormous understanding and unflinching encouragement during the seemingly unending hours spent in preparing and writing this text.

Chi Pok Cheung would like to express his deep gratitude to his wife, Christine, who has been very patient during the long hours of preparing and writing this text.

Jiyuan Tu would like to acknowledge his wife, Xue, and his son, Tian, who gave their boundless support to this project.

Special thanks are given to Jonathan Simpson, who led the project in the initial stages, and to Fiona Geraghty, who guided us for the rest of the project. Both of them have provided much encouragement in the writing of this book. The authors are especially grateful to Elsevier Science & Technology, and the Publisher, who have offered immense assistance both in academic elucidation and professional skills in the publication process.

To those we have failed to mention but have been involved in one way or another, the authors extend our deepest heartfelt appreciation.

Guan Heng Yeoh
Chi Pok Cheung
Jiyuan Tu

Introduction

Multiphase systems dealing with bubbles, drops, and particles are frequently encountered in many natural, biological, and industrial systems. The principal objective of this book is to provide applied scientists, practicing engineers, undergraduate and graduate students, and researchers concerned with multiphase phenomena, with a comprehensive critical review of the literature on particle–particle and particle–fluid interactions; a particle may be taken more specifically to be in the form of a gas bubble, liquid drop, or solid particle. There are important similarities as well as significant differences when dealing with these three types of particle. Because of the parallel relevancies, this book aims to examine the different treatments of gas bubbles, liquid drops, and solid particles in order to produce a unified text on the subject.

Guan Heng Yeoh
Chi Pok Cheung
Jiyuan Tu

CHAPTER 1

Introduction

1.1 Classification and Application of Multiphase Flows

Multiphase flows can be viewed as a fluid flow system comprising two or more distinct phases simultaneously co-flowing in a mixture where the level of separation between phases is at a scale well above the molecular level. In principle, multiphase flows can be classified on the basis of *number of phases*, *types of phases*, *size of phases* and *interaction between phases.* Depending on the combination of different phases, multiphase flows can be further classified according to: *dispersed phase flows*, *separated phase flows*, *gas—liquid flows*, *liquid—liquid flows*, *gas—particle flows*, *liquid—particle flows,* and *three-phase flows.*

Dispersed phase flows can be considered as flows whereby one of the phases exists as discrete particles. Two classical examples are the motion of bubbles in a liquid flow and the motion of liquid droplets in a gas. In such flows, there is no connection between these discrete fluid particles in the liquid. It is thus taken as a mixture of different sized bubbles or droplets that are being dispersed in a continuous medium.

Separated phase flows are different from dispersed phase flows due to the prevalence of a distinct line of contact separating two phases. One example is annular flow where there is a liquid layer along the pipe wall and a gaseous inner core. Categorically, one phase in this type of flow is distinctly separated from another in the same medium.

Gas—liquid flows having one phase being in gaseous form and the other in liquid state can assume many forms. Since the gas phase is permitted to freely deform within the liquid phase, several different geometrical shapes are possible, which include spherical, elliptical, distorted, toroidal and cap. Such flows also often exhibit other complex interfacial structures, namely, mixed or transitional flows that depict the transition between the dispersed phase flows and separated phase flows. Change of interfacial structures occurs through the occurrence of bubble—bubble or droplet—droplet interactions due to coalescence and break up and the presence of any phase change process.

Liquid—liquid flows belong to the special category where two immiscible fluids are co-flowing within the medium. One typical example is the presence of oil droplets in water or vice versa.

Gas—particle flows are concerned with the motion of suspended solid particles in the gas phase. For small particle number density, the gas flow exerts the main effect on the particles. Dilute gas—particle flows are predominantly governed by the surface and body forces acting on the particles. For very dilute gas–particle flows, the solid particles are treated as passive tracers, which do not have an effect on the gas flow. For large particle number density, particle–particle interactions become more important than the forces due to interstitial gas. Such two-phase flows are referred to as dense gas–particle flows. Collisions inherently exist between the solid particles and significantly influence their movement and migration in the gas phase. For substantially bigger particles, such flows are categorically known as granular flows.

Liquid–particle flows, in contrast to gas–particles flows, consist of the transport of solid particles in liquid flow instead. Here, the solid particles do not have a distinct velocity field but generally follow the liquid velocity field. Also known as slurry flows, they fall into the same category as dispersed phase flows in which the liquid now represents the continuous medium. The different liquid and solid phases are mainly driven by, and largely respond to, the presence of pressure gradients since the density ratio between phases is low and the drag between phases is generally rather high in such flows.

Three-phase flows are encountered in a number of engineering applications of technical relevance, for example, bubbles in slurry flows resulting in three phases co-flowing together in the same medium. In this particular category, solid particles and gas bubbles co-flow with the continuous liquid phase. The coexistence of three phases considerably complicates the computational modeling of the flow physics due to the required understanding associated with the phenomena of particle–particle, bubble–bubble, particle–bubble, particle–fluid, and bubble–fluid interactions.

Based on the above classifications, multiphase flows are also widely featured in many diverse natural, biological, and industrial systems. Some examples of multiphase flows that can be distinguished in these three different systems are illustrated in Table 1.1. It should be noted that the list of examples presented in this table is by no means exhaustive but is intended to provide the reader a broad overview of the range of applications and types of multiphase flows that can be found in these systems.

Multiphase flows are inherently complex. The physical understanding of flows where more than one phase is involved offers problems of complexity that are immeasurably far greater than in single-phase flows. This is because the phases do not, in general, uniformly mix and because small-scale interactions between the phases can have profound effects on the macroscopic properties of the fluid flow. This clearly reflects the ubiquitous challenges that still exist when dealing with the complex nature of multiphase flows.

1.2 Complexity of Multiphase Flows

Dispersed flows of gas bubbles, liquid drops and solid particles are central to the analysis of multiphase flows and development of generic computational approaches. The motion of these

Table 1.1: Examples of multiphase flows in natural, biological and industrial systems.

	Multiphase flows				
	Gas–liquid	Liquid–liquid	Gas–particle	Liquid–particle	Three-phase
Natural	Rain droplets, ocean waves, mist formation	Oil–water mixture, liquid bridges	Sand storms, volcanoes, avalanches	Sedimentation of sand in rivers and sea, soil erosion, mud slides, debris flows	Oil–water–sand mixture
Biological	Aerosols	Water and hydrophobic phases (lipids or membranes)	Dust particles, smoke (soot particles)	Nanoparticles in blood flow	Plasma, red blood cells and leukocytes in blood flow
Industrial	Boiling water and pressurized water nuclear reactors, boilers, heat exchangers, chemical reactor desalination systems	Emulsifiers, fuel-cell systems, micro-channel applications	Pneumatic conveyers, dust collectors, fluidized beds, pulverized solid particles, spray drying	Slurry transportation, flotation, fluidized beds, water jet cutting, sewage treatment plants	Air lift pumps, air lift bio-reactors, fluidized beds

particles is strongly affected by the relative motion of the phases within the mixture. As the relative motion becomes large enough, inertial instabilities can give rise to *mesoscale* structures (Agrawal et al., 2001). The complex nature of such flows is manifested by the appearance of clusters of bubbles, drops or particles (Serizawa, 2003).

In dispersed flows of bubbles or drops, the existence of dynamically changing interface can cause individual gas bubbles or liquid drops to deform, coalesce through the breaking down of the interfaces between the phases to form larger bubbles or drops, and break up into two smaller bubbles or drops due to the shearing of the fluid. As the volume flow of gases or drops increases, coalescence causes the appearance of caplike gas bubbles or liquid drops. For small diameter tubes, a pattern is exhibited whereby slugs of highly aerated liquid fill the whole tube. These so-called Taylor bubbles or drops have characteristics of a spherical cap nose and are somewhat abruptly terminated at the bottom edge. With increasing volume fractions of gases or drops, a churn pattern persists resulting in a haphazard flow of gas–liquid or liquid–liquid mixture. In contrast to small diameter tubes, churn flow can be rather different in large diameter pipes where Taylor bubbles are not formed. At very high velocities of the dispersed phase, an annular pattern is observed whereby part of the liquid flows along the wall and the remainder is the dispersed phase. The complexity of flow regime transitions remains an important problem whereby an improved physical understanding of one flow regime to another is required to establish a sound physical theory for the prediction of the different flow patterns.

In dispersed flows of solid particles, the entire range of particle volume fractions from dilute to dense conditions results in an array of complex physical processes. For dilute particle laden flows with high mass loading ratio, the motion of fluid has a significant effect on the particle motion and vice versa, when present. Interparticle collisions, which may be significant especially in "fluid-like" suspensions, can lead to particles interacting through interstitial fluid and ephemeral impulsive interactions. For conditions leading to transition from fluid-like to "solid-like" behavior, such as in a random close packing environment, not only particles will interact with each other through enduring contact between each other but also aggregation between individual solid particles may occur if the net interparticle force is attractive and strong enough to overcome the hydrodynamic forces. Conversely, breakage of the aggregate may be affected by the fluid shear overcoming the attractive interparticle force, which holds the individual solid particles together within the aggregate.

Principally, the difficulty in analyzing multiphase flows lies in the unconstrained behavior of the phases, which can assume a large number of complicated configurations. The main issue toward the understanding of why the phases configure in a particular way requires the identification of the microphysics controlling the organization of the phases of specific multiphase flows. For gas–liquid or liquid–liquid flows, the microphysics problem is the physical understanding of the formation or destruction of the changing interfaces between the fluid and gas bubbles or liquid drops. For gas–particle or liquid–particle flows, since the interfaces remain intact, the microphysics problem predominantly concerns the interfacial forces and interaction behaviors between particles. In turbulent multiphase flows, the possible interactions between turbulent eddies and interfacial structures as well as exchanges between individual phases introduce additional complexities to the flow phenomena. Under some circumstances, the discrete gas bubbles, liquid drops or solid particles can assume large-scale turbulent motions that do not directly reflect the turbulence patterns being observed in the fluid.

One recurring theme throughout the study of multiphase flows is the requirement to model and predict the detailed behavior of such flows and the phenomena that they continue to manifest. Computation of multiphase flows serves many essential roles: (1) as a tool to develop an understanding of the basic physics such as clarifying the importance of physical effects such as surface tension and gravity by adding or removing them, (2) as an aid in closing the averaged transport equations based upon a macroscopic formulation such as the development of closure relationships and testing against numerical simulations, and (3) as a means to solve actual problems such as small-scale problems that can be tackled through direct numerical simulation and large-scale problems via reduced formulations such as averaged transport equations. Computational techniques are becoming powerful tools to resolve a range of multiphase flows. The effective use of a variety of computational approaches to aptly handle different classifications of multiphase flows is demonstrated in Section 1.3.

1.3 Multiscale Characteristics of Multiphase Flows

Multiphase flow physics are inherently multiscale in nature. Figure 1.1 describes the many different physical characteristics that can be observed at different length scales. At *microscale*, it is essential to understand the interaction of the gas bubbles, liquid drops and solid particles with the continuum fluid through tracking the motion of the individual discrete particles in space and time. With increasing length scale, interaction between the discrete particles may become significant resulting in local structural changes due to agglomeration/ coalescence and breakage/break up processes of gas bubbles, liquid drops and solid particles that are prevalent at the *mesoscale*. At the device scale, the influence of the *macroscale* hydrodynamic behavior of the background fluid on the clusters of gas bubbles, liquid drops and solid particles results in the existence of large-scale flow structures encompassing the different individual phases within the multiphase flow.

As will be illustrated through some examples below, computational techniques can be employed to reveal details of peculiar flow physics that otherwise could not be visualized by experiments or to clarify particular accentuating mechanisms that are consistently being manifested in complex multiphase flows. Such an approach, based on the utilization of advanced numerical methods and models, usually contains very detailed information, producing an accurate realization of the fluid flow.

Fujita and Yamaguchi (2007) have investigated the self-organization of nanoparticles in a dense suspension medium. Here, the motion of nanoparticles, which are treated as rigid spheres, is solved by the linear and angular momentum equations based on Newton's law in a

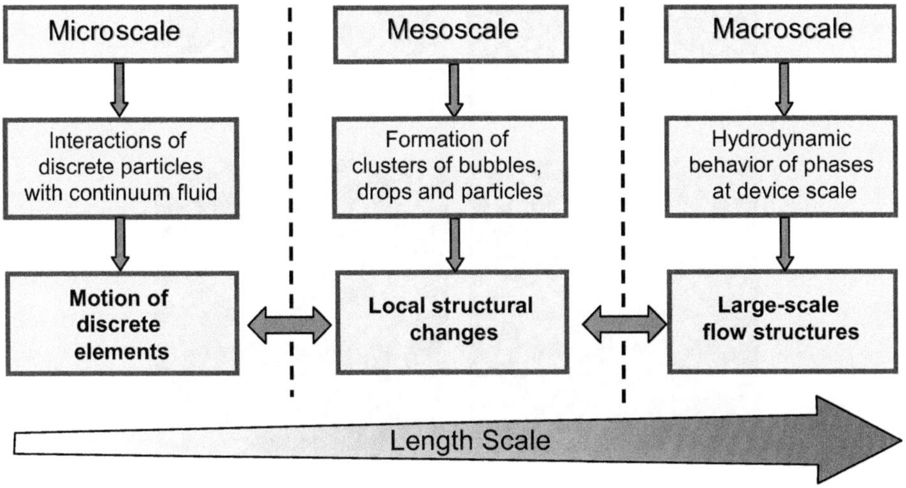

Figure 1.1
Multiscale characteristics of multiphase flows. (For color version of this figure, the reader is referred to the online version of this book.)

Lagrangian framework. The soft-sphere model based on the discrete element method is adopted to resolve the solid contact forces between particles. Additional consideration of other forces acting on the nanoparticles includes the capillary force, electrostatic force, van der Waals force, Brownian random force and fluid force. The flow of solvent is solved by the Stokes equation in the Eulerian framework. This provides the influence of the fluid force acting on the nanoparticles. Based on a nanoparticle diameter of 50 nm (5×10^{-8} m), the numerical simulation reveals that locally distributed nanoparticles form chainlike clusters shortly after the start of the simulation at time $t = 2.5$ μs, which is mainly due to Brownian motion and the van der Waals attractive force exerted between nanoparticles close to each other. These clusters continue to further attract neighboring nanoparticles and they increase in size at time $t = 5$ μs. As time progresses, the chainlike structures aggregate with one another and form an organized network structure throughout the computational domain at time $t = 10$ μs, such as illustrated in Figure 1.2.

In a similar investigation, Marshall (2009) has also treated the motion of aerosols through a Lagrangian scheme based on the discrete element method. Being of micron-sized particles (10^{-6} m), the Brownian random force is not considered. Nonetheless, the reduced gravity force is accounted for along with fluid forces comprising drag, lift, Magnus and added mass. The soft-sphere model based on the discrete element method is employed to resolve the solid

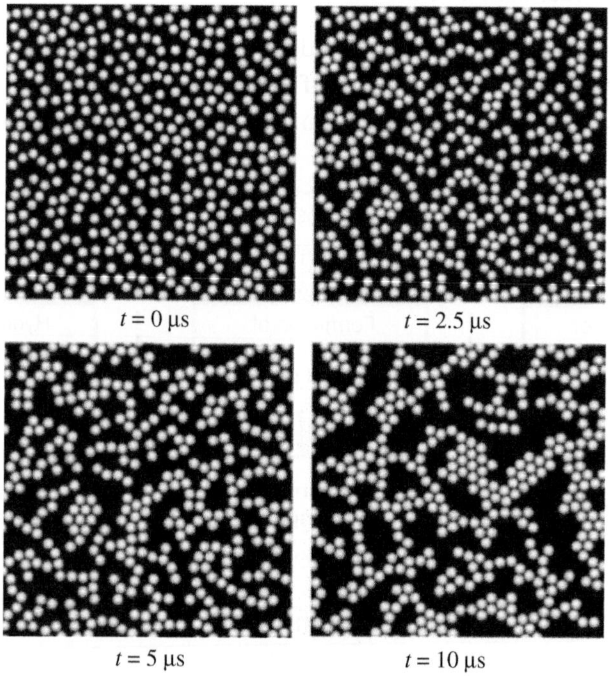

Figure 1.2
Self-organization process of nanoparticles in the dense suspension medium.
After Fujita and Yamaguchi (2007).

interactions between aerosols. In order to account for the effect of adhesion, the van der Waals force is considered only within the flattened contact region. Numerical simulation is performed by examining the aggregation behavior of particles advected in a circular pipe flow with a parabolic laminar velocity profile such as exemplified in Figure 1.3. Assuming that the aerosols are being initially spaced within the channel and shortly after the start of the simulation, they begin to collide and form aggregates with the channel. These small aggregates begin colliding with each other to progressively form larger aggregates. On one hand, some aggregates near the walls become attached to the walls, forming fingerlike projections sticking out into the flow. On the other hand, some of the projecting aggregates break off due to the imposed shear, leaving a part attached to the wall and a part that returns to the flow. Other aggregates that are initially projected out into the flow subsequently bend via rolling to be attached to the wall at several different locations. As aggregates near the wall tumble due to the imposed shear, they capture other particles from the flow and transport them to the wall, such that most of the aerosols are eventually attached to the wall. Those aerosols that are contained in the aggregate layer lining the walls settle to a state with zero velocity, so that they become trapped at the wall.

Lu and Tryggvason (2008) have carried out direct numerical simulation of buoyant bubbles in turbulent flow. The immiscible phase boundary separating the fluids inside and outside the bubble is tracked through the so-called front tacking method, which considers explicit marker points following the fluid interface through space and time. The Navier–Stokes equation is solved for the entire computation domain with appropriate consideration of the jump condition, such as surface tension at the interface as a source term to the equation. Dynamics of freely evolving arrays of nearly spherical and deformable bubbles of size between 1 mm (10^{-3} m) and 2 mm (2×10^{-3} m) in a vertical channel are investigated. The void fraction profiles across the channel (shown in Figure 1.4) are fundamentally different for both systems. Bubble deformability, rather than size, is considered to be the dominant effect where large and small bubbles result in a different void fraction distribution. This is not entirely surprising as the lift force of a bubble, and therefore its lateral migration, is dependent on how deformed it is since bubble deformation can result in the reversal of the sign of the lift force. For strongly deformed bubbles, any lateral lift is relatively weak and the dispersion of the bubbles by the turbulent flow and interactions with each other overcomes any tendency of the bubbles to "bunch" together in the middle of the channel. For nearly spherical bubbles, the bubble distribution consists of a core where the mixture is in hydrostatic equilibrium and the flow is essentially homogeneous and a wall layer exists with a large number of bubbles sliding along the wall. Figure 1.5 shows the bubbles and isocontours of the vertical velocity. Most of the nearly spherical bubbles are found to flow near the walls, and the flow in the middle of the channel is relatively quiescent.

Quan et al. (2009) have recently demonstrated the capability of adopting a moving mesh interface tracking coupled with local mesh adaption to handle interface merging. Numerical

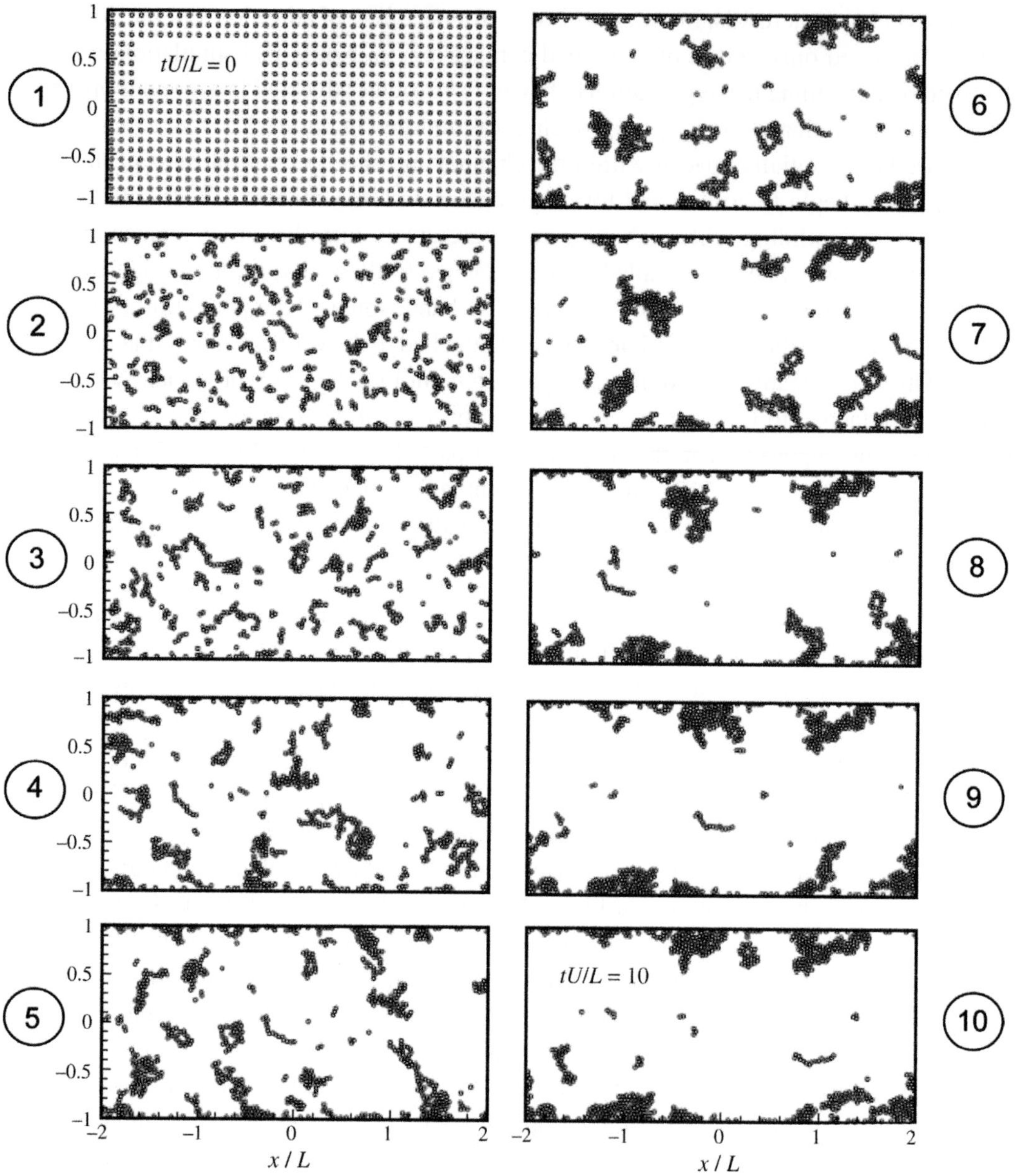

Figure 1.3

Evolution of particle locations within the fluid flow, depicting the development of aggregates and adhesion of aggregates to the channel walls due to van der Waals forces, at nondimensional times $tU/L = 0–10$. *After Marshall (2009).*

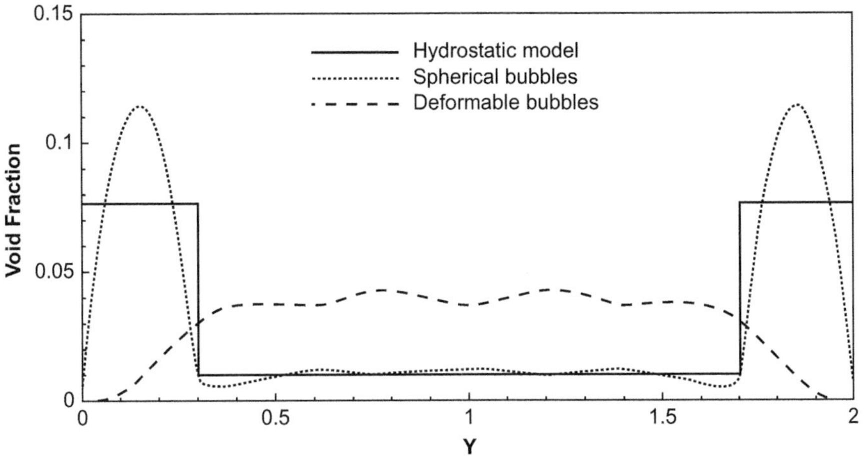

Figure 1.4
Average void fraction profiles across channel at steady state for nearly spherical and deformable bubbles. *After Lu and Tryggvason (2008).*

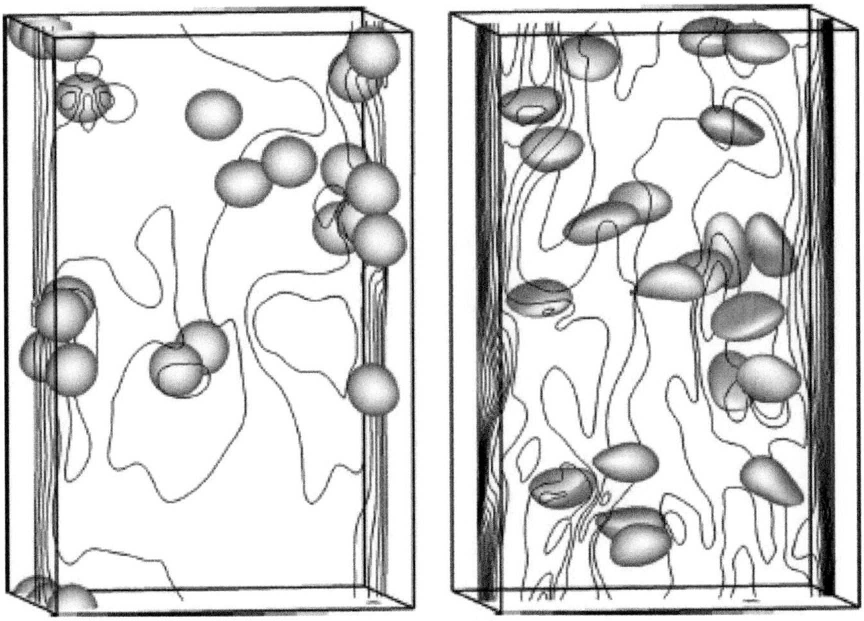

Figure 1.5
Bubbles and flow field in a turbulent vertical channel flow. Left figure - near spherical bubbles. Right figure - deformable bubbles. *After Lu and Tryggvason (2008).*

simulations are performed on droplet pair collisions. Figure 1.6 illustrates the shape evolution of the off-center coalescence of two droplets with high surface tension coefficient. It can be seen that at the time when a bridge is formed between the two droplets (see droplet shapes c and d), the two ends continue to move in opposite directions due to the momentum, and the droplet is stretched (see droplet shapes of g and h). A ligament with two bulbous ends is formed (see sequence of droplet shapes of i to l); the two ends move toward the center due to the prevalence of high surface tension forces. The coalesced droplet eventually becomes a sphere. Figure 1.6 also depicts the shape evolution of the off-center coalescence of two droplets but with a smaller surface tension coefficient. In this particular case, the bridge region can be seen to move outward while the two ends continue to move in opposite directions. A stretched droplet is formed with two bulbous ends. With the ligament being continuously stretched due to the momentum, the bulbous ends become spherical, and a neck region is created near the right end. Then this end is pinched off. A necking is also created near the left end afterward, and then the left end is pinched off resulting in the formation of two end droplets. These two ends of the middle section contract due to the high surface tension forces at the ends, and a bulbous ligament is again created. More significantly,

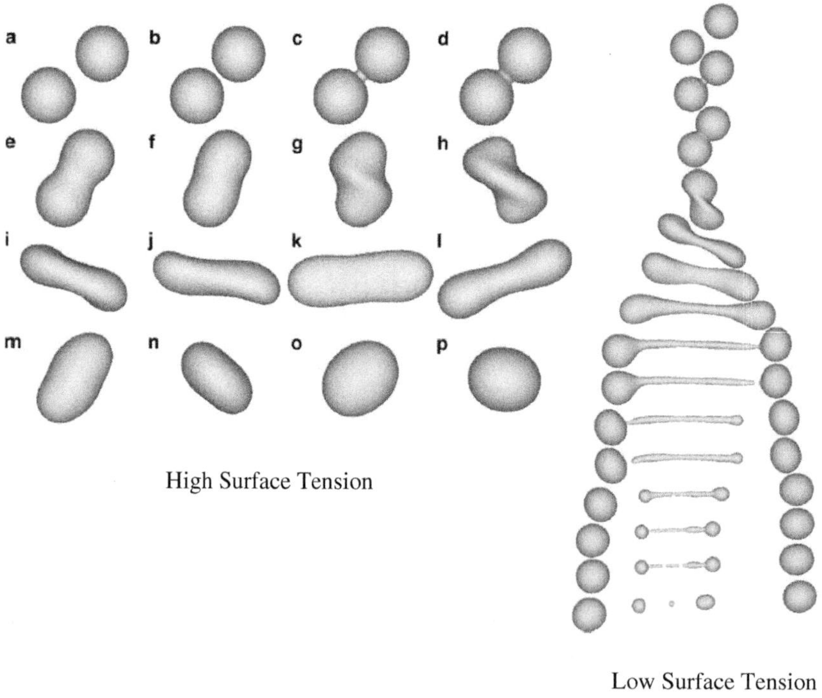

Figure 1.6
Evolution of the off-center coalescence of two droplets with high and low surface tension coefficients. *After Quan et al. (2009).*

the middle section experiences break up that results in three satellite droplets being formed in the middle region.

As demonstrated by the examples above, it is becoming evermore possible to directly solve the transport equations governing the conservation of mass and momentum for each of the phases and to compute every detail of the multiphase flow, the motion of all the fluid around every gas bubble, liquid droplet and solid particle and the position of every interface. The role of such computations in answering the fundamental questions in multiphase flows will undoubtedly increase through time, and it must be recognized that such simulations demonstrate, in some sense, what types of problems can be possibly solved based on current availability of computational resources. Nevertheless, it is in the examination of very complex large-scale systems, where it is necessary to follow the complex evolution of an enormous range of scales over a long time period that such comprehensive treatment through the means of direct numerical simulation is still prohibitive and only restricted to turbulent flows of low Reynolds number and dynamics of a limited amount of individual gas bubbles, liquid droplets and solid particles.

There is, thus, still a great need to adopt a macroscopic formulation based on averaging the transport equations, which yields a multiphase flow continuum formulation, thereby effectively eliminating the interfacial discontinuities for large-scale multiphase flows that are highly turbulent at the device scale. Some typical industrial systems where multiphase flows are prevalent include spray dryers, cyclone separators, bubble columns, stirred tank reactors, nuclear reactors, fluidized beds and sludge removal. In such systems, the gross features of the fluid flow are only predicted. Hence, the reliance on ever more realistic models encompassing the turbulent effects for momentum and energy as well as for interfacial exchanges for mass, momentum and energy transfer in order to provide the necessary closure to the modeling of multiphase flows is paramount. Such simulations, in which if it is possible to access the complete data and to accurately control every aspect of the systems, will not only revolutionize the predictive capability but also open up new ways of better controlling the behavior of such systems with greater efficiencies and significant production outputs.

Development of the correct microphysics in the required closure laws for the averaged transport equations remains the key challenge for a wide range of conditions. In the framework of computational fluid dynamics, it is hard to imagine another form of practical engineering simulation tool except through the consideration of the multifluid approach, which treats all phases as continuum mediums. New insights and wealth of information (as exemplified by the above examples) that can be offered through detailed simulations of the flow physics and to possibly construct a quantitative model, allowing other similar flows to be computed, can prove to be instrumental in developing better closures for the models of the averaged transport equations. These new advances, which capitalize on the ongoing development of state-of-the-art numerical capabilities, can lead to farther-reaching modeling tools for practical applications.

1.4 Need of Population Balance Modeling for Multiphase Flows

In comparison to single-phase flows, the presence of gas bubbles, liquid drops and solid particles (regardless whether they are inherently present within or deliberately introduced into the flow system) in almost all multiphase flows of major importance significantly affect the behavior of the fluid flow. Because of mounting interest in determining the influence of gas bubbles, liquid drops and solid particles on the fluid flow, the need of population balance modeling to resolve the microphysics which occurs at the mesoscale is becoming ever more important to better synthesize the behavior and dynamic evolution of the population of these discrete particles. As a result, much consideration has been concentrated toward describing the spatial and temporal evolution of the geometrical structures through the formation and destruction of clusters through interactions among and between the discrete particles and turbulent eddies in turbulent flows. The overarching issue, especially using the macroscopic formulation based on averaging the transport equations, is the determination of the interfacial transfer terms to provide the appropriate closure to the equations, which can be realized through the consideration of suitable mechanistic models accounting for the physical interaction between the different phases. One such approach is population balance modeling, especially to determine the local size distribution of the discrete elements in space and time.

In essence, the population balance of any system is a record for the number of gas bubbles, liquid drops and solid particles whose presence or occurrence governs the overall behavior of the multiphase system under consideration. The record of these discrete elements is dynamically dependent on the *birth* and *death* processes that *create* new discrete elements and *terminate* existing particulates within a finite or defined space. Since practical multiphase flows generally contain millions or billions of discrete particles that are simultaneously varying in space and over time, the feasibility of direct numerical simulation in resolving such flows is still far beyond the capacity of existing computational resources. Population balance, which records the number of discrete particles as an averaged function through the population balance equation, has been shown to be very extremely promising in handling the flow complexity because of its comparatively lower computational requirements.

The development of the population balance model can be traced back to as early as the end of the eighteenth century when the Boltzmann equation, devised by Ludwig Boltzmann, could be regarded as the first population balance equation that can be expressed in terms of a statistical distribution of molecules or particles in a state space. However, the derivation of a generic population balance concept was actually initiated from the middle of the nineteenth century. In the 1960s, Hulburt and Katz (1964) and Randolph and Larson (1964), based on the statistical mechanics and continuum mechanical frameworks respectively, presented the population balance concept to solve size variation due to nucleation, growth and agglomeration processes of solid particles. A series of research development was thereafter presented by Fredrickson et al. (1967), Ramkrishna and Borwanker (1973) and Ramkrishna (1979, 1985), where the treatment of population balance equations were successfully generalized with various internal coordinates. A number of textbooks mainly concerning population balance of aerocolloidal systems have also appeared (Hidy and Brock, 1970; Pandis and Seinfeld, 1998; Friedlander, 2000). Nevertheless, the flexibility and capability of population balance in solving practical engineering problems has not been fully exposed, until recently, when Ramkrishna (2000) published a textbook focusing on the generic issues of population balance for various applications.

Although the concept of population balance was formulated over many decades, implementation of population balance modeling has only been realized in very recent times. Such a dramatic breakthrough has only been made possible by the rapid development of computational fluid dynamics and in situ experimental measuring techniques. The capacity to measure the sizes of discrete elements or other population balance variables in the fluid flow from experiments remains an important component in the population balance framework. These experimental data not only allow the knowledge of sizes and their evolution within the systems to be realized but also provide a scientific basis for model calibrations and validations. In the context of computational fluid dynamics, external variables of the population balance equation can be readily acquired by decoupling the equation from the *external* coordinates (referring to the physical space), which can then enable solution algorithms to be developed within the *internal* coordinates (referring to the property space).

1.5 Scope of this Book

Computational investigations of multiphase flows are increasingly being instigated not only to arrive at a better understanding of the behavior of particles interacting with other surrounding particles and the fluid but also to provide practical solutions to many existing and emerging systems of natural and technological significance. On the latter, averaged transport equations are still widely applied in most computational analysis of multiphase flows. The appropriate microphysics that are required to be accommodated via the interphase mass, momentum and energy exchanges between the different phases in these equations greatly depend on the size distribution of discrete gas bubbles, liquid drops or solid particles in the fluid. In order to adequately describe the temporal as well as spatial changes of the size distribution, population balance modeling is progressively being adopted. This therefore brings us to the underlying purpose of this book. The basic concepts and formulation of mechanistic models, as well as mathematical and theoretical illustrations of computational techniques that can be effectively applied and utilized in feasibly solving the many classifications of multiphase flows, are presented based on the combined frameworks of computational multiphase fluid dynamics and population balance.

The scope of this book is as follows:

Chapter 2 deals with the foundation of multiphase flow formulation in the Eulerian and Lagrangian frames of reference. The former describes the local instant formulation of the multiphase flow based on the *interpenetrating media framework*; the explicit existence of the interface separating the phases is described. This subsequently leads to the derivation of effective equations governing the conservation mass, momentum and energy. The latter is based on the consideration of *discrete element framework*. The motion of a particle is deduced from the linear and angular momentum (*Newton's second law*) which is subjected to the action of possible forces acting on the particle as well as eddy interaction with the particle in turbulent flow. The hard sphere model for particle–particle collision is presented.

Fundamental development of the population balance equation is presented in Chapter 3. The concept of population of particles being distributed in physical space as well as property space is described. Various integrated forms of the population balance equation through moment transformation are discussed. Formulation of relevant source/sink terms, due to aggregation/coalescence and breakage/break up processes for the population balance equation are generically presented.

Mechanistic models for gas–liquid/liquid–liquid flows are presented in Chapter 4. Various mechanisms and wide diversity of available kernel models for the coalescence and break up of fluid particles are described.

Mechanistic models for gas–particle/liquid–particle flows are presented in Chapter 5. Specific mechanisms and kernel models for the aggregation and breakage of solid particles are described. In addition, the soft-sphere model for particle–particle collision is introduced. Forces acting on particles with and without adhesion as well as in some cases subjected to external electrical, magnetic and temperature fields, are discussed.

Chapter 6 focuses on the solution methods and turbulence modeling. For the solution methods for Eulerian models, mesh system based on unstructured mesh generation, numerical discretization based on the finite volume method and specific numerical techniques to solve the discretized equations including simultaneous solution of nonlinearly coupled equations (SINCE), which is dedicated for sequential iterative calculations for multiphase problems, are described. The application of the interphase slip algorithm (IPSA) and its variant interphase slip algorithm–coupled (IPSA-C) and a coupled solver approach for the coupling between the pressure and velocity are also described for the specific consideration in the context of multiphase calculations. For the solution methods for the population balance equation, efficient methods based on discrete classes, standard of moments and numerical quadrature are described. For the solution methods for the Lagrangian models, different particle methods, which only differ by the different particle interaction laws as well as by the imposition of random forcing in some models to mimic collisions or interaction with molecules (or particles) of the surrounding fluid via molecular dynamics, Brownian dynamics and discrete element method, are described. Last, the consideration of turbulence models based on Reynolds averaging and spatial filtering through the concept of large eddy simulation is presented in view of achieving the necessary closure to the mathematical system of equations.

In Chapter 7, some practical examples are provided to enhance the understanding of the complex modeling concept and other important considerations that are required to best tackle different classifications of multiphase flows based upon the combined frameworks of computational multiphase fluid dynamics and population balance.

The future of the population balance approach is introduced in Chapter 8. Some interesting features through the use of population balance to resolve problems in emerging areas such as natural systems and biological systems, bulk attrition, crystallization and synthesis of nanoparticles, are briefly discussed.

CHAPTER 2

Computational Multiphase Fluid Dynamics Framework

2.1 Eulerian Formulation Based on Interpenetrating Media Framework

In the Eulerian formulation of a multiphase flow, the description of a fluid in such a flow may be regarded as a clearly identifiable portion of a particular phase occupying the flow system. For two-phase flows of gas and liquid and liquid and particle, one fluid may be characterized by the continuous liquid which occupies a connected region of space, while the other fluid represents the flow of finely dispersed gas bubbles and solid particles, respectively. Analogously, for the dispersed flow of gas and particle, one fluid can be taken to be the continuous gas, while the other fluid accounts for the flow of dispersed solid particles in the gas flow.

Within a typical multiphase flow system, multiple phases that co-exist simultaneously in the flow often exhibit relative motion among the phases, as well as heat transfer across the phase boundary. In general, the microscopic motions and thermal characteristics of the individual constituents are very complex, and the solution to the microlevel evolutionary equations is difficult due to the uncertainty of the exact locations of the particular constituents at any particular time. For most practical purposes, the exact prediction on the evolution of the details within the multiphase flow system is neither possible nor deemed desirable. On the other hand, the gross features of the fluid flow and heat transfer are of more significant importance. Owing to the complexities of interfaces and resultant discontinuities in fluid properties, as well as from physical scaling issues, it is thereby customary to apply some sort of averaging process to the conservation equations, which leads to the derivation of the effective conservation equations in the interpenetrating media framework.

In the framework of computational multiphase fluid dynamics, the averaging process allows the feasibility of solving the multiphase flow through suitable numerical techniques in the Eulerian framework and the ease of comparison with the experimental data. There have been numerous averaging approaches that have been proposed in a multitude of multiphase flow investigations. Details concerning the adoption of an appropriate choice of averaging may be found in Vernier and Delhaye (1968); Ishii (1975); Yadigaroglu and Lahey (1976); Delhaye and Achard (1976); Panton (1968); Agee et al. (1978); Banerjee and Chan (1980); Drew

Multiphase Flow Analysis Using Population Balance Modeling.

(1983); Lahey and Drew (1988); Besnard and Harlow (1988); Joseph et al. (1990); Drew and Passman (1999); Kolev (2005); and Ishii and Hibiki (2006). In retrospect, averaging may be performed in time, space, over an ensemble, or in some combination of these. By defining an instantaneous field to be ϕ (x, y, z, t), the commonly used averaging approaches in multiphase flow can be mathematically described as follows:

Time average:

$$\overline{\phi}(x,y,z) = \lim_{T\to\infty} \frac{1}{T}\int \phi(x,y,z,t)\mathrm{d}t \tag{2.1}$$

Space average:

$$\langle\phi\rangle_V(t) = \lim_{V\to\infty} \frac{1}{V}\iiint \phi(x,y,z,t)\mathrm{d}V \tag{2.2}$$

Ensemble average:

$$\langle\phi\rangle_E(x,y,z,t) = \lim_{N\to\infty} \frac{1}{N}\sum_{n=1}^{N} \phi_n(x,y,z,t) \tag{2.3}$$

where T is an averaging time scale, V is the volume based on an averaging length scale, and N is the total number of realizations. While averaging allows the mathematical solution of the problem to be more tractable, there is a judicious requirement to recover lost information regarding the local gradients between each phase, which have to be resupplied in the form of semiempirical closure relationships, also known as constitutive equations, for the various interphase interaction properties such as the interfacial mass, momentum, and energy exchanges in the conservation equations of mass, momentum, and energy.

On the basis of the continuum assumption where each phase behaves like a continuous fluid, the formulation of the local instantaneous conservation equations of mass, momentum, and energy can be achieved through classical consideration, which is similar to the derivation of conservation equations governing the single-phase fluid flow. A phase indicator function is introduced to allow a clear physical description of the jump or interphase interaction conditions, which express the conservation of mass, momentum, and energy at the interface. Through the use of a phase indicator equation and applying appropriate averaging, the effective conservation equations of mass, momentum, and energy can be subsequently derived. It should be noted that either the space (volume) averaging or ensemble averaging could be employed to formulate the effective conservation equations since both of them result essentially in the same form of equations. For succinctness, we denote $\langle\rangle$ as a generic averaging process of representing either space (volume) averaging $\langle\rangle_V$ or ensemble averaging $\langle\rangle_E$ so that $\langle\phi\rangle$ corresponds to a generic averaged field.

2.1.1 Mass Conservation

The first basic equation of fluid motion for each continuous phase is the conservation of mass.

This conservation law, which states "matter can neither be created nor destroyed", implying that mass must always be conserved, is important to the derivation of the conservation equation of mass. From the classical physics consideration of single-phase fluid flow, the instantaneous conservation equation of mass may be derived from the consideration of a suitable model of the flow for a continuum fluid, which is also deemed to be valid for the continuous flow in each phase. Adopting the infinitesimal fluid element approach, such an approach prescribes an infinitesimal element in the multiphase fluid flow as shown in Figure 2.1. An elemental volume dV inside each fractional volume containing different phases allows the fundamental physical principles to be applied to the infinitesimal small fluid element. Considering that the fluid element is fixed in space and the fluid is permitted to flow through it, this approach leads directly to the fundamental equation in partial differential form.

Consider the enlarged elemental volume dV containing any kth ($= 1, 2, 3, \ldots, N$) phase as shown in Figure 2.1. The fundamental physical principle governing the conservation of mass is

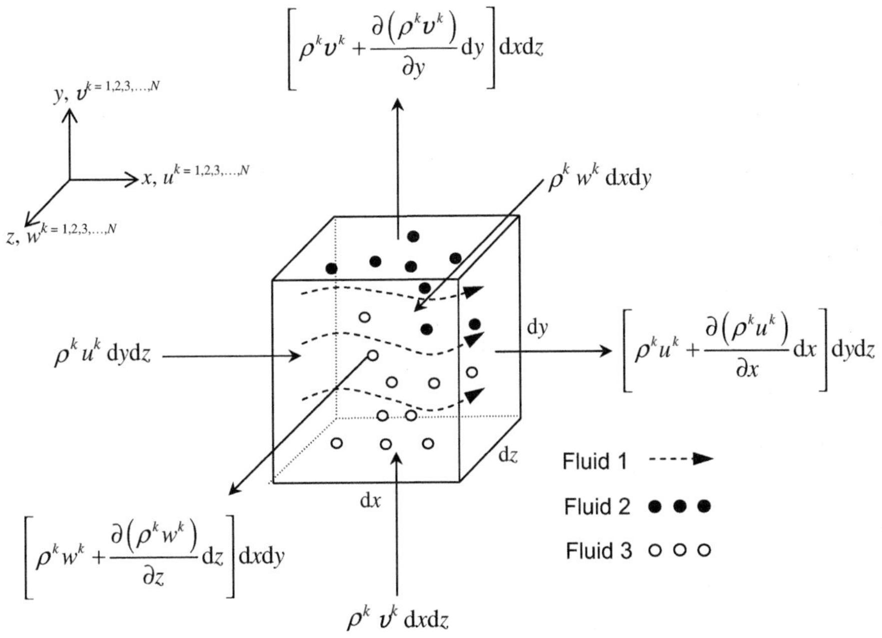

Figure 2.1
The conservation of mass based on the elemental volume for the kth phase. Fluid 1, fluid 2, and fluid 3 are taken as continuous fluids within the elemental volume.

$$\begin{array}{ccc} \textit{Rate of increase of} & & \textit{Net rate at which} \\ \textit{mass within the fluid} & = & \textit{mass enters the} \\ \textit{element} & & \textit{elemental volume} \end{array} \tag{2.4}$$

For a Cartesian coordinate system, the kth phase density and velocity are functions of (x, y, z) space and time t, which are denoted as ρ^k and u^k, v^k, and w^k. Since the mass of any fluid m^k is the product of the kth phase density and the elemental volume, $\rho^k\, \mathrm{d}V\,(=\mathrm{d}x\, \mathrm{d}y\, \mathrm{d}z)$, the rate of increase of mass within the fluid element is thus given by

$$\frac{\partial m^k}{\partial t} = \frac{\partial \rho^k}{\partial t}\mathrm{d}x\mathrm{d}y\mathrm{d}z \tag{2.5}$$

To account for the mass flow across each of the faces of the element, it can be seen from Figure 2.1 that the net rate at which mass enters the elemental volume can be expressed as

$$\underbrace{(\rho^k u^k)\mathrm{d}y\mathrm{d}z - \left[(\rho^k u^k) + \frac{\partial(\rho^k u^k)}{\partial x}\mathrm{d}x\right]\mathrm{d}y\mathrm{d}z}_{\text{along } x} +$$

$$\underbrace{(\rho^k v^k)\mathrm{d}x\mathrm{d}z - \left[(\rho^k v^k) + \frac{\partial(\rho^k v^k)}{\partial y}\mathrm{d}y\right]\mathrm{d}x\mathrm{d}z}_{\text{along } y} +$$

$$\underbrace{(\rho^k w^k)\mathrm{d}x\mathrm{d}y - \left[(\rho^k w^k) + \frac{\partial(\rho^k w^k)}{\partial z}\mathrm{d}z\right]\mathrm{d}x\mathrm{d}y}_{\text{along } z}$$

or

$$-\left[\frac{\partial(\rho^k u^k)}{\partial x} + \frac{\partial(\rho^k v^k)}{\partial y} + \frac{\partial(\rho^k w^k)}{\partial z}\right]\mathrm{d}x\mathrm{d}y\mathrm{d}z \tag{2.6}$$

Referring to Eqn (2.4), the partial differential equation of the conservation of mass can be obtained by equating Eqns (2.5)–(2.6) and, after dividing by the control volume $\mathrm{d}V$ $(=\mathrm{d}x\ \mathrm{d}y\ \mathrm{d}z)$, as

$$\frac{\partial \rho^k}{\partial t} = -\left[\frac{\partial(\rho^k u^k)}{\partial x} + \frac{\partial(\rho^k v^k)}{\partial y} + \frac{\partial(\rho^k w^k)}{\partial z}\right] \tag{2.7}$$

Defining $\mathbf{U}^k \equiv U_j^k \equiv (u^k, v^k, w^k)$, Eqn (2.7) can be written in a compact form, and, noting that the term in brackets is simply the divergence of the mass flux, it becomes

$$\frac{\partial \rho^k}{\partial t} + \frac{\partial\left(\rho^k U_j^k\right)}{\partial x_j} = \frac{\partial \rho^k}{\partial t} + \nabla \cdot \left(\rho^k \mathbf{U}^k\right) = 0 \tag{2.8}$$

Equation (2.8) essentially represents the local instantaneous equation for the conservation of mass of the kth phase.

The phase indicator function $\mathscr{F}^k(x, y, z, t)$ is introduced to distinguish the phases that are present within the fluid flow. By definition,

$$\mathscr{F}^k(x, y, z, t) = \begin{cases} 1 & \text{if } (x, y, z) \text{ is in } k\text{th phase at time } t \\ 0 & \text{otherwise} \end{cases} \tag{2.9}$$

It can be demonstrated from Drew and Passman (1999) that the topological equation reflecting the material derivatives of $\mathscr{F}^k (D\mathscr{F}^k/Dt)$ for each kth phase following the interface velocity $\mathbf{U}^{\text{int}} \equiv U_j^{\text{int}} \equiv (u^{\text{int}}, v^{\text{int}}, w^{\text{int}})$ vanish as

$$\underbrace{\frac{\partial \mathscr{F}^k}{\partial t} + U_j^{\text{int}} \frac{\partial \mathscr{F}^k}{\partial x_j}}_{D\mathscr{F}_k/Dt} = \underbrace{\frac{\partial \mathscr{F}^k}{\partial t} + \mathbf{U}^{\text{int}} \cdot \nabla \mathscr{F}^k}_{D\mathscr{F}_k/Dt} = 0 \tag{2.10}$$

Figure 2.2 describes the distinct fields of different phases that are separated by an interface, and each phase experiences different velocity on the interface. On the basis of Eqn (2.10), it can be observed that both partial derivatives of $\mathscr{F}^k$ vanish away from the interface; the left-hand side of Eqn (2.10) vanishes identically according to the topological equation. The phase indicator function $\mathscr{F}^k$ on the interface can be regarded as a jump condition that remains constant so that the material derivatives following the interface vanish. It should be noted that if mass transfer exists across the interface from one fluid to the other, the interface moves not

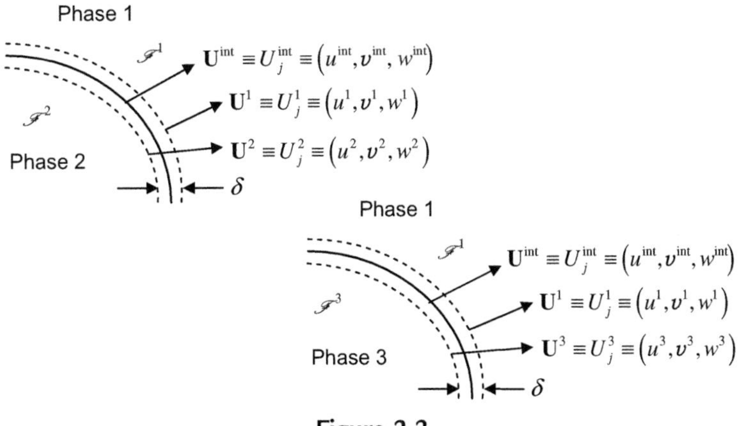

Figure 2.2
Definition of the interface characteristics.

only by convection but also by the amount of mass being transferred between the fields. In such a case, the interface velocity is not equivalent to the neighboring velocities.

The averaged form of the equation for the conservation of mass can be derived by first multiplying Eqn (2.8) by $\mathscr{F}^k$, which is written as

$$\mathscr{F}^k \frac{\partial \rho^k}{\partial t} + \mathscr{F}^k \nabla \cdot \left(\rho^k \mathbf{U}^k\right) = 0 \tag{2.11}$$

Using Eqn (2.10) and applying the following identities below:

$$\mathscr{F}^k \frac{\partial \rho^k}{\partial t} = \frac{\partial\left(\rho^k \mathscr{F}^k\right)}{\partial t} - \rho^k \frac{\partial \mathscr{F}^k}{\partial t} = \frac{\partial\left(\rho^k \mathscr{F}^k\right)}{\partial t} + \rho^k \mathbf{U}^{\text{int}} \cdot \nabla \mathscr{F}^k \tag{2.12}$$

and

$$\mathscr{F}^k \nabla \cdot \left(\rho^k \mathbf{U}^k\right) = \nabla \cdot \left(\mathscr{F}^k \rho^k \mathbf{U}^k\right) - \rho^k \mathbf{U}^k \cdot \nabla \mathscr{F}^k \tag{2.13}$$

yields

$$\frac{\partial\left(\mathscr{F}^k \rho^k\right)}{\partial t} + \nabla \cdot \left(\mathscr{F}^k \rho^k \mathbf{U}^k\right) = \rho^k \left(\mathbf{U}^k - \mathbf{U}^{\text{int}}\right) \cdot \nabla \mathscr{F}^k \tag{2.14}$$

Assuming the averaging process on the field ϕ satisfies the following rules:

$$\left.\begin{aligned} \langle\langle a \rangle\rangle &= \langle a \rangle && (2.15)\\ \langle a + b \rangle &= \langle a \rangle + \langle b \rangle && (2.16)\\ \langle\langle a \rangle b\rangle &= \langle a \rangle \langle b \rangle && (2.17) \end{aligned}\right\} \textit{Reynolds rule}$$

$$\left\langle \frac{\partial a}{\partial t} \right\rangle = \frac{\partial \langle a \rangle}{\partial t} \quad \textit{Leibnitz rule} \tag{2.18}$$

$$\left\langle \frac{\partial a}{\partial x_j} \right\rangle = \frac{\partial \langle a \rangle}{\partial x_j} = \nabla \langle a \rangle \quad \textit{Gauss rule} \tag{2.19}$$

By applying the above averaging rules to Eqn (2.14), the instantaneous averaged equation for the conservation of mass is given by

$$\begin{aligned} &\left\langle \frac{\partial\left(\mathscr{F}^k \rho^k\right)}{\partial t} \right\rangle + \left\langle \nabla \cdot \left(\mathscr{F}^k \rho^k \mathbf{U}^k\right) \right\rangle = \left\langle \rho^k \left(\mathbf{U}^k - \mathbf{U}^{\text{int}}\right) \cdot \nabla \mathscr{F}^k \right\rangle \Rightarrow \\ &\frac{\partial \left\langle \mathscr{F}^k \rho^k \right\rangle}{\partial t} + \nabla \cdot \left\langle \mathscr{F}^k \rho^k \mathbf{U}^k \right\rangle = \underbrace{\left\langle \rho^k \left(\mathbf{U}^k - \mathbf{U}^{\text{int}}\right) \cdot \nabla \mathscr{F}^k \right\rangle}_{\Gamma^k} \end{aligned} \tag{2.20}$$

The term on the right-hand side in Eqn (2.20) represents the interfacial mass source.

2.1.2 Momentum Conservation

The second basic equation of fluid motion for each continuous phase is the conservation of momentum. In order to derive the instantaneous equation for the kth phase, consider the fluid element such has been described in Figure 2.1 for the conservation of mass. Newton's second law of motion states that

$$\begin{array}{lcl} \textit{Rate increase of} & & \textit{Sum of forces} \\ \textit{momentum of the} & = & \textit{acting on the fluid} \\ \textit{fluid element} & & \textit{element} \end{array} \tag{2.21}$$

For the derivation of the equation governing the conservation of momentum, there are effectively three scalar relations along the x, y, and z directions of the Cartesian frame of which this particular fundamental law can be invoked. The x-component of Newton's second law is expressed as

$$m^k a_x^k = \sum F_x^k \tag{2.22}$$

where F_x^k and a_x^k are the force and acceleration along the x direction for the kth phase. The acceleration a_x^k from the above equation is simply the time rate change of u^k, which is given by the material derivative of u^k

$$a_x^k = \frac{Du^k}{Dt} \tag{2.23}$$

Given that the mass of the fluid element m^k is $\rho^k\,\mathrm{d}V$ ($=\mathrm{d}x\,\mathrm{d}y\,\mathrm{d}z$), the rate of increase of momentum of the fluid element is simply

$$\rho^k \frac{Du^k}{Dt}\mathrm{d}x\mathrm{d}y\mathrm{d}z \tag{2.24}$$

On the left-hand side of Eqn (2.22), the sum of forces acting on the fluid element normally comprises two sources: surface forces and body forces. These forces are normally incorporated as additional source/sink terms in the momentum equations. The surface forces for the velocity component u^k as illustrated in Figure 2.3 deforming the fluid element are specifically due to the normal stress σ_{xx}^k and tangential stresses τ_{yx}^k and τ_{zx}^k acting on the surfaces of the fluid element. In the x direction, the net force is the sum of the force components acting on the fluid element. Considering the velocity component u^k, the surface forces are due to the normal viscous stress σ_{xx}^k and tangential viscous stresses τ_{yx}^k and τ_{zx}^k acting on the surfaces of the fluid element. The total net force per unit volume on the fluid due to these surface stresses should be equal to the sum of the normal and tangential forces. Hence, the total net force per unit volume along the x direction is

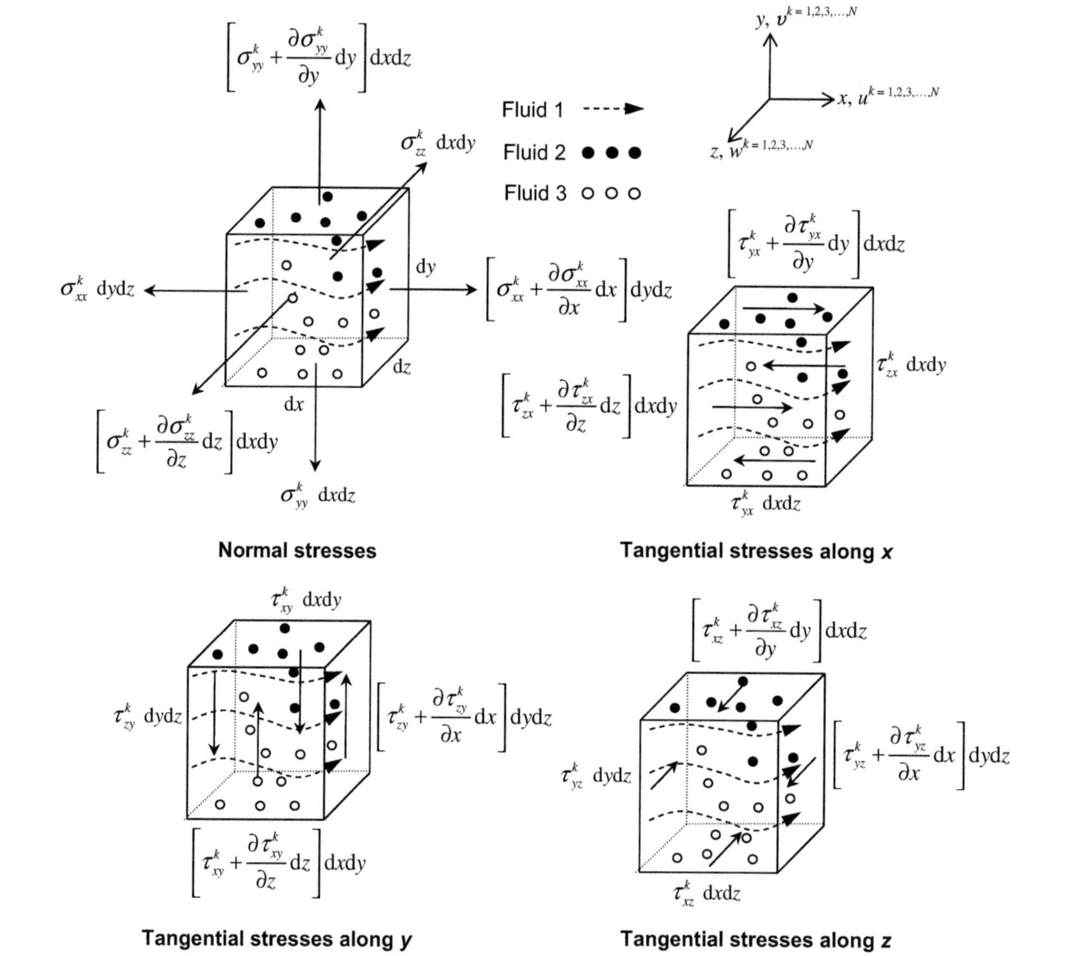

Figure 2.3

Normal and tangential stresses acting on infinitesimal control volumes for velocity components u^k, v^k, and w^k along the Cartesian directions of x, y, and z. Fluid 1, fluid 2, and fluid 3 are taken as continuous fluids within the elemental volume.

$$\left[\frac{\partial \sigma_{xx}^k}{\partial x} + \frac{\partial \tau_{yx}^k}{\partial y} + \frac{\partial \tau_{zx}^k}{\partial z}\right] dxdydz \tag{2.25}$$

Along the y direction and z direction, the total net forces per unit volume on the control volume surfaces can also be similarly derived as

$$\left[\frac{\partial \tau_{xy}^k}{\partial x} + \frac{\partial \sigma_{yy}^k}{\partial y} + \frac{\partial \tau_{zy}^k}{\partial z}\right] dxdydz \tag{2.26}$$

$$\left[\frac{\partial \tau_{xz}^k}{\partial x}+\frac{\partial \tau_{yz}^k}{\partial y}+\frac{\partial \sigma_{zz}^k}{\partial z}\right]\mathrm{d}x\mathrm{d}y\mathrm{d}z \tag{2.27}$$

Combining Eqn (2.25) with the time rate of change of the velocity component u^k given by Eqn (2.24) and the presence of body forces, the x-momentum equation is given by

$$\begin{aligned}\rho^k\frac{Du^k}{Dt}\mathrm{d}x\mathrm{d}y\mathrm{d}z &= \left[\frac{\partial \sigma_{xx}^k}{\partial x}+\frac{\partial \tau_{yx}^k}{\partial y}+\frac{\partial \tau_{zx}^k}{\partial z}+\sum F_x^{k,\text{body forces}}\right]\mathrm{d}x\mathrm{d}y\mathrm{d}z \Rightarrow \\ \rho^k\frac{Du^k}{Dt} &= \frac{\partial \sigma_{xx}^k}{\partial x}+\frac{\partial \tau_{yx}^k}{\partial y}+\frac{\partial \tau_{zx}^k}{\partial z}+\sum F_x^{k,\text{body forces}}\end{aligned} \tag{2.28}$$

Analogously, the y-momentum and z-momentum equations, using Eqns (2.26) and (2.27), can be written as

$$\rho^k\frac{Dv^k}{Dt}=\frac{\partial \tau_{xy}^k}{\partial x}+\frac{\partial \sigma_{yy}^k}{\partial y}+\frac{\partial \tau_{zy}^k}{\partial z}+\sum F_y^{k,\text{body forces}} \tag{2.29}$$

$$\rho^k\frac{Dw^k}{Dt}=\frac{\partial \tau_{xz}^k}{\partial x}+\frac{\partial \tau_{yz}^k}{\partial y}+\frac{\partial \sigma_{zz}^k}{\partial z}+\sum F_z^{k,\text{body forces}} \tag{2.30}$$

Defining $\mathbf{T}^k \equiv (\sigma_{xx}^k, \tau_{xy}^k, \ldots, \tau_{yz}^k, \sigma_{zz}^k)$ to be the Cauchy stress tensor representing the respective stresses along the Cartesian coordinate directions of x, y, and z and $\sum \mathbf{F}^{k,\text{body forces}} \equiv (\sum F_x^{k,\text{body forces}}, \sum F_y^{k,\text{body forces}}, \sum F_z^{k,\text{body forces}})$, the conservation momentum equation can be written in a compact form as

$$\rho^k\frac{\partial \mathbf{U}^k}{\partial t}+\rho^k\mathbf{U}^k\cdot\nabla\mathbf{U}^k=\nabla\cdot\mathbf{T}^k+\sum \mathbf{F}^{k,\text{body forces}} \tag{2.31}$$

In many fluid flows, it is customary to represent $\mathbf{T}^k$ in terms of pressure and extra stresses. Describing the normal stresses as σ_{xx}^k, σ_{yy}^k, and σ_{zz}^k in terms of the kth phase pressure p^k and normal viscous stress components τ_{xx}^k, τ_{yy}^k, and τ_{zz}^k acting perpendicular to the control volume as

$$\sigma_{xx}^k=-p^k+\tau_{xx}^k \quad \sigma_{yy}^k=-p^k+\tau_{yy}^k \quad \sigma_{zz}^k=-p^k+\tau_{zz}^k \tag{2.32}$$

the conservation momentum equation can be alternatively written as

$$\rho^k\frac{\partial \mathbf{U}^k}{\partial t}+\rho^k\mathbf{U}^k\cdot\nabla\mathbf{U}^k=-\nabla p^k+\nabla\cdot\boldsymbol{\tau}^k+\sum \mathbf{F}^{k,\text{body forces}} \tag{2.33}$$

where $\boldsymbol{\tau}^k$ denotes the extra stresses, i.e. $\boldsymbol{\tau}^k \equiv (\tau_{xx}^k, \tau_{xy}^k, \ldots, \tau_{yz}^k, \tau_{zz}^k)$. Equation (2.31) represents the local instantaneous equation for the conservation of momentum of the kth phase.

Similar to the derivation of the averaged equation governing the conservation of mass, the averaged equation governing the conservation of momentum can be derived by first multiplying Eqn (2.31) with the phase indictor function, and subsequently applying the Reynolds, Leibnitz, and Gauss rules. In other words,

$$\mathscr{F}^k \rho^k \frac{\partial \mathbf{U}^k}{\partial t} + \mathscr{F}^k \rho^k \mathbf{U}^k \cdot \nabla \mathbf{U}^k = -\mathscr{F}^k \nabla p^k + \mathscr{F}^k \nabla \cdot \tau^k + \mathscr{F}^k \sum \mathbf{F}^{k,\text{body forces}} \tag{2.34}$$

For the first term (local acceleration term) on the left-hand side of Eqn (2.34),

$$\mathscr{F}^k \rho^k \frac{\partial \mathbf{U}^k}{\partial t} = \frac{\partial \left(\mathscr{F}^k \rho^k \mathbf{U}^k\right)}{\partial t} - \mathbf{U}^k \frac{\partial \left(\mathscr{F}^k \rho^k\right)}{\partial t} \tag{2.35}$$

Equation (2.35) can also be rewritten by employing Eqn (2.14) for the time derivative term $\partial(\mathscr{F}^k \rho^k)/\partial t$ according to

$$\mathscr{F}^k \rho^k \frac{\partial \mathbf{U}^k}{\partial t} = \frac{\partial \left(\mathscr{F}^k \rho^k \mathbf{U}^k\right)}{\partial t} + \mathbf{U}^k \cdot \nabla \left(\mathscr{F}^k \rho^k \mathbf{U}^k\right) - \rho^k \mathbf{U}^k \left(\mathbf{U}^k - \mathbf{U}^{\text{int}}\right) \cdot \nabla \mathscr{F}^k \tag{2.36}$$

For the second term (advection term) on the left-hand side of Eqn (2.34),

$$\mathscr{F}^k \rho^k \mathbf{U}^k \cdot \nabla \mathbf{U}^k = \nabla \cdot \left(\mathscr{F}^k \rho^k \mathbf{U}^k \otimes \mathbf{U}^k\right) - \mathbf{U}^k \cdot \nabla \left(\mathscr{F}^k \rho^k \mathbf{U}^k\right) \tag{2.37}$$

in which

$$\mathbf{U}^k \otimes \mathbf{U}^k = \begin{pmatrix} u^k \mathbf{i} \\ v^k \mathbf{j} \\ w^k \mathbf{k} \end{pmatrix} \begin{pmatrix} u^k \mathbf{i} & v^k \mathbf{j} & w^k \mathbf{k} \end{pmatrix} = \begin{pmatrix} u^k u^k \mathbf{ii} & u^k v^k \mathbf{ij} & u^k w^k \mathbf{ik} \\ v^k u^k \mathbf{ji} & v^k v^k \mathbf{jj} & v^k w^k \mathbf{jk} \\ w^k u^k \mathbf{ki} & w^k v^k \mathbf{kj} & w^k w^k \mathbf{kk} \end{pmatrix}$$

where **i**, **j**, and **k** are unit vectors. On the right-hand side of Eqn (2.34), the pressure and extra stresses terms can be expressed as

$$\mathscr{F}^k \nabla p^k = \nabla \left(\mathscr{F}^k p^k\right) - p^k \nabla \mathscr{F}^k \tag{2.38}$$

$$\mathscr{F}^k \nabla \cdot \tau^k = \nabla \cdot \left(\mathscr{F}^k \tau^k\right) - \tau^k \cdot \nabla \mathscr{F}^k \tag{2.39}$$

Substituting Eqns (2.35)–(2.39) into Eqn (2.34) yields

$$\begin{aligned} \frac{\partial \left(\mathscr{F}^k \rho^k \mathbf{U}^k\right)}{\partial t} + \nabla \cdot \left(\mathscr{F}^k \rho^k \mathbf{U}^k \otimes \mathbf{U}^k\right) &= -\nabla \left(\mathscr{F}^k p^k\right) + \nabla \cdot \left(\mathscr{F}^k \tau^k\right) \\ &\quad + p^k \nabla \mathscr{F}^k k \mathbf{U}^k \left(\mathbf{U}^k - \mathbf{U}^{\text{int}}\right) \cdot \nabla \mathscr{F}^k - \tau^k \cdot \nabla \mathscr{F}^k \\ &\quad + \mathscr{F}^k \sum \mathbf{F}^{k,\text{body forces}} \end{aligned} \tag{2.40}$$

Applying the Reynolds, Leibnitz, and Gauss rules as previously described in section 2.1.1, the averaged equation governing the conservation of momentum is given by

$$\frac{\partial\langle \mathscr{F}^k\rho^k\mathbf{U}^k\rangle}{\partial t} + \nabla\cdot\langle \mathscr{F}^k\rho^k\mathbf{U}^k\otimes\mathbf{U}^k\rangle$$

$$= -\nabla\langle \mathscr{F}^k p^k\rangle + \nabla\cdot\left\langle \mathscr{F}^k\tau_{ij}^k\right\rangle + \langle \mathscr{F}^k\rangle\left\langle \sum \mathbf{F}^{k,\text{body forces}}\right\rangle \tag{2.41}$$

$$+ \underbrace{\langle \rho^k\mathbf{U}^k(\mathbf{U}^k - \mathbf{U}^{\text{int}})\cdot\nabla\mathscr{F}^k\rangle + \langle p^k\rangle\langle\nabla\mathscr{F}^k\rangle - \langle \boldsymbol{\tau}^k\cdot\nabla\mathscr{F}^k\rangle}_{\boldsymbol{\Omega}^k}$$

The term $\boldsymbol{\Omega}^k$ on the right-hand side in Eqn (2.41) represents the interfacial momentum sources.

2.1.3 Energy Conservation

The third basic equation of fluid motion for each continuous phase is the conservation of energy. Based on the first law of thermodynamics, the energy equation can be derived for the continuous fluid assuming no addition or removal of heat due to external heat sources as

$$\begin{array}{c} \textit{Rate increase of} \\ \textit{energy of the} \\ \textit{fluid element} \end{array} = \underbrace{\begin{array}{c} \textit{Net rate of heat} \\ \textit{added to the} \\ \textit{fluid element} \end{array}}_{\sum\dot{Q}} + \underbrace{\begin{array}{c} \textit{Net rate of work} \\ \textit{done on the fluid} \\ \textit{element} \end{array}}_{\sum\dot{W}} \tag{2.42}$$

Analogous to the consideration of the conservation of momentum, the time rate of change of energy for the moving fluid element of a kth phase is simply the product between the density ρ^k and the substantial derivative of the energy E^k. The rate of increase of energy of the fluid element is given by

$$\rho^k\frac{DE^k}{Dt}\mathrm{d}x\mathrm{d}y\mathrm{d}z \tag{2.43}$$

Referring to Figure 2.4, the rate of work done on the elemental volume in the x direction is the product between the surface forces (caused by the normal stress σ_{xx}^k and tangential stresses τ_{yx}^k and τ_{zx}^k) and the velocity component u^k. The net rate of work done by these surface forces acting along the x direction can be written as

$$\left[\frac{\partial(u^k\sigma_{xx}^k)}{\partial x} + \frac{\partial\left(u^k\tau_{yx}^k\right)}{\partial y} + \frac{\partial(u^k\tau_{zx}^k)}{\partial z}\right]\mathrm{d}x\mathrm{d}y\mathrm{d}z \tag{2.44}$$

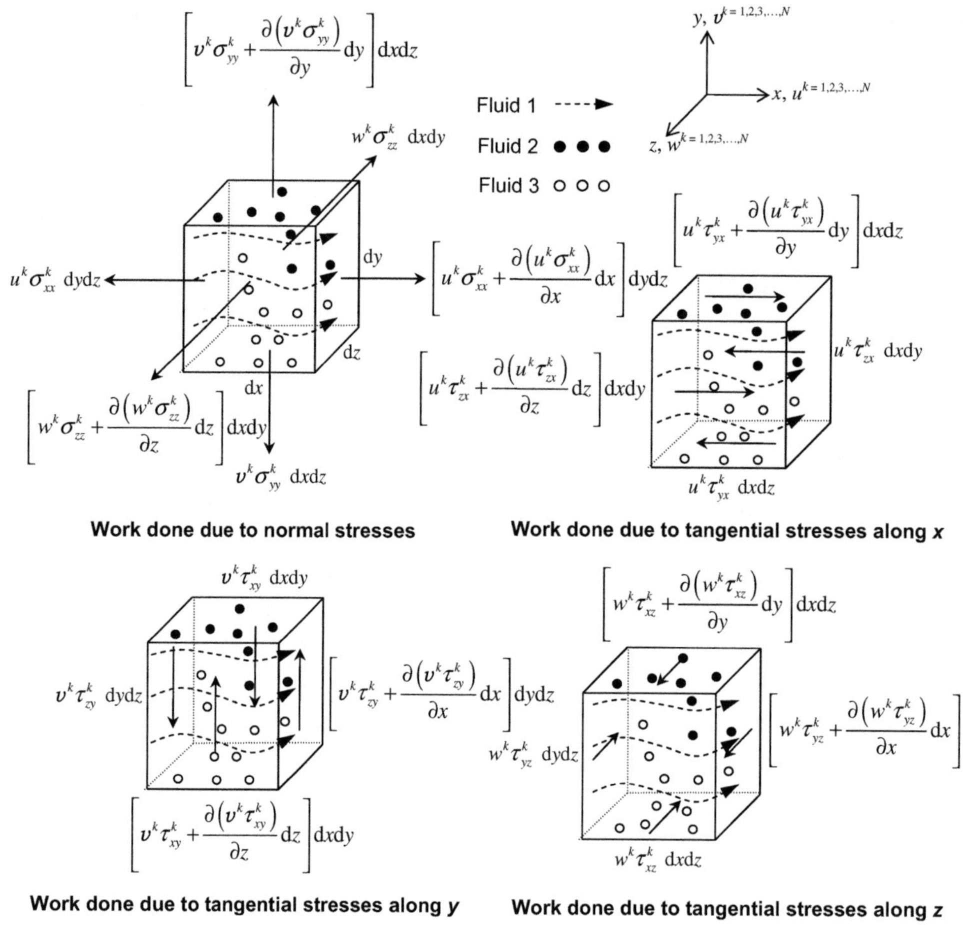

Figure 2.4

Work done due to normal and tangential stresses along the Cartesian directions of *x*, *y*, and *z*. Fluid 1, fluid 2, and fluid 3 are taken as continuous fluids within the elemental volume.

Work done due to surface stress components along the *y* direction and *z* direction is similarly derived; these additional rates of work done on the fluid are

$$\left[\frac{\partial\left(v^k\tau_{xy}^k\right)}{\partial x}+\frac{\partial\left(v^k\sigma_{yy}^k\right)}{\partial y}+\frac{\partial\left(v^k\tau_{zy}^k\right)}{\partial z}\right]dxdydz \qquad (2.45)$$

$$\left[\frac{\partial\left(w^k\tau_{xz}^k\right)}{\partial x}+\frac{\partial\left(w^k\tau_{yz}^k\right)}{\partial y}+\frac{\partial\left(w^k\sigma_{zz}^k\right)}{\partial z}\right]dxdydz \qquad (2.46)$$

In addition to the work done by surface forces on the fluid element, possible work done due to body forces is also considered. Hence, the net rate work done is given by

$$\sum \dot{W} = \left[\frac{\partial(u^k\sigma^k_{xx})}{\partial x} + \frac{\partial\left(u^k\tau^k_{yx}\right)}{\partial y} + \frac{\partial(u^k\tau^k_{zx})}{\partial z} + \frac{\partial\left(v^k\tau^k_{xy}\right)}{\partial x} + \frac{\partial\left(v^k\sigma^k_{yy}\right)}{\partial y} + \frac{\partial\left(v^k\tau^k_{zy}\right)}{\partial z} + \frac{\partial(w^k\tau^k_{xz})}{\partial x} \right.$$
$$+ \frac{\partial\left(w^k\tau^k_{yz}\right)}{\partial y} + \frac{\partial(w^k\sigma^k_{zz})}{\partial z} + \sum F_x^{k,\text{body force}} u^k + \sum F_y^{k,\text{body force}} v^k$$
$$\left. + \sum F_z^{k,\text{body force}} w^k \right] dxdydz \tag{2.47}$$

Substituting the normal stress relationships for σ^k_{xx}, σ^k_{yy}, and σ^k_{zz} in terms of the kth phase pressure and normal viscous stress components as previously described in the derivation of the conservation of momentum from Eqn (2.47), the net rate work done on the fluid element becomes

$$\sum \dot{W} = \left[-\frac{\partial(p^k u^k)}{\partial x} - \frac{\partial(p^k v^k)}{\partial y} - \frac{\partial(p^k w^k)}{\partial z} + \frac{\partial(u^k\tau^k_{xx})}{\partial x} + \frac{\partial\left(u^k\tau^k_{yx}\right)}{\partial y} + \frac{\partial(u^k\tau^k_{zx})}{\partial z} + \frac{\partial\left(v^k\tau^k_{xy}\right)}{\partial x} \right.$$
$$+ \frac{\partial\left(v^k\tau^k_{yy}\right)}{\partial y} + \frac{\partial\left(v^k\tau^k_{zy}\right)}{\partial z} + \frac{\partial(w^k\tau^k_{xz})}{\partial x} + \frac{\partial\left(w^k\tau^k_{yz}\right)}{\partial y} + \frac{\partial(w^k\tau^k_{zz})}{\partial z} + \sum F_x^{k,body\ force} u^k$$
$$\left. + \sum F_y^{k,body\ force} v^k + \sum F_z^{k,body\ force} w^k \right] dxdydz \tag{2.48}$$

For the consideration of heat addition, the net rate of heat transfer to the fluid due to the heat flow along the x direction, y direction, and z direction is given by the difference between the heat entering and leaving the control volume such as is illustrated in Figure 2.5. The total rate of heat added is

$$\sum \dot{Q} = \left[-\frac{\partial q^k_x}{\partial x} - \frac{\partial q^k_y}{\partial y} - \frac{\partial q^k_z}{\partial z} \right] dxdydz \tag{2.49}$$

Taking the contributions based on Eqns (2.43), (2.48), and (2.49), and substituting them into Eqn (2.42), the conservation energy equation after dividing by the control volume $dx\,dy\,dz$ can be written as

$$\rho^k \frac{DE^k}{Dt} = -\frac{\partial(p^k u^k)}{\partial x} - \frac{\partial(p^k v^k)}{\partial y} - \frac{\partial(p^k w^k)}{\partial z} - \frac{\partial q^k_x}{\partial x} - \frac{\partial q^k_y}{\partial y} - \frac{\partial q^k_z}{\partial z} + \sum F_x^{k,\text{body force}} u^k$$
$$+ \sum F_y^{k,\text{body force}} v^k + \sum F_z^{k,\text{body force}} w^k + \Phi^k_E \tag{2.50}$$

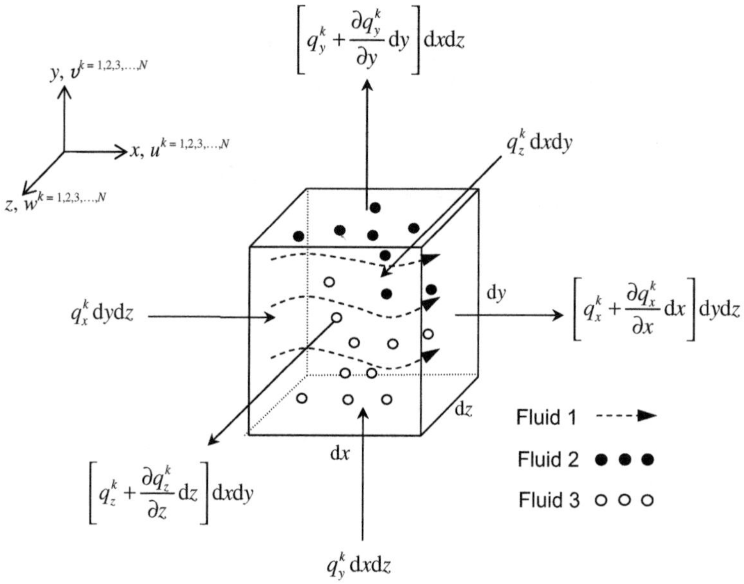

Figure 2.5

Heat added to the fluid along the Cartesian directions of *x*, *y*, and *z*. Fluid 1, fluid 2, and fluid 3 are taken as continuous fluids within the elemental volume.

The effects due to the viscous extra stresses in the above conservation energy equation are described by Φ_E^k:

$$\Phi_E^k = \frac{\partial\left(u^k\tau_{xx}^k\right)}{\partial x} + \frac{\partial\left(u^k\tau_{yx}^k\right)}{\partial y} + \frac{\partial\left(u^k\tau_{zx}^k\right)}{\partial z} + \frac{\partial\left(v^k\tau_{xy}^k\right)}{\partial x} + \frac{\partial\left(v^k\tau_{yy}^k\right)}{\partial y} + \frac{\partial\left(v^k\tau_{zy}^k\right)}{\partial z} + \frac{\partial\left(w^k\tau_{xz}^k\right)}{\partial x} + \frac{\partial\left(w^k\tau_{yz}^k\right)}{\partial y} + \frac{\partial\left(w^k\tau_{zz}^k\right)}{\partial z} \tag{2.51}$$

Equation (2.51) effectively represents a source of energy due to work done deforming the fluid element. This work is extracted from the mechanical energy that causes fluid movement, which is subsequently converted into heat.

The specific energy E^k of a fluid can often be defined as the sum of the specific internal energy and kinetic energy:

$$E^k = \underbrace{e^k}_{\substack{\text{specific internal energy}\\ \text{for the } k\text{th phase}}} + \underbrace{\frac{1}{2}\left(u^k u^k + v^k v^k + w^k w^k\right)}_{\substack{\text{kinetic energy}\\ \text{for the } k\text{th phase}}} \tag{2.52}$$

Substituting Eqn (2.52) into Eqn (2.50) results in the local instantaneous equation for the conservation of energy in terms of the specific internal energy and kinetic energy. Rather than using this form directly, it is common practice to extract the changes of the kinetic energy in order to obtain an equation for the specific internal energy alone. This can be accomplished by multiplying u^k to Eqn (2.28), v^k to Eqn (2.29), and w^k to Eqn (2.30) and then adding the resultant equations together. Expressing in terms of the kth phase pressure and normal viscous stress components via Eqn (2.32), the conservation equation for the kinetic energy can be written as

$$\begin{aligned}\rho^k\frac{D}{Dt}\left(\frac{1}{2}\left(u^ku^k+v^kv^k+w^kw^k\right)\right) =& -u^k\frac{\partial p^k}{\partial x}-v^k\frac{\partial p^k}{\partial y}-w^k\frac{\partial p^k}{\partial z}\\ &+\sum F_x^{k,\text{body force}}u^k+\sum F_y^{k,\text{body force}}v^k\\ &+\sum F_z^{k,\text{body force}}w^k+u^k\left(\frac{\partial\tau_{xx}^k}{\partial x}+\frac{\partial\tau_{yx}^k}{\partial y}+\frac{\partial\tau_{zx}^k}{\partial z}\right)\\ &+v^k\left(\frac{\partial\tau_{xy}^k}{\partial x}+\frac{\partial\tau_{yy}^k}{\partial y}+\frac{\partial\tau_{zy}^k}{\partial z}\right)+w^k\left(\frac{\partial\tau_{xz}^k}{\partial x}+\frac{\partial\tau_{yz}^k}{\partial y}+\frac{\partial\tau_{zz}^k}{\partial z}\right)\end{aligned} \tag{2.53}$$

Subtracting Eqn (2.53) from Eqn (2.50) yields the specific internal energy equation, which is given by

$$\rho^k\frac{De^k}{Dt}=-p^k\left(\frac{\partial u^k}{\partial x}+\frac{\partial v^k}{\partial y}+\frac{\partial w^k}{\partial z}\right)-\frac{\partial q_x^k}{\partial x}-\frac{\partial q_y^k}{\partial y}-\frac{\partial q_z^k}{\partial z}+\Phi_e^k \tag{2.54}$$

where

$$\Phi_e^k=\tau_{xx}^k\frac{\partial u^k}{\partial x}+\tau_{yx}^k\frac{\partial u^k}{\partial y}+\tau_{zx}^k\frac{\partial u^k}{\partial z}+\tau_{xy}^k\frac{\partial v^k}{\partial x}+\tau_{yy}^k\frac{\partial v^k}{\partial y}+\tau_{zy}^k\frac{\partial v^k}{\partial z}+\tau_{xz}^k\frac{\partial w^k}{\partial x}+\tau_{yz}^k\frac{\partial w^k}{\partial y}+\tau_{zz}^k\frac{\partial w^k}{\partial z} \tag{2.55}$$

Equation (2.54) represents another form of the equation governing the conservation of energy. Here, the kinetic energy and body force terms have dropped out. It is important to note that this form of energy equation, in terms of the specific internal energy alone, does not explicitly account for the body force.

Equation (2.50) can also be rearranged to give an equation for the enthalpy. The sensible enthalpy h_s^k and the total enthalpy H^k of a fluid can be defined as

$$h_s^k=e^k+\frac{p^k}{\rho^k}\quad\text{and}\quad H^k=h_s^k+\frac{1}{2}\left(u^ku^k+v^kv^k+w^kw^k\right)$$

Combining these two definitions with the specific energy E^k yields

$$H^k = e^k + \frac{p^k}{\rho^k} + \frac{1}{2}\left(u^k u^k + v^k v^k + w^k w^k\right) = E^k + \frac{p^k}{\rho^k} \tag{2.56}$$

Substituting Eqn (2.54) into Eqn (2.48), the total enthalpy equation can be written as

$$\begin{aligned}\rho^k \frac{DH^k}{Dt} = & -\frac{\partial p^k}{\partial t} - \frac{\partial q_x^k}{\partial x} - \frac{\partial q_y^k}{\partial y} - \frac{\partial q_z^k}{\partial z} \\ & + \sum F_x^{k,\text{body force}} u^k + \sum F_y^{k,\text{body force}} v^k \\ & + \sum F_z^{k,\text{body force}} w^k + \Phi_H^k\end{aligned} \tag{2.57}$$

where $\Phi_H^k = \Phi_E^k$.

Among the many different forms governing the conservation of energy as derived above, the total enthalpy equation represents a convenient form that is most frequently used in multiphase flow investigations. Henceforth, we will concentrate on the derivation of the averaged form of the total enthalpy equation. Expressing Eqn (2.57) in a compact form:

$$\rho^k \frac{\partial H^k}{\partial t} + \rho^k \mathbf{U}^k \cdot \nabla H^k = -\frac{\partial p^k}{\partial t} - \nabla \cdot \mathbf{q}^k + \sum \mathbf{F}^{k,\text{body force}} \cdot \mathbf{U}^k + \Phi_H^k \tag{2.58}$$

where $\Phi_H^k = \nabla \cdot (\mathbf{U}^k \cdot \boldsymbol{\tau}^k)$ and multiplying the phase indicator function gives

$$\begin{aligned}\mathscr{F}^k \rho^k \frac{\partial H^k}{\partial t} + \mathscr{F}^k \rho^k \mathbf{U}^k \cdot \nabla H^k = & -\mathscr{F}^k \frac{\partial p^k}{\partial t} - \mathscr{F}^k \nabla \cdot \mathbf{q}^k \\ & + \mathscr{F}^k \sum \mathbf{F}^{k,\text{body force}} \cdot \mathbf{U}^k + \mathscr{F}^k \Phi_H^k\end{aligned} \tag{2.59}$$

By applying the following general entities:

$$\mathscr{F}^k \rho^k \frac{\partial \phi}{\partial t} = \frac{\partial(\mathscr{F}^k \rho^k \phi)}{\partial t} + \phi \nabla \cdot (\mathscr{F}^k \rho^k \mathbf{U}^k) - \rho^k \phi (\mathbf{U}^k - \mathbf{U}^{\text{int}}) \cdot \nabla \mathscr{F}^k \tag{2.60}$$

$$\mathscr{F}^k \rho^k \mathbf{U} \cdot \nabla \phi = \nabla \cdot (\mathscr{F}^k \rho^k \mathbf{U}^k \phi) - \phi \nabla \cdot (\mathscr{F}^k \rho^k \mathbf{U}^k) \tag{2.61}$$

$$\mathscr{F}^k \frac{\partial \phi}{\partial t} = \frac{\partial(\mathscr{F}^k \phi)}{\partial t} - \phi \frac{\partial \mathscr{F}^k}{\partial t} \tag{2.62}$$

$$\mathscr{F}^k \nabla \cdot \phi = \nabla \cdot (\mathscr{F}^k \phi) - \phi \cdot \nabla \mathscr{F}^k \tag{2.63}$$

Equation (2.59) can be alternatively expressed by

$$\frac{\partial(\mathscr{F}^k\rho^kH^k)}{\partial t}+\nabla\cdot\left(\mathscr{F}^k\rho^k\mathbf{U}^kH^k\right)=-\frac{\partial(\mathscr{F}^kp^k)}{\partial t}-\nabla\cdot\left(\mathscr{F}^k\mathbf{q}^k\right)+\mathscr{F}^k\sum\mathbf{F}^{k,\text{body force}}\cdot\mathbf{U}^k$$
$$+\underbrace{\mathscr{F}^k\Phi_H^k}_{\Phi_H'^k+\Phi^{\text{int}}}+\rho^kH^k\left(\mathbf{U}^k-\mathbf{U}^{\text{int}}\right)\cdot\nabla\mathscr{F}^k+\mathbf{q}^k\cdot\nabla\mathscr{F}^k+p^k\frac{\partial\mathscr{F}^k}{\partial t}\tag{2.64}$$

In the above equation, the term $\mathscr{F}^k\Phi_H^k$ may be written in the form of $\Phi_H'^k+\Phi^{\text{int}}$:

$$\Phi_H'^k=\frac{\partial(\mathscr{F}^ku^k\tau_{xx}^k)}{\partial x}+\frac{\partial\left(\mathscr{F}^ku^k\tau_{yx}^k\right)}{\partial y}+\frac{\partial(\mathscr{F}^ku^k\tau_{zx}^k)}{\partial z}+\frac{\partial\left(\mathscr{F}^kv^k\tau_{xy}^k\right)}{\partial x}+\frac{\partial\left(\mathscr{F}^kv^k\tau_{yy}^k\right)}{\partial y}$$
$$+\frac{\partial\left(\mathscr{F}^kv^k\tau_{zy}^k\right)}{\partial z}+\frac{\partial(\mathscr{F}^kw^k\tau_{xz}^k)}{\partial x}+\frac{\partial\left(\mathscr{F}^kw^k\tau_{yz}^k\right)}{\partial y}+\frac{\partial(\mathscr{F}^kw^k\tau_{zz}^k)}{\partial z}\tag{2.65}$$

$$\Phi^{\text{int}}=-u^k\tau_{xx}^k\frac{\partial\mathscr{F}^k}{\partial x}-u^k\tau_{yx}^k\frac{\partial\mathscr{F}^k}{\partial y}-u^k\tau_{zx}^k\frac{\partial\mathscr{F}^k}{\partial z}-v^k\tau_{xy}^k\frac{\partial\mathscr{F}^k}{\partial x}-v^k\tau_{yy}^k\frac{\partial\mathscr{F}^k}{\partial y}$$
$$-v^k\tau_{zy}^k\frac{\partial\mathscr{F}^k}{\partial z}-w^k\tau_{xz}^k\frac{\partial\mathscr{F}^k}{\partial x}-w^k\tau_{yz}^k\frac{\partial\mathscr{F}^k}{\partial y}-w^k\tau_{zz}^k\frac{\partial\mathscr{F}^k}{\partial z}\tag{2.66}$$

Averaging Eqn (2.64) according to Reynolds, Leibnitz, and Gauss rules as previously described in Section 2.1.1, the averaged conservation equation for the total enthalpy is given by

$$\frac{\partial\langle\mathscr{F}^k\rho^kH^k\rangle}{\partial t}+\nabla\cdot\langle\mathscr{F}^k\rho^k\mathbf{U}^kH^k\rangle$$
$$=-\frac{\partial\langle\mathscr{F}^kp^k\rangle}{\partial t}-\nabla\cdot\langle\mathscr{F}^k\mathbf{q}^k\rangle+\langle\mathscr{F}^k\rangle\left\langle\sum\mathbf{F}^{k,\text{body force}}\cdot\mathbf{U}^k\right\rangle+\langle\Phi_H'^k\rangle$$
$$+\underbrace{\langle\rho^kH^k\left(\mathbf{U}^k-\mathbf{U}^{\text{int}}\right)\cdot\nabla\mathscr{F}^k\rangle+\langle\mathbf{q}^k\cdot\nabla\mathscr{F}^k\rangle+\langle p^k\rangle\left\langle\frac{\partial\mathscr{F}^k}{\partial t}\right\rangle+\langle\Phi^{\text{extra}}\rangle}_{\Pi_H^k}\tag{2.67}$$

where

$$\langle\Phi_H'^k\rangle=\nabla\cdot\langle\mathscr{F}^k\mathbf{U}^k\cdot\boldsymbol{\tau}^k\rangle\tag{2.68}$$

$$\langle\Phi^{\text{extra}}\rangle=-\langle\mathbf{U}^k\cdot\boldsymbol{\tau}^k\cdot\nabla\mathscr{F}^k\rangle\tag{2.69}$$

The term Π_H^k, on the right-hand side of Eqn (2.67), represents the interfacial energy terms. Also, the energy fluxes along the Cartesian coordinate directions, q_x, q_y and q_z in the term

$\nabla \cdot \langle \mathscr{F}^k \mathbf{q}^k \rangle$ on the right-hand side of Eqn (2.67) can be formulated by applying Fourier's law of heat conduction that relates heat flux to the local temperature gradient:

$$q_x^k = -\lambda^k \frac{\partial T^k}{\partial x} \quad q_y^k = -\lambda^k \frac{\partial T^k}{\partial y} \quad q_z^k = -\lambda^k \frac{\partial T^k}{\partial z} \tag{2.70}$$

where λ_k is the thermal conductivity for the kth phase. It is worth noting that the term $\partial \mathscr{F}^k / \partial t$ in Eqn (2.67) represents the local acceleration term of the material derivative of the phase indicator function as stipulated by the topological equation Eqn (2.10). It may be rewritten as $\partial \mathscr{F}^k / \partial t = -\mathbf{U}^{\text{int}} \cdot \nabla \mathscr{F}^k$.

2.1.4 Physical Description of Interfacial Exchange Terms

In this section, the physical significance of the various interfacial sources in the averaged equations governing the conservation of mass (Section 2.1.1), momentum (Section 2.1.2), and energy (Section 2.1.3) of the multifluid model is described.

For mass conservation, it is apparent that the gradient $\nabla \mathscr{F}^k$ is zero everywhere except at the interface. As has been demonstrated by Drew (1983), the gradient $\nabla \mathscr{F}^k$appearing in the interfacial mass source Γ^k of Eqn (2.20) behaves like a delta function δ_s and is aligned with the surface unit normal vector pointing to the kth phase according to

$$\nabla \mathscr{F}^k = \mathbf{n}^k \delta_s \left(\mathbf{x} - \mathbf{x}^{\text{int}}, t \right) \tag{2.71}$$

where $\mathbf{n}^k$ is the unit external to the component k, which is shown in Figure 2.6, and $\mathbf{x} \equiv (x, y, z)$ and $\mathbf{x}^{\text{int}} \equiv (x^{\text{int}}, y^{\text{int}}, z^{\text{int}})$. The physical significance of Γ^k that is given by

$$\Gamma^k = \left\langle \rho^k \left(\mathbf{U}^k - \mathbf{U}^{\text{int}} \right) \cdot \mathbf{n}^k \delta_s \right\rangle \tag{2.72}$$

represents the mass flux to the kth phase from the other phases via the interface. Since there cannot be any storage or accumulation of mass at the interface, the averaged interfacial mass balance constraint (jump condition) must be

$$\sum_{k=1}^{2} \Gamma^k = \sum_{k=1}^{2} \left\langle \rho^k \left(\mathbf{U}^k - \mathbf{U}^{\text{int}} \right) \cdot \mathbf{n}^k \delta_s \right\rangle = 0 \tag{2.73}$$

For momentum conservation, focusing on the interfacial pressure at any point along the interface such as is depicted in Figure 2.6, it can usually be expressed by the sum of pressure in the kth phase of either side of the interface and the difference between the pressure at the interface and the pressure in the kth phase, in other words,

$$p_{\text{int}}^k = p^k + \Delta p_{\text{int}}^k \tag{2.74}$$

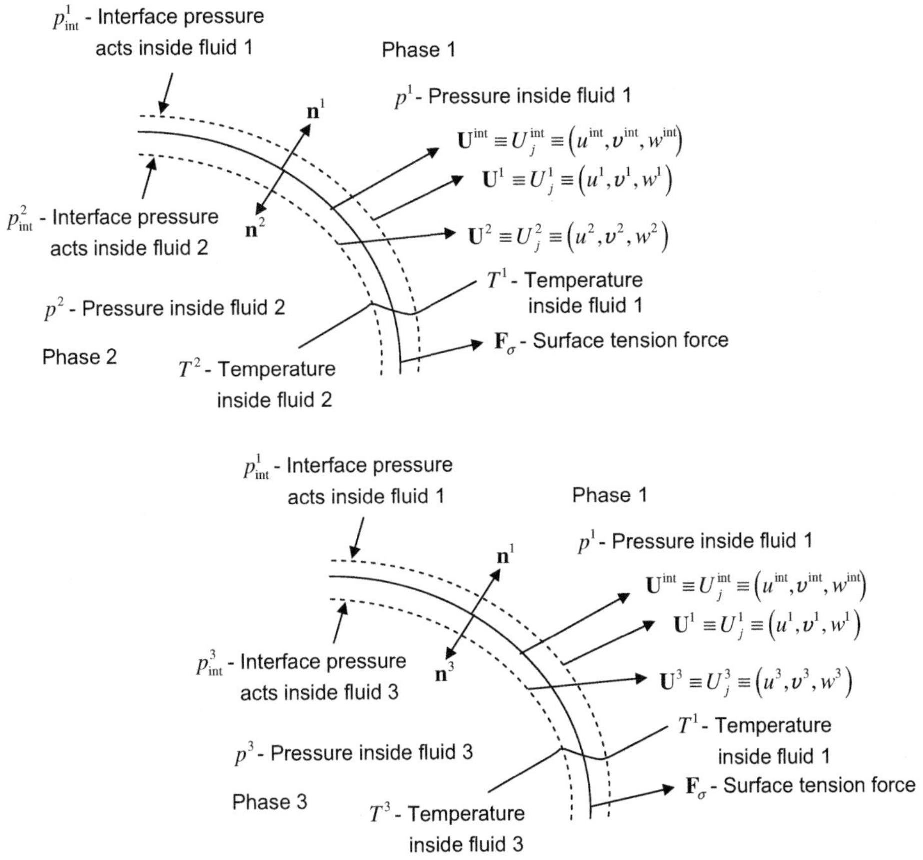

Figure 2.6
Interfacial characteristics for the conservation of mass, momentum, and energy.

The additional term denoted by Δp^k_{int} on the left-hand side of the above equation denotes a nonzero contribution, especially when considering the case of the pressure at the surface of a bubble or drop as it moves through a continuous fluid. Owing to the curvature of the interface, the continuous phase accelerates in a layer in the immediate area surrounding the bubble or drop giving a pressure lower than the surrounding pressure. Expressing $\mathbf{\Omega}^k$ of Eqn (2.41) in terms of the interfacial pressure and difference between the phase and interfacial pressures, the interfacial momentum sources can be immediately written as

$$\mathbf{\Omega}^k \equiv \left\langle \rho^k \mathbf{U}^k \left(\mathbf{U}^k - \mathbf{U}^{int}\right) \cdot \nabla \mathcal{F}^k \right\rangle + \left\langle p^k_{int} \right\rangle \left\langle \nabla \mathcal{F}^k \right\rangle + \left\langle \left(p^k - p^k_{int}\right) \nabla \mathcal{F}^k - \boldsymbol{\tau}^k \cdot \nabla \mathcal{F}^k \right\rangle \quad (2.75)$$

Physically, the first term on the right-hand side of the above equation represents the interfacial momentum due to mass exchange across the interface, while the last two terms are usually lumped together to represent the interfacial force density, such as that suggested by Drew (1983). The interfacial force density contains forces acting on the dispersed phase due to viscous drag, wake, and boundary layer formations, and the unbalanced pressure distributions leading to the well-known effects of added or virtual mass, lift, and Bassett history contribution. Similar to mass conservation, the averaged interfacial momentum balance constraint (jump condition), written in terms of the interfacial pressure and difference in pressure, is

$$\sum_{k=1}^{2}\mathbf{\Omega}^k=\sum_{k=1}^{2}\Big[\langle\rho^k\mathbf{U}^k(\mathbf{U}^k-\mathbf{U}^{\text{int}})\cdot\nabla\mathscr{F}^k\rangle+\langle p_{\text{int}}^k\rangle\langle\nabla\mathscr{F}^k\rangle \\ +\langle(p^k-p_{\text{int}}^k)\nabla\mathscr{F}^k-\boldsymbol{\tau}^k\cdot\nabla\mathscr{F}^k\rangle\Big]\equiv\mathbf{F}_\sigma \tag{2.76}$$

The term $\mathbf{F}_\sigma$ in Eqn (2.76) is a notable interfacial momentum source. For example, it is the contribution to the total force acting on the mixture specifically due to the surface tension at the interface for gas–liquid flow. For a constant surface tension coefficient, this singular force can be expressed as

$$\mathbf{F}_\sigma=\sigma\langle\kappa\,\nabla\mathscr{F}^1\rangle=\sigma\langle\kappa\,\mathbf{n}^{\text{int}}\delta_s\rangle \tag{2.77}$$

where σ is the surface tension coefficient, κ is the interface curvature, and the normal vector to the interface is $\mathbf{n}^{\text{int}}=\mathbf{n}^1=-\mathbf{n}^2$ or $\mathbf{n}^{\text{int}}=\mathbf{n}^1=-\mathbf{n}^3$.

For energy conservation, the interfacial energy transfer terms of the averaged conservation energy equations for the total enthalpy of Eqn (2.65) is rewritten as

$$\Pi_H^k\equiv\langle\rho^kH^k(\mathbf{U}^k-\mathbf{U}^{\text{int}})\cdot\nabla\mathscr{F}^k\rangle+\langle\mathbf{q}^k\cdot\nabla\mathscr{F}^k\rangle+\langle p^k\rangle\left\langle\frac{\partial\mathscr{F}^k}{\partial t}\right\rangle+\langle\Phi^{\text{extra}}\rangle \tag{2.78}$$

From physical perspective, the first term on the right-hand side of above equation represents the interfacial energy due to mass exchange across the interface while the term $\langle\mathbf{q}^k\cdot\nabla\mathscr{F}^k\rangle\equiv\langle\mathbf{q}^k\cdot\mathbf{n}^k\delta_s\rangle$ characterizes the flux of heat being transferred into the kth phase from the other phases normal to the interface. The last two terms on the right-hand side of the above equation denote the interfacial work done due to the pressure and extra stresses acting on the interface. In summing all the interfacial energy transfer contributions, the averaged interfacial energy balance constraints (jump conditions) for the total enthalpy are

$$\sum_{k=1}^{2}\Pi_H^k=\sum_{k=1}^{2}\left[\langle\rho^kH^k(\mathbf{U}^k-\mathbf{U}^{\text{int}})\cdot\nabla\mathscr{F}^k\rangle+\langle\mathbf{q}^k\cdot\nabla\mathscr{F}^k\rangle+\langle p^k\rangle\left\langle\frac{\partial\mathscr{F}^k}{\partial t}\right\rangle+\langle\Phi^{\text{extra}}\rangle\right]\equiv\zeta \tag{2.79}$$

The term ζ represents the interfacial energy source contributed by the total work done due to the surface tension at the interface, which is simply given by the product of the surface tension force with the interfacial velocity. If a constant surface tension coefficient is assumed, the interfacial energy source ζ is thus given by

$$\zeta = \mathbf{F}_\sigma \cdot \mathbf{U}^{\text{int}} = \sigma \langle \kappa \mathbf{U}^{\text{int}} \cdot \mathbf{n}^{\text{int}} \delta_s \rangle \tag{2.80}$$

2.1.5 Effective Conservation Equations

In most practical flows of interest, turbulence is associated with the existence of random fluctuations in the fluid. The presence of the random nature of the fluid flow generally precludes computations based on equations that describe the fluid motion proceeding at the desired accuracy. It is therefore preferable that some means of practically resolving the random transient distribution of the instantaneous field ϕ with time is realized for practical computations.

One approach that has been commonly adopted is that the instantaneous field ϕ can be decomposed to a steady mean motion $\overline{\phi}$ and a fluctuating motion ϕ' according to

$$\phi = \overline{\phi} + \phi' \tag{2.81}$$

This particular flow decomposition presents an attractive way of characterizing a turbulent flow by the mean values of flow properties with corresponding statistical fluctuating property. By definition, the time averaged of the fluctuating motion ϕ' is zero, in other words,

$$\overline{\phi'} = \lim_{T\to\infty} \frac{1}{T} \int \phi' \mathrm{d}t = 0 \tag{2.82}$$

By applying volume averaging or ensemble averaging, the instantaneous averaged field is similarly given by

$$\langle \phi \rangle = \overline{\langle \phi \rangle} + \phi'' \tag{2.83}$$

In Eqn (2.83), the time average of the fluctuating component ϕ'' is also, by definition, zero:

$$\overline{\phi''} = \lim_{T\to\infty} \frac{1}{T} \int \phi'' \mathrm{d}t = 0 \tag{2.84}$$

In multiphase flow analysis, preference is normally given to the Favre-averaging approach in order to alleviate the complication of modeling additional correlation terms containing averages of fluctuating quantities. Two types of averaged variables are employed, namely phase-weighted averages and mass-weighted averages. The phase-weighted average for the variable ϕ can be defined by

$$\overline{\langle \phi \rangle} = \frac{\overline{\langle \mathscr{F}^k \phi \rangle}}{\overline{\langle \mathscr{F}^k \rangle}} \tag{2.85}$$

while the mass-weighted average of the variable ψ can also be defined in accordance with

$$\overline{\langle \psi \rangle} = \frac{\overline{\langle \rho^k \psi \rangle}}{\overline{\langle \rho^k \rangle}} \tag{2.86}$$

Hence, the instantaneous volume-averaged variables or ensemble-averaged variables of ϕ and ψ may now be written as

$$\langle \phi \rangle = \overline{\langle \phi \rangle} + \phi'' \tag{2.87}$$

$$\langle \psi \rangle = \overline{\langle \psi \rangle} + \psi'' \tag{2.88}$$

where ϕ'' and ψ'' are the superimposed fluctuations. It can be shown that by multiplying Eqn (2.87) by the averaged phase indicator function $\langle \mathscr{F}^k \rangle$ and Eqn (2.88) by the averaged density $\langle \rho^k \rangle$:

$$\langle \mathscr{F}^k \phi \rangle = \langle \mathscr{F}^k \rangle \overline{\langle \phi \rangle} + \langle \mathscr{F}^k \rangle \phi'' \tag{2.89}$$

$$\langle \rho^k \psi \rangle = \langle \rho^k \rangle \overline{\langle \psi \rangle} + \langle \rho^k \rangle \psi'' \tag{2.90}$$

Time averaging the above equations, noting that the *Reynolds* rule $\overline{\overline{a}} = \overline{a}$ applies, yields

$$\overline{\langle \mathscr{F}^k \phi \rangle} = \overline{\langle \mathscr{F}^k \rangle}\ \overline{\langle \phi \rangle} + \overline{\langle \mathscr{F}^k \rangle \phi''} \tag{2.91}$$

$$\overline{\langle \rho^k \psi \rangle} = \overline{\langle \rho^k \rangle}\ \overline{\langle \psi \rangle} + \overline{\langle \rho^k \rangle \psi''} \tag{2.92}$$

On the basis of the definitions of the phase-weighted averages and mass-weighted averages in Eqns (2.85) and (2.86), it follows that

$$\overline{\langle \mathscr{F}^k \rangle \phi''} = 0 \tag{2.93}$$

$$\overline{\langle \rho^k \rangle \psi''} = 0 \tag{2.94}$$

Note also that $\overline{\phi''} = 0$ and $\overline{\psi''} = 0$, by definition of Eqn (2.84).

The local volume fraction (or volumetric concentration, or relative residence time) represents an important parameter in multiphase flow investigations. It can be defined as the fraction of time in which the continuous or dispersed phase occupies a particular given point in space. Strictly speaking, the local volume fraction α^k can be regarded as the ratio of the fractional volume V_k of the kth phase in an arbitrary small region over the total volume $V$of the region in question within the multiphase flow. Incidentally, it also corresponds to the volume-averaged variable of the phase indicator function, i.e. $\alpha^k = V_k/V = \langle \mathscr{F}^k \rangle$. In the event where the governing equations are volume-averaged variables and subsequently time averaged, suitable forms of equations governing the conservation of mass, momentum, and energy via the phase-weighted variables and mass-weighted averages, i.e. Equations (2.85) and (2.86), for the multifluid model, can be formulated. Dropping the bars and parentheses, which by default denote the Favre-averaging and volume-averaging processes, the effective conservation equations written in terms of the local volume fraction and products of averages are as follows:

Mass conservation:

$$\frac{\partial(\alpha^k \rho^k)}{\partial t} + \nabla \cdot (\alpha^k \rho^k \ \mathbf{U}^k) = \Gamma'^k \tag{2.95}$$

Momentum conservation:

$$\frac{\partial(\alpha^k \rho^k \mathbf{U}^k)}{\partial t} + \nabla \cdot (\alpha^k \rho^k \mathbf{U}^k \otimes \mathbf{U}^k) = -\alpha^k \nabla p^k - p^k \nabla \alpha^k + \nabla \cdot (\alpha^k \boldsymbol{\tau}^k) - \nabla \cdot (\alpha^k \boldsymbol{\tau}^{k''}) + \alpha^k \sum \mathbf{F}^{k,\text{body forces}} + \boldsymbol{\Omega}'^k \tag{2.96}$$

Energy conservation:

$$\frac{\partial(\alpha^k \rho^k H^k)}{\partial t} + \nabla \cdot (\alpha^k \rho^k \mathbf{U}^k H^k) = p^k \frac{\partial \alpha^k}{\partial t} + \alpha^k \frac{\partial p^k}{\partial t} - \nabla \cdot (\alpha^k \lambda^k \nabla T^k) - \nabla \cdot (\alpha^k \mathbf{q}_H^{k''}) + \alpha^k \sum \mathbf{F}^{k,\text{body forces}} \cdot \mathbf{U}^k + \Phi_H''^k + \Pi_H'^k \tag{2.97}$$

From the above, the mean total enthalpy is given by $H^k = h_s^k + \frac{1}{2}\mathbf{U}^k\mathbf{U}^k + \frac{1}{2}\mathbf{U}''^k\mathbf{U}''^k$, the extra stresses term $\Phi_H''^k$ signifies the Favre-averaged approach, and volume-averaged variables processes of the instantaneous term Φ_H^k and the interfacial terms Γ'^k, $\boldsymbol{\Omega}'^k$ and $\Pi_H'^k$ represent the Favre-averaged variables that are subsequently performed on top of the volume-averaged variables terms Γ^k, $\boldsymbol{\Omega}^k$ and Π_H^k. It is worthwhile noting that if ensemble-averaged variables are performed on the governing equations, and fluctuating quantities are subsequently introduced into the equations, the final forms of the governing equations are no different from those of the twice-averaged conservation equations. More details on the ensemble averaging concept for developing averaged conservation equations can be found in Kashima and Rauenzahn (1994); Lhuillier (1996); Brackbill et al. (1997), and Drew and Passman (1999).

Comments

1. In the momentum and energy equations, the normal and shear stress $\boldsymbol{\tau}^k$ can be taken to be proportional to the time rate of strain, i.e. velocity gradients. Such fluids are usually designated as Newtonian fluids. Fluids which do not behave like Newtonian fluids are generally classified as non-Newtonian fluids. In most multiphase flow problems, the fluid

is normally assumed to be Newtonian. The normal and shear viscous stress components for the kth phase according to Newton's law of viscosity are:

$$\begin{aligned}
&\tau_{xx} = 2\mu^k \frac{\partial u^k}{\partial x} - \frac{2}{3}\mu^k \nabla \cdot \mathbf{U}^k \quad \tau_{yy} = 2\mu^k \frac{\partial v^k}{\partial y} - \frac{2}{3}\mu^k \nabla \cdot \mathbf{U}^k \\
&\tau_{zz} = 2\mu^k \frac{\partial w^k}{\partial z} - \frac{2}{3}\mu^k \nabla \cdot \mathbf{U}^k \\
&\tau_{xy} = \tau_{yx} = \mu^k \left(\frac{\partial v^k}{\partial x} + \frac{\partial u^k}{\partial y} \right) \quad \tau_{xz} = \tau_{zx} = \mu^k \left(\frac{\partial w^k}{\partial x} + \frac{\partial u^k}{\partial z} \right) \\
&\tau_{yz} = \tau_{zy} = \mu^k \left(\frac{\partial w^k}{\partial y} + \frac{\partial v^k}{\partial z} \right)
\end{aligned} \tag{2.98}$$

where μ^k is the dynamic viscosity for the kth phase.

2. Examples of possible body forces that can be significant in engineering applications are gravity and electromagnetic forces. It should be noted that body force due to gravity appears to be the most common force in multiphase analysis. We shall therefore restrict ourselves to describing of only this force in the multifluid model, which is given by

$$\sum \mathbf{F}^{k,\text{body forces}} = \rho^k \mathbf{g} \tag{2.99}$$

3. Additional turbulent flux terms such as the Reynolds stress ($\boldsymbol{\tau}^{k\prime\prime}$) and Reynolds flux ($\mathbf{q}_H^{k\prime\prime}$) appearing in Eqns (2.96) and (2.97) are considered to be fluctuating quantities due to flow decomposition, which can be taken to be equivalent to the turbulent flux terms in single-phase turbulence problems. Consideration of turbulence closure via standard models for single-phase fluid flows in resolving these terms can be adopted for the multifluid modeling approach. This will be further discussed in Chapter 6. In the absence of these terms, laminar consideration of multiphase flow is resorted to.

4. Much success of the multifluid model in handling various forms of multiphase flow centers on the formulation of suitable constitutive equations for the interfacial exchange terms Γ'^k, $\boldsymbol{\Omega}'^k$, and $\Pi_H'^k$. Nonetheless, construction of these equations is nontrivial since no universally applicable methodologies that are independent of the topology of the flow or flow pattern currently exist in multiphase flow modeling. In order to cater for the various forms of multiphase flow in question, the best that can be achieved with the present state of knowledge is to attempt to formulate heuristic models of Γ'^k, $\boldsymbol{\Omega}'^k$, and $\Pi_H'^k$ given for a particular flow structure.

5. Assuming that the Reynolds stress and flux as well as the interfacial exchange terms can be properly ascertained, there are five equations governing the conservation of mass, momentum, and energy which can be solved in determining the local volume fraction α^k,

velocity components u^k, v^k, and w^k, and total enthalpy H^k for each fluid or phase. To evaluate the density ρ^k and temperature T^k, the consideration of the algebraic equation of state for density and the constitutive equation for total enthalpy can be respectively introduced accordingly:

$$\rho^k = \rho^k\left(T^k, p^k\right) \quad H^k = H^k\left(T^k, p^k\right)$$

with the algebraic constraint for the local volume fraction in the multifluid model satisfying $\sum_{k=1}^{N} \alpha^k = 1$. The unknown, pressure p^k, is usually given by algebraic constraints on the pressure. If the relative velocity, also referred to as the slip velocity—the difference between dispersed and continuous phase velocities—is small and there exists no appreciable dispersed phase expansion/contraction, the simplest assumption commonly adopted in most multiphase calculations is that all phases share the same pressure field: $p^k = p$. This supposes that there is instantaneous microscopic pressure equilibration. Other constraints on the pressures are also possible, which may include, depending on the particular type of multiphase flow, the effects of surface tension or solid compression.

6. Most practical multiphase flow applications of interest are of low Mach number or low speed flows. They can be regarded as weakly compressible flows. The term weakly refers to the consequence of density change being affected mainly by the substantial temperature variations but not the pressure variations since the pressure remains relatively unperturbed within the surroundings. On the basis of the weakly compressible assumption, and the kinetic energy $\frac{1}{2}\mathbf{U}^k\mathbf{U}^k$ as well as the mean flow kinetic energy $\frac{1}{2}\mathbf{U}''^k\mathbf{U}''^k$ in the definition of enthalpy, the pressure work term $\partial p/\partial t$ and the sources of energy due to work done by the body forces and extra stresses deforming the fluid element and on the interface are ignored.

In computational multiphase fluid dynamics, the interfacial exchange terms Γ'^k, $\boldsymbol{\Omega}'^k$, and Π'^k_H can normally be linearized for numerical treatment according to

$$\Gamma'^k \equiv \sum_{l=1}^{N} (\dot{m}_{lk} - \dot{m}_{kl}) \tag{2.100}$$

$$\boldsymbol{\Omega}'^k \equiv \sum_{l=1}^{N} \left(\dot{m}_{lk}\mathbf{U}^l - \dot{m}_{kl}\mathbf{U}^k\right) + p^k_{\text{int}}\nabla\alpha^k + \mathbf{F}^k_{\text{D}} \tag{2.101}$$

$$\Pi'^k_H \equiv \sum_{l=1}^{N} \left(\dot{m}_{lk}H^l - \dot{m}_{kl}H^k\right) + Q^{\text{int}}_H \tag{2.102}$$

where $\dot{m}_{lk}$ and $\dot{m}_{kl}$ characterize the mass transfer from the lth phase to kth phase and from the kth phase to lth phase, respectively. From mass conservation, $\dot{m}_{kk} = \dot{m}_{ll} = 0$. On the basis of Eqns (2.98)–(2.102), the multifluid model for the governing equations of conservation of mass, momentum, and energy can be written for a turbulent mixture in the form

Mass conservation:

$$\frac{\partial\left(\alpha^k\rho^k\right)}{\partial t}+\nabla\cdot\left(\alpha^k\rho^k\mathbf{U}^k\right)=\sum_{l=1}^{N}(\dot{m}_{lk}-\dot{m}_{kl}) \tag{2.103}$$

Momentum conservation:

$$\begin{aligned}&\frac{\partial\left(\alpha^k\rho^k\mathbf{U}^k\right)}{\partial t}+\nabla\cdot\left(\alpha^k\rho^k\mathbf{U}^k\otimes\mathbf{U}^k\right)\\&=-\alpha^k\nabla p^k+\left(\nabla\cdot\alpha^k\left[\mu^k\left(\nabla\mathbf{U}^k+\left(\nabla\mathbf{U}^k\right)^T\right)-\frac{2}{3}\mu^k\nabla\cdot\mathbf{U}^k\delta\right]\right)\\&\quad-\nabla\cdot\left(\alpha^k\boldsymbol{\tau}^{k\prime\prime}\right)+\alpha^k\rho^k\mathbf{g}+\sum_{l=1}^{N}\left(\dot{m}_{lk}\mathbf{U}^l-\dot{m}_{kl}\mathbf{U}^k\right)\\&\quad+\underbrace{\left(p^k_{\text{int}}-p^k\right)\nabla\alpha^k+\mathbf{F}_{\text{D}}^{k,\text{drag}}+\mathbf{F}_{\text{D}}^{k,\text{non-drag}}}_{\mathbf{F}_D^k}\end{aligned} \tag{2.104}$$

Energy conservation:

$$\begin{aligned}\frac{\partial\left(\alpha^k\rho^kH^k\right)}{\partial t}+\nabla\cdot\left(\alpha^k\rho^k\mathbf{U}^kH^k\right)&=\nabla\cdot\left(\alpha^k\lambda^k\nabla T^k\right)-\nabla\cdot\left(\alpha^k\mathbf{q}_H^{k\prime\prime}\right)\\&\quad+\sum_{l=1}^{N}\left(\dot{m}_{lk}H^l-\dot{m}_{kl}H^k\right)+Q_H^{\text{int}}\end{aligned} \tag{2.105}$$

In Eqn (2.104), both phases have been taken to share the same pressure field, i.e. $p^k=p$. The interfacial force $\mathbf{F}_{\text{D}}^k$ in Eqn (2.104) is normally split in terms of the drag force $\mathbf{F}_{\text{D}}^{k,\text{drag}}$ and any other interfacial nondrag forces in $\mathbf{F}_{\text{D}}^{k,\text{non-drag}}$. The interfacial drag force $\mathbf{F}_{\text{D}}^{k,\text{drag}}$ and interfacial heat source Q_H^{int} can be expressed usually in linear forms according to

$$\mathbf{F}_{\text{D}}^{k,\text{drag}}\equiv\sum_{l=1}^{N}B_{kl}\left(\mathbf{U}^l-\mathbf{U}^k\right) \tag{2.106}$$

$$Q_H^{\text{int}}\equiv\sum_{l=1}^{N}C_{kl}\left(T^l-T^k\right) \tag{2.107}$$

where B_{kl} and C_{kl} are the interphase drag and heat transfer terms. Through appropriate modeling considerations, closure to the interfacial exchange terms is generally attained through prescribed algebraic functions of the governing flow parameters. In the momentum and energy equations, turbulent fluxes $\boldsymbol{\tau}^{k\prime\prime}$ and $\mathbf{q}_H^{k\prime\prime}$ can be resolved via the consideration of suitable turbulence models.

2.2 Lagrangian Description on Discrete Element Framework

The multifluid model, also known as the Eulerian—Eulerian approach, solves the disperse phase as an ensemble of individual discrete particles flowing like another fluid in the flow system. An alternate strategy in handling the disperse phase is via the consideration of the Eulerian—Lagrangian approach. The Eulerian component of this approach consists of solving the surrounding fluid (continuous phase) through the governing Eqns (2.103)—(2.105). Nevertheless, the particles—gas bubbles, liquid drops, or solid particles—are now tracked independently through the surrounding fluid through a trajectory model in the Lagrangian component. Interphase interaction effects in the continuous phase equations governing the mass, momentum, and heat exchanges between the particles and surrounding fluid are hereby represented by the summation of all sources and sinks of representative (or computed) trajectories in altering the flow field.

2.2.1 Equations of Motion

In the time-driven discrete element method, the trajectory of a particle can be determined by calculating all forces and moments acting on it. Depending on the particle characteristics and its surrounding, a variety of forces are considered. The instantaneous particle velocity $\mathbf{V}_\mathrm{p}$ and particle rotation rate $\mathbf{\Omega}^\mathrm{p}$ can be obtained through solution of the particle linear and angular momentum equations (Newton's second law), given by

$$m_\mathrm{p}\frac{D\mathbf{V}_\mathrm{p}}{Dt} = \sum \mathbf{F} \tag{2.108}$$

$$I_\mathrm{p}\frac{D\mathbf{\Omega}_\mathrm{p}}{Dt} = \sum \mathbf{M} \tag{2.109}$$

where m_p is the particle mass, and I_p is the particle moment of inertia. On the right hand side, the Lagrangian time derivatives are essentially the material derivatives of the particle velocity as well as the rotation rate while on the left hand side, the source terms as stipulated in Eqns (2.108) and (2.109) represent the sum of forces and moments acting on the particle.

A range of complex phenomena associated with heat and mass transfer processes can also be handled within the Lagrangian framework. In addition to the particle linear and angular momentum Eqns (2.108) and (2.109), particle heat and mass transfers are tracked along the discrete particle trajectories and solved by the particle conservation equations of mass and energy, which, written in terms of the material derivatives of the particle mass m_p and particle temperature T_p, read as follows:

$$\frac{Dm_\mathrm{p}}{Dt} = S_{m_\mathrm{p}} \tag{2.110}$$

$$m_p C_p \frac{DT_p}{Dt} = S_{T_p} \tag{2.111}$$

From Eqns (2.110) and (2.111), the source term S_{m_p} represents the mass transfer between the particle and surrounding fluid while the source term S_{T_p} is governed primarily by three modes of heat transfer:

1. Convective heat transfer;
2. Latent heat transfer associated with mass transfer; and
3. Net radiative power absorbed by the particle.

In the absence of mass transfer, the source term S_{T_p} is affected by only the two modes of heat transfer represented by (1) and (3), respectively. The product of the particle mass, specific heat of constant pressure (C_p), and material derivative of the particle temperature denotes the sensible heating term of the particle energy equation.

In turbulent flows, the dispersion of particles due to turbulent eddies generally requires the full time history of the complex flow characteristics. It should be noted that Eqns (2.108)–(2.111) yield the instantaneous particle velocity, mass, and temperature while the Favre-averaged form of the transport equations for the continuous phase results in the mean values of these fields. In order to determine the particle source terms of the mass, linear, and angular momentum and energy equations, the recovery of omitted statistical fluctuations of the surrounding fluid is needed for the prediction of the turbulent dispersion characteristics of particles in the fluid. Suitable approaches to estimate the fluctuating components are discussed in Section 2.2.3.

2.2.2 Fluid–Particle Interaction (Forces Related to Fluid Acting on Particle One-Way, Two-Way Coupling)

The importance of the fluid forces acting on particles can be determined according to the definition of the Stokes number:

$$St = \frac{\tau_p}{\tau_f} \tag{2.112}$$

where τ_p is the particle relaxation time, while τ_f is the fluid integral scale. For a small Stokes number, $St << 1$, the particles can be considered to follow the fluid streamline with a small drift velocity relative to the fluid velocity. At moderate values of Stokes number, $St = O(1)$, large particle dispersion relative to the fluid velocity becomes more significant. For a large Stokes number, $St >> 1$, the particles are no longer in equilibrium with the surrounding fluid phase and they divert rather substantially from the fluid stream path leading to significant momentum transfer from the particle to the fluid. Here, the inertia effect of particles becomes more prevalent and will, therefore, exert a significant influence on the background fluid.

There are a wide variety of different types of fluid forces that act on particles. According to Clift et al. (1978); Shirolkar et al. (1996); Crowe et al. (1998); and Gouesbet and Berlemont (1999), the fluid forces include the drag, virtual or added mass, Basset history, lift, Magnus, pressure gradient, and reduced gravity.

In most practical flows of engineering interest, the most important force exerted on the particles by the surrounding fluid is the drag force. For a spherical particle, the expression is given by

$$\mathbf{F}_{\text{Drag}} = \frac{\pi}{8}\rho^{c} d_{p}^{2} C_{D} (\mathbf{V}_{f} - \mathbf{V}_{p}) |\mathbf{V}_{f} - \mathbf{V}_{p}| \tag{2.113}$$

where ρ^c is the continuous phase fluid density. The drag coefficient denoted by C_D is normally a function of the particle Reynolds number. Suitable relationships for the drag coefficient need to be specified depending on the different types of flows and configurations being solved. In Eqn (2.113), $\mathbf{V}_f$ represents the instantaneous fluid velocity.

The virtual or added mass force originates because of the difference in acceleration between the fluid and the particle. This force becomes dominant when there is significant difference in the density of the fluid and the particle. For an inviscid flow, it can be written as

$$\mathbf{F}_{\text{Added}} = K_{A} m_{p} \left(\frac{d\mathbf{V}_{f}}{dt} - \frac{d\mathbf{V}_{p}}{dt} \right) \tag{2.114}$$

where the added mass coefficient for a sphere is taken to be $K_A = 0.5$. It has been demonstrated by Rivero et al. (1991) that Eqn (2.114) with an added mass coefficient of a constant value of 0.5 also holds for a sphere in viscous flows over a wide range of Reynolds and fluid accelerations.

Owing to the transitory nature of the particle's boundary layer especially in an oscillatory flow field, the Basset history force arises, which is mainly influenced by the history of the particle trajectory. This force can be expressed as

$$\mathbf{F}_{\text{Basset}} = K_{B} d_{p}^{2} \sqrt{\pi \rho^{c} \mu^{c}} \int_{t_0}^{t} \left(\frac{d\mathbf{V}_{f}}{dt} - \frac{d\mathbf{V}_{p}}{dt} \right) \frac{ds}{\sqrt{t - s}} \tag{2.115}$$

where the Basset history force coefficient K_B can range between 1.5 and 6.0 and μ^c is the continuous phase dynamic viscosity.

At very low Reynolds number, Saffman (1965) demonstrated that a small rotating particle moving in a uniform shear flow experiences a lift force due to both pressure difference between the top and bottom of the particle when it rotates with the fluid and local gradients of transitional fluid velocities. The Saffman lift force is

$$\mathbf{F}_{\text{Lift}} = K_{\text{L}}\chi m_{\text{p}}\frac{(\mathbf{V}_{\text{f}} - \mathbf{V}_{\text{p}}) \times \boldsymbol{\omega}_{\text{f}}}{\text{Re}_{\text{p}}^{1/2}\alpha_{\text{L}}^{1/2}} \tag{2.116}$$

In Eqn (2.116), $\chi = \rho^{\text{c}}/\rho^{\text{p}}$ is the density ratio with ρ^{p} being the particle density, $\boldsymbol{\omega}_{\text{f}}$ is the fluid vorticity at a particular location which can be determined according to $\boldsymbol{\omega}_{\text{f}} = \nabla \times \mathbf{V}_{\text{f}}$, $\alpha_{\text{f}} \equiv |\boldsymbol{\omega}_{\text{f}}|d_{\text{p}}/(2|\mathbf{V}_{\text{f}} - \mathbf{V}_{\text{p}}|)$, and $\text{Re}_{\text{p}} \equiv |\mathbf{V}_{\text{f}} - \mathbf{V}_{\text{p}}|d_{\text{p}}/\nu^{\text{c}}$, where ν^{c} represents the continuous phase kinematic viscosity. The lift coefficient K_{L} takes on a constant value of 2.18. For flows of finite Reynolds number, particle rotation appears to have little influence on the lift force. Further increasing the Reynolds number has shown that the lift force decreases significantly for both rotating and nonrotating conditions. According to Bagchi and Balachandar (2002), the lift on a nonrotating sphere decreases more rapidly than that on a rotating sphere.

The Magnus force results when a rotating particle is subjected to a nonrotating fluid, especially at high Reynolds number. It can be written as

$$\mathbf{F}_{\text{Magnus}} = \frac{3}{4}\chi m_{\text{p}}\left(\frac{1}{2}\boldsymbol{\omega}_{\text{f}} - \boldsymbol{\Omega}_{\text{p}}\right) \times (\mathbf{V}_{\text{f}} - \mathbf{V}_{\text{p}}) \tag{2.117}$$

The Magnus force can be added to the lift on a particle rotating at the same rate as the fluid flow or, to obtain to high accuracy, the total lift force acting on a particle traveling through the fluid with arbitrary rotation rate. The corresponding viscous torque acting on the particle due to the differential fluid rotation rate can be obtained by

$$\mathbf{M}_{\text{Fluid}} = \pi\mu^{\text{c}}d_{\text{p}}^{3}\left(\frac{1}{2}\boldsymbol{\omega}_{\text{f}} - \boldsymbol{\Omega}_{\text{p}}\right) \tag{2.118}$$

It should be noted that the pressure gradient force may be required because of the force required to accelerate the fluid which would occupy the particle volume if the particle were absent. The associated force can be written as

$$\mathbf{F}_{\text{Pressure}} = \chi m_{\text{p}}\frac{D\mathbf{V}_{\text{f}}}{Dt} \tag{2.119}$$

Also, the reduced gravity force can be included, which is given by

$$\mathbf{F}_{\text{Gravity}} = m_{\text{p}}(1 - \chi)\mathbf{g} \tag{2.120}$$

Equation (2.120) includes both the gravitational force and the corresponding fluid buoyancy force acting on the particle.

In addition to the various forces listed above, the consideration of a random force that acts on a particle may be deemed to be necessary in some flows. For a very small particle (especially of nanometer diameter), this random force results from the Brownian motion induced by individual molecular collision with the particle. The amplitudes of the Brownian force components at each time step may be determined according to Li and Ahmadi (1992) as

$$\mathbf{F}_{\mathrm{Brownian},i} = G_i\sqrt{\frac{\pi S_{\mathrm{o}}}{\Delta t}} \tag{2.121}$$

where G_i represent the zero mean, unit-variance, independent Gaussian random numbers, and

$$S_{\mathrm{o}} = \frac{216\nu^{\mathrm{c}}\sigma T^{\mathrm{c}}}{\pi^2\rho^{\mathrm{c}}d_{\mathrm{p}}^5\left(\frac{\rho^{\mathrm{p}}}{\rho^{\mathrm{c}}}\right)^2 C_{\mathrm{c}}} \tag{2.122}$$

In the above equation, σ is the Stefan–Boltzmann constant, T^{c} is the continuous phase fluid temperature, and C_{c} is the Cunningham correction to Stokes' drag law, which can be calculated by the following expression:

$$C_{\mathrm{c}} = 1 + \frac{2\lambda}{d_{\mathrm{p}}}\left(1.257 + 0.4e^{-\left(1.1d_{\mathrm{p}}/2\lambda\right)}\right) \tag{2.123}$$

where λ represents the molecular mean free path. For turbulent flows, a random force is often employed to model the effects of subgrid scale turbulence on the dispersion of particles. A discussion of relevant models for particle flows will be given in Section 2.2.3.

As the fluid flow exerts a force on the particles, in a similar way the particles may exert a force back on the fluid flow. The particle-induced force on the overall fluid flow can be assessed by the significance of the particle mass loading. The mass-loading ratio can be defined by $m_{\mathrm{p}}/m_{\mathrm{f}}$ where m_{p} and m_{f} are the mass of the particle and fluid phases respectively. When the mass-loading ratio is small, the particles can be treated as passive contaminants. In the simplest case, we can safely deal with a one-way coupling problem. Nevertheless, when the mass-loading ratio is increased, particularly for particles in turbulent flows, global turbulence modifications may be induced. In contrast to the particles behaving like passive contaminants, turbulence modification can be expected to significantly affect the energy distribution of the surrounding fluid since the particle diameter is much larger than the Kolmogorov scale. For sufficiently large mass-loading ratio, the existence of relative motion between particles and the carrier fluid results in an extra dissipation of the turbulence energy. Because of this, we are now confronted with a two-way coupling problem: turbulence modifies the behavior of particles and, in return, modifies the fluid turbulence.

In addition, the importance of particle–particle interactions in turbulent flows brings forth the consideration of a four-way coupling problem. The terminology "four-way" comes from the fact that if a particle *A* influences a particle *B*, then, reciprocally, particle *B* must influence particle *A*, by action and reaction. The importance of interparticle collisions can be determined based on the ratio of the particle relaxation time (τ_p) and the characteristics time of collisions (τ_c). In the framework of kinetic theory, τ_c, for a statistically homogeneous distribution of dynamically identical particles, depends on the particle volume fraction, particle diameter and particle kinetic energy. Two distinct regimes can be identified. The dilute regime is denoted by $\tau_p/\tau_c << 1$ while the dense regime is given by $\tau_p/\tau_c >> 1$.

Flows of dilute applications are controlled by the surface and body forces acting on the particles. For particle volume fractions less than 10^{-6}, the particles are expected to exert negligible influence on the turbulence of the fluid phase. This corresponds to the description of very dilute flows. The particles can be treated as passive contaminants which are carried along by the fluid phase and they do not alter the bulk fluid flow characteristics. For particle volume fractions between 10^{-6} and 10^{-3}, particles augment the turbulence. This corresponds now to dilute flows. The effect of the particle trajectories on the continuum can be accounted for by feeding appropriate sources or sinks back into the balance equations of the continuous fluid phase so as to modify the fluid flow equations for turbulence intensity and dissipation due to the presence of the particle phase. For particle volume fractions greater than 10^{-3}, which indicate dense flows, the motion of particles is significantly controlled by particle—particle collisions or interactions. The averaged time between two collisions is smaller than the particle relaxation time so that the particles do not have sufficient time to recover their own behavior between two collisions. Figure 2.7 exemplifies the one-way coupling problem, two-way coupling problem, and four-way coupling problem of the fluid—particle flow through the proposed map of particle-turbulence modulation (Elghobashi, 1994). It should be noted that dense flows in the absence of turbulence reduce to a two-way coupling problem.

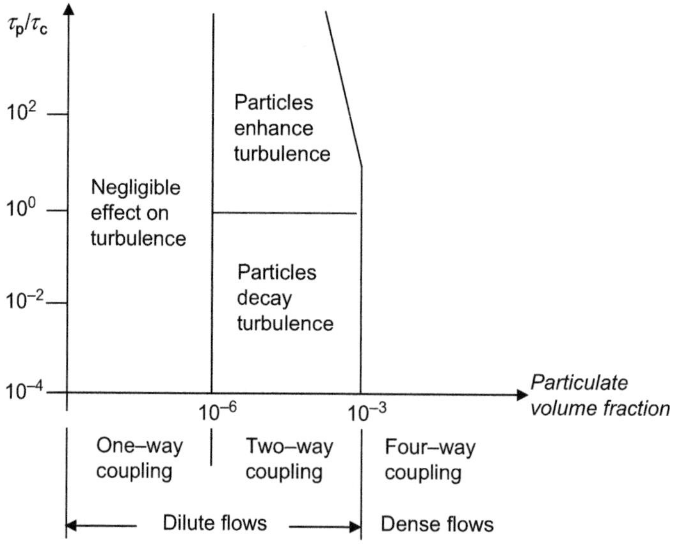

Figure 2.7
Proposed map for particle-turbulence modulation. *After Elghobashi, 1994.*

2.2.3 Particle–Particle Interaction (Four-Way Coupling Concept: Collisions and Turbulent Dispersion of Particles)

In the continuum mechanics framework, contact of particles can, in general, be studied in detail. Numerical methods such as the finite element method may be utilized, yielding detailed insight into the temporal evolution of stresses and strains in the volume or on the surface of the particle. Integration of the stresses leads to the determination of the forces that are required for the solution of the equations of motion, as in Eqns (2.108) and (2.109). Nevertheless, even for single binary collision, the finite element method is highly time consuming. Owing to the specific consideration of effectively using the discrete element method for large particle assemblies, there is certainly a need to model the contact forces by much simpler approaches. The hard-sphere model can be applied in the framework of the discrete element method, which assumes that the particles are rigid and contact forces are subsequently derived from a point on the bodies for such a model.

Hard-sphere model

The hard-sphere model may be applied for simulating particle–particle collision. Main assumptions of the model, which are concerned with the particle shape, the deformation history during collision, and the nature of collisions, are the following: the particles are generally taken to be spherical and quasi rigid and the shape of these particles is retained after impact; the dynamics of idealized binary collision and the collisions between particles are taken to be instantaneous; the contact of particles during collision occurs at a point; and the interaction forces are taken to be impulsive and all other finite forces are negligible during collisions. These assumptions are believed to be sufficiently realistic for collisions of relatively coarse particles ($>100\ \mu$m). One characteristic feature of the hard-sphere model is the ability to process a sequence of collisions one at a time. Another important feature is that the simulations can be readily performed with realistic values of key parameters such as the restitution and friction coefficients.

At any instant during the impact of two particles such as shown in Figure 2.8, the motions of the particles are governed by the linear and angular impulse momentum laws which yield the following set of equations for a binary collision of two spheres:

$$
\begin{aligned}
m_k(\mathbf{V}_k - \mathbf{V}_k^0) &= \mathbf{J} \\
m_l(\mathbf{V}_l - \mathbf{V}_l^0) &= -\mathbf{J} \\
\frac{I_k}{R_k}(\boldsymbol{\omega}_k - \boldsymbol{\omega}_k^0) &= \mathbf{J} \times \mathbf{n} \\
\frac{I_l}{R_l}(\boldsymbol{\omega}_l - \boldsymbol{\omega}_l^0) &= \mathbf{J} \times \mathbf{n}
\end{aligned}
\qquad (2.124)
$$

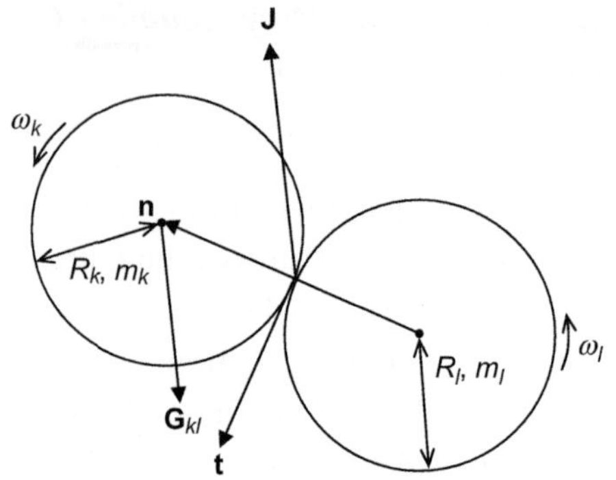

Figure 2.8
Contact between two particles for the hard-sphere model.

where the superscript 0 denotes conditions just before collision, m_k and m_l are the mass, R_k and R_l are the radii, $\boldsymbol{\omega}_k$ and $\boldsymbol{\omega}_l$ are the angular velocities, and I_k and I_l are the moment of inertia of particle k and particle l. The velocities prior to collisions are the velocities at the last time step just before collision (the corresponding time difference is not larger than 10^{-4} s). By definition, $I = mR_{\text{gyration}}^2$ where R_{gyration} is the radius of gyration of the particle ($R_{\text{gyration}}^2 = \frac{2}{3}R_{\text{p}}^2$ for spherical particles) and $\mathbf{J}$ is the impulse vector. By adopting the vector relation $(\mathbf{J} \times \mathbf{n}) \times \mathbf{n} = \mathbf{J} - (\mathbf{J} \cdot \mathbf{n})\mathbf{n}$, we obtain

$$\mathbf{V}_{kl} - \mathbf{V}_{kl}^0 = B_k\mathbf{J} - (B_k - B_l)(\mathbf{J} \cdot \mathbf{n})\mathbf{n} \tag{2.125}$$

where $\mathbf{V}_{kl}$ is the relative velocity at the contact point between two particles with velocities $\mathbf{V}_k$ and $\mathbf{V}_l$ defined as

$$\mathbf{V}_{kl} = \mathbf{G}_{kl} - (R_k\boldsymbol{\omega}_k - R_l\boldsymbol{\omega}_l) \times \mathbf{n} \tag{2.126}$$

and the collision constants B_1 and B_2 are

$$\begin{aligned} B_1 &= \frac{1}{m_k} + \frac{1}{m_l} + \frac{R_k^2}{I_k} + \frac{R_l^2}{I_l} \\ B_2 &= \frac{1}{m_k} + \frac{1}{m_l} \end{aligned} \tag{2.127}$$

In Eqn (2.126), the relative velocity of particle centroids $\mathbf{G}_{kl}$ is given by

$$\mathbf{G}_{kl} = \mathbf{V}_k - \mathbf{V}_l \tag{2.128}$$

The normal and tangential unit vectors that define the collision coordinate system in Figure 2.8 are

$$\mathbf{n} = \frac{\mathbf{x}_k - \mathbf{x}_l}{|\mathbf{x}_k - \mathbf{x}_l|} \tag{2.129}$$

$$\mathbf{t} = \frac{\mathbf{V}_{kl} - (\mathbf{G}_{kl}^0 \cdot \mathbf{n})\mathbf{n}}{|\mathbf{V}_{kl} - (\mathbf{G}_{kl}^0 \cdot \mathbf{n})\mathbf{n}|} \tag{2.130}$$

Some parameters are established to relate the velocities before and after collisions. The first collision parameter is the coefficient of normal restitution, e_n:

$$\mathbf{V}_{kl} \cdot \mathbf{n} = -e_\mathrm{n}(\mathbf{V}_{kl}^0 \cdot \mathbf{n}) \tag{2.131}$$

By combining the above equation with Eqn (2.125), the normal component of the impulse vector can be written as

$$J_\mathrm{n} = (1 + e_\mathrm{n})\frac{(\mathbf{V}_{kl}^0 \cdot \mathbf{n})}{B_2} \tag{2.132}$$

The second and third collision parameters comprise the coefficient of tangential restitution, e_t, and the coefficient of friction, μ_f. These two parameters concern the two kinds of collisions—particle sticking and sliding in the tangential impact process. The case where the tangential component of the impact velocity is sufficiently high or the friction coefficient is small by comparison is exemplified by

$$\mu_\mathrm{f} < \frac{(1 + e_\mathrm{t})}{J_\mathrm{n}}\frac{(\mathbf{V}_{kl}^0 \cdot \mathbf{t})}{B_1} \tag{2.133}$$

where gross sliding occurs throughout the whole duration of the contact. Applying Coulomb's law, the tangential component of the impulse is then given by

$$J_{t,\mathrm{sliding}} = -\mu_\mathrm{f} J_\mathrm{n} \tag{2.134}$$

On the other hand, if the friction coefficient is sufficiently high,

$$\mu_\mathrm{f} \geq \frac{(1 + e_\mathrm{t})}{J_\mathrm{n}}\frac{(\mathbf{V}_{kl}^0 \cdot \mathbf{t})}{B_1} \tag{2.135}$$

in which sticking collisions occur after an initial sliding phase—the relative tangential velocity between two colliding particles becomes zero—the tangential impulse for this case is

$$J_{t,\mathrm{sticking}} = -(1 + e_\mathrm{t})\frac{(\mathbf{V}_{kl}^0 \cdot \mathbf{t})}{B_1} \tag{2.136}$$

where the coefficient of tangential restitution, e_t is defined as

$$\mathbf{V}_{kl} \cdot \mathbf{t} = -e_\mathrm{t}(\mathbf{V}_{kl}^0 \cdot \mathbf{t}) \tag{2.137}$$

Once all the impulse vectors are known, the postcollision velocities can now be calculated from Eqn (2.124).

Based on the work of Hoomans et al. (1996), a two-step approach is generally adopted to solve the hard-sphere particle dynamics. The first step consists of a fixed interaction process in which the particles are taken to be fixed in space and particle velocities are calculated via Newton's equation of motion to account for fluid forces acting on the particle. In the second step, possible collision events between particles are recorded and collision dynamics is thereby executed for each collision event. It is noted that in this step the particles are assumed to be in free flight before collisions. This step signifies the movement and collision process because each particle would have moved to the correct position before the occurrence of collisions.

Turbulent transport of particle

Besides particle–particle collision, particles that are transported in a turbulent flow may also be greatly influenced by their interaction with turbulent eddies. Figure 2.9 depicts a schematic illustration of the transport of particles subject to collision and the motion of different sized eddies interacting with particles of various sizes. In determining the outcome of the eddy-particle interaction, the size of the particle with respect to the eddy size is an important parameter in determining the dispersive motion of the particle.

In summary, the important factors contributing to the prediction of turbulent transport of particle are

1. Particle size with respect to the turbulent length scale (eddy size) in the fluid
2. Relative density between the particle and the fluid
3. Fluctuating fluid velocity surrounding the particle

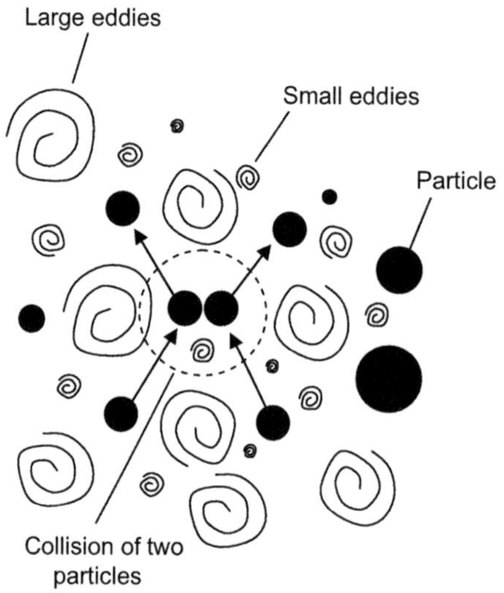

Figure 2.9
Schematic illustration of particle–particle and eddy–particle interactions in turbulent flow.

4. Particle relaxation time
5. Eddy lifetime and particle Lagrangian time scale
6. Cross trajectory effect phenomenon.

If the particle density is taken to be much greater than the surrounding fluid density, the inertia force at the fluid–particle interface will act to damp the fluctuations in its velocity comparing with the fluctuations of the surrounding fluid. This so called dense particle will have less fluctuating velocity as compared to that of the fluid. The reduction in the fluctuating velocity is known as the inertia effect and is characterized by a time scale called the particle relaxation time. It should be noted that when a small particle is introduced in a turbulent flow, it will remain trapped inside an eddy for some duration of time before it is influenced by another eddy. The maximum time for which a particle can remain under the influence of a particular eddy is the *eddy lifetime*. Eddy properties are taken to remain constant or uniform during the entire eddy lifetime; it is, therefore, assumed that if a particle is trapped inside an eddy, it will experience a uniform velocity field during its residence time within an eddy.

Another interesting inertia effect is that of the time interval over which the particle velocity is correlated with its initial velocity and the increase of particle inertia. The initial particle velocity can be viewed as the velocity in one particular eddy just before the particle migrates to another eddy. The final velocity of the particle with a resulting higher inertia in the previous eddy will exert a greater influence on the velocity outcome of the same particle in the present eddy. This correlated time known as the *particle integral time scale* or *particle Lagrangian time scale*, is roughly the time for which the particle maintains its initial velocity before it undergoes a turbulent collision and changes its velocity.

Consideration of the *crossing trajectory effect* involves the particle experiencing a premature migration from one eddy to another before the decay due to the turbulence of the original eddy. In this instance, the particle does not remain trapped inside an eddy for the entire eddy lifetime. An obvious outcome is the reduction in the particle Lagrangian time scale, which is due to an abrupt change in the fluid conditions surrounding the particle. In order to estimate the interaction time between an eddy and a particle, a particle drift velocity—the difference between surrounding fluid velocity and particle velocity—is employed to appropriately determine the time the particle would take to cross a given eddy.

In turbulent flows, Eqns (2.108) and (2.109) may be solved ignoring the fluctuating components of the fluid velocity and rotation. Such an approach accounts only for the advection of particles due to mean flow and thus conveniently neglects the dispersion of particle due to turbulent velocity fluctuations. Particles with the same physical properties and initial conditions will have identical trajectories. These *deterministic* models can nonetheless be applied for flows having very low turbulence levels. For highly turbulent flows, the dispersive effect of turbulent fluctuations on the particle motion becomes significant and is necessary to be incorporated as such models will not yield identical trajectories even for particles with the same physical properties and initial conditions. In order to obtain reliable statistics that characterize the particle behavior, Monte Carlo simulations need to be performed; such models are known as *stochastic* models. One core predicament in such models is the determination of the appropriate fluctuating fluid velocity at the particle location needed to solve the equation of motion of the particle. In many practical problems, this can be achieved through the time-averaged component of the instantaneous fluid velocity along the particle trajectory, calculated via a suitable turbulent model (e.g. the two-equation k-ε turbulence model), along with a simulated fluctuating velocity component which is then superimposed to calculate the instantaneous fluid velocity. Stochastic models are, therefore, normally used to estimate the unknown fluctuating fluid component by accounting for the eddy—particle interaction.

For the models based on the eddy lifetime concept, a fluid eddy is assigned a fluctuating velocity ($\mathbf{V}_{\mathrm{f}}''$), which is randomly sampled from a probability density function (PDF) and assumed to stay constant during the interaction time, namely the minimum of the eddy lifetime (t_{e}) and the eddy transit time (t_{c}). The eddy—particle interaction time at each particle location can be determined from the eddy lifetime and eddy size which can be estimated from local turbulence properties available from a turbulence model, such as the two equation k-ε model in the Reynolds-averaged framework. With the knowledge of the interaction time and randomly sampled fluctuating fluid velocity, the velocity and position of the particle are subsequently determined. At the end of each time step, a new fluctuating fluid velocity is sampled from a new PDF, which is generated using local turbulence properties, and the next interaction time is subsequently determined from local properties at the new particle location. This essentially represents the basic idea of the stochastic models based on the eddy lifetime concept. The fluctuating or eddy velocity can be calculated according to

$$\mathbf{V}_{\mathrm{f}}'' = \psi v_{\mathrm{rms}} \tag{2.138}$$

where v_{rms} is the root mean square (rms) of the fluid fluctuating velocity in the relevant direction and ψ is a random number drawn from a prescribed Gaussian probability distribution of zero mean and unity standard deviation. The fluctuating velocity's standard deviation in isotropic flows can be estimated from the local turbulent kinetic energy k^{c} of the continuous phase as

$$\underbrace{\sqrt{\left(u_{\mathrm{f}}''\right)^2}}_{v_{\mathrm{rms},1}} = \underbrace{\sqrt{\left(v_{\mathrm{f}}''\right)^2}}_{v_{\mathrm{rms},2}} = \underbrace{\sqrt{\left(w_{\mathrm{f}}''\right)^2}}_{v_{\mathrm{rms},3}} = \sqrt{\frac{2}{3}k^{\mathrm{c}}} \tag{2.139}$$

where the three-dimensional fluctuating velocities are taken to be equal to each other. A mathematical expression for the Gaussian probability distribution of the fluctuating fluid velocity is expressed by

$$P(v_{\mathrm{rms},i}) = \frac{1}{\sqrt{2\pi}\sqrt{\frac{2}{3}k^{\mathrm{c}}}}\exp\left(\frac{v_{\mathrm{rms},i}^2}{\frac{4}{3}k^{\mathrm{c}}}\right) \text{ for } i = 1, 2, 3 \tag{2.140}$$

Owing to the assumption of isotropy, independent fluctuating velocities are sampled from the above PDF for each coordinate direction at every time step.

The eddy lifetime and eddy size that are required at each particle location to determine the next interaction time can be obtained from local turbulence properties. Different investigators have employed different expressions for the eddy lifetime and eddy size in isotropic stochastic models. These two scales can be determined according to

$$t_{\mathrm{e}} = A\frac{k^{\mathrm{c}}}{\varepsilon^{\mathrm{c}}} \quad \text{and} \quad l_{\mathrm{e}} = B\frac{(k^{\mathrm{c}})^{3/2}}{\varepsilon^{\mathrm{c}}} \tag{2.141}$$

where A and B are two dependent constants and ε^{c} is the dissipation of the turbulent kinetic energy of the continuous phase. Gosman and Ioannides (1983) have prescribed values of A and B constants as 0.37 and 0.3 while Chen and Crowe (1984) have employed higher values of A and B constants as 0.45 and 0.457. Nevertheless, values of A and B constants as 0.2 and 0.164 have been more commonly adopted such as can be found in Faeth (1983); Shuen et al. (1983); Kallio and Stock (1986); Mostafa and Mongia (1987, 1988); and Adeniji-Fashola and Chen (1990). Milojevic (1990) has nonetheless applied values of A and B constants as 0.3 and 0.245. Based on the appropriate expressions of the eddy lifetime and eddy size, the interaction time is, henceforth, determined. The mathematical expression for the interaction time t_{int} take to be the equivalent to the time step Δt is

$$t_{\mathrm{int}} = \Delta t = \min(t_{\mathrm{e}}, t_{\mathrm{c}}) \tag{2.142}$$

where t_{c} is the minimum crossing time. The particle characteristic dimension l_{e} crossing an eddy is given by

$$t_{\mathrm{c}} = \frac{l_{\mathrm{e}}}{|\mathbf{V}_{\mathrm{f}} - \mathbf{V}_{\mathrm{p}}|} \tag{2.143}$$

Note that if t_{c} calculated using the above equation is smaller than t_{e}, the particle would jump to another eddy. Since the drift velocity in Eqn (2.143) is generally not known in advance, the drift velocity is required to be approximated at the beginning of the new interaction time. A different expression for the transit time, based on a simplified and linearized equation of motion for a particle in a uniform flow, is recommended. Thus,

$$t_{\mathrm{c}} = -\tau_{\mathrm{p}}\left(1 - \frac{l_{\mathrm{e}}}{\tau_{\mathrm{p}}|\mathbf{V}_{\mathrm{f}} - \mathbf{V}_{\mathrm{p}}|}\right) \tag{2.144}$$

The particle relaxation time τ_p in Eqn (2.144), which is the rate of response of particle acceleration to the relative velocity between the particle and the surrounding fluid, can be written as

$$\tau_p = \frac{4}{3}\frac{d_p^2}{\mu^c}\frac{\rho^p}{C_D}\frac{1}{Re_p} \tag{2.145}$$

Re_p is particle Reynolds number defined as $Re_p = \rho^c|\mathbf{V}_f - \mathbf{V}_p|\, d_p/\mu^c$. The magnitude of the relative velocity during the particle–eddy interaction is approximated by its value at the beginning of the new interaction. When $l_e > \tau_p|\mathbf{V}_f - \mathbf{V}_p|$, the linearized stopping distance of the particle is smaller than the eddy size, and Eqn (2.144) does not possess a solution. In such a case, the eddy has captured the particle and the interaction time is the eddy lifetime. Through these models, the time correlation between the fluctuating fluid velocities in the eddy lifetime is accounted for in a simple manner. The process of randomly sampling a set of independent fluid velocities and assuming them to be constant over random time steps implicitly produces a linear decrease in the Lagrangian autocorrelation. However, the method does not account for the fluctuating fluid velocity cross-correlations or anisotropy due to the isotropic assumption undertaken. For other more sophisticated stochastic models, the use of time-correlated stochastic models and PDF propagation models to account for both the fluid fluctuating time and cross-correlations is described next.

In the time-correlated stochastic models, both temporal fluid correlations and directional fluid correlations can be accounted for since a fluid particle is now simultaneously tracked along with the discrete particle trajectory to estimate the fluctuating velocity at each time step. By following the fluid and discrete particles simultaneously and moving down the flow field, the fluid fluctuating velocity can be determined at the discrete particle location and at the same time account for the temporal directional fluid correlations.

The first step in relation to the time-correlated stochastic models is to solve the fluid particle trajectory through accounting for both temporal fluid correlations and cross-correlations of fluctuating fluid velocity. A fluid particle can be constructed using a Markov-chain model as

$$\begin{aligned}
\mathbf{V}_f(t) &= \mathbf{V}_f^{mean}(t) + \mathbf{V}_f''(t) \\
\mathbf{V}_f''(t) &= R_{f,ij}^{L}(\Delta t)\mathbf{V}_f''(t)(t - \Delta t) + \mathbf{d}_t \\
\mathbf{x}_f(t) &= \mathbf{x}_f(t - \Delta t) + \left(\frac{\mathbf{V}_f(t) + \mathbf{V}_f(t - \Delta t)}{2}\right)\Delta t
\end{aligned} \tag{2.146}$$

where $i, j = x, y, z$ and $\mathbf{d}_t$ is the zero mean normal vector independent of the velocity vector $\mathbf{V}_f''$. The time $t - \Delta t$ denotes the present time and Eqn (2.146) demonstrates the computation of the location and fluctuating velocity of the fluid at the next time step. A commonly used approximation of the correlation tensor $R^{\mathrm{L}}_{\mathrm{f},ij}$ is the Frenkiel function:

$$R^{\mathrm{L}}_{\mathrm{f},ij}(\Delta t) = \exp\left[\frac{-\Delta t}{(m^2+1)t_{\mathrm{e}_{ij}}}\right]\cos\left[\frac{-m\Delta t}{(m^2+1)t_{\mathrm{e}_{ij}}}\right] \tag{2.147}$$

in which m is a modeling parameter, referred to as a negative loop parameter, set normally to a value of unity. Equation (2.147) assumes that the correlation tensor is a function of the present time ($t - \Delta t$) and the time step (Δt). This implies a Markovian approximation. The Lagrangian fluid time scale tensor $t_{\mathrm{e}_{ij}}$ can be estimated from the fluid properties according to

$$t_{\mathrm{e}_{ij}}(t-\Delta t) = C_{\mathrm{L}}\frac{\overline{v''_{\mathrm{f},i}(t-\Delta t)v''_{\mathrm{f},j}(t-\Delta t)}}{2\varepsilon}\ \text{for } i,j = 1,2,3 \tag{2.148}$$

where $(v''_{\mathrm{f},1}, v''_{\mathrm{f},2}, v''_{\mathrm{f},3}) \equiv (u''_{\mathrm{f}}, v''_{\mathrm{f}}, w''_{\mathrm{f}})$, C_{L} is a constant which ranges between 0.2 and 0.6 and ε, here again, is the dissipation of the turbulent kinetic energy of the fluid.

The second step of the time-correlated stochastic models is to determine the fluctuating particle velocity and thereafter obtain the instantaneous discrete particle velocity and location. Burry and Bergeles (1993) proposed that the fluctuating particle velocity vector is connected to the fluctuating fluid velocity vector according to

$$\mathbf{V}''_{\mathrm{p}} = R^{\mathrm{L}}_{ij}(r)\mathbf{V}''_{\mathrm{f}} + \mathbf{e}_{\mathrm{t}} \tag{2.149}$$

where $\mathbf{e}_{\mathrm{t}}$ is the randomness due to turbulence. Correlation term $R^{\mathrm{L}}_{ij}(r)$ characterizes the spatial correlation function which is given in terms of the distance (r) between the fluid and discrete particle locations and the Eulerian fluid length scale ($l_{\mathrm{e}_{ij}}$), in a similar form of the Frenkiel function:

$$R^{\mathrm{L}}_{ij}(r) = \exp\left[\frac{-r}{(m^2+1)l_{\mathrm{e}_{ij}}}\right]\cos\left[\frac{-mr}{(m^2+1)l_{\mathrm{e}_{ij}}}\right] \tag{2.150}$$

where $l_{\mathrm{e}_{ij}} \propto t_{\mathrm{e}_{ij}}\sqrt{\overline{v''_{\mathrm{f},i}v''_{\mathrm{f},j}}}$ and $t_{\mathrm{e}_{ij}}$ is evaluated from Eqn (2.148). The instantaneous particle velocity vector and position vector can then be determined by

$$\begin{aligned} \mathbf{V}_{\mathrm{p}}(t) &= \mathbf{V}^{\mathrm{mean}}_{\mathrm{p}}(t) + \mathbf{V}''_{\mathrm{p}}(t) \\ \mathbf{x}_{\mathrm{p}}(t) &= \mathbf{x}_{\mathrm{p}}(t-\Delta t) + \left(\frac{\mathbf{V}_{\mathrm{p}}(t) + \mathbf{V}_{\mathrm{p}}(t-\Delta t)}{2}\right)\Delta t \end{aligned} \tag{2.151}$$

where the mean particle velocity $\mathbf{V}^{\mathrm{mean}}_{\mathrm{p}}$ is calculated through the linear momentum equation using the mean fluid phase velocity:

$$\frac{D\mathbf{V}^{\mathrm{mean}}_{\mathrm{p}}}{Dt} = \frac{1}{\tau_{\mathrm{p}}}(\mathbf{V}^{\mathrm{mean}}_{\mathrm{f}} - \mathbf{V}^{\mathrm{mean}}_{\mathrm{p}}) + \left(1 - \frac{\rho^{\mathrm{f}}}{\rho^{\mathrm{p}}}\right)\mathbf{g} + \frac{1}{m_{\mathrm{p}}}\mathrm{OF} \tag{2.152}$$

where the "lumped" term "OF" represents the other forces that act on the particle. The reader is encouraged to consult Burry and Bergeles (1993) and Shirolkar et al. (1996) for more in-depth derivation of such models.

Comments

1. Isotropic stochastic models based on the eddy lifetime concept and anisotropic time-correlated stochastic models, which require the evaluation of two turbulent parameters—turbulent kinetic energy k^c and dissipation of the turbulent kinetic energy ε^c of the continuous phase—have been principally developed for the so-called Reynolds-averaged Navier–Stokes (RANS) framework. In this framework, only mean velocity fields are obtained thereby removing the complexity of actually solving for the fluctuating components governing the fluid flow behavior. Hence, the basic idea of the isotropic stochastic models and anisotropic stochastic models for the particle is to determine or recover the fluctuating components of the fluid velocity in order that it can be superimposed on the mean fluid velocity to yield the instantaneous fluid velocity. The formulation of the Reynolds-averaged equations along with suitable turbulent models for the fluid phase is described in Chapter 6.

2. Realistically speaking, the proper prediction of the turbulent transport of particle should be realized through capturing or fully resolving all scales of motion as well as all interfacial configurations of the fluid flow. One approach that calculates such flows at sufficiently high enough spatial and temporal resolution subject to the availability of computational resources is the direct numerical simulation (DNS). It is well-known that the required computational resources for DNS are comparatively very large. Because of this, DNS simulations of many multiphase flows have been mainly restricted to laminar multiphase systems or low Reynolds number turbulent flows. Alternatively, another approach based upon the concept of large eddy simulation (LES) treats the turbulent flow structure as the distinct transport of large- and small-scale motions. On this basis, the large-scale motion is directly simulated on a scale that the underlying computational mesh will allow while the small-scale motion is modeled accordingly. Since the large-scale motion is generally much more energetic and by far the most effective transporter compared with the small-scale ones, such an approach, which treats the large eddies precisely but approximates the small eddies so long as the small-scale turbulence physically exhibits isotropic turbulence, can be feasibly realized for turbulent multiphase systems. The computational requirement for LES is, in general, more intensive than RANS but still not as costly when compared to DNS. In the context of turbulent transport of particle, Vreman et al. (2009) have applied LES to investigate a vertical turbulent channel flow laden with a multitude of solid particles and that considers both particle–fluid and particle–particle interactions. In LES, the flow equations are spatially filtered in order to reduce dynamical complexity. Fundamentally, subgrid contributions arising from the filtering process need to

be integratted into the particle translation and rotation momentum equations in order to determine the particle trajectory and position. For a relatively coarse particle which is slightly larger than the Kolmogorov length scale, and where the Stokes response time is an order of magnitude larger than the Kolmogorov time, the subgrid contributions can be neglected. The motion of particle is, therefore, assumed to be influenced only by the large-scale eddies in the fluid flow. These have been confirmed in the study performed by Fede and Simonin (2006). The formulation of the filtered equations of the fluid phase along with suitable LES subgrid models for the small-scale eddies is described accordingly in Chapter 6.

2.3 Differential, Generic and Integral Form of the Transport Equations for Multiphase Flow

On the basis of the derivation of the conservation equations described, the set of partial differential equations can be collated governing the time-dependent three-dimensional fluid flow and heat transfer of a Newtonian fluid for the multifluid model in Cartesian coordinates.

Mass

$$\frac{\partial(\alpha^k\rho^k)}{\partial t}+\frac{\partial(\alpha^k\rho^k u^k)}{\partial x}+\frac{\partial(\alpha^k\rho^k v^k)}{\partial y}+\frac{\partial(\alpha^k\rho^k w^k)}{\partial z}=S_{m^k}^{\text{int}} \tag{2.153}$$

x-Momentum

$$\frac{\partial(\alpha^k\rho^k u^k)}{\partial t}+\frac{\partial(\alpha^k\rho^k u^k u^k)}{\partial x}+\frac{\partial(\alpha^k\rho^k v^k u^k)}{\partial y}+\frac{\partial(\alpha^k\rho^k w^k u^k)}{\partial z}$$
$$=\frac{\partial}{\partial x}\left[\alpha^k\mu^k\frac{\partial u^k}{\partial x}\right]+\frac{\partial}{\partial y}\left[\alpha^k\mu^k\frac{\partial u^k}{\partial y}\right]+\frac{\partial}{\partial z}\left[\alpha^k\mu^k\frac{\partial u^k}{\partial z}\right]+S_{u^k}^k \tag{2.154a}$$

y-Momentum

$$\frac{\partial(\alpha^k\rho^k v^k)}{\partial t}+\frac{\partial(\alpha^k\rho^k u^k v^k)}{\partial x}+\frac{\partial(\alpha^k\rho^k v^k v^k)}{\partial y}+\frac{\partial(\alpha^k\rho^k w^k v^k)}{\partial z}$$
$$=\frac{\partial}{\partial x}\left[\alpha^k\mu^k\frac{\partial v^k}{\partial x}\right]+\frac{\partial}{\partial y}\left[\alpha^k\mu^k\frac{\partial v^k}{\partial y}\right]+\frac{\partial}{\partial z}\left[\alpha^k\mu^k\frac{\partial v^k}{\partial z}\right]+S_{v^k}^k \tag{2.154b}$$

z-Momentum

$$\frac{\partial(\alpha^k\rho^k w^k)}{\partial t}+\frac{\partial(\alpha^k\rho^k u^k w^k)}{\partial x}+\frac{\partial(\alpha^k\rho^k v^k w^k)}{\partial y}+\frac{\partial(\alpha^k\rho^k w^k w^k)}{\partial z}$$
$$=\frac{\partial}{\partial x}\left[\alpha^k\mu^k\frac{\partial w^k}{\partial x}\right]+\frac{\partial}{\partial y}\left[\alpha^k\mu^k\frac{\partial w^k}{\partial y}\right]+\frac{\partial}{\partial z}\left[\alpha^k\mu^k\frac{\partial w^k}{\partial z}\right]+S_{w^k}^k \tag{2.154c}$$

Enthalpy

$$\frac{\partial(\alpha^k\rho^kH^k)}{\partial t}+\frac{\partial(\alpha^k\rho^ku^mH^m)}{\partial x}+\frac{\partial(\alpha^k\rho^kv^kH^k)}{\partial y}+\frac{\partial(\alpha^k\rho^kw^kH^k)}{\partial z}=$$
$$\frac{\partial}{\partial x}\left[\alpha^k\lambda^k\frac{\partial T^k}{\partial x}\right]+\frac{\partial}{\partial y}\left[\alpha^k\lambda^k\frac{\partial T^k}{\partial y}\right]+\frac{\partial}{\partial z}\left[\alpha^k\lambda^k\frac{\partial T^k}{\partial z}\right]+S^k_{H^k} \tag{2.155}$$

For Eqns (2.153)−(2.155), there are commonalities that exist between them. Employing the general variable ϕ^k, the generic form of the governing equations can be written for the multifluid model as

$$\frac{\partial(\alpha^k\rho^k\phi^k)}{\partial t}+\frac{\partial(\alpha^k\rho^ku^k\phi^k)}{\partial x}+\frac{\partial(\alpha^k\rho^kv^k\phi^k)}{\partial y}+\frac{\partial(\alpha^k\rho^kw^k\phi^k)}{\partial z}$$
$$=\frac{\partial}{\partial x}\left[\alpha^k\Gamma^k_{\phi^k}\frac{\partial\phi^k}{\partial x}\right]+\frac{\partial}{\partial y}\left[\alpha^k\Gamma^k_{\phi^k}\frac{\partial\phi^k}{\partial y}\right]+\frac{\partial}{\partial z}\left[\alpha^k\Gamma^k_{\phi^k}\frac{\partial\phi^k}{\partial z}\right]+S^k_{\phi^k} \tag{2.156}$$

Equation (2.156) is principally known as the transport equations for any variable ϕ^k. It essentially illustrates the various physical transport processes occurring in the fluid flow: the rate of change ϕ^k representing the local acceleration term accompanied by the advection term on the left hand side, is equivalent to the diffusion term (where $\Gamma^k_{\phi^k}$ is designated as the diffusion coefficients) and the source term $S^k_{\phi^k}$ on the right-hand side. In order to bring forth the common features, terms that are not shared between the equations are placed into the source terms.

By setting the transport variable ϕ^k equal to 1, u^k, v^k, w^k, and H^k and selecting appropriate values for the diffusion coefficients Γ_{ϕ^k} and source terms S_{ϕ^k}, we obtain special forms, presented in Table 2.1, for each of the partial differential equations for the conservation of mass, momentum, and energy for the multifluid model. Although we have systematically walked through the derivation of the complete set of governing equations in detail from basic conservation principles, the final general forms pertaining to the fluid motion, heat transfer, and turbulent scalars to be defined in Chapter 6 conform simply to the generic form of Eqn (2.156). These equations are therefore of enormous significance within the computational fluid dynamics framework as the inclusion of these increasingly complex physical processes associated with multiphase flows can be appropriately accommodated and correctly solved. In Table 2.1, $S^{\text{int}}_{u^k}$, $S^{\text{int}}_{v^k}$, $S^{\text{int}}_{w^k}$, and $S^{\text{int}}_{H^k}$ represent the interfacial source terms for the respective phases.

There are commonalities that are also shared between the various equations for the particle in the trajectory model based on a discrete element framework. In general, the form of the governing equations can simply be written in terms of the material derivative and source term as

$$\frac{D\phi^d}{Dt}=S_{\phi^d} \tag{2.157}$$

Table 2.1: General form of governing equations for the multifluid model.

ϕ^k	$\Gamma^k_{\phi^k}$	$S^k_{\phi^k}$
1	0	$S^{\text{int}}_{m^k}$
u^k	μ^k	$-\alpha^k \frac{\partial p'^k}{\partial x} + \alpha^k \rho^k g_x + \frac{\partial}{\partial x}\left[\alpha^k \mu^k \frac{\partial u^k}{\partial x}\right] + \frac{\partial}{\partial y}\left[\alpha^k \mu^k \frac{\partial v^k}{\partial x}\right] + \frac{\partial}{\partial z}\left[\alpha^k \mu^k \frac{\partial w^k}{\partial x}\right] + \frac{\partial}{\partial x}\left[\alpha^k \tau^{k''}_{xx}\right] + \frac{\partial}{\partial y}\left[\alpha^k \tau^{k''}_{xy}\right] + \frac{\partial}{\partial z}\left[\alpha^k \tau^{k''}_{xz}\right] + S^{\text{int}}_{u^k}$
v^k	μ^k	$-\alpha^k \frac{\partial p'^k}{\partial y} + \alpha^k \rho^k g_y \frac{\partial}{\partial x}\left[\alpha^k \mu^k \frac{\partial u^k}{\partial y}\right] + \frac{\partial}{\partial y}\left[\alpha^k \mu^k \frac{\partial v^k}{\partial y}\right] + \frac{\partial}{\partial z}\left[\alpha^k \mu^k \frac{\partial w^k}{\partial y}\right] + \frac{\partial}{\partial x}\left[\alpha^k \tau^{k''}_{yx}\right] + \frac{\partial}{\partial y}\left[\alpha^k \tau^{k''}_{yy}\right] + \frac{\partial}{\partial z}\left[\alpha^k \tau^{k''}_{yz}\right] + S^{\text{int}}_{v^k}$
w^k	μ^k	$-\alpha^k \frac{\partial p'^k}{\partial z} + \alpha^k \rho^k g_z + \frac{\partial}{\partial x}\left[\alpha^k \mu^k \frac{\partial u^k}{\partial z}\right] + \frac{\partial}{\partial y}\left[\alpha^k \mu^k \frac{\partial v^k}{\partial z}\right] + \frac{\partial}{\partial z}\left[\alpha^k \mu^k \frac{\partial w^k}{\partial z}\right] + \frac{\partial}{\partial x}\left[\alpha^k \tau^{k''}_{zx}\right] + \frac{\partial}{\partial y}\left[\alpha^k \tau^{k''}_{zy}\right] + \frac{\partial}{\partial z}\left[\alpha^k \tau^{k''}_{zz}\right] + S^{\text{int}}_{w^k}$
H^k	0	$\frac{\partial}{\partial x}\left(\alpha^k \lambda^k \frac{\partial T^k}{\partial x}\right) + \frac{\partial}{\partial y}\left(\alpha^k \lambda^k \frac{\partial T^k}{\partial y}\right) + \frac{\partial}{\partial z}\left(\alpha^k \lambda^k \frac{\partial T^k}{\partial z}\right) + \frac{\partial}{\partial x}\left[\alpha^k q^{k''}_{H_x}\right] + \frac{\partial}{\partial y}\left[\alpha^k q^{k''}_{H_y}\right] + \frac{\partial}{\partial z}\left[\alpha^k q^{k''}_{H_z}\right] + S^{\text{int}}_{H^k}$

Note: p'^k is the modified pressure defined by $p'^k = p^k + \frac{2}{3}\mu^k \nabla \cdot \mathbf{U}^k$.

where ϕ^d represents the instantaneous transport variable accompanied by the instantaneous forcing term S_{ϕ^d}. Table 2.2 summarizes the special forms of each equation governing the conservation of mass, momentum, and energy for the particle.

In order to numerically solve the approximate form of Eqn (2.156), it is convenient to consider the integral form of this generic transport equation over a finite control volume. Integration of the equation for the mixture over a three-dimensional control volume V yields

Table 2.2: General form of governing equations for the trajectory model.

ϕ^d	S_{ϕ^d}
m_P	S_{m_P}
u_P	$\frac{1}{\rho^P V^P}\sum F_x$
v_P	$\frac{1}{\rho^P V^P}\sum F_y$
w_P	$\frac{1}{\rho^P V^P}\sum F_z$
$\Omega_{x,P}$	$\frac{1}{I_P}\sum M_x$
$\Omega_{y,P}$	$\frac{1}{I_P}\sum M_y$
$\Omega_{z,P}$	$\frac{1}{I_P}\sum M_z$
T_P	$\frac{1}{m_P C_P} S_{T_P}$

$$\int_V \frac{\partial(\alpha^k\rho^k\phi^k)}{\partial t}\mathrm{d}V + \int_V \left\{\frac{\partial(\alpha^k\rho^k u^k\phi^k)}{\partial x} + \frac{\partial(\alpha^k\rho^k v^k\phi^k)}{\partial y} + \frac{\partial(\alpha^k\rho^k w^k\phi^k)}{\partial z}\right\}\mathrm{d}V$$
$$= \int_V \left\{\frac{\partial}{\partial x}\left[\alpha^k\Gamma^k_{\phi^k}\frac{\partial\phi^k}{\partial x}\right] + \frac{\partial}{\partial y}\left[\alpha^k\Gamma^k_{\phi^k}\frac{\partial\phi^k}{\partial y}\right] + \frac{\partial}{\partial z}\left[\alpha^k\Gamma^k_{\phi^k}\frac{\partial\phi^k}{\partial z}\right]\right\}\mathrm{d}V + \int_V S^k_{\phi^k}\mathrm{d}V \quad (2.158)$$

By applying the Gauss divergence theorem to the volume integral of the advection and diffusion terms, Eqn (2.158) can now be expressed in terms of their elemental projected areas along the Cartesian coordinate directions dA^x, dA^y, and dA^z as

$$\int_V \frac{\partial(\alpha^k\rho^k\phi^k)}{\partial t}\mathrm{d}V + \int_A \{(\alpha^k\rho^k u^k\phi^k)\mathrm{d}A^x + (\alpha^k\rho^k v^k\phi^k)\mathrm{d}A^y + (\alpha^k\rho^k w^k\phi^k)\mathrm{d}A^z\}$$
$$= \int_A \left\{\left[\alpha^k\Gamma^k_{\phi^k}\frac{\partial\phi^k}{\partial x}\right]\mathrm{d}A^x + \left[\alpha^k\Gamma^k_{\phi^k}\frac{\partial\phi^k}{\partial y}\right]\mathrm{d}A^y + \left[\alpha^k\Gamma^k_{\phi^k}\frac{\partial\phi^k}{\partial z}\right]\mathrm{d}A^z\right\} + \int_V S^k_{\phi^k}\mathrm{d}V \quad (2.159)$$

Note that the projected areas are positive if their outward normal vector from the volume surface are directed in the same direction along the Cartesian coordinate system; otherwise they are negative.

2.4 Boundary Conditions for Multiphase Flow

Conservation equations of the mass, momentum, and enthalpy equations for the multifluid model govern the fluid flow, heat transfer, and mass transfer in a multiphase flow system. With regard to additional closure required via appropriate turbulence models and constitutive equations for the interfacial terms, suitable boundary conditions, and sometimes initial conditions form another important component in the computation of multiphase flow since they dictate the particular solutions that can be obtained from these governing equations. This is of great significance especially in multiphase flow modeling as any numerical solutions of the governing equations must result in a strong and compelling representation of the actual physical flow.

In order to impose relevant boundary conditions in multiphase flow systems, consider for the purpose of illustration a three-phase gas-liquid-solid flow in, for example, a conduit, as described in Figure 2.10. Let us first consider the boundary conditions for velocity. For the multifluid model, the x-component of the superficial velocity u^k_s can be related to the actual velocity u^k according to

$$u^k_s = \alpha^k u^k \quad \text{at the inflow boundary} \quad (2.160)$$

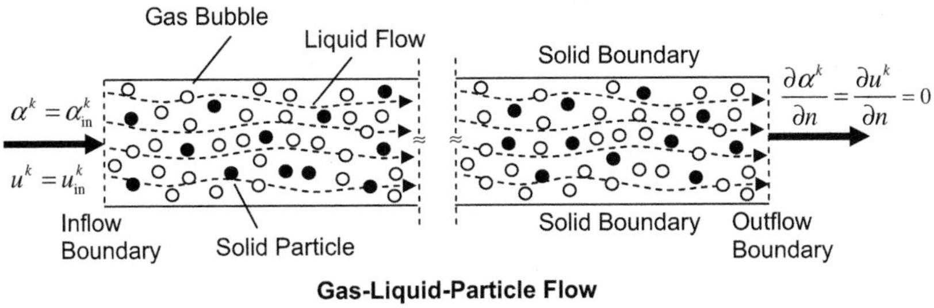

Figure 2.10
Specification of the velocity boundary conditions for a gas-liquid-particle flow.

By definition, superficial velocity, sometimes referred as the volumetric flux, is effectively the volume flow of the fluid divided by the cross-sectional area of the flow domain. The fraction of the area occupied by the fluid is assumed to be equivalent to the volume fraction occupied by the fluid. From Eqn (2.160), it is evident that the knowledge of both distributions of the volume fraction and actual velocity is required for the transport field ϕ^k. The *Dirichlet* boundary conditions for the volume fraction and actual velocity are:

$$\alpha^k = \alpha^k_{\text{in}}, u^k = u^k_{\text{in}} \quad \text{and} \quad v^k = w^k = 0 \quad \text{at the inflow boundary} \tag{2.161}$$

where α^k_{in} and u^k_{in} can either be specified as constant values or prescribed profiles at the surface. From a computational perspective, *Dirichlet* boundary conditions can be applied rather accurately so long as α^k_{in}, u^m_{in}, and u^k_{in} are continuous. Similar boundary conditions in the x direction may also be imposed at the inflow boundaries in the y and z directions.

For outflow boundaries, it is rather important that they should be located where the flows are approximately unidirectional and where surface stresses can take known values. In a fully-developed flow, the velocity component in the direction across the boundary remains unchanged and by satisfying stress continuity, the shear forces along the surface are taken to be zero. The outflow boundary conditions are given by

$$\frac{\partial \alpha^k}{\partial n} = \frac{\partial u^k}{\partial n} = \frac{\partial v^k}{\partial n} = \frac{\partial w^k}{\partial n} = 0 \quad \text{at the outflow boundary} \tag{2.162}$$

where n is the direction normal to the surface. This condition which is commonly known as the *Neumann* boundary condition is, nonetheless, equally applicable for the phase velocity as well as the local volume fraction. Equation (2.162) also implies that the normal gradient of the superficial velocity u^k_{s} is equal to zero, i.e. $\partial u^k_{\text{s}}/\partial n = 0$ at the outflow boundary.

The *no-slip* condition is generally imposed for the solid walls of the conduit. A zero relative velocity is assumed between the surface and the fluid immediately at the surface. In the multifluid model, the physical boundary conditions for the disperse phase (gas–solid flow) are distinguished from the continuous phase (liquid flow). Noting that the superscript

c denotes the continuous phase, the *no-slip* condition is invoked and the velocity components are prescribed as

$$u^{\mathrm{c}} = v^{\mathrm{c}} = w^{\mathrm{c}} = 0 \quad \text{at the solid boundary} \tag{2.163}$$

For the dispersed phase, it is more preferable to consider a slip condition. Hence, the velocity at the wall is a finite, nonzero value. In the x direction, the wall boundary condition, noting that the superscript p denotes the dispersed particle phase, is given as

$$u^{\mathrm{p}} = u^{\mathrm{p}}_{\mathrm{wall}},\, v^{\mathrm{p}} = v^{\mathrm{p}}_{\mathrm{wall}} \quad \text{and} \quad w^{\mathrm{p}} = w^{\mathrm{p}}_{\mathrm{wall}} \quad \text{at the solid boundary} \tag{2.164}$$

where $u^{\mathrm{p}}_{\mathrm{wall}}$, $v^{\mathrm{p}}_{\mathrm{wall}}$, and $w^{\mathrm{p}}_{\mathrm{wall}}$ are the velocity components of the dispersed particle–fluid velocity immediately adjacent to the wall. The Neumann boundary condition is imposed for the local volume fraction:

$$\frac{\partial \alpha^k}{\partial n} = 0 \quad \text{at the solid boundary} \tag{2.165}$$

For a multiphase flow system with heat transfer, it is also common to impose the Dirichlet and Neumann boundary conditions at the inflow and outflow boundaries. With reference to Figure 2.10, the boundary conditions for T^k are

$$T^k = T^k_{\mathrm{in}} \quad \text{at the inflow boundary} \tag{2.166}$$

$$\frac{\partial T^k}{\partial n} = 0 \quad \text{at the outflow boundary} \tag{2.167}$$

where T^k_{in} can either be specified as constant values or prescribed profiles at the surface. A *no-slip* condition can also be analogously applied for the temperature at the walls of the conduit. For a given problem where the wall temperature T^k_{wall} is known, the Dirichlet boundary condition applies and the boundary condition is given by

$$T^k = T^k_{\mathrm{wall}} \quad \text{at the solid boundary} \tag{2.168}$$

Nevertheless, if the wall temperature is changing as a function of time due to the heat transfer to or from the surface, then Fourier's law of heat conduction can be applied to provide the necessary boundary condition at the surface. Denoting the wall heat flux as q^k_{wall} then according to Fourier's law

$$q^k_{\mathrm{wall}} = -\left(\alpha^k \lambda^k \frac{\partial T^k}{\partial n}\right)_{\mathrm{wall}} \quad \text{at the solid boundary} \tag{2.169}$$

From above, the changing surface mixture temperature T^k_{wall} is now responding to the thermal response of the wall material through the heat transfer to the wall q^k_{wall}. For the case where there is no heat transfer to the surface, the proper boundary condition comes from Eqn (2.169) with $q^k_{\text{wall}} = 0$, in other words,

$$\left(\frac{\partial T^k}{\partial n}\right)_{\text{wall}} = 0 \quad \text{at the solid boundary} \tag{2.170}$$

This condition corresponds with the Neumann boundary condition for the phase temperature at the outflow boundary. At this stage, the only physical boundary conditions for the continuum viscous flow are the boundary conditions associated with the velocity and temperature. Other flow variables, such as density and pressure, fall out as part of the solution. In the absence of interfacial mass exchange, it is evident that the boundary conditions (2.161)–(2.164) close the system mathematically and satisfy local and overall mass conservation for the multifluid model.

The boundary conditions for the turbulent scalars may be written in general forms for the different boundaries of the multiphase flows as

$$\phi^n = \phi^n_{\text{in}} \quad \text{at the inflow boundary} \tag{2.171}$$

$$\frac{\partial \phi^n}{\partial n} = 0 \quad \text{at the outflow boundary} \tag{2.172}$$

$$\phi^n = \phi^n_{\text{wall}} \quad \text{or} \quad \frac{\partial \phi^n}{\partial n} = 0 \quad \text{at the solid boundary} \tag{2.173}$$

where the superscript n denotes the phase quantities and ϕ^n_{in} and ϕ^n_{wall} can either be specified as constant values or prescribed profiles at the inflow and solid boundaries, respectively. For turbulent multiphase flows, experimentally verified quantities, whenever possible, should always be applied at the inflow boundaries for the turbulent scalars. Adopting some sensible engineering assumptions, specification of the inflow turbulent kinetic energy may be realized through relating the inflow turbulence to the turbulence intensity, which is defined as the ratio of the fluctuating component of the velocity to the mean velocity, as well as the upstream inflow conditions. Approximate values for the turbulent kinetic energy, dissipation, and frequency can be determined according to the following relationships:

$$k^n_{\text{in}} = \frac{3}{2}\left(u^n_{\text{in}} I\right)^2 \tag{2.174}$$

$$\varepsilon^n_{\text{in}} = C_\mu^{3/4} \frac{\left(k^n_{\text{in}}\right)^{3/2}}{l} \tag{2.175}$$

$$\omega^n_{\text{in}} = \frac{\sqrt{k^n_{\text{in}}}}{C_\mu^{1/4} l} \tag{2.176}$$

where I is the turbulence intensity level and l is the characteristic length scale. Different levels of I and length scales of l are considered depending on whether the multiphase systems are internal or external flows. In cases where problems arise in specifying appropriate turbulence quantities, the inflow boundaries for the application of all turbulence models should be moved sufficiently far away from the region of interest such that the inflow boundary layer and subsequently the turbulence are allowed to be developed naturally. Near the solid walls, universal wall functions which are employed alongside with the two-equation *k-ε model* avoid the necessity of resolving the flow structure within the viscous sublayer. In these models, the dissipation ε^n is normally empirically evaluated at the first computational mesh point adjacent to the solid boundary.

Other boundary conditions that are also of importance and often required for multiphase modeling include the symmetry and periodic boundary conditions. The symmetric boundary condition can be employed to take advantage of special geometrical features of the solution region. This boundary condition can be imposed by prescribing the normal velocity at the surface and the normal gradients of the other velocity components to be zero. The *Neumann* boundary condition is subsequently applied for the rest of the variables. For the periodic boundary condition, the transport property of one of the surface ϕ_1^n can be taken to be equivalent to the transport property of the second surface ϕ_2^n depending on which of the two surfaces of the flow domain experience periodicity, i.e. $\phi_1^n = \phi_2^n$.

Comments

1. For particle flow in a confined conduit, there are a number of events the particle experiences when it strikes a wall. As an example, a solid particle may be reflected via an elastic collision or inelastic collision. In the Lagrangian framework, this particular boundary condition concerns the solid particle trajectory as it rebounds off the boundary with a change of momentum according to its normal and tangential coefficients of restitution. These coefficients, which define the amount of momentum in the directions normal and tangential to the wall being retained by the solid particle after collision with the boundary, are generally required to be known *a priori*. The proper specification of this boundary condition is, however, not trivial. They are strongly influenced by the topology, material properties and conditioning at the surface. Currently, there are no universal coefficients of restitution that can be readily adopted for all types of surfaces. In addition, there is the possibility where the solid particle may slide along the wall depending on the particle properties and impact angle and/or "stick" to the wall, which increases the complexity of determining the actual behavior of the solid particle after colliding with the boundary. In the Eulerian description of the solid particle flow, the possible occurrences of individual particles impacting at the wall, sliding along the wall or sticking on the wall are treated via the specification of the slip boundary condition. Such consideration of this particular boundary condition should not be construed as to reflect the actual

behaviors of the solid particles at the wall; possible events occurring at the wall are grossly approximated through prescribing feasible averaged values of the wall velocity components. This slip boundary condition, pending a more rigorous formulation, is the best which can be applied in the present state of multiphase modeling.

2. For the flow gas bubbles in a liquid, a degassing boundary condition is typically employed to feasibly model a free surface from which the dispersed gas bubbles are permitted to escape, but not the continuous liquid. The problem can thus be treated crudely by assuming that the continuous and disperse phases view this boundary as a slip wall and an outlet though the main issue still remains in ensuring the total conservation of mass is appropriately satisfied for the two-phase flow. This boundary condition can also be applied for a boiling liquid in an open vessel where the top surface is exposed to the surroundings only if the volume of gas being evaporated is negligible in comparison to the volume of liquid that is contained in the vessel. In reality, the possible loss of liquid in the vessel due to the evaporation occurring at the top surface should be accounted for, which could possibly be achieved via tracking the shrinkage of the volume of liquid as the level subsides.

2.5 Summary

The basic equations that form the foundation of the Eulerian–Eulerian interpenetrating continua and Eulerian–Lagrangian on discrete element frameworks for multiphase flow calculations have been described. By means of a general purpose model of fluid flow, and heat and mass transfer stemming from the basic principles of conservation of mass, momentum, and energy, the consideration of an infinitesimal small control volume, the introduction of a phase indicator function, and averaging being performed in time, space, over an ensemble, or in some combination of these, the effective conservation equations as well as the interfacial sources, are derived for the multifluid model. Although the Newtonian model of viscous stresses has been assumed to tie up the system of equations, accommodation of fluids having non-Newtonian characteristics could have been easily incorporated within the framework of these equations. With regard to the conservation equations that have been derived, significant commonalities can be found between them. This subsequently leads to the formulation of the generic form of the governing equations of the multifluid model. Appropriate boundary conditions and their physical significance on different typical forms of multiphase flows, as well as some comments on the specification of the wall boundary conditions for complex multiphase problems, are provided. In the next chapter, the generic framework for population balance is explored and described.

CHAPTER 3

Population Balance Approach—A Generic Framework

3.1 What is a Population Balance Approach?

Particles, regardless of whether they are naturally present within or engineered into the multiphase flow system, can significantly affect the behavior of such flow a system. The size distribution of the secondary phase, which may comprise gas bubbles, liquid drops, or solid particles, can evolve in conjunction with the transport of the secondary phase in this type of flow system. These evolutionary processes may have a combination of different phenomena associated with nucleation, growth, dispersion, dissolution, aggregation or coalescence, and breakage or break up producing the dispersion. To determine the extent of particles influencing the fluid flow, a balance equation based on the *population balance approach* can be adopted to better synthesize the behavior of the population of particles and understand its dynamical evolution subject to the system environments. In essence, the population balance approach is specifically utilized to consider the balances in the population of the particles, in addition to the consideration of momentum, mass, and energy balances.

The viability of the *population balance approach* to handle a wide variety of particle processes has certainly received unprecedented attention. The coupling of the computational multiphase fluid dynamics model and appropriate models of population balance is increasingly being employed to predict the evolution of the size distribution of particles in order to improve the crystallization, precipitation and polymerization processes of materials for a wide variety of applications including pharmaceutical, agriculture, and specialty chemical products as well as the many aggregation–breakage processes in engineering applications associated with coagulation and rupture of flocs, addition and degradation of polymers, and coalescence and break up of liquid drops. Mounting interest in population balance in the petrochemical and mining industries has also resulted in significant design improvements to the widely used bubble column reactors to promote increasing rates of mass transport of gas to the liquid and more efficient mixing of competing gas–liquid reactions. In retrospect, a *population balance* applying to any flow system is mainly concerned with tracking the number of entities—solid particles, bubbles, or droplets—present within the system, whose development of these entities ultimately dictates the overall behavior of the system under consideration.

Multiphase Flow Analysis Using Population Balance Modeling.

The field equations that are derived for the multifluid model in the previous chapter can normally be expressed by six or nine conservation equations consisting of mass, momentum and energy for each phase in a two-phase flow system or three-phase flow system. Within these balance equations, the existence of interphase transfer terms—mass, momentum, and energy exchanges through the interface between each phase, which are obtained from an appropriate averaging of local instantaneous balance equations—indicates an important characteristic of the multifluid model formulation. These terms in the respective conservation equations determine the rate of phase changes and the degree of mechanical and thermal nonequilibrium between phases; they are thus required to be modeled accurately for multiphase flows.

3.2 Basic Definitions

3.2.1 Coordinate System and Density Function

The population balance of any system can be conceptually described as the evolution of the population of countable entities that exist within the flow system. In this framework, the population of particles can be treated not only as being distributed in the physical space but also in an abstract property space. Dependent variables of these entities can be taken to exist in two different coordinates: *internal* and *external* coordinates (Ramkrishna, 2000). Mathematically, the former is taken to be the property coordinates, the latter as the spatial coordinates. An example of the internal and external coordinates involved in the population balance for gas–liquid–particle flow is illustrated in Figure 3.1. The joint space comprising the internal and external coordinates is referred to as the *particle phase space*. In this space, the quantity of basic interest which is useful to the characterization of distinct gas bubbles, solid particles, or liquid drops is the *density function*. The population balance may, therefore, be regarded as the transport for the *number density* and the *number balance for particles of a particular state*.

Evolution of the density function within the system must consider the various ways in which particles of a specific state can either be created or eliminated from the system. Change of density function with respect to external coordinates refers to motion through the physical space, that of internal coordinates through the abstract property space. In other words, the external coordinates refer to the spatial location of each particle in the physical space while the internal coordinates are concerned primarily with the internal properties of particles such as size, surface area, volume, composition, and so forth. The physical processes that can be described based on this format are collectively denoted as the advection or convection processes since they result from motion in the particle phase space. At the same time, this group of physical processes may contribute to the creation and termination of the specific particle types. The number of particles of a particular type

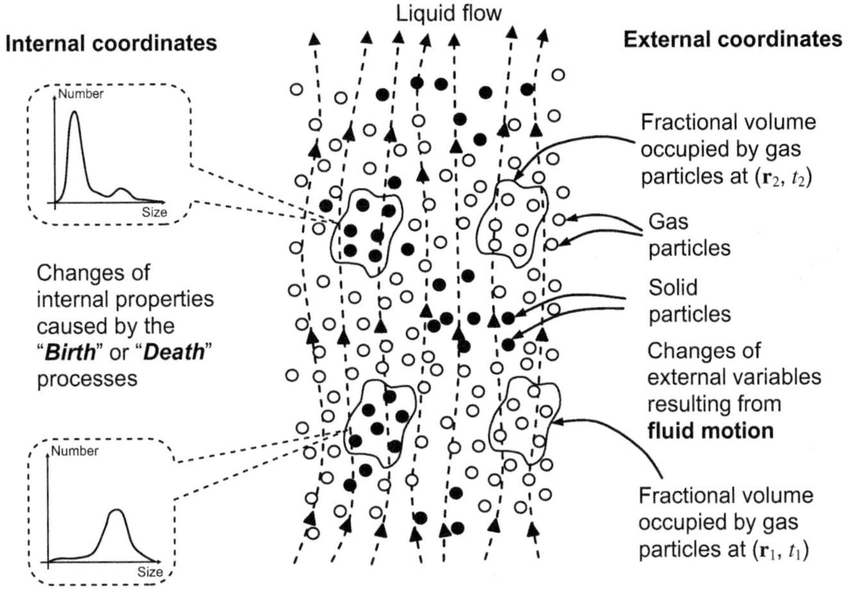

Figure 3.1
An example of the internal and external coordinates of population balance for gas–liquid–particle flow.

can also be altered through the creation of new particles, known as *birth processes* and termination of existing particles, known as *death processes*. Phenomenologically, population balance comprises convection processes as well as the birth and death processes of the population of particles.

3.2.2 Particle State Vector

The choice of particle state can be established through the consideration of variables that are required to satisfy the rate of change of those parameters of direct interest to a particular application and the birth and death processes. This assumption implies that memory effects can be neglected for the particle processes on the current particle in question. In practice, the particle state vector can be described by a finite dimensional vector. This vector can accommodate the description of particles with considerable internal structure as well as particles with spatial, internal morphology where several distinct components may be located at the centroid of the particle.

Following the notation of Ramkrishna (2000), it is convenient to define the density function as $f_1(\mathbf{x},\mathbf{r},t)$, which is denoted as the number of particles per unit volume of the particle phase at time t, at the internal coordinates $\mathbf{x} \equiv (x_1, x_2, \ldots, x_n)$, where n represents the number of different quantities associated with the particle and at the external coordinates $\mathbf{r} \equiv (r_1, r_2, r_3)$

which may be employed to indicate the position vector of the particle as determined by its centroid. This number density $f_1(\mathbf{x},\mathbf{r},t)$ is taken to be smooth in order that it can be differentiated with respect to any of its arguments as many times as desired. The particle state vector accounts for both internal and external coordinates; the domain of internal coordinate is taken to be $V_\mathbf{x}$, and the domain of external coordinates to be $V_\mathbf{r}$, which represents a set of points in the physical space in which particles are present.

3.2.3 Continuous Phase Vector

The behavior of each particle being affected by the continuous phase may also be collated into a finite dimensional vector field. The continuous phase vector, which can be defined by $\mathbf{Y}(\mathbf{r},t)\equiv(\alpha^k,u^k,v^k,w^k,\rho^k,\ldots)$, is a function of only the external coordinates $\mathbf{r}$ and time t and is calculated from the governing transport equations and the boundary conditions associated with the particular problem. It should be noted that a continuous phase may not be necessarily considered for some applications where there is interaction between the population of particles, and the continuous phase does not result in a substantial change in the continuum phase vector. Analysis of the population thus reduces to only the equation describing the population balance.

3.2.4 Rate of Change of Particle State Vector and Particle State Continuum

Particle states can vary significantly in time. Of interest, smooth changes in particle state, which can be described by separate velocity vectors, are defined over the particle state space for both the internal and external coordinates. While change of internal coordinates refers to motion through an abstract property space, change of external coordinates refers to motion through physical space. In order to distinguish the different convection processes, the velocities for the internal and external coordinates are defined respectively as $\mathbf{v}_\mathbf{x}(\mathbf{x},\mathbf{r},\mathbf{Y},t)$ and $\mathbf{v}_\mathbf{r}(\mathbf{x},\mathbf{r},\mathbf{Y},t)$. Based on these velocities, which are assumed to be as smooth as is necessary, it is also possible to identify particle fluxes (number of particles per unit time per unit area normal to the direction of the velocity). Hence, $f_1(\mathbf{x},\mathbf{r},t)\mathbf{v}_\mathbf{x}(\mathbf{x},\mathbf{r},\mathbf{Y},t)$ is the particle flux through the internal coordinate space while $f_1(\mathbf{x},\mathbf{r},t)\mathbf{v}_\mathbf{r}(\mathbf{x},\mathbf{r},\mathbf{Y},t)$ represents the particle flux through the physical space; both of these fluxes are determined at time t and point $(\mathbf{x},\mathbf{r})$ in particle state space.

For the ease of deriving the population balance equation, it is convenient to designate a particle space continuum which permeates the space of internal and external coordinates. In retrospect, this continuum may be viewed as deforming in space and time in accordance with the field—$\mathbf{v}_\mathbf{x}(\mathbf{x},\mathbf{r},\mathbf{Y},t)$ and $\mathbf{v}_\mathbf{r}(\mathbf{x},\mathbf{r},\mathbf{Y},t)$—relative to the fixed coordinates. Based on this continuum assumption, each particle phase shall be deemed to behave like a continuous fluid.

3.3 Fundamentals of Population Balance Equation

3.3.1 Basic Consideration

Considering that the particles are firmly embedded in the deforming particle state continuum, the integral formulation of the population balance states that the number of particles can only change through the birth and death processes. Denoting the net rate of generation of particles by $S_{f_1}(\mathbf{x},\mathbf{r},\mathbf{Y},t)$ and $dV_\mathbf{x}$ and $dV_\mathbf{r}$ as infinitesimal volumes in abstract property space of arbitrary domain $V_\mathbf{x}$ and physical space of arbitrary domain $V_\mathbf{r}$, the system balance deforming in time and space may be given by

$$\frac{D}{Dt}\int\limits_{V_\mathbf{x}(t)}\int\limits_{V_\mathbf{r}(t)} f_1 dV_\mathbf{x} dV_\mathbf{r} = \int\limits_{V_\mathbf{x}(t)}\int\limits_{V_\mathbf{r}(t)} S_{f_1} dV_\mathbf{x} dV_\mathbf{r} \tag{3.1}$$

where the notation D/Dt denotes the substantial derivative operator. Adopting the notation $V(t) = V_\mathbf{x}(t) \cup V_\mathbf{r}(t)$, $dV = dV_\mathbf{x} dV_\mathbf{r}$ and $\int_{V(t)} = \int_{V_\mathbf{x}(t)}\int_{V_\mathbf{r}(t)}$, Eqn (3.1) can be rewritten in a compact form:

$$\frac{D}{Dt}\int\limits_{V(t)} f_1 dV = \int\limits_{V(t)} S_{f_1} dV \tag{3.2}$$

For general vector spaces, the conventional Reynolds theorem is written as

$$\frac{D}{Dt}\int\limits_{V(t)} f_1 dV = \frac{\partial}{\partial t}\int\limits_{V(t)} f_1 dV + \int\limits_{S(t)} (\mathbf{v}f_1)\cdot \mathbf{n} dS \tag{3.3}$$

where $\mathbf{v}$ $(\equiv \mathbf{v}_\mathrm{x} + \mathbf{v}_\mathrm{r})$ represents the combined phase velocity vector, dS $(\equiv dS_\mathbf{x} + dS_\mathbf{r})$ denotes infinitesimal surface areas in $S \equiv S_\mathbf{x} + S_\mathbf{r}$ and $\mathbf{n}$ $(\equiv \mathbf{n}_\mathbf{x} + \mathbf{n}_\mathbf{r})$ denotes a combined normal unit vector. Also, Gauss's divergence theorem can be applied, which equates the volume integral of a divergence $(\nabla\cdot)$ of a vector into an area integral over the surface:

$$\int\limits_{V(t)} \nabla\cdot(\mathbf{v}f_1) dV = \int\limits_{S(t)} (\mathbf{v}f_1)\cdot \mathbf{n} dS \tag{3.4}$$

Using the above theorem, the surface integral in Eqn (3.3) may be replaced by a volume integral and the equation can subsequently be rewritten as a volume integral. In other words,

$$\int\limits_{V(t)} \left(\frac{\partial f_1}{\partial t} + \nabla\cdot(\mathbf{v}f_1) - S_{f_1}\right) dV = 0 \tag{3.5}$$

Since (3.5) is valid for any size of combined domain *V*, the implication is that the integrand must vanish everywhere in the particle state space. This leads to the differential form of the population balance equation:

$$\frac{\partial f_1}{\partial t} + \nabla \cdot (\mathbf{v} f_1) = S_{f_1} \tag{3.6}$$

From another continuum perspective, consider the schematic representation of a flow of particles where the particle space continuum may be viewed as an ensemble of elastic strings deforming everywhere, with the imbedded particles in the particle state space, as illustrated in Figure 3.2. Defining the combined phase velocity vector as $\mathbf{v} \equiv (v_1, v_2, v_3)$, an infinitesimal small control volume $\mathrm{dV} = \mathrm{d}x\,\mathrm{d}y\,\mathrm{d}z$ fixed in the Cartesian space (this is enlarged within the figure) is analyzed with respect to the (v_1, v_2, v_3) flow field. To account for the combined phase particle fluxes across each of the faces of the element, the net flow of f_1 can be expressed as:

$$\underbrace{\left[(v_1 f_1) + \frac{\partial (v_1 f_1)}{\partial x}\mathrm{d}x\right]\mathrm{d}y\mathrm{d}z - (v_1 f_1)\mathrm{d}y\mathrm{d}z}_{along\ x} +$$

$$\underbrace{\left[(v_2 f_1) + \frac{\partial (v_2 f_1)}{\partial y}\mathrm{d}y\right]\mathrm{d}x\mathrm{d}z - (v_2 f_1)\mathrm{d}x\mathrm{d}z}_{along\ y} +$$

$$\underbrace{\left[(v_3 f_1) + \frac{\partial (v_3 f_1)}{\partial z}\mathrm{d}z\right]\mathrm{d}x\mathrm{d}y - (v_3 f_1)\mathrm{d}x\mathrm{d}y}_{along\ z}$$

or

$$\left[\frac{\partial (v_1 f_1)}{\partial x} + \frac{\partial (v_2 f_1)}{\partial y} + \frac{\partial (v_3 f_1)}{\partial z}\right]\mathrm{d}x\mathrm{d}y\mathrm{d}z \tag{3.7}$$

Time rate of increase of f_1 in the elemental volume is subsequently given by

$$\frac{\partial f_1}{\partial t}\mathrm{d}x\mathrm{d}y\mathrm{d}z \tag{3.8}$$

The system balance requires that, for unsteady flow, *the time rate of increase of f_1 and the net flow of f_1 must be equal to the net generation of f_1 within the elemental volume*. Denoting the net generation of f_1 occurring inside the elemental volume as $S_{f_1}\mathrm{d}x\mathrm{d}y\mathrm{d}z$, the statement can be expressed in terms of Eqns (3.7) and (3.8) as

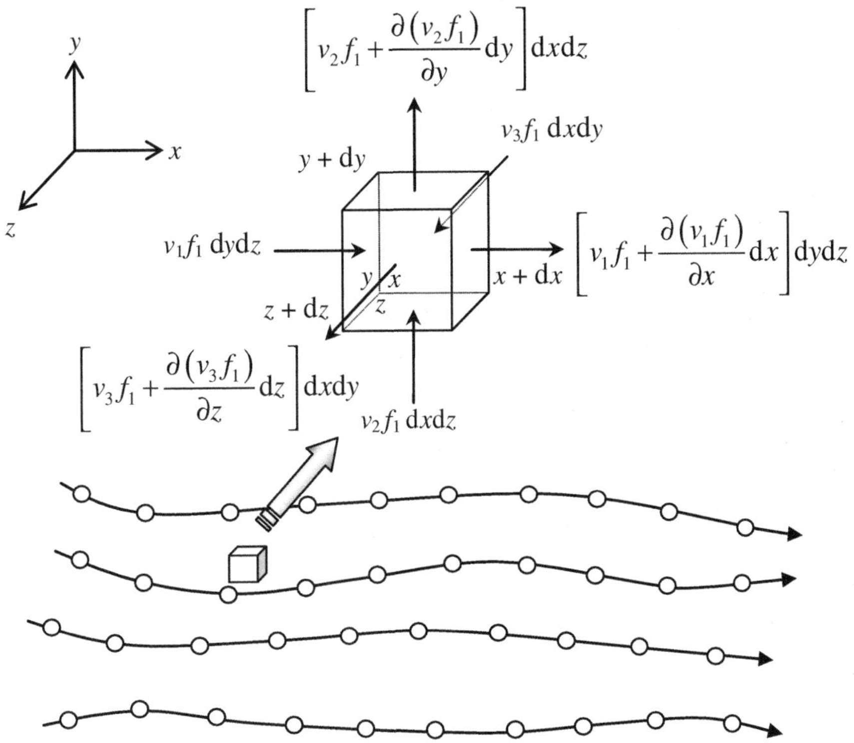

Figure 3.2
Consideration of an infinitesimal control volume of the particle space continuum.

$$\frac{\partial f_1}{\partial t} dxdydz + \left[\frac{\partial(v_1 f_1)}{\partial x} + \frac{\partial(v_2 f_1)}{\partial y} + \frac{\partial(v_3 f_1)}{\partial z}\right] dxdydz = S_{f_1} dxdydz$$

or

$$\frac{\partial f_1}{\partial t} + \left[\frac{\partial(v_1 f_1)}{\partial x} + \frac{\partial(v_2 f_1)}{\partial y} + \frac{\partial(v_3 f_1)}{\partial z}\right] = S_{f_1} \tag{3.9}$$

Equation (3.9) is the partial differential equation for the number density and has been derived based on the consideration of an infinitesimal, small element fixed in space. In vector form, Eqn (3.9) has exactly the same form as that derived in Eqn (3.6).

It is worthwhile noting that the population balance equation may be derived through the consideration of averaging the statistical Boltzmann-type equation. Here, the description of the multiphase flow is developed based on the Boltzmann theory of gases. The fundamental variable is the particle distribution function along with an appropriate choice of internal coordinates pertaining to a particular problem being solved. By defining $p(\mathbf{x}, \mathbf{r}, \mathbf{c}, t)\mathbf{dxdrdc}$ to be the particle number density distribution function, which is assumed to be continuous and specifies the probable number density of particles with internal coordinates about $\mathbf{x}$ in the range of $\mathbf{dx}$, about position $\mathbf{r}$ in the range of $\mathbf{dr}$, and about velocity $\mathbf{c}$ in the range $\mathbf{dc}$, at about time t, an equation can be written accordingly

$$p(\mathbf{x} + \mathbf{dx}, \mathbf{r} + \mathbf{dr}, \mathbf{c} + \mathbf{dc}, t + dt)\mathbf{dxdrdc} - p(\mathbf{x}, \mathbf{r}, \mathbf{c}, t)\mathbf{dxdrdc} = S_p \mathbf{dxdrdc}dt \qquad (3.10)$$

where $\mathbf{dr} = \mathbf{c}dt$, $\mathbf{dc} = \mathbf{F}dt$ in which $\mathbf{F}$ is the force per unit mass acting on a particle and S_p comprises the rates of change of $p(\mathbf{x},\mathbf{r},\mathbf{c},t)$ due to breakage/break up and aggregation/coalescence as well as other sources or sinks due to particle interactions (for example, rebounding of particles). Assuming that the change of particle velocity within the time interval t to $t + dt$ is negligible, Eqn (3.10) reduces to

$$\frac{\partial f}{\partial t} + \nabla_{\mathbf{r}} \cdot (\mathbf{v}_{\mathbf{r}} f) + \nabla_{\mathbf{x}} \cdot (\mathbf{v}_{\mathbf{x}} f) = S_f \qquad (3.11)$$

Equation (3.11) is now analogous to a Boltzmann-type equation in describing the temporal and spatial rate of change of the distribution function:

$$f(\mathbf{x}, \mathbf{r}, t) = \int_{-\infty}^{\infty} p\mathbf{dc} \qquad (3.12)$$

which denotes the probable number density of particles with internal coordinates about $\mathbf{x}$ in the range of $\mathbf{dx}$, about position $\mathbf{r}$ in the range of $\mathbf{dr}$ and at about time t. Note that this function is identical to the basic definition of the density function $f_1(\mathbf{x},\mathbf{r},t)$. The velocity independent source/sink term on the left hand side of Eqn (3.11) is

$$S_f = \int_{-\infty}^{\infty} S_p \mathbf{dc} \qquad (3.13)$$

In Eqn (3.11), by integrating the force $\mathbf{F}$ over the whole velocity space, no net contribution due to this force appears explicitly in the population balance equation since the distribution vanishes as the velocity approaches infinity ($\pm\infty$). Also, the source/sink term consists of only the net generation rate of particles due to breakage/break up and aggregation/coalescence processes while other sources or sinks, due to particle interactions, vanish as the number of particles is conserved, in particular for the case during the rebounding processes in the system. Defining the sum of $\mathbf{v}_{\mathbf{x}}$ and $\mathbf{v}_{\mathbf{x}}$ to be the combined phase velocity vector $\mathbf{v}$, Eqn (3.11) has exactly the same form as that derived in Eqn (3.6).

3.3.2 Various Integrated Forms of Transport Equations

The local population balance of the number density $f_1(\mathbf{x},\mathbf{r},t)$ can be written as

$$\frac{\partial f_1(\mathbf{x},\mathbf{r},t)}{\partial t} + \nabla_{\mathbf{r}} \cdot (\mathbf{v}_{\mathbf{r}}(\mathbf{x},\mathbf{r},\mathbf{Y},t) f_1(\mathbf{x},\mathbf{r},t)) + \nabla_{\mathbf{x}} \cdot (\mathbf{v}_{\mathbf{x}}(\mathbf{x},\mathbf{r},\mathbf{Y},t) f_1(\mathbf{x},\mathbf{r},t)) = S_{f_1}(\mathbf{x},\mathbf{r},\mathbf{Y},t) \tag{3.14}$$

In Eqn (3.14), the physical significance of the first term on the left hand side of the equation is the local change of the particle number density with time. The second term represents the change of the number density due to advection in the external coordinates, and the third term denotes the change of the number density due to advection in the internal coordinates indicating several particle growth phenomena. However, the physical significance of the term on the right hand side of the equation is the net particle source/sink rate, representing the change in the density function which can be attributed to particle break up/breakage and coalescence/aggregation.

For a multiphase system of nonreactive, isothermal, particle mixtures, it is customary to assume that all relevant internal variables can be calculated from the consideration of the particle volume or diameter. If the internal coordinate is taken to be the particle volume V_p or diameter d_p in describing incompressible fluid particle and solid particle dispersions, the population balance Eqn (3.14) is given by

$$\frac{\partial f_1(V_\text{p},\mathbf{r},t)}{\partial t} + \nabla_{\mathbf{r}} \cdot \left(\mathbf{v}_{\mathbf{r}}(V_\text{p},\mathbf{r},\mathbf{Y},t) f_1(V_\text{p},\mathbf{r},t)\right) + \frac{\partial\left(d\dot{V}_\text{p} f_1(V_\text{p},\mathbf{r},t)\right)}{\partial V_\text{p}} = S_{f_1}(V_\text{p},\mathbf{r},\mathbf{Y},t) \tag{3.15}$$

or

$$\frac{\partial f_1(d_\text{p},\mathbf{r},t)}{\partial t} + \nabla_{\mathbf{r}} \cdot \left(\mathbf{v}_{\mathbf{r}}(d_\text{p},\mathbf{r},\mathbf{Y},t) f_1(d_\text{p},\mathbf{r},t)\right) + \frac{\partial\left(\dot{d}_\text{p} f_1(d_\text{p},\mathbf{r},t)\right)}{\partial d_\text{p}} = S_{f_1}(d_\text{p},\mathbf{r},\mathbf{Y},t) \tag{3.16}$$

where $\dot{V}_\text{p}$ denotes the time rate of change of particle volume while $\dot{d}_\text{p}$ represents the time rate of change of particle diameter. For multiphase flow where the compressibility effect is important, the use of particle mass as an internal coordinate may prove to be more advantageous. This requirement is particularly important in the gas phase because this quantity is conserved under pressure changes. The population balance equation Eqn (3.14) can now be written in the following way

$$\frac{\partial f_1(m_\text{p},\mathbf{r},t)}{\partial t} + \nabla_{\mathbf{r}} \cdot \left(\mathbf{v}_{\mathbf{r}}(m_\text{p},\mathbf{r},\mathbf{Y},t) f_1(m_\text{p},\mathbf{r},t)\right) + \frac{\partial\left(\dot{m}_\text{p} f_1(m_\text{p},\mathbf{r},t)\right)}{\partial m_\text{p}} = S_{f_1}(m_\text{p},\mathbf{r},\mathbf{Y},t) \tag{3.17}$$

where $\dot{m}_p$ represents the time rate of change of particle mass m_p. It should be noted in the case of describing incompressible fluids that the use of particle mass as an internal coordinate is essentially equivalent to utilizing those population balance formulations based on the use of the particle diameter as an internal coordinate. To close the population balance problem, closure is required for the particle growth as well as the break up/breakage and coalescence/aggregation kernels through appropriate models in the source/sink term S_{f_1}. These kernels are required to be consistent with the internal coordinate used.

Nevertheless, Eqns (3.15)–(3.17) are generally still too detailed to be employed and solved via direct numerical simulations. Practical multiphase flows that are encountered in natural and technological systems generally contain millions of particles that are simultaneously varying along the internal coordinates. Hence, the use of direct numerical simulations in resolving such flows is still far beyond the capacity of existing computer resources. One simpler alternative that can be derived is through use of some presumed particle size distribution. This distribution is retained during the process under investigation. Under these assumptions, the complete particle size distribution can be represented by a limited number of parameters, and the population balance equation can thus be reformulated in terms of these parameters.

One practical approach to solve the population balance of particles being distributed in the flow is applying moment transformation of the population balance equation. Adopting the particle volume V_p as the internal coordinate, the mth moments of the number density function in terms of volume are defined as

$$m_k(\mathbf{r}, t) = \int_0^\infty f_1(V_p, \mathbf{r}, t) V_p^k \mathrm{d}V_p \tag{3.18}$$

Various mth moments in Eqn (3.18) contribute to physically important moments of the number density function. In retrospect, particle number density, particle mass density, interfacial area concentration and local volume fraction essentially correspond to zero-order moments through third-order moments. If the volume V_p is treated as the independent internal coordinate, the particle number density N, average particle volume $\overline{v}$, interfacial area concentration a_i and local volume fraction α^p are as follows:

Zeroth moment

$$N(\mathbf{r}, t) = \int_0^\infty f_1(V_p, \mathbf{r}, t) \mathrm{d}V_p \tag{3.19}$$

First moment

$$\overline{v}(\mathbf{r}, t) = \left[\int_0^\infty f_1(V_p, \mathbf{r}, t) V_p \mathrm{d}V_p \right] \Big/ N(\mathbf{r}, t) \tag{3.20}$$

Second moment

$$a_{\mathrm{i}}(\mathbf{r},t)=\int_{0}^{\infty}\pi D_{e}^{2}f_{1}\left(V_{\mathrm{p}},\mathbf{r},t\right)\mathrm{d}V_{\mathrm{p}} \tag{3.21}$$

Third moment

$$\alpha^{\mathrm{p}}(\mathbf{r},t)=\int_{0}^{\infty}\frac{\pi}{6}D_{e}^{3}f_{1}\left(V_{\mathrm{p}},\mathbf{r},t\right)\mathrm{d}V_{\mathrm{p}} \tag{3.22}$$

In Eqns (3.21) and (3.22), the particle surface area and volume can be expressed in terms of equivalent diameters.

Considering the special case, where there is no mass change of the particle (absence of particle growth), the transport equation for the particle number density reduces to

$$\frac{\partial N(\mathbf{r},t)}{\partial t}+\nabla_{\mathbf{r}}\cdot\left(\mathbf{v}_{\mathrm{pn}}(\mathbf{r},\mathbf{Y},t)N(\mathbf{r},t)\right)=S_{\mathrm{N}}(\mathbf{r},\mathbf{Y},t) \tag{3.23}$$

where $\mathbf{v}_{\mathrm{pn}}(\mathbf{r},\mathbf{Y},t)$ denotes the average local particle velocity weighted by the number density defined by

$$\mathbf{v}_{\mathrm{pn}}(\mathbf{r},\mathbf{Y},t)\equiv\frac{\int_{0}^{\infty}\mathbf{v}_{\mathbf{r}}\left(V_{\mathrm{p}},\mathbf{r},\mathbf{Y},t\right)f_{1}\left(V_{\mathrm{p}},\mathbf{r},t\right)\mathrm{d}V_{\mathrm{p}}}{\int_{0}^{\infty}f_{1}\left(V_{\mathrm{p}},\mathbf{r},t\right)\mathrm{d}V_{\mathrm{p}}} \tag{3.24}$$

and

$$S_{\mathrm{N}}(\mathbf{r},\mathbf{Y},t)\equiv\int_{0}^{\infty}S_{f_{1}}\left(V_{\mathrm{p}},\mathbf{r},\mathbf{Y},t\right)\mathrm{d}V_{\mathrm{p}} \tag{3.25}$$

In order to derive the transport equation for the interfacial area concentration, Eqn (3.23) can be modified by the following geometric relation:

$$N(\mathbf{r},t)=\frac{\alpha^{P}(\mathbf{r},t)}{\overline{v}(\mathbf{r},t)} \tag{3.26}$$

Equation (3.26) can thus be rewritten in the form:

$$N(\mathbf{r},t)=\Psi\frac{a_{\mathrm{i}}^{3}(\mathbf{r},t)}{\alpha^{2}(\mathbf{r},t)} \tag{3.27}$$

where Ψ represents the factor characterizing the shape of the particles. For spherical particles, $\Psi=1/36\pi$. Substituting Eqn (3.27) into Eqn (3.23) yields the transport equation for the interfacial area concentration:

$$\frac{\partial a_{\mathrm{i}}(\mathbf{r},t)}{\partial t}+\nabla_{\mathbf{r}}\cdot\left(\mathbf{v}_{\mathrm{pi}}(\mathbf{r},\mathbf{Y},t)a_{\mathrm{i}}(\mathbf{r},t)\right)=\frac{1}{3\Psi}\left(\frac{\alpha(\mathbf{r},t)}{a_{\mathrm{i}}(\mathbf{r},t)}\right)^{2}S_{\mathrm{N}}(\mathbf{r},\mathbf{Y},t)$$
$$+\left(\frac{2}{3}\frac{a_{\mathrm{i}}(\mathbf{r},t)}{\alpha(\mathbf{r},t)}\right)\left[\frac{\partial\alpha(\mathbf{r},t)}{\partial t}+\nabla_{\mathbf{r}}\cdot\left(\mathbf{v}_{\mathrm{pi}}(\mathbf{r},\mathbf{Y},t)\alpha(\mathbf{r},t)\right)\right] \tag{3.28}$$

where $\mathbf{v}_{\mathrm{pi}}(\mathbf{r},\mathbf{Y},t)$ denotes the average local particle velocity weighted by the interfacial area concentration defined by

$$\mathbf{v}_{\mathrm{pi}}(\mathbf{r},\mathbf{Y},t)\equiv\frac{\int_{0}^{\infty}\mathbf{v}_{\mathbf{r}}(V_{\mathrm{p}},\mathbf{r},\mathbf{Y},t)\pi D_{e}^{2}f_{1}(V_{\mathrm{p}},\mathbf{r},t)\mathrm{d}V_{\mathrm{p}}}{\int_{0}^{\infty}\pi D_{e}^{2}f_{1}(V_{\mathrm{p}},\mathbf{r},t)\mathrm{d}V_{\mathrm{p}}} \tag{3.29}$$

Analogously, the transport equation for the average particle volume can also be expressed through the use of Eqn (3.26) as

$$\frac{\partial\overline{v}(\mathbf{r},t)}{\partial t}+\nabla_{\mathbf{r}}\cdot\left(\mathbf{v}_{pv}(\mathbf{r},\mathbf{Y},t)a_{\mathrm{i}}(\mathbf{r},t)\right)=-\frac{\overline{v}^{2}(\mathbf{r},t)}{\alpha(\mathbf{r},t)}S_{\mathrm{N}}(\mathbf{r},\mathbf{Y},t)$$
$$+\frac{\overline{v}(\mathbf{r},t)}{\alpha(\mathbf{r},t)}\left[\frac{\partial\alpha(\mathbf{r},t)}{\partial t}+\nabla_{\mathbf{r}}\cdot\left(\mathbf{v}_{pv}(\mathbf{r},\mathbf{Y},t)\alpha(\mathbf{r},t)\right)\right] \tag{3.30}$$

where $\mathbf{v}_{pv}$ $(\mathbf{r},\mathbf{Y},t)$ denotes the average local particle velocity weighted by the local volume fraction defined by

$$\mathbf{v}_{pv}(\mathbf{r},\mathbf{Y},t)\equiv\frac{\int_{0}^{\infty}\mathbf{v}_{\mathbf{r}}(V_{\mathrm{p}},\mathbf{r},\mathbf{Y},t)\ \frac{\pi}{6}D_{e}^{3}f_{1}(V_{\mathrm{p}},\mathbf{r},t)\mathrm{d}V_{\mathrm{p}}}{\int_{0}^{\infty}\frac{\pi}{6}D_{e}^{3}f_{1}(V_{\mathrm{p}},\mathbf{r},t)\mathrm{d}V_{\mathrm{p}}} \tag{3.31}$$

3.3.3 Breakage/Break up Processes

As defined by Ramkrishna (2000), the term breakage or break up in this instance can be applied not only to systems in which particles undergo random breakage but also new particles that arise from existing particles by other mechanisms. The net generation rate $S_{\mathrm{B}}(\mathbf{x},\mathbf{r},\mathbf{Y},t)$ due to breakage/break up processes may be assumed to decompose into a source term $S_B^+(\mathbf{x},\mathbf{r},\mathbf{Y},t)$ and sink term $S_B^-(\mathbf{x},\mathbf{r},\mathbf{Y},t)$. Formulation of these source and sink terms are described as follows:

By assuming that the physical breakages/breakups of the particles occur independently of each other, the average rate of particles *lost* by breakage or break up of state (**x**,**r**) per unit time or the sink term can be written as

$$S_B^-(\mathbf{x},\mathbf{r},\mathbf{Y},t) = b(\mathbf{x},\mathbf{r},\mathbf{Y},t)f_1(\mathbf{x},\mathbf{r},t) \quad (3.32)$$

The term $b(\mathbf{x},\mathbf{r},\mathbf{Y},t)$ on the right hand side of Eqn (3.32) represents the specific breakage rate of particles, more commonly known as the breakage frequency, which has dimension of reciprocal time. This term is also sometimes referred to as a transition probability function for breakage. Modeling of the breakage frequency must be considered in the premise of examining the events on the time scale in which they occur before either leading to particle breakage or leaving the particle intact. Because of the breakage processes being random in nature, probabilistic theory is adopted for the modeling of the breakage frequency.

To characterize the source term, the average production rate of particles originating from the breakage/break up of particles of state (**x**,**r**) per unit time applying to all particle states—with both internal and external coordinates—is given by

$$S_B^+(\mathbf{x},\mathbf{r},\mathbf{Y},t) = \int_{V_\mathbf{r}}\int_{V_\mathbf{x}} \nu(\mathbf{x}',\mathbf{r}',\mathbf{Y},t)b(\mathbf{x}',\mathbf{r}',\mathbf{Y},t)P(\mathbf{x},\mathbf{r}|\mathbf{x}',\mathbf{r}',\mathbf{Y},t)f_1(\mathbf{x}',\mathbf{r}',t)dV_\mathbf{r}dV_\mathbf{x} \quad (3.33)$$

In Eqn (3.33), the breakage/break up functions $\nu(\mathbf{x}',\mathbf{r}',\mathbf{Y},t)$ and $P(\mathbf{x},\mathbf{r}|\mathbf{x}',\mathbf{r}',\mathbf{Y},t)$ are designated as the average number of particles and probability density function for particles from the breakage/break up of a single particle of state $(\mathbf{x}',\mathbf{r}')$ in an environment of **Y** at time t. The probability density function $P(\mathbf{x},\mathbf{r}|\mathbf{x}',\mathbf{r}',\mathbf{Y},t)$ in Eqn (3.33) is a continuously distributed fraction over particle state space and is commonly associated with the daughter particle size distribution function denoting the size distribution of daughter particles produced upon the breakage/break up of a parent particle. This quantity needs to be determined through either detailed modeling of the breakage/break up process or from experimental observation. This function inherits certain properties which constrain the breakage/break up process. In other words, it must satisfy the normalization condition in the internal coordinates:

$$\int_{V_\mathbf{x}} P(\mathbf{x},\mathbf{r}|\mathbf{x}',\mathbf{r}',\mathbf{Y},t)dV_\mathbf{x} = 1 \quad (3.34)$$

as well as the conservation of mass:

$$\int_{V_\mathbf{x}} P(\mathbf{x},\mathbf{r}|\mathbf{x}',\mathbf{r}',\mathbf{Y},t)dV_\mathbf{x} = 1 \quad (3.35)$$

With regard to the integrand on the right hand side of Eqn (3.33), new particles resulting from the breakage/break up processes $\nu(\mathbf{x}', \mathbf{r}', \mathbf{Y}, t)b(\mathbf{x}', \mathbf{r}', \mathbf{Y}, t)f_1(\mathbf{x}', \mathbf{r}', t)\mathrm{d}V_{\mathbf{r}}\mathrm{d}V_{\mathbf{x}}$ are generated of which a fraction $P(\mathbf{x}, \mathbf{r}|\mathbf{x}', \mathbf{r}', \mathbf{Y}, t)\mathrm{d}V_{\mathbf{r}}\mathrm{d}V_{\mathbf{x}}$ represents particles of state (**x**,**r**) considering that the number of particles of state $(\mathbf{x}', \mathbf{r}')$ that undergo breakage/break up per unit of time is $b(\mathbf{x}', \mathbf{r}', \mathbf{Y}, t)f_1(\mathbf{x}', \mathbf{r}', t)\mathrm{d}V_{\mathbf{r}}\mathrm{d}V_{\mathbf{x}}$.

As stipulated in Jakonsen (2008), the integral of Eqn (3.33) in physical space is based on the assumption that a parent particle is breaking up at one location in physical space and producing daughter particles that can occur elsewhere. Obviously, the location of the birth of these daughter particles must coincide with that of the parent particle. Given that breakage/break up is a local phenomenon in physical space, the source term has to be rewritten for problems in the external coordinates. When introducing a Dirac delta function $(\mathbf{x}', \mathbf{r}')$ within the physical space integral argument, an appropriate modification can be obtained leading to the breakage/break up source term simplifying to

$$S_{\mathrm{B}}^{+}(\mathbf{x}, \mathbf{r}, \mathbf{Y}, t) = \int_{V_{\mathbf{x}}} \nu(\mathbf{x}', \mathbf{r}, \mathbf{Y}, t)b(\mathbf{x}', \mathbf{r}, \mathbf{Y}, t)P(\mathbf{x}, \mathbf{r}|\mathbf{x}', \mathbf{r}, \mathbf{Y}, t)f_1(\mathbf{x}', \mathbf{r}, t)\mathrm{d}V_{\mathbf{x}} \quad (3.36)$$

where $P(\mathbf{x}, \mathbf{r}|\mathbf{x}', \mathbf{r}, \mathbf{Y}, t)$ is now the daughter size distribution function.

The net generation of particles of state (**x**,**r**) per unit time due to breakage/break up is henceforth given by

$$S_{\mathrm{B}}(\mathbf{x}, \mathbf{r}, \mathbf{Y}, t) = S_{\mathrm{B}}^{+}(\mathbf{x}, \mathbf{r}, \mathbf{Y}, t) - S_{\mathrm{B}}^{-}(\mathbf{x}, \mathbf{r}, \mathbf{Y}, t) \quad (3.37)$$

or

$$S_{\mathrm{B}}(\mathbf{x}, \mathbf{r}, \mathbf{Y}, t) = \int_{V_{\mathbf{x}}} \nu(\mathbf{x}', \mathbf{r}, \mathbf{Y}, t)b(\mathbf{x}', \mathbf{r}, \mathbf{Y}, t)P(\mathbf{x}, \mathbf{r}|\mathbf{x}', \mathbf{r}, \mathbf{Y}, t)f_1(\mathbf{x}', \mathbf{r}, t)\mathrm{d}V_{\mathbf{x}}$$
$$- b(\mathbf{x}, \mathbf{r}, \mathbf{Y}, t)f_1(\mathbf{x}, \mathbf{r}, t) \quad (3.38)$$

3.3.4 Aggregation/Coalescence Processes

As described by Ramkrishna (2000), the process of aggregation or coalescence must occur between two particles. Although it is conceivable that adjacent particles could simultaneously aggregate in very crowded systems, attention is primarily directed toward significant binary aggregation/coalescence. The current framework also includes coagulation which represents features of *floc* of particles held together by surface forces. Similarly to the breakage/break up processes, the net generation rate $S_{\mathrm{C}}(\mathbf{x},\mathbf{r},\mathbf{Y},t)$ due to aggregation/coalescence processes may also be taken to decompose into a source term $S_{\mathrm{C},f_1}^{+}(\mathbf{x}, \mathbf{r}, \mathbf{Y}, t)$ and sink term $S_{\mathrm{C},f_1}^{-}(\mathbf{x}, \mathbf{r}, \mathbf{Y}, t)$. Formulation of these source and sink terms is described below.

The aggregation/coalescence frequency is first defined: $a(\tilde{\mathbf{x}},\tilde{\mathbf{r}};\mathbf{x},\mathbf{r},\mathbf{Y},t)$, which represents the fraction of pairs of particles of states $(\tilde{\mathbf{x}},\tilde{\mathbf{r}})$ and $(\mathbf{x}',\mathbf{r}')$ that aggregate/coalesce per unit of time. This frequency is defined for an ordered pair of particles. From a physical viewpoint, the ordering of pairs of particles should not change the value of the frequency. In other words, $a(\tilde{\mathbf{x}},\tilde{\mathbf{r}};\mathbf{x},\mathbf{r},\mathbf{Y},t)$ should satisfy the symmetry property $a(\tilde{\mathbf{x}},\tilde{\mathbf{r}};\mathbf{x},\mathbf{r},\mathbf{Y},t) = a(\mathbf{x},\mathbf{r};\tilde{\mathbf{x}},\tilde{\mathbf{r}},\mathbf{Y},t)$; hence only the order of a given pair of particles is considered. It is also essential to identify the state of the particle as a result of aggregation/coalescence. Assuming that the particle state of one of the aggregating/coalescing pairs given those of the other aggregating/coalescing particles and the new particle, can be solved, with the knowledge of state ($\mathbf{x}$,$\mathbf{r}$) of the new particle and state $(\mathbf{x}',\mathbf{r}')$ of one of the two aggregating/coalescing particles, the state of the other aggregating/coalescing particle can be known and can be designated by $[\tilde{\mathbf{x}}(\mathbf{x},\mathbf{r}|\mathbf{x}',\mathbf{r}'),\tilde{\mathbf{r}}(\mathbf{x},\mathbf{r}|\mathbf{x}',\mathbf{r}')]$.

Defining the pair density function $f_2(\mathbf{x},\mathbf{r},t)$ to represent the average number of distinct pairs of particles at time t within the state spaces located about ($\mathbf{x}$,$\mathbf{r}$) and $(\mathbf{x}',\mathbf{r}')$ respectively, the source term due to aggregation/coalescence may be written in accordance with the form stipulated by Ramkrishna (2000) as

$$S_C^+(\mathbf{x},\mathbf{r},\mathbf{Y},t) = \int_{V_\mathbf{r}}\int_{V_\mathbf{x}} \frac{1}{\delta} a(\tilde{\mathbf{x}},\tilde{\mathbf{r}};\mathbf{x}',\mathbf{r}',\mathbf{Y},t)\, f_2(\tilde{\mathbf{x}},\tilde{\mathbf{r}};\mathbf{x}',\mathbf{r}',t)\, \frac{\partial(\tilde{\mathbf{x}},\tilde{\mathbf{r}})}{\partial(\mathbf{x},\mathbf{r})} \mathrm{d}V_\mathbf{r}\mathrm{d}V_\mathbf{x} \tag{3.39}$$

In Eqn (3.39), δ denotes the number of times identical pairs have been considered in the interval of integration where $1/\delta$ corrects for the redundancy. As normally adopted in population balance derivation, a coarse approximation of the following pair density function is assumed: $f_2(\tilde{\mathbf{x}},\tilde{\mathbf{r}};\mathbf{x}',\mathbf{r}',t) \approx \tilde{f}_1(\tilde{\mathbf{x}},\tilde{\mathbf{r}},t) f_1'(\mathbf{x}',\mathbf{r}',t)$. This assumption implies the absence of any statistical correlation between particles of state spaces ($\mathbf{x}$,$\mathbf{r}$) and $(\mathbf{x}',\mathbf{r}')$ at any instant of time t. The term $\partial(\tilde{\mathbf{x}},\tilde{\mathbf{r}})/\partial(\mathbf{x},\mathbf{r})$ represents the Jacobian determinant, which is given by

$$\begin{vmatrix} \frac{\partial \tilde{x}_1}{\partial x_1} & \cdots & \frac{\partial \tilde{x}_1}{\partial x_n} & \frac{\partial \tilde{x}_1}{\partial r_1} & \frac{\partial \tilde{x}_1}{\partial r_2} & \frac{\partial \tilde{x}_1}{\partial r_3} \\ \vdots & \vdots & \vdots & \vdots & \vdots & \vdots \\ \frac{\partial \tilde{x}_n}{\partial x_1} & \cdots & \frac{\partial \tilde{x}_n}{\partial x_n} & \frac{\partial \tilde{x}_n}{\partial r_1} & \frac{\partial \tilde{x}_n}{\partial r_2} & \frac{\partial \tilde{x}_n}{\partial r_3} \\ \frac{\partial \tilde{r}_1}{\partial x_1} & \cdots & \frac{\partial \tilde{r}_1}{\partial x_n} & \frac{\partial \tilde{r}_1}{\partial r_1} & \frac{\partial \tilde{r}_1}{\partial r_2} & \frac{\partial \tilde{r}_1}{\partial r_3} \\ \frac{\partial \tilde{r}_2}{\partial x_1} & \cdots & \frac{\partial \tilde{r}_2}{\partial x_n} & \frac{\partial \tilde{r}_2}{\partial r_1} & \frac{\partial \tilde{r}_2}{\partial r_2} & \frac{\partial \tilde{r}_2}{\partial r_3} \\ \frac{\partial \tilde{r}_3}{\partial x_1} & \cdots & \frac{\partial \tilde{r}_3}{\partial x_n} & \frac{\partial \tilde{r}_3}{\partial r_1} & \frac{\partial \tilde{r}_3}{\partial r_2} & \frac{\partial \tilde{r}_3}{\partial r_3} \end{vmatrix} \tag{3.40}$$

Also stipulated in Jakonsen (2008), the integral of Eqn (3.39) in physical space is not suitable for the source term based on the assumption that two particles aggregating/coalescing at a particular location can produce a new larger particle elsewhere. Here again, the birth of the particle should be at the same location where the parent particles aggregate/coalesce. Introducing a Dirac delta function $(\mathbf{x}', \mathbf{r}')$ within the physical space integral argument in the given source term definition, the aggregation/coalescence source term reduces to

$$S_C^+(\mathbf{x}, \mathbf{r}, \mathbf{Y}, t) = \int_{V_x} \frac{1}{\delta} a(\tilde{\mathbf{x}}, \mathbf{r}; \mathbf{x}', \mathbf{r}, \mathbf{Y}, t)\, f_2(\tilde{\mathbf{x}}, \mathbf{r}; \mathbf{x}', \mathbf{r}, t)\, \frac{\partial(\tilde{\mathbf{x}}, \mathbf{r})}{\partial(\mathbf{x}, \mathbf{r})} dV_{\mathbf{x}} \tag{3.41}$$

The sink term due to aggregation/coalescence is more readily found to be

$$S_C^-(\mathbf{x}, \mathbf{r}, \mathbf{Y}, t) = \int_{V_r} \int_{V_x} a(\mathbf{x}', \mathbf{r}'; \mathbf{x}, \mathbf{r}, \mathbf{Y}, t) f_2(\mathbf{x}', \mathbf{r}'; \mathbf{x}, \mathbf{r}, t) dV_{\mathbf{r}} dV_{\mathbf{x}} \tag{3.42}$$

Analogously, the integral of Eqn (3.42) in physical space is not appropriate as it is based on two particles aggregating/coalescing at a particular location producing a new larger particle elsewhere. Following the same procedure as for the source term, the sink term due to aggregation/coalescence can be redefined as

$$S_C^-(\mathbf{x}, \mathbf{r}, \mathbf{Y}, t) = \int_{V_x} a(\mathbf{x}', \mathbf{r}'; \mathbf{x}, \mathbf{r}, \mathbf{Y}, t) f_2(\mathbf{x}', \mathbf{r}'; \mathbf{x}, \mathbf{r}, t) dV_{\mathbf{x}} \tag{3.43}$$

The net generation of particles of state $(\mathbf{x},\mathbf{r})$ per unit time due to aggregation/coalescence is henceforth given by

$$S_C(\mathbf{x}, \mathbf{r}, \mathbf{Y}, t) = S_C^+(\mathbf{x}, \mathbf{r}, \mathbf{Y}, t) - S_C^-(\mathbf{x}, \mathbf{r}, \mathbf{Y}, t) \tag{3.44}$$

or

$$\begin{aligned} S_C(\mathbf{x}, \mathbf{r}, \mathbf{Y}, t) = & \int_{V_x} \frac{1}{\delta} a(\tilde{\mathbf{x}}, \mathbf{r}; \mathbf{x}', \mathbf{r}, \mathbf{Y}, t)\, f_2(\tilde{\mathbf{x}}, \mathbf{r}; \mathbf{x}', \mathbf{r}, t)\, \frac{\partial(\tilde{\mathbf{x}}, \mathbf{r})}{\partial(\mathbf{x}, \mathbf{r})} dV_{\mathbf{x}} \\ & - \int_{V_x} a(\mathbf{x}', \mathbf{r}'; \mathbf{x}, \mathbf{r}, \mathbf{Y}, t) f_2(\mathbf{x}', \mathbf{r}'; \mathbf{x}, \mathbf{r}, t) dV_{\mathbf{x}} \end{aligned} \tag{3.45}$$

3.3.5 Net Generation of Particles

The relevant terms due to breakage/break up and aggregation/coalescence in the source/sink term for the population balance equation have been generically presented in the previous sections for particles being present within the multiphase system. In this system, the distribution of these particles is a result of breakage/break up and aggregation/coalescence processes occurring *simultaneously*. With regard to this, the net rate of generation of particles $S_{f_1}(\mathbf{x},\mathbf{r},\mathbf{Y},t)$ in Eqn (3.14) can be obtained by algebraically summing the net generation of particles due to breakage/break up $S_B(\mathbf{x},\mathbf{r},\mathbf{Y},t)$ and the net generation of particles due to aggregation/coalescence $S_C(\mathbf{x},\mathbf{r},\mathbf{Y},t)$. In other words,

$$S_{f_1}(\mathbf{x},\mathbf{r},\mathbf{Y},t) = S_B(\mathbf{x},\mathbf{r},\mathbf{Y},t) + S_C(\mathbf{x},\mathbf{r},\mathbf{Y},t) \tag{3.46}$$

or

$$\begin{aligned} S_{f_1}(\mathbf{x},\mathbf{r},\mathbf{Y},t) = & \int_{V_\mathbf{x}} \nu(\mathbf{x}',\mathbf{r},\mathbf{Y},t) b(\mathbf{x}',\mathbf{r},\mathbf{Y},t) P(\mathbf{x},\mathbf{r}|\mathbf{x}',\mathbf{r},\mathbf{Y},t) f_1(\mathbf{x}',\mathbf{r},t) \mathrm{d}V_\mathbf{x} \\ & - b(\mathbf{x},\mathbf{r},\mathbf{Y},t) f_1(\mathbf{x},\mathbf{r},t) + \int_{V_\mathbf{x}} \frac{1}{\delta} a(\tilde{\mathbf{x}},\mathbf{r};\mathbf{x}',\mathbf{r},\mathbf{Y},t)\, f_2(\tilde{\mathbf{x}},\mathbf{r};\mathbf{x}',\mathbf{r},t)\, \frac{\partial(\tilde{\mathbf{x}},\mathbf{r})}{\partial(\mathbf{x},\mathbf{r})} \mathrm{d}V_\mathbf{x} \\ & - \int_{V_\mathbf{x}} a(\mathbf{x}',\mathbf{r}';\mathbf{x},\mathbf{r},\mathbf{Y},t) f_2(\mathbf{x}',\mathbf{r}';\mathbf{x},\mathbf{r},t) \mathrm{d}V_\mathbf{x} \end{aligned} \tag{3.47}$$

The physical significance of the processes due to breakage/break up and aggregation/ coalescence in Eqn (3.47) can be described as follows. The breakage/break up of the particles has two outcomes. First, the number of particles in a particle fraction increases because of larger particles breaking up. The integral yields the rate of breakage/break up of all larger particles. Second, the number of particles in a particle fraction decreases because of these particles breaking up. Each particle that broke apart will no longer belong to its previous cluster of particles. Analogously, the aggregation/coalescence of the particles has two outcomes. First, new particles are formed due to merging of smaller particles. Second, the number of particles in a particle fraction subsequently decreases since some of the particles merge with other particles.

3.4 Practical Considerations of Population Balance Framework

Owing to the complexities associated with the particle growth term as well as the birth and death processes in the source/sink term, appropriate constitutive relationships are inevitably required to achieve closure of the population balance equation. Strictly speaking, this parameterization process represents the *weakest* component, which should demand the greatest attention. In most cases, the detailed functionality of the closure relationships and even the physical insights of the birth and death processes are generally unknown or unresolved. Also, formulation of the growth terms that may be dependent on the choice of internal coordinates or particle properties leads to the additional complexity of suitably characterizing the particle phase in the population balance framework. Modeling of these terms for particular multiphase flows remains extremely challenging because of the complexity of the physics involved. In some flows, the modeling is less familiar and remains in its early stage of development.

Focusing on the particle growth term, if a relation for the contact area as a function of the internal coordinates can be properly identified, the modeling of this term due to interfacial mass transfer can be readily modeled in accordance with a number of existing theories. For example, the film theory (albeit being semiempirical) and the ideal gas law may be applied for the gas phase. For the source/sink term, phenomenological models are generally proposed to describe the particle processes due to breakage/break up and aggregation/coalescence. Based on these models, suitable breakage/break up and aggregation/coalescence rate functions are developed and employed to solve the general population balance equation. Parameters of the models are evaluated by comparison with experimental data on particle size distributions obtained over a range of flow conditions. As these models have been derived from fundamental concepts, such models can be generalized to mechanistically describe the basic hydrodynamics and physical properties of the multiphase system and circumvent the reliance of empirical analysis. Some significant development and formulation of the rate functions based on the continuum framework of particle breakage/break up and aggregation/coalescence, considering breakage of drops, coalescence and break up of fluid particles, aggregation of solid particles and other similar processes, have been dealt with by Tsouris and Tavlarides (1994), Kocamustafaogullari and Ishii (1995), Ramkrishna (2000) and Friedlander (2000).

In practice, the transport of particle number density function is solved within the framework of computational multiphase fluid dynamics. By taking the volume V_p as the independent internal coordinate, the derivation of the net generation of particles becomes

$$\begin{aligned} S_{f_1}(V_p, \mathbf{r}, \mathbf{Y}, t) = & \int_{V_p}^{\infty} \nu\left(V_p', \mathbf{r}, \mathbf{Y}\right) b\left(V_p', \mathbf{r}, \mathbf{Y}\right) P\left(V_p, \mathbf{r} \middle| V_p', \mathbf{r}, \mathbf{Y}, t\right) f_1\left(V_p', \mathbf{r}, t\right) dV_p' \\ & - b(V_p, \mathbf{r}, \mathbf{Y}) f_1(V_p, \mathbf{r}, t) \\ & + \frac{1}{2} \int_0^{V_p} a\left(\tilde{V}_p, \tilde{\mathbf{r}}; V_p', \mathbf{r}', \mathbf{Y}\right) f_1(\tilde{V}_p, \tilde{\mathbf{r}}, t) f_1(V', \mathbf{r}', t) dV_p' \\ & - \int_0^{\infty} a\left(V_p', \mathbf{r}'; V_p, \mathbf{r}, \mathbf{Y}\right) f_1\left(V_p', \mathbf{r}', t\right) f_1(V_p, \mathbf{r}, t) dV_p' \end{aligned} \tag{3.48}$$

Following the formulation proposed by Prince and Blanch (1990), Luo (1993) and Luo and Svendsen (1996), which can be generalized to any forms of multiphase flows, the continuous size range of particles represented by the number density function can be discretized into a series number of discrete size classes. For each particle class, the particle number density is solved to accommodate the population changes caused by particle breakage/break up and aggregation/coalescence. Substituting Eqn (3.48) into Eqn (3.15), the population balance equation for the particle number density defined by

$$N_i(\mathbf{r}, t) = \int_{v_i}^{v_{i+1}} f_1\left(V_\mathrm{p}, \mathbf{r}, t\right) \mathrm{d}v \quad i = 0, 1, 2, \ldots, M \tag{3.49}$$

and in the absence of particle growth can be written according to

$$\begin{aligned}
&\frac{\partial N_i(\mathbf{r}, t)}{\partial t} + \nabla_\mathrm{r} \cdot \left(\mathbf{v}_{\mathrm{r},i}(\mathbf{r}, \mathbf{Y}, t) N_i(\mathbf{r}, t)\right) = S_{N_i}(\mathbf{r}, \mathbf{Y}, t) \\
&= \int_{v_i}^{v_{i+1}} \mathrm{d}v \left[\sum_{j=0}^{M} \int_{v_j}^{v_{j+1}} \nu\left(V'_\mathrm{p}, \mathbf{r}, \mathbf{Y}\right) b\left(V'_\mathrm{p}, \mathbf{r}, \mathbf{Y}\right) P\left(V_\mathrm{p}, \mathbf{r} \middle| V'_\mathrm{p}, \mathbf{r}, \mathbf{Y}, t\right) f_1\left(V'_\mathrm{p}, \mathbf{r}, t\right) \mathrm{d}v' \right. \\
&\quad - b\left(V_\mathrm{p}, \mathbf{r}, \mathbf{Y}\right) f_1\left(V_\mathrm{p}, \mathbf{r}, t\right) + \frac{1}{2} \sum_{j=0}^{i-1} \int_{v_j}^{v_{j+1}} a\left(\tilde{V}_\mathrm{p}, \tilde{\mathbf{r}}; V'_\mathrm{p}, \mathbf{r}', \mathbf{Y}\right) f_1\left(\tilde{V}_\mathrm{p}, \tilde{\mathbf{r}}, t\right) f_1\left(V', \mathbf{r}', t\right) \mathrm{d}v' \\
&\quad \left. - \sum_{j=0}^{M} \int_{v_j}^{v_{j+1}} a\left(V'_\mathrm{p}, \mathbf{r}'; V_\mathrm{p}, \mathbf{r}, \mathbf{Y}\right) f_1\left(V'_\mathrm{p}, \mathbf{r}', t\right) f_1\left(V_\mathrm{p}, \mathbf{r}, t\right) \mathrm{d}v' \right]
\end{aligned} \tag{3.50}$$

On the basis of our early work in Cheung et al. (2007a, b), the feasibility of predicting the bubble size distribution in isothermal vertical bubbly flows has been realized through solving a single transport equation for the average number density. This so-called "one-group" approach is applicable to bubbly flows which are dominated primarily by spherical shape bubbles. Nevertheless, the "two-group" approach which has been adopted in our latest works, i.e. of Qi et al. (2012) and Cheung et al. (2012), is based on the characterization of spherical shape and cap shape bubbles in two different groups. Two transport equations of the average number density in the form similar to Eqn (3.50) along with the formulation of appropriate source and sink terms are solved in order to predict the transition between bubbly and cap flows. This will be further discussed in the next chapter.

3.5 Comments on the Coupling Between Population Balance and Computational Multiphase Fluid Dynamics

In order to be consistent with the generic form of Eqn (2.156) and variables used in the framework of computational multiphase fluid dynamics in Chapter 2, the transport equation for the particle number density (Eqn (3.48)) in terms of size fraction is normally adopted. By definition, $\alpha_i^p = N_i v_i$, expressing in terms of size fraction f_i, $\alpha^p f_i = N_i v_i$. Subsequently in terms of mass $m_{p,i}$,

$$\alpha^p \rho_i^p f_i = N_i m_{p,i} \tag{3.51}$$

Taking the substantive derivative of Eqn (3.51) leads to

$$\frac{D\left(\alpha^p \rho_i^p f_i\right)}{Dt} = \frac{D\left(N_i m_{p,i}\right)}{Dt} = m_{p,i}\frac{DN_i}{Dt} + N_i\frac{Dm_{p,i}}{Dt} \tag{3.52}$$

Since the mass remains invariant within the particle number density, the transport equation for the size fraction can be expressed in terms of the rate of increase and advection of f_i on the right-hand side and a source/sink term on the left-hand side as

$$\frac{\partial\left(\alpha^p \rho_i^p f_i\right)}{\partial t} + \nabla \cdot \left(\alpha^p \rho_i^p \mathbf{v}_{p,i} f_i\right) = m_{p,i} S_{N_i} \tag{3.53}$$

It is still not common to solve Eqn (3.53) by direct numerical simulations (which attempt to resolve the whole spectrum of possible turbulent length scales in the flow and provide the propensity of describing the complex flow structures within the external coordinates). Practical multiphase flows that are encountered in natural and technological systems generally contain millions of particles that are simultaneously varying along the internal coordinates. Hence, the feasibility of direct numerical simulations in resolving such flows is still far beyond the capacity of existing computer resources. In Chapter 2, averaging has been introduced so that the feasibility of solving the multiphase flow through suitable numerical techniques can be realized with reasonable time and space resolutions. To be consistent with the governing equations of the multifluid model, an averaged form of the population balance equation is derived. Applying volume-averaging (or ensemble-averaging) and introducing Favre-averaging to Eqn (3.53) to alleviate the complication of modeling additional correlation terms containing averages of fluctuating quantities, the phase-weighted average and mass-weighted average for the transport of size fraction can be formulated. Dropping the bars and parentheses, which by default denote the Favre-averaging and volume-averaging (or ensemble-averaging) processes, the effective equation for the size fraction can now be written as

$$\frac{\partial\left(\alpha^p \rho_i^p f_i\right)}{\partial t} + \nabla \cdot \left(\alpha^p \rho_i^p \mathbf{v}_{p,i} f_i\right) = -\nabla \cdot \left(\alpha^p \mathbf{q}_{f_i}^{p''}\right) + m_{p,i} S_{N_i} \tag{3.54}$$

where $\mathbf{q}_{f_i}^{p''}$ is the turbulent flux of the size fraction. The physics of $\mathbf{q}_{f_i}^{p''}$ in Eqn (3.54) is yet to be fully understood, which makes modeling this term in the computational multiphase fluid dynamic framework challenging. There is still some dispute concerning the diffusion of particles in the equations governing the conservation of mass and momentum; the diffusion of particles is thus included in the population balance equations of the particle and average number density for consideration of large spatial gradients over the swarm of particles present in the multiphase flow. In the absence of these large gradients within the flow system, the terms are usually neglected in practice.

As the population balance equation is being solved in conjunction with the multifluid model, relevant boundary conditions for either the size fraction or average number density also form another important consideration in the computation of multiphase flow. At the inflow and outflow boundaries, *Dirichlet* and *Neumann* boundary conditions are adopted. In other words,

$$f_i = f_{i,\text{in}} \text{ or } N = N_{i,\text{in}} \quad \text{at the inflow boundary} \tag{3.55}$$

$$\frac{\partial f_i}{\partial n} = 0 \text{ or } \frac{\partial N_i}{\partial n} = 0 \quad \text{at the outflow boundary} \tag{3.56}$$

At the inflow boundary, $f_{i,\text{in}}$ have values ranging between zero and unity, while $N_{i,\text{in}}$ is usually determined based on the volume fraction and particle volume. Both of these variables may be specified as either constant values or prescribed profiles at the surface. For a given problem where $f_{i,\text{wall}}$ and $N_{i,\text{wall}}$ are known at solid walls of the computational domain, the *Dirichlet* boundary condition applies. The boundary condition is given by

$$f_i = f_{i,\text{wall}} \text{ or } N_i = N_{i,\text{wall}} \quad \text{at the solid boundary} \tag{3.57}$$

Otherwise, the *Neumann* boundary condition is imposed:

$$\frac{\partial f_i}{\partial n} = 0 \text{ or } \frac{\partial N_i}{\partial n} = 0 \quad \text{at the solid boundary} \tag{3.58}$$

This condition also corresponds with the *Neumann* boundary condition for the size fraction and average number density at the outflow boundary. Other boundary conditions such as symmetry and periodic boundary conditions may also be adopted for the size fraction and the average number density to take advantage of special geometrical features of the solution domain.

3.6 Summary

The basic consideration of the population balance equation for multiphase flow calculations, having a secondary phase consisting of solid particles, bubbles, or droplets, has been discussed and described. Through the consideration of integral formulation, discrete infinitesimal small control volume in the fluid flow, or averaging the statistical Boltzmann-type equation, the population balance equation can be derived which essentially results in the transport of density function in time and space subject to the net generation rate of particles due to breakage/break up and aggregation/coalescence processes. In practice, the population balance equation in terms of density function is still too detailed to be employed and solved via direct numerical simulations. Instead, the transport of particle number density is normally considered within the computational multiphase fluid dynamics framework. This also forms the premise whereby the fundamental development of suitable source or sink terms will be described in Chapters 4 and 5 for different types of multiphase flows. Similarly to the governing equations in Chapter 2, appropriate boundary conditions and their physical significance for different typical forms of multiphase flows as well as specification of the wall boundary conditions are discussed and provided.

CHAPTER 4

Mechanistic Models for Gas–Liquid/ Liquid–Liquid Flows

4.1 Introduction

Motion of individual gas bubbles or liquid droplets traveling in a liquid phase, better known as gas–liquid or liquid–liquid flows, is characterized by the existence of interfaces between deforming phases and discontinuity of associated properties. Complex interphase mass, momentum, and energy transfers are often encountered through these interfaces. In a number of industry applications, namely chemical reactors, boiler and condenser equipment, and nuclear reactors, the extent of particle–particle interaction can profoundly influence the overall performance of these applications; such interaction can significantly alter the interfacial area between the phases via the distribution of different sizes of gas bubbles or liquid droplets. In addition to the understanding of fundamental flow physics, the study of fluid particle dynamics—comprising coalescence and break up processes as well as any phase changes occurring—represents another important consideration, particular in the rational design and optimization of these multiphase systems.

The phenomenon of fluid particle coalescence and break up has been the subject of extensive investigations. There are two important aspects in comprehending the essential fluid particle coalescence and break up processes involved. From a physical perspective, the exchange of mass, momentum, and energy between the dispersed phase—gas bubbles or liquid drops, and the continuous phase—the liquid flow, is significantly influenced by the local concentration of the particle interfacial area or corresponding parameters such as the diameter, volume, or number density of fluid particles. This fluid particle interfacial area, size, or density represents a key parameter in the study of gas–liquid mixtures or liquid–liquid dispersion flows. From a modeling perspective, the interphase mass, momentum, and energy transfer terms as a result of averaging performed on the local instantaneous conservation equations are important considerations in the multifluid model. These terms are generally required to be modeled and solved as accurately as possible in order to obtain the necessary phasic parameters (for example, the volume fraction of the dispersed phase) of such flows.

For the treatment of fluid particle size distributions in multiphase systems, the population balance equation introduced in Chapter 3 can be applied. In order to achieve the desired closure of the conservation equation, reliable models are needed for the growth as well as the birth and death source and sink terms which are dependent on local events and fluid properties. In the case of the birth and death source and sink terms, which principally describe the creation and destruction of a particle population due to coalescence and break up, the conservation equation will require fluid particle coalescence and break up rates and detailed phenomenological models able to predict the nature of the interactions of the fluid particles with the surrounding continuous field and between the fluid particles themselves. Strictly speaking, the modeling of birth and death rates for these classifications of multiphase flows remains extremely challenging because of the complexity of the physics involved. In some cases, the detailed functionality of the closure relations and even the physical insights of the birth and death processes are yet to be discovered. The formulation of the growth term that may be dependent upon the choice of internal coordinates or particle properties also leads to the additional complexity of suitably characterizing the particle phase in the frameworks of computational multiphase fluid dynamics and population balance.

In the subsequent sections, various mechanisms and available kernel models for the coalescence and break up of fluid particles are described. It should be noted that the concept of bubble, drop, or particle is exchangeable in this chapter because the physical effects relating to coalescence and break up of gas bubbles and coalescence and break up of liquid drops in liquid are comparable in nature.

4.2 Mechanisms and Kernels of Fluid Particle Coalescence

Coalescence, together with break up and mass transfer due to particle growth, is responsible for the evolution of fluid particle sizes in gas–liquid or liquid–liquid flows. Focusing initially on the process of coalescence, it is generally considered to be complex because it involves not only the interaction of particles with the surrounding fluid, but also the interaction between fluid particles themselves once these particles are brought together due to the action of the external flow or body forces. It is worthwhile mentioning that gas bubbles or liquid drops behave rather differently than solid particles because they are deformable, elastic, and may agglomerate or eventually coalesce after random collisions. However, the basic assumptions where solid particles can be treated, according to the kinetic theory, as hard spheres and the collisions to be perfectly elastic and obey the classical conservation laws do not hold for real fluid particles. Because of the complicated physics, a practical consideration of the birth and death rates due to fluid particle coalescence is proposed where appropriate source and sink terms are developed directly on the averaging scales performing analysis of fluid particle coalescence, particularly in turbulent gas–liquid or liquid–liquid flows.

The population balance equation of the particle number density, which was derived in Chapter 3 (Eqn (3.50)), is commonly adopted in conjunction with the multifluid model. The macroscopic population balance formulation of the discrete birth and death rates due to fluid particle coalescence is presented herein. First, the birth of fluid particles of volume ν_i due to coalescence results from the coalescence between all fluid particles smaller than ν_i. Therefore, the birth rate for particles that form a fluid particle volume ν_i can be written as

$$S_{C,i}^{+}(\mathbf{r}, \mathbf{Y}, t) = \int_{\nu_i}^{\nu_{i+1}} d\nu \left[\frac{1}{2} \sum_{j=0}^{i-1} \int_{\nu_j}^{\nu_{j+1}} \underbrace{a\left(\tilde{V}_{\mathrm{p}}, \tilde{\mathbf{r}}; V_{\mathrm{p}}', \mathbf{r}', \mathbf{Y}\right)}_{\text{coalescence frequency}} f_1\left(\tilde{V}_{\mathrm{p}}, \tilde{\mathbf{r}}, t\right) f_1(V', \mathbf{r}', t) d\nu' \right] \quad (4.1)$$

Second, the death of fluid particles due to coalescence between two particles of volume ν_i or between one fluid particle of volume ν_i can be determined by

$$S_{C,i}^{-}(\mathbf{r}, \mathbf{Y}, t) = \int_{\nu_i}^{\nu_{i+1}} d\nu \left[\sum_{j=0}^{M} \int_{\nu_j}^{\nu_{j+1}} \underbrace{a\left(V_{\mathrm{p}}', \mathbf{r}'; V_{\mathrm{p}}, \mathbf{r}, \mathbf{Y}\right)}_{\text{coalescence frequency}} f_1\left(V_{\mathrm{p}}', \mathbf{r}', t\right) f_1\left(V_{\mathrm{p}}, \mathbf{r}, t\right) d\nu' \right] \quad (4.2)$$

For coalescence between fluid particles, the coalescence frequency in Eqns (4.1) and (4.2) could be derived as the product of a collision frequency (or collision density) and a coalescence efficiency (or coalescence probability). The general expression for the coalescence frequency in gas—liquid or liquid—liquid flows, which is defined as the rate of particle formation as a result of binary collisions of fluid particles with volumes ν_i and ν_j, can be written as

$$a(\nu_i, \nu_j) = h(\nu_i, \nu_j)\lambda(\nu_i, \nu_j) \quad (4.3)$$

A variety of mechanisms for the collision frequency $h(\nu_i, \nu_j)$ that promotes collisions among particles in a turbulent flow is reviewed in Liao and Lucas (2010). They can be distinguished by the following:

1. Motion induced by turbulent fluctuations in the surrounding continuous liquid
2. Wake interactions or helical/zig-zag trajectories
3. Motion induced by mean-velocity gradients in the flow
4. Different particle rise velocities induced by buoyancy or body forces
5. Bubble capture in a turbulent eddy.

The coalescence efficiency $\lambda(\nu_i, \nu_j)$ can be developed according to three postulated theories or criteria (Liao and Lucas, 2010).

The first theory that was widely adopted was the film drainage model. According to Shinnar and Church (1960), after collision, two particles may come together and be prevented from undergoing coalescence by the presence of a thin film of liquid trapped between them. It is the attractive forces between these two particles that subsequently drive the film to drain out until the two surfaces collapse and coalescence proceeds. This type of coalescence is usually divided into three subprocesses. First, as two particles collide, a small amount of liquid is trapped between them. Second, the two particles remain in contact until the liquid film drains out to a critical thickness. Third, the thin film separating the two fluid particles ruptures, resulting in the coalescence of these particles. In retrospect, the duration of collisions is constricted due to the prevailing fluctuations, and the process of coalescence will occur only if the interaction time between fluid particles is sufficient for the intervening film to drain out to the critical rupture thickness. The evolution of coalescence of two droplets based on this theory was confirmed in Chapter 1 through the adoption of a moving mesh interface tracking methodology coupled with local mesh adaptation to handle the interface merging of the two droplets.

The second theory, which was proposed by Howarth (1964) focusing on the energy model where coalescence will proceed, depends on the impact of colliding fluid particles. During energetic collisions, when the approach velocity of two colliding fluid particles exceeds a critical value, immediate coalescence without liquid film capturing and thinning will become the main mechanism. In this situation, the turbulence force in controlling the coalescence probability is considered to be far greater than the attraction force between two colliding interfaces of the fluid particles.

The third theory, based on the work by Lehr and Mewes (1999) and Lehr et al. (2002), introduced the critical approach velocity model, which stems from an empirical theory derived from the experimental observation of Doubliez (1991) and Duineveld (1994). Here, they consider that small approach velocities result in high coalescence probability.

In the next three sections, various physical models proposed for the collision frequency are outlined followed by the description of models for the film drainage, energy, and critical approach velocity for coalescence efficiency.

4.2.1 Collision Frequency Due to Turbulent Fluctuation and Random Collision

In gas–liquid or liquid–liquid flows, fluid particles may collide due to the fluctuating turbulent velocity of the surrounding liquid. Because of the random movement of these fluid particles, the motion has been taken to be similar to the random movement of molecules in the classical kinetic theory of gases (Figure 4.1). Obviously, this is a rough assumption, but it forms the basis for developing the collision frequency models.

According to the kinetic gas theory of Kennard (1938), the collision frequency for binary collision or the effective swept volume rate of sizes of $d_{p,i}$, and $d_{p,j}$ can be expressed as the product of the collision-sectional area A_{ij} and the relative velocity u' according to

$$h(v_i, v_j) = A_{ij}u' \tag{4.4}$$

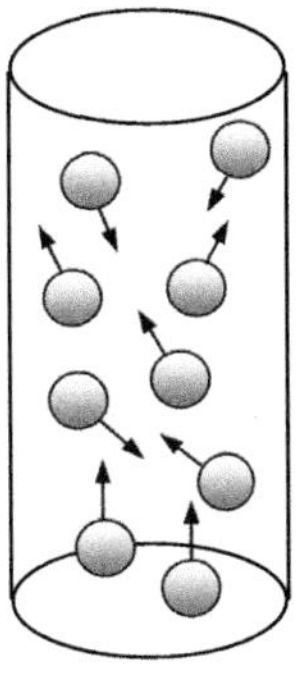

Figure 4.1
A schematic representation of the random movement of fluid particles.

where A_{ij} can be calculated as

$$A_{ij} = \frac{\pi}{4}\left(d_{p,i} + d_{p,j}\right)^2 \tag{4.5}$$

In order to determine the characteristic velocity, it is normally assumed that the characteristic velocity of the colliding fluid particles is that of an equally-sized eddy, as proposed by Coulaoglou and Tavlarides (1977), Lee et al. (1987), Prince and Blanch (1990), and Luo (1993). It should be recognized that very small eddies possess insufficient energy to significantly affect the motion of the fluid particle, and much larger eddies only convey the fluid particles and have no effect on the relative motion. The relative velocity between two fluid particles can be approximated based on the kinetic theory of gasses by the root mean square of two equivalent turbulent velocities:

$$u' = \left(u_{t,i}^2 + u_{t,j}^2\right)^{1/2} \tag{4.6}$$

in which $u_{t,i}$ and $u_{t,j}$ correspond to the turbulent velocities with sizes of $d_{p,i}$, and $d_{p,j}$. The discrete particle diameters $d_{p,i}$, and $d_{p,j}$ may be evaluated based on the spherical bubble assumption as

$$d_{p,i} = \left(\frac{6}{\pi}v_i\right)^{1/3} \quad d_{p,j} = \left(\frac{6}{\pi}v_j\right)^{1/3} \tag{4.7}$$

For fluid particles, estimation of the turbulent velocity has been obtained by considering certain expressions developed in the classical theory on isotropic turbulence due to Kolmogorov (1941). The theory states that within the inertial subrange of turbulence, where the distance between two points in the flow field is much smaller than the turbulent characteristic length scale but much larger than the Kolmogorov microscale, the second-order velocity structure function can be taken to be a function of the dissipation of turbulent kinetic energy and the distance separating the two points. Analogous with the theory of Kolmogorov (1941), if the second-order velocity structure function is interpreted as an absolute velocity squared and the distance separating the two points is taken to be equivalent to the particle diameter, the turbulent velocity $u_{t,i}$ for particle diameter $d_{p,i}$ is thus approximated as

$$u_{t,i}^2 = C\left(\varepsilon^c d_{p,i}\right)^{2/3} \tag{4.8}$$

where C is a constant and ε^c is the dissipation of the turbulent kinetic energy in the continuous phase.

On the basis of Eqns (4.5), (4.6) and (4.8), the collision frequency can be written as

$$h(v_i, v_j) = C_1\left(d_{p,i} + d_{p,j}\right)^2\left(d_{p,i}^{2/3} + d_{p,j}^{2/3}\right)^{1/2}\left(\varepsilon^c\right)^{1/3} \tag{4.9}$$

where C_1 lies in the range between 0.28 and 1.11 (Liao and Lucas, 2010).

Some modifications have nonetheless been proposed to improve the calculation of the collision frequency as depicted in the form of Eqn (4.9).

The first modification is the elucidation of the effect of size ratio between fluid particles and eddies; in the original formulation, all fluid particles have been assumed to be inertial subrange and have the same velocity as the equal-sized eddies. Colin et al. (2004) proposed that for fluid particles larger than the turbulent characteristic length scale, turbulent eddies are not efficient in transporting the fluid particles. The relative velocity is, therefore, mainly due to the mean shear flow. Denoting the turbulent characteristic length scale as l_e, turbulent collisions occur only in the following:

$$u' = \frac{C_t}{\sqrt{1.61}}\left(\varepsilon^c \frac{d_{p,i} + d_{p,j}}{2}\right)^{1/3} \quad \left(d_{p,i} < l_e, d_{p,j} < l_e\right) \tag{4.10a}$$

$$u' = \frac{C_t}{\sqrt{1.61}}\left(\varepsilon^c d_{p,i}\right)^{1/3} \quad \left(d_{p,i} < l_e, d_{p,j} > l_e\right) \tag{4.10b}$$

In the above two equations, the adjustable coefficient C_t accounts for the difference between the velocity of fluid particles and the liquid eddies, while the factor $1/\sqrt{1.61}$ considers the deceleration of the approaching fluid particles due to an increase in virtual mass. Based on Eqn (4.5) and the above equations, the collision frequency can be written as

$$h(v_i, v_j) = C_2\left(d_{p,i} + d_{p,j}\right)^{7/3}\left(\varepsilon^c\right)^{1/3} \quad \left(d_{p,i} < l_e, d_{p,j} < l_e\right) \tag{4.11a}$$

$$h(v_i, v_j) = C_3\left(d_{p,i} + d_{p,j}\right)^2\left(d_{p,i}\right)^{1/3}\left(\varepsilon^c\right)^{1/3} \quad \left(d_{p,i} < l_e, d_{p,j} < l_e\right) \tag{4.11b}$$

where $C_2 = \frac{1}{2}\left(\frac{8\pi}{3}\right)^{1/2}\frac{C_t}{\sqrt{1.61}}\frac{1}{\sqrt[3]{2}}$ and $C_3 = \frac{1}{2}\left(\frac{8\pi}{3}\right)^{1/2}\frac{C_t}{\sqrt{1.61}}$.

The second modification to the calculation of the collision frequency is to consider the reduction of the free space movement of fluid particles. In this instance, the collision frequency will increase due to the increased likelihood of collisions among the fluid particles. A factor γ can be multiplied with the collision frequency to describe the effect. Different expressions for factor γ have been found in the literature, and these are summarized as follows:

Wu et al. (1998)

$$\gamma = \frac{1}{\sqrt[3]{\alpha_{\max}^{\mathrm{p}}}\left(\sqrt[3]{\alpha_{\max}^{\mathrm{p}}} - \sqrt[3]{\alpha^{\mathrm{p}}}\right)} \tag{4.12}$$

Hibiki and Ishii (2000a, b)

$$\gamma = \frac{1}{\alpha_{\max}^{\mathrm{p}} - \alpha^{\mathrm{p}}} \tag{4.13}$$

Lehr et al. (2002)

$$\gamma = \exp\left[\left(\frac{\sqrt[3]{\alpha_{\max}^{\mathrm{p}}} - \sqrt[3]{\alpha^{\mathrm{p}}}}{\sqrt[3]{\alpha_{\max}^{\mathrm{p}}}}\right)^2\right] \tag{4.14}$$

Wang et al. (2005a, b)

$$\gamma = \frac{\alpha_{\max}^{\mathrm{p}}}{\alpha_{\max}^{\mathrm{p}} - \alpha^{\mathrm{p}}} \tag{4.15}$$

The maximum volume fraction of the fluid particle $\alpha_{\max}^{\mathrm{p}}$ as depicted in Eqns (4.12)–(4.15) varies between 0.52 and 0.8.

The third modification is to consider the ratio of mean distance between fluid particles to their average turbulent path length. When the distance between fluid particles is larger than the turbulent path length, no collision should be counted; hence, a decreasing factor Π should be included with collision frequency. By assuming that the average size of eddies is of the same order as the size of a fluid particle, the expression for Π derived by Wu et al. (1998) is given by

$$\Pi = 1 - \exp\left(-C'\frac{\sqrt[3]{\alpha_{\max}^{\mathrm{p}}}\sqrt[3]{\alpha^{\mathrm{p}}}}{\sqrt[3]{\alpha_{\max}^{\mathrm{p}}} - \sqrt[3]{\alpha^{\mathrm{p}}}}\right) \tag{4.16}$$

where C' is an adjustable constant. For an air–water system, it has been set to a value of 3. Alternatively, Wang et al. (2005a, b) proposed a different expression by considering that the factor Π should approach unity when the ratio of the mean distance between fluid particles ($h_{p,ij}$) with sizes of $d_{p,i}$, and $d_{p,j}$ to the mean relative turbulent path length scale ($l_{t,ij}$) of fluid particles with sizes of $d_{p,i}$, and $d_{p,j}$—$h_{p,ij}/l_{t,ij}$ is small and approaches zero at large ratios. The expression can be written as

$$\Pi = \exp\left[-\left(\frac{h_{p,ij}}{l_{t,ij}}\right)^6\right] \tag{4.17}$$

Based on the second and third modifications, Wang et al. (2005a, b) adopted the modified form of Eqn (4.9) according to

$$h(\nu_i, \nu_j) = C_4 \gamma \Pi \left(d_{p,i} + d_{p,j}\right)^2 \left(d_{p,i}^{2/3} + d_{p,j}^{2/3}\right)^{1/2} (\varepsilon^c)^{1/3} \tag{4.18}$$

In principle, the modification factors γ and Π play a similar role because both are related to the volume fraction of the fluid particles. The influence of the volume fraction is obvious and has been shown to be important. Nevertheless, the necessity for γ and Π demands further investigation.

4.2.2 Collision Frequency Due to Wake Entrainment

Besides collisions due to random movement of the particles in a turbulent flow, the free-rise of fluid particles through a liquid will inevitably carry a certain amount of liquid, and wakes are subsequently formed behind these fluid particles. When fluid particles enter the so-called wake region of a leading fluid particle, they will accelerate and may collide with the preceding fluid particle. As observed in the experiment performed by Stewart (1995), the wake behind the traveling fluid particles was observed to be the driving force and sole mechanism for particle–particle interaction. In particular, wake-induced collisions could result in coalescence between pairs of fluid particles of large cap shape where the flow regions behind the fluid particles are sufficiently viscous to keep their wakes laminar.

In accordance with Kalkach-Navarro et al. (1994), the collision frequency between the trailing fluid particle in the wake and the leading fluid particle can be defined as the volume that has to be occupied at time t–dt to permit combination with the leading fluid particle at time t, per unit time ($\dot{V}_c$), multiplied by the probability for the fluid particle to be in the wake of the leading fluid particle (P), i.e. $\dot{V}_c P$. As can be seen in Figure 4.2, the volume which a fluid particle of radius $r_{p,i}$ has to occupy at time t–dt to combine with a fluid particle of radius $r_{p,j}$, in terms of equivalent radii, is given by

$$\dot{V}_{c,i} = 2\pi \int_0^{r_s} r \Delta u_l \left(r_s^2 - r^2\right)^{1/2} \mathrm{d}r \tag{4.19}$$

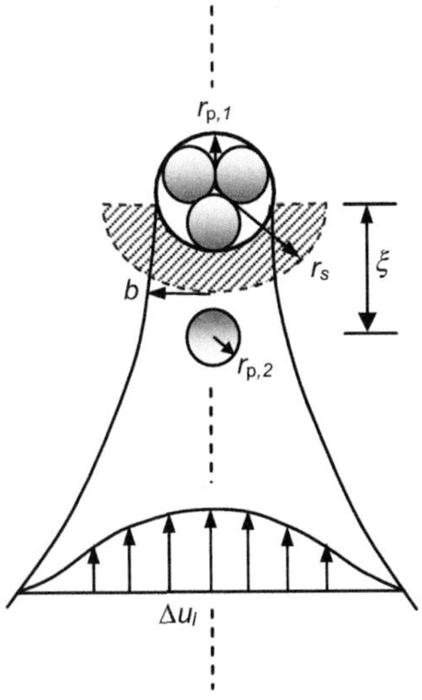

Figure 4.2
A schematic representation of the geometric approximation used for the wake model by Kalkach-Navarro et al. (1994).

where $r_s = r_{p,i} + r_{p,j}$. The integration domain is the shaded area depicted in Figure 4.2, while the velocity difference in the wake is determined according to the correlation suggested by Schlichting (1979) as

$$\Delta u_l = k_1 \left(\frac{g\sigma}{\rho^c}\right)^{1/4} \left(\frac{r_{p,i}}{\xi}\right)^{2/3} C_D^{1/3} \tag{4.20}$$

where g is the gravity, σ is the surface tension, ρ^c is the continuous phase density, and ξ is the distance between the centers of the leading and trailing fluid particles.

The drag coefficient C_D in Eqn (4.20) can be evaluated based on the model proposed by Harmathy (1960) for distorted fluid particles:

$$C_D = \frac{4}{3}\left[\frac{g(\rho^c - \rho^p)}{\sigma(1-\alpha^p)}\right]^{1/2} r_{p,i} \tag{4.21}$$

where ρ^p is the particle density.

The probability of a fluid particle being in the wake of the leading fluid particle can be assumed to be proportional to the area occupied by the wake divided by the area of the conduit:

$$P = k_2 \left(\frac{b}{D_c}\right)^2 \tag{4.22}$$

where D_c is the diameter of the conduit and b is the width of the wake at the location of the combined radius r_s which is given by

$$b = k_3 C_D^{1/3}\left(r_{p,i} + r_{p,j}\right)^{1/3} r_{p,i}^{2/3} \tag{4.23}$$

Substituting Eqns (4.20) and (4.21) into Eqn (4.19) and integrating the collision frequency along with Eqns (4.22) and (4.23), as well as considering two possibilities that the fluid particle with radius $r_{p,i}$ is the leading fluid particle and that the fluid particle with radius $r_{p,j}$ is the trailing fluid particle, we get:

$$h(v_i, v_j) = \left(\frac{4}{3}\pi\right)^{5/3} C_5\left(r_{p,i}^3 + r_{p,j}^3\right)\left(r_{p,i} + r_{p,j}\right)^2 \tag{4.24}$$

where C_5 contains the nondimensional proportionality constants of k_1, k_2, and k_3:

$$C_5 = 2\pi \frac{k_1 k_2 k_3^2}{D_c}\left[\frac{g^3(\rho^c - \rho^p)^2}{\sigma \rho^l (1 - \alpha^p)^2}\right]^{1/4}\left(\frac{3}{4\pi}\right)^{5/3} \tag{4.25}$$

Expressing the above equation in terms of volumes of the fluid particle yields

$$h(v_i, v_j) = C_5\left(v_i + v_j\right)\left(v_i^{1/3} + v_j^{1/3}\right)^2 \tag{4.26}$$

Alternatively, Collela et al. (1999) developed a novel model for wake-induced collision frequency by accounting for the wake interactions, swarm velocity, and shape of the fluid particles. The resulting expression is given by

$$h(v_i, v_j) = u_{rel}\frac{V_i^{BOX}}{\zeta} \tag{4.27}$$

where u_{rel} is the relative velocity between two coalescing fluid particles, V_i^{BOX} is the volume influenced by the wake of a fluid particle of size $d_{p,i}$, and ζ is the average distance between fluid particles. The model devised by Nevers and Wu (1971) is employed to determine u' and V_i^{BOX}. Referring to Figure 4.3, the volume influenced by the wake of a fluid particle is assumed to be of a conical shape in which the base of the cone is represented by the cross-sectional area of the leading fluid particle. The height of the cone, according to experimental results (Nevers and Wu, 1971; Miyahara et al., 1991; Stewart, 1995), is normally taken to be five times the base diameter. Along this height, the velocity of the trailing fluid particle can be expressed by the sum of its terminal velocity and the relative velocity between the leading and trailing fluid particles:

$$u_p = u_{terminal} + u_{rel} \tag{4.28}$$

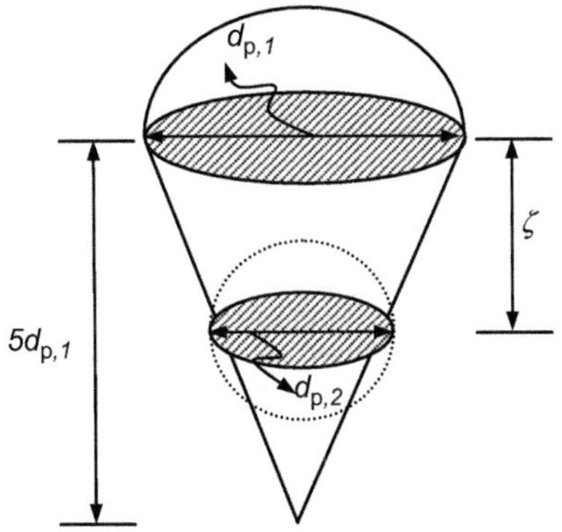

Figure 4.3
A schematic representation of the geometric approximation used for the wake model by Collela et al. (1999).

By considering only the drag and buoyancy forces,

$$F_{\text{drag}} = \frac{1}{2} C_D \rho^c u_p^2 A_{\text{projected}}$$
$$F_{\text{buoyancy}} = \rho^c g \left(\frac{2}{3} \pi r^3 \right) \tag{4.29}$$

where the projected area is $A_{\text{projected}} = \pi(r^2 - \beta^2)$. Equating these forces on the trailing fluid particle:

$$\rho^c g \left(\frac{2}{3} \pi r^3 \right) = \frac{1}{2} C_D \rho^c u_p^2 \pi (r^2 - \beta^2) \tag{4.30}$$

and substituting the above into Eqn (4.28) for u_p gives

$$u_{\text{terminal}} + u_{\text{rel}} = \left(\frac{4r^3 g}{3(r^2 - \beta^2) C_D} \right)^{1/2} \tag{4.31}$$

Since $u_{\text{terminal}} = (4rg/3C_D)^{1/2}$, the relative velocity can thus be obtained according to the above expression. It should be noted that the geometrical quantities below are used to determine the relative velocity:

$$\frac{r_{p,i}}{\beta} = \frac{5r_{p,i}}{5r_{p,i} - \zeta - h}$$
$$r_{p,j}^2 = \beta^2 + (r - h)^2 \tag{4.32}$$
$$r = r_{p,i} - r_{p,j}$$

The swarm effect can be accounted for via the equation proposed by Richardson and Zaki (1954).

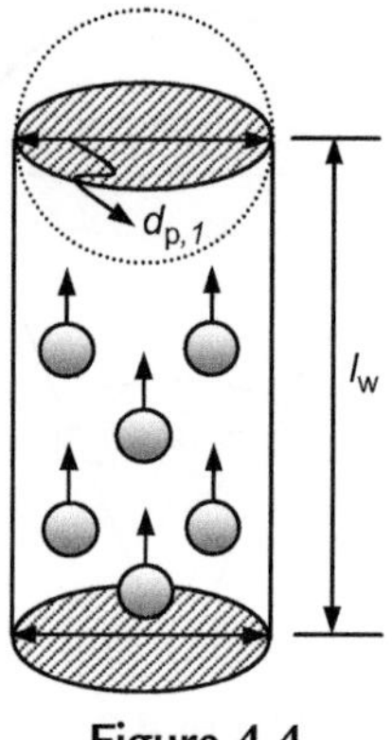

Figure 4.4
A schematic representation of the geometric approximation used for the wake model by Hibiki and Ishii (2000b).

Nevertheless, Wang et al. (2005a) adopted the model which was originally proposed by Wu et al. (1998) and later extended by Hibiki and Ishii (2000b) in their two-group formulation. Here, the collision frequency is related to the number and the rise velocity of the trailing fluid particles in the effective wake region of the leading fluid particle. Assuming that all trailing fluid particles collide with the leading fluid particle, which is approximated to be a large spherical fluid particle, an analytical form of the velocity distribution in the wake region can be obtained (Figure 4.4). With the volume influenced by the wake of the leading fluid particle of size $d_{p,i}$, given by

$$V_i^{\mathrm{WAKE}} = \frac{\pi}{4} d_{p,i}^2 \left(l_w - \frac{d_{p,i}}{2} \right) \tag{4.33}$$

the number of trailing particles of size $d_{p,j}$ in the wake region of a leading large bubble of size $d_{p,i}$ is determined by

$$N_{wi,j} = V_i^{\mathrm{WAKE}} N_j = \frac{\pi}{4} d_{p,i}^2 \left(l_w - \frac{d_{p,i}}{2} \right) N_j \tag{4.34}$$

The collision frequency can be obtained by assuming that all trailing fluid particles collide with the leading fluid particle within an average time interval ΔT:

$$h(v_i, v_j) = k_4 \frac{N_{wi,j}}{\Delta T N_j} \tag{4.35}$$

Substituting Eqn (4.34) into Eqn (4.35) yields

$$h(v_i, v_j) = k_4 \frac{\pi}{4} d_{p,i}^2 \underbrace{\left(\frac{l_w - d_{p,i}/2}{\Delta T} \right)}_{u_{wi,j}} \tag{4.36}$$

where l_w is the effective length of the wake region and $u_{wi,j}$ is the fluid particle rise velocity relative to the leading fluid particle. According to Schlichting (1979) and Bilicki and Kestin (1987),

$$u_{wi,j} = \frac{k_5}{l_w/d_{p,i} - 1/2}\left[\left(\frac{l_w}{d_{p,i}/2}\right)^{1/3} - 1\right]u_{rise,i} \tag{4.37}$$

where $u_{rise,i}$ is the rise velocity of the leading fluid particle, which can be calculated according to Fan and Tsuchiya (1990) using the following expression:

$$u_{rise,i} = 0.71\sqrt{gd_{p,i}} \tag{4.38}$$

Combining Eqns (4.37) and (4.38):

$$h(v_i, v_j) = C_6 d_{p,i}^2 u_{rise,i} \tag{4.39}$$

where C_6 contains the nondimensional proportionality constants of k_4 and k_5:

$$C_6 = \frac{\pi}{4}\frac{k_4 k_5}{l_w/d_{p,i} - 1/2}\left[\left(\frac{l_w}{d_{p,i}/2}\right)^{1/3} - 1\right] \tag{4.40}$$

In Eqn (4.40), $l_w/d_{p,i}$ has been assumed to be independent of size $d_{p,i}$. The effective length l_w can be taken as described above to be five times that of the base diameter of the leading fluid particle; therefore, it is reasonable to treat $l_w/d_{p,i}$ as a constant depending on the fluid properties.

4.2.3 Collision Frequency Due to Other Mechanisms

Collision of fluid particles in a turbulent flow may also be promoted by a variety of other mechanisms. This is in addition to the mechanisms that have been described in previous sections. In the literature, three possible mechanisms have been proposed for collision frequency. These are: (1) velocity gradient-induced collisions, (2) buoyancy-induced collision, and (3) capture in a turbulent eddy.

As pointed out by Friedlander (1977) and Williams and Loyalka (1991), particles in a uniform, laminar shear flow may collide with each other because of the prevalence of velocity gradients in the bulk of a turbulent flow. Assuming the flow has straight streamlines and the particle motion is rectilinear, the functional form of the collision frequency due to laminar shear can be expressed by

$$h(v_i, v_j) = C_7\left(\frac{d_{p,i} + d_{p,j}}{2}\right)^3 \overline{\left(\frac{du_{shear}}{dr}\right)} \tag{4.41}$$

where C_7 is a constant (a value of 4/3 prescribed by Friedlander, 1977) and $\overline{\left(\frac{du_{\text{shear}}}{dr}\right)}$ denotes the average shear rate. In theory, the model can be applied to any collision case resulting from a velocity gradient. For a turbulent gas—liquid or liquid—liquid flow, fluid particles collisions induced by the velocity shear of the mean flow can be described by Eqn (4.41).

According to Prince and Blanch (1990), fluid particle collisions may occur due to the difference in rise velocities of fluid particles having different sizes. A typical model for the collision frequency due to buoyancy collision as proposed by Friedlander (1977) is given by

$$h(v_i, v_j) = \frac{\pi}{4}\left(d_{p,i} + d_{p,j}\right)\left|u_{\text{rise},i} - u_{\text{rise},j}\right| \tag{4.42}$$

Although Eqn (4.42) has the same form as Eqn (4.41), the main difference is that the relative velocity is calculated from the rise velocity caused by body forces (for example, buoyancy). Wang et al. (2005a) has adopted the correlation as depicted in Eqn (4.38), while Prince and Blanch (1990) adopted the expression developed by Clift et al. (1978):

$$u_{\text{rise},i} = \sqrt{\frac{2.14\sigma}{\rho^{\text{l}} d_{p,i}} + 0.505 g d_{p,i}} \tag{4.43}$$

Analogously, the rise velocity for fluid particle with size $d_{p,j}$ is also determined according to Eqn (4.43).

As stipulated by Chesters (1991), when the drop size is considerably smaller than the size of the energy dissipating eddies in a turbulent flow, the force governing the collision can be taken to be predominantly viscous driven. In this instance, the fluid particle velocities will be very close to the velocity of the continuous phase flow field. Under this circumstance, the collision frequency will be determined by only the local shear of the flow in turbulent eddies similar to uniform laminar shear. The collision frequency due to capture in a turbulent eddy can be written as

$$h(v_i, v_j) = C_8\left(\frac{d_{p,i} + d_{p,j}}{2}\right)^3 \sqrt{\frac{\varepsilon^{\text{c}}}{\nu^{\text{c}}}} \tag{4.44}$$

where C_8 is a constant, and $\sqrt{\varepsilon^{\text{c}}/\nu^{\text{c}}}$ is often referred as the turbulent shear rate where ν^{c} is the kinematic viscosity of the continuous phase. Kocamustafaogullari and Ishii (1995) have specified $C_8 = 0.618$ for a fluid particle smaller than turbulent eddy size, while Colin et al. (2004) have prescribed $C_8 = 4\pi/3\sqrt{1/61}$ for a fluid particle larger than the characteristic length of turbulence.

It is rather difficult to ascertain which mechanism plays the most important role. Hence, the consideration of when to apply particular mechanism is dependent upon the considered multiphase system in question. In general, there is a rough guideline that can be adopted.

If the size of the fluid particles is in the inertial subrange of turbulence, they will be exposed continuously to the stresses exerted by turbulent eddies from all directions, and the random motion will therefore be the most important. Nevertheless, if the fluid particles have dimensions that are smaller than the Kolmogorov length scale, slip velocity will become negligible, and the relative velocity will be determined primarily by the local turbulence characteristics such as turbulent shear while the mean-velocity gradient affects the relative motion of all fluid particles in laminar shear. The buoyancy mechanism becomes more dominant in the presence of increasing density differences while the mechanism of wake entrainment is only considered for fluid particles of large size.

4.2.4 Coalescence Efficiency Due to Film Drainage Model

A number of experiments have shown that not all collisions lead to the coalescence of fluid particles. In most cases, only a fraction of collisions result in actual coalescence while most of the colliding fluid particles separate from each other after a collision event. In addition to the collision frequency, coalescence efficiency is, therefore, additionally introduced to aptly describe the coalescence process.

The film drainage model is described in this section. Here, the binary coalescence process is assumed to occur in three consecutive stages, which can be seen in the schematic drawing in Figure 4.5. First, as fluid particles collide, a small amount of liquid is trapped between them. Second, this liquid is drained out over a period of time from an initial thickness until this liquid film separating the fluid particles reaches a critical thickness, under the action of the film hydrodynamics. These hydrodynamics of the film depend on whether the film surface is immobile or mobile. Third, film rupture occurs at this point of film instability resulting in instantaneous coalescence. The final result of the binary coalescence is a new larger fluid particle.

For this model, the coalescence efficiency can be described according to two characteristic time scales, namely the contact time due to particle–particle collision and the drainage time,

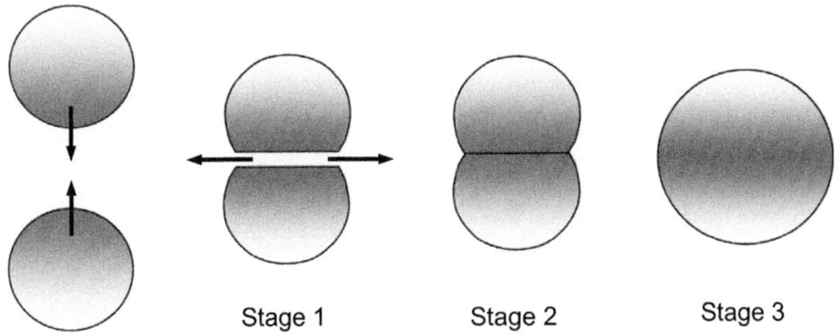

Figure 4.5
A schematic representation of the three consecutive stages of binary coalescence of fluid particles.

which is the time required for the intervening film between the fluid particles to drain to the critical thickness so that film rupture and coalescence can occur. According to Ross (1971), the probability of coalescence of a normal distribution of the coalescence efficiency by assuming that the coalescence and contact time are random variables can be written as

$$\lambda(v_i, v_j) = \frac{1}{2}\exp\left(-\frac{t_{\text{drainage}}}{t_{\text{contact}}}\right)\exp\left(\frac{1}{2}\frac{\sigma^2_{t_{\text{drainage}}}}{t^2_{\text{contact}}}\right) \times \text{erfc}\left(\frac{\sqrt{2}}{2}\frac{\sigma^2_{t_{\text{drainage}}} - t_{\text{drainage}}t_{\text{contact}}}{t_{\text{contact}}\sigma_{t_{\text{drainage}}}}\right) \tag{4.45}$$

Equation (4.45) is further simplified by Coulaloglou (1975) by assuming that the coalescence time is not distributed but the contact time remains a random variable. With $\sigma_{t_{\text{drainage}}} = 0$, the coalescence efficiency is henceforth determined by

$$\lambda(v_i, v_j) \approx \exp\left(-\frac{t_{\text{drainage}}}{t_{\text{contact}}}\right) \tag{4.46}$$

The above functional relationship is often used and has become the starting point of almost all subsequent models. However, the main difference is the formulation of modified relations for estimating the drainage and contact time.

In order to derive the drainage time, two regimes of film drainage may be considered:

1. The rigidity of particles surface—deformable or nondeformable and
2. The mobility of the contact interfaces—immobile, partially mobile, or fully mobile.

Figures 4.6 and 4.7 illustrate the regimes of the film drainage model. The postulation of various models is subsequently discussed thereafter.

Nondeformable rigid particles

When bubbles or drops are very viscous when compared to the continuous liquid phase, or they are very small in size (<1 mm), their interfaces at sufficiently large distances are

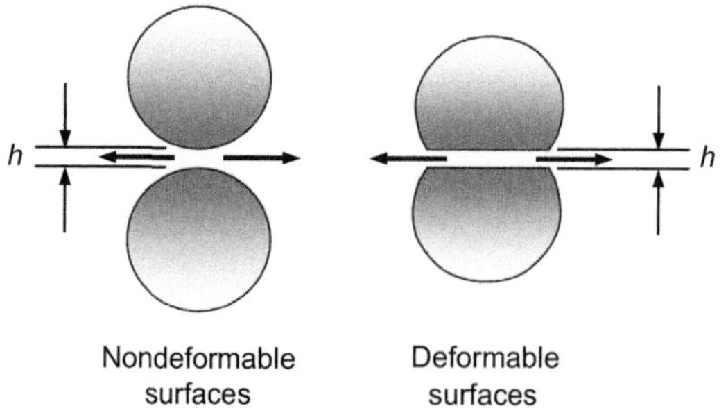

Figure 4.6

A schematic representation of nondeformable surfaces and deformable surfaces of fluid particles.

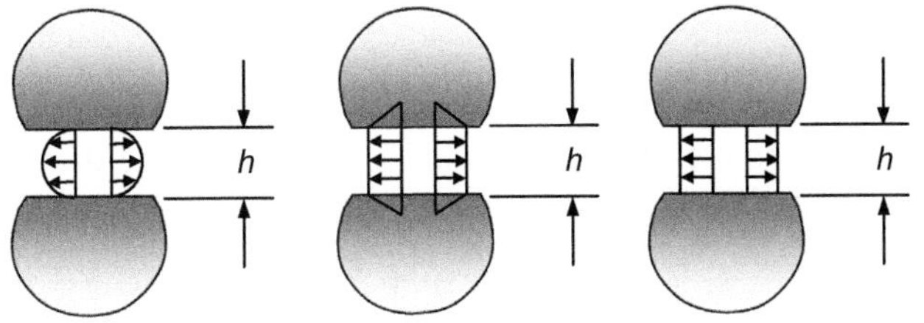

Figure 4.7
A schematic representation of immobile, partially mobile, and fully mobile contact interfaces of fluid particles.

only slightly deformed and they thus behave as nearly rigid spherical particles. Chesters (1991) has derived the drainage time for two nondeformable spheres with equal sizes via the consideration of the Poiseuille relation as

$$t_{\text{drainage}} = \frac{3\pi\mu^{c}}{2F} r_{\text{p}}^{2} \ln\left(\frac{h_{\text{i}}}{h_{\text{f}}}\right) \tag{4.47}$$

where F is the compressing force, h_{i} is the initial film thickness, and h_{f} is the critical film thickness. Replacing the radius r_{p} in Eqn (4.47) with an equivalent radius, $r_{\text{p,eq}} = 2r_{\text{p},i}r_{\text{p},j}/(r_{\text{p},i} + r_{\text{p},j})$ such as suggested by Chesters and Hofman (1982), the drainage time can be extended to describe the case of unequal sizes of fluid particles by

$$t_{\text{drainage}} = \frac{6\pi\mu^{c}}{F}\left(\frac{r_{\text{p},i}r_{\text{p},j}}{r_{\text{p},i} + r_{\text{p},j}}\right)^{2} \ln\left(\frac{h_{\text{i}}}{h_{\text{f}}}\right) \tag{4.48}$$

Nevertheless, the assumption of nondeformable rigid particles is only strictly valid for very small fluid particles. In most applications where large fluid particles exist within the flow, the deformation of the fluid particle surface during collision needs to be accounted for. Based on the proposal by Simon (2004), the so-called parallel model can be utilized to describe the film drainage between deformable interfaces. Here, the model assumes that the surfaces of coalescing fluid particles deform into two parallel disks such as seen in Figure 4.7. Based on this, the subject of deformable particles with immobile interfaces, partially mobile interfaces, and fully mobile interfaces is discussed next.

Deformable particles with immobile interfaces

For this particular consideration, the film drainage is governed by a viscous thinning behavior. In other words, the liquid is expelled from between the two rigid surfaces by laminar flow. This results in the consideration of the velocity profile in the film being parabolic with no slip at the surface. Therefore, the interaction between the film drainage and the circulation inside the particles is not a coupled mechanism.

Assuming the case of a constant force, Chesters (1991) derived the drainage time as

$$t_{\text{drainage}} = \frac{3\mu^{\text{c}}F}{16\pi\sigma^2} r_{\text{p}}^2 \left(\frac{1}{h_{\text{f}}^2} - \frac{1}{h_{\text{i}}^2} \right) \tag{4.49}$$

If the radius r_{p} in Eqn (4.49) is replaced with an equivalent radius $r_{\text{p,eq}}$, the drainage time for the case of two unequal-sized particles becomes

$$t_{\text{drainage}} = \frac{3\mu^{\text{c}}F}{4\pi\sigma^2} \left(\frac{r_{\text{p},i} r_{\text{p},j}}{r_{\text{p},i} + r_{\text{p},j}} \right)^2 \left(\frac{1}{h_{\text{f}}^2} - \frac{1}{h_{\text{i}}^2} \right) = \frac{3\mu^{\text{c}}F}{16\pi\sigma^2} \left(\frac{d_{\text{p},i} d_{\text{p},j}}{d_{\text{p},i} + d_{\text{p},j}} \right)^2 \left(\frac{1}{h_{\text{f}}^2} - \frac{1}{h_{\text{i}}^2} \right) \tag{4.50}$$

Note that the approximation of immobility of the film surface in deriving the drainage time above is applicable only to multiphase systems with extremely high viscosities for the dispersed phase or containing a certain concentration of soluble surfactant.

Deformable particles with partially mobile interfaces

For liquid–liquid flow in particular, drainage is predominantly controlled by the motion of the film surface. The contribution of the additional flow is now due to the prevailing pressure gradient being much smaller within the film.

Assuming a quasi-steady creeping flow, such as that suggested by Chesters (1991), the drainage time for partially mobile interfaces can be determined according to

$$t_{\text{drainage}} = \frac{\pi \mu^{\text{p}} F^{1/2}}{2(2\pi\sigma/r_{\text{p}})^{3/2}} \left(\frac{1}{h_{\text{f}}} - \frac{1}{h_{\text{i}}} \right) \tag{4.51}$$

Lee et al. (1987) considered the model of Sagert and Quinn (1976) for the partially mobile case to derive the time for film thinning, which is given by

$$t_{\text{thinning}} = -3M\mu^{\text{c}}R_{\text{a}}^2 \int_{h_{\text{i}}}^{h_{\text{f}}} \frac{\text{d}h}{8h^3 \left[2\sigma/r_{\text{p}} + H(6\pi h^3) \right]} \tag{4.52}$$

where M is the surface mobility parameter, which usually takes a value between 0 and 4, μ^{c} is the dynamic viscosity of the continuous phase, R_{a} is the radius of the disk being separated by the thin film in between the fluid particles, and H is the Hamaker constant (10^{-19}–10^{-20} J). Lee et al. (1987) further considered that the time for rupture combining with the time for film thinning yields the drainage time: $t_{\text{drainage}} = t_{\text{thinning}} + t_{\text{rupture}}$. Based on the proposal by

Ruckenstein and Jain (1974), a general expression for the breaking of the thin film can be simplified to

$$t_{\text{rupture}} = \frac{24\pi^2 M\sigma\mu^{\text{c}} h_{\text{f}}^5}{H^2} \tag{4.53}$$

Tsouris and Tavlarides (1994) employed the approximate relationship of the resisting force by Davis et al. (1989) to derive the drainage time for partially mobile interfaces:

$$\mathrm{d}t = \frac{6\pi\mu^{\text{c}} r_{\text{p,eq}}^2}{F} \frac{1+0.38\gamma}{1+1.69\gamma+0.43\gamma^2} \frac{\mathrm{d}h}{h} \tag{4.54}$$

where $\gamma = \mu^{\text{c}}/\mu^{\text{p}}\sqrt{r_{\text{p,eq}}/h}$ characterizes the interface mobility. Assuming a constant film drainage rate, the drainage time between two deformable particles from h_{i} to h_{f} is

$$t_{\text{drainage}} = \frac{6\pi\mu^{\text{c}} r_{\text{p,eq}}^2}{F}\zeta \tag{4.55}$$

where

$$\zeta = 1.872 \ln\left(\frac{\sqrt{h_{\text{i}}}+1.378q}{\sqrt{h_{\text{f}}}+1.378q}\right) + 0.127 \ln\left(\frac{\sqrt{h_{\text{i}}}+0.312q}{\sqrt{h_{\text{f}}}+0.312q}\right)$$
$$q = \mu^{\text{c}}\sqrt{r_{\text{p,eq}}}/\mu^{\text{p}} \tag{4.56}$$

Deformable particles with fully mobile interfaces

In order to derive the drainage time for fully mobile interfaces, it should be recognized that the drainage process is governed by not only the inertia force but also the viscous force. Chesters (1975) proposed a relationship to characterize a parallel-film model for fully mobile interfaces which can be expressed as

$$\frac{\mathrm{d}H'}{\mathrm{d}t} = \left(\frac{\sigma}{3\mu^{\text{c}} r_{\text{p}}} \frac{\mathrm{d}H'}{\mathrm{d}t}\right) \exp\left(-\frac{12\mu^{\text{c}} t}{\rho^{\text{c}} R_{\text{a}}^2}\right) - \frac{\sigma}{3\mu^{\text{c}} r_{\text{p}}} \tag{4.57}$$

where $H' = 1/2 \ln h$. For highly viscous liquids, the drainage velocity can be taken to be independent of the film size and, subsequently, the force as well. Because the film is thinning viscously, the drainage time at this limit is

$$t_{\text{drainage}} = \frac{3\mu^{\text{c}} r_{\text{p}}}{2\sigma} \ln\left(\frac{h_{\text{i}}}{h_{\text{f}}}\right) \tag{4.58}$$

The above equation can be used for unequal sizes of fluid particles by replacing r_{p} with $r_{\text{p,eq}}$. In the inertia-controlled limit, Eqn (4.57) can be manipulated to be reduced to

$$t_{\text{drainage}} = \frac{1}{2}\frac{\rho^{c} u_{t} r_{p}^{2}}{\sigma} \tag{4.59}$$

Luo (1993) extended the above model for fluid particles of unequal sizes

$$t_{\text{drainage}} = \frac{1}{2}\frac{\rho^{c} u' r_{p,eq}^{2}}{\sigma} = \frac{2\rho^{c} u'}{\sigma}\left(\frac{r_{p,i} r_{p,j}}{r_{p,i} + r_{p,j}}\right)^{2} \tag{4.60}$$

From the above equation, the inertia thinning is proportional to the relative velocity between two fluid particles. This implies that when the drainage time is small, the coalescence efficiency is high when the relative velocity is low.

On the other hand, Lee et al. (1987) proposed that the thinning process can be taken to be inertially controlled for mobile interfaces according to Sagert and Quinn (1976):

$$t_{\text{thinning}} = \frac{R_{a}}{4}\left(\frac{\rho^{c} r_{p}}{\sigma}\right)^{1/2} \ln\left(\frac{h_{i}}{h_{f}}\right) \tag{4.61}$$

For the case of unequal sizes of fluid particles, the equivalent radius $r_{p,eq}$ is determined instead and, combined with the time for rupture from Eqn (4.53), yields the drainage time.

Oolman and Blanch (1986) derived an alternative expression for the liquid film drainage model which includes the Hamaker contribution at very low film thicknesses and changes in the concentration of surfactant species. It can be written as

$$\frac{dh}{dt} = \left\{\frac{8}{\rho^{c} R_{a}^{2}}\left[-\frac{4c}{RT}\left(\frac{d\sigma}{dc}\right)^{2} + h^{2}\left(\frac{2\sigma}{r_{p}} + \frac{H}{6\pi h^{3}}\right)\right]\right\}^{1/2} \tag{4.62}$$

where R is the gas constant, T is the temperature, and c is the concentration of surfactant species. In the work of Prince and Blanch (1990), assuming that there are no Hamaker contributions, no surface impurities, simplified interface geometry, and constant initial and critical film thicknesses, Eqn (4.62) can be integrated to

$$t_{\text{drainage}} = \left(\frac{\rho^{c} r_{p,eq}^{3}}{16\sigma}\right)^{1/2} \ln\left(\frac{h_{i}}{h_{f}}\right) \tag{4.63}$$

Compressing force and contact time

In the models which have been derived in the previous sections for film drainage, an interaction force F is required to be determined. This force is normally not constant. Appropriate relationships are thus needed to describe the complex nature of the film drainage process.

According to Coulaloglou and Tavlarides (1977) and Tsouris and Tavlarides (1994), the force may be assumed to be proportional to the mean square of the relative velocity at the ends of the eddy with a size of the equivalent diameter. Following Ross (1971),

$$F \approx \rho^{c}(\varepsilon^{c})^{2/3}(d_{p,i}+d_{p,j})^{2/3}\left(\frac{d_{p,i}d_{p,j}}{d_{p,i}+d_{p,j}}\right)^{2} \tag{4.64}$$

Chesters (1991) suggested that the collision force should be based on both viscous and inertial collisions in turbulent flows. For the viscous regime where the fluid particles are much smaller than the Kolmogorov scale, a typical force between two colliding fluid particles can be expected to be proportional to the turbulent shear rate. In other words,

$$F \approx 6\pi\mu^{c} r_{p}^{2}\sqrt{\frac{\varepsilon^{c}}{\nu^{c}}} \tag{4.65}$$

Nevertheless, the interaction force for inertial collisions that is exerted by one fluid particle onto an other should be greater than that calculated by Eqn (4.65). The force F is essentially the force exerted by the external flow:

$$F \approx \pi R_{a}^{2}\left(\frac{2\sigma}{r_{p}}\right) \tag{4.66}$$

As also demonstrated through Eqn (4.46), the calculation for the coalescence efficiency requires the interaction or contact time between fluid particles. Levich (1962) proposed a relationship based on dimensional analysis in a turbulent system, which can be written as

$$t_{\text{contact}} \approx \left(\frac{d_{p}^{2}}{\varepsilon^{c}}\right)^{1/3} \tag{4.67}$$

Nevertheless, Chesters (1991) argued that the collision force and its duration should be controlled by the bulk external flow. Using the analogy of the flow of solid particles in viscous shear, the contact time of fluid particles at viscous collisions in turbulent flow can be taken to be inversely proportional to the start rate of flow in the smallest eddies:

$$t_{\text{contact}} \approx \sqrt{\frac{\nu^{c}}{\varepsilon^{c}}} \tag{4.68}$$

During inertial collisions, Chesters (1991) considered a conversion process between kinetic energy and surface energy, and the actual contact time for the inertial system should be less than that inherent in Eqn (4.67) proposed by Levich (1962). Adopting an energy balance, the contact time is derived as

$$t_{\text{contact}} \approx \sqrt{\frac{\left(\frac{4\rho^{\text{p}}}{3\rho^{\text{c}}}+1\right)\rho^{\text{c}} r_{\text{p}}^3}{2\sigma}} \tag{4.69}$$

In contrast to the expressions proposed by Chesters (1991), Luo (1993) derived a more reasonable and fundamental expression for the contact time based on a simple parallel model. The contact time can be written for unequal sizes of fluid particles in the form of

$$t_{\text{contact}} \approx \left(1+\frac{d_{\text{p},i}}{d_{\text{p},j}}\right)\sqrt{\frac{\left(\frac{\rho^{\text{p}}}{\rho^{\text{c}}}+C_{\text{vm}}\right)\rho^{\text{c}} d_{\text{p},i}^3}{3\left(1+\frac{d_{\text{p},i}^2}{d_{\text{p},j}^2}\right)\left(1+\frac{d_{\text{p},i}^3}{d_{\text{p},j}^3}\right)\sigma}} \tag{4.70}$$

where C_{vm} represents the virtual mass coefficient, which is taken to be a constant between 0.5 and 0.8 (Jeelani and Hartland, 1991).

Kamp and Chesters (2001) considered the contact time of fluid particles as the interval between the onset of film formation and the moment at which the fluid particles begin to rebound. Assuming a balance between the increasing surface free energy and the corresponding reduction of the kinetic energy in the system, the contact time is given by

$$t_{\text{contact}} \approx \sqrt{\frac{\rho^{\text{c}} C_{\text{vm}}}{3\sigma}\left(\frac{2d_{\text{p},i}d_{\text{p},j}}{d_{\text{p},i}+d_{\text{p},j}}\right)^3} \tag{4.71}$$

4.2.5 Coalescence Efficiency Due to Energy Model

Aside from the popular film drainage model, the development of other models to determine the coalescence efficiency for fluid particles has also been realized. One which will be described herein is the energy model. Based on the original proposal by Howarth (1964, 1967) and later confirmed by optical records of coalescence in a dispersion by Park and Blair (1975) and Kuboi et al. (1972), which demonstrated that a significant fraction of collisions result in immediate coalescence and the probability increases with increasing energy of collision, the coalescence efficiency of fluid particles could be derived based on such considerations. Sovova (1981) developed a model to represent the energetic collisions according to

$$\lambda^{\text{energy}}(v_i, v_j) = \exp\left(-C_9\frac{E_{\text{int}}}{E_{\text{kin}}}\right) \tag{4.72}$$

where E_{kin} is the kinetic collision energy and E_{int} is the interfacial energy. Constant C_9 in Eqn (4.72) can be specified between 0.00068 and 0.0151 (Sovova, 1981). The kinetic energy is assumed to be proportional to the equivalent volume $v_{\text{p,eq}}(=v_i v_j/v_i + v_j)$ and the relative velocity of two colliding fluid particles:

$$E_{\text{kin}} = \frac{1}{2}\rho^{\text{p}} v_{\text{p,eq}} u'^2 \tag{4.73}$$

Noting the relative velocity that is given by Eqn (4.6), Eqn (4.73) becomes

$$\begin{aligned} E_{\text{kin}} &\propto \rho^{\text{p}} v_{\text{p,eq}} (\varepsilon^{\text{c}})^{2/3} \left(d_{\text{p},i}^{2/3} + d_{\text{p},j}^{2/3}\right) \Rightarrow \\ E_{\text{kin}} &\propto \rho^{\text{p}} (\varepsilon^{\text{c}})^{2/3} v_{\text{p,eq}} \left(v_{\text{p},i}^{2/9} + v_{\text{p},j}^{2/9}\right) \end{aligned} \tag{4.74}$$

The interfacial energy can be taken to be proportional to the interfacial tension and surface area:

$$E_{\text{int}} = \sigma \left(v_{\text{p},i}^{2/3} + v_{\text{p},j}^{2/3}\right) \tag{4.75}$$

Based on Eqns (4.74) and (4.75), the coalescence efficiency is subsequently evaluated.

Simon (2004) also proposed a similar expression for the kinetic energy. Considering the momentum balance during the collision, the kinetic energy can be determined as

$$E_{\text{kin}} \propto \rho^{\text{p}} (\varepsilon^{\text{c}})^{2/3} \left(v_{\text{p},i}^{11/9} + v_{\text{p},j}^{11/9}\right) \tag{4.76}$$

Using Eqns (4.75) and (4.76), the coalescence efficiency is henceforth determined.

Sovova (1981) also proposed combining the energy model with the film drainage model to determine the combined effect of the coalescence efficiency. This approach was adopted by Chatzi et al. (1989) and Lafi and Reyes (1994). The overall coalescence efficiency can be evaluated as

$$\lambda(v_i, v_j) = \lambda^{\text{energy}}(v_i, v_j) + \lambda^{\text{film}}(v_i, v_j) - \lambda^{\text{energy}}(v_i, v_j)\lambda^{\text{film}}(v_i, v_j) \tag{4.77}$$

where λ^{film} is the coalescence efficiency due to the film drainage model determined through Eqn (4.46).

4.2.6 Coalescence Efficiency Due to Critical Approach Velocity Model

Another model to determine the coalescence is the consideration of the critical approach velocity. In the energy model, it has been assumed that coalescence will occur immediately when the approach velocity exceeds a critical level at the instant of collision. Nevertheless, experimental observations by Doubliez (1991) and Duineveld (1994) depicted that coalescence will occur in a more gentle manner. Lehr and Mewes (1999) and Lehr et al. (2002) proposed a simple relationship for the coalescence efficiency considering the approach velocity, of which the correct form should be

$$\lambda(v_i, v_j) = \min\left(\frac{u_{\text{critical}}}{u_{\text{approach}}}, 1\right) \tag{4.78}$$

In the above equation, the critical velocity u_{critical} can be determined according to

$$u_{\text{critical}} = \sqrt{\frac{\text{We}_{\text{critical}}\sigma}{\rho^c d_{\text{p,eq}}}} \tag{4.79}$$

where $\text{We}_{\text{critical}}$ is the critical Weber number and $d_{\text{p,eq}}$ is the equivalent diameter $(=2d_{\text{p},i}d_{\text{p},j}/d_{\text{p},i}+d_{\text{p},j})$. The approach velocity u_{approach} is evaluated by taking the maximum of the relative velocity and the slip velocity of the two colliding fluid particles.
In other words,

$$u_{\text{approach}} = \max\left(u', \left|\mathbf{U}_{\text{p},i} - \mathbf{U}_{\text{p},j}\right|\right) \tag{4.80}$$

4.3 Mechanisms and Kernels of Fluid Particle Break up

Similar to the coalescence of fluid particles in turbulent gas–liquid or liquid–liquid flows, the break up of fluid particles in turbulent dispersions is significantly influenced by the continuous phase hydrodynamics and interfacial interactions. Physically, the break up mechanism can be viewed as a balance between external stresses from the continuous phase that attempt to obliterate the fluid particle and the surface stresses of the particle, in addition to the viscous stresses of the fluid inside the particle that restore its form. Therefore, the break up of a fluid particle is determined by the hydrodynamic conditions in the surrounding liquid and the characteristics of the particle itself. In accordance with the fluid particle coalescence, a practical consideration of the birth and death rates due to fluid particle break up is also adopted. Appropriate source and sink terms are henceforth developed directly on the averaging scales, performing analyses of fluid particle break up.

The macroscopic population balance formulation of the discrete birth and death rates due to fluid particle break up are outlined. First, the birth of particles of volume v_i due to break up results from the break up of all fluid particles smaller than v_i. The birth rate for particles of volume v_i can be obtained by use of

$$S_{B,i}^{+}(\mathbf{r},\mathbf{Y},t) = \int_{v_i}^{v_{i+1}} dv \left[\sum_{j=0}^{M} \int_{v_j}^{v_{j+1}} v\left(V_p',\mathbf{r},\mathbf{Y}\right) \underbrace{b\left(V_p',\mathbf{r},\mathbf{Y}\right)}_{\text{break up frequency}} \times \underbrace{P\left(V_p,\mathbf{r}\middle|V_p',\mathbf{r},\mathbf{Y},t\right)}_{\text{daughter size distribution function}} f_1\left(V_p',\mathbf{r},t\right) dv' \right] \tag{4.81}$$

Second, the death of volume v_i due to break up stems from the break up of the particles within this class. In other words,

$$S_{B,i}^{-}(\mathbf{r},\mathbf{Y},t) = \int_{v_i}^{v_{i+1}} dv \left[\underbrace{b(V_p,\mathbf{r},\mathbf{Y})}_{\text{break up frequency}} f_1\left(V_p,\mathbf{r},t\right) \right] \tag{4.82}$$

For break up between particles, general expressions for the break up frequency and daughter size distribution function in gas–liquid or liquid–liquid flows can normally be expressed in terms of volumes v_i and v_j as $b(v_i)$ and $P(v_j, v_i)$ or $\beta(f_{BV}, 1)$, respectively.

A variety of mechanisms for the fluid particle break up was identified by Liao and Lucas (2009). They can be classified into four main categories of break up: due to

1. Turbulent fluctuation and collision or turbulent shearing
2. Viscous shear force
3. Interfacial instability
4. Shearing-off process.

Various physical models for the above four categories of particle break up are further elaborated upon in the following sections.

4.3.1 Break up Due to Turbulent Shearing

In turbulent multiphase flow, the break up of fluid particles is primarily caused by the presence of turbulent pressure fluctuations along the surface or by the collisions between turbulent eddy and fluid particle. The shape of the fluid particle may be modified from its original spherical form with the fluctuation of the surrounding fluid or due to eddy collision with the particle. When the amplitude of oscillation is sufficient to cause the surface of the fluid particle to be unstable, it will begin to deform and stretch leading to a necking that subsequently contracts and fragments into two or more daughter fluid particles. From the perspective of considering the force balance on the fluid particle, the external force that is initiating the oscillation is the dynamic pressure difference around the fluid particle and the break up mechanism can be expressed as a balance between the dynamic pressure and surface stress. This balance leads to the prediction of a critical Weber number, above which the fluid particle is no longer stable:

$$\mathrm{We}_{\mathrm{cr}} = \frac{\tau_i}{\tau_{\mathrm{s}}} \tag{4.83}$$

where τ_i is the dynamic pressure and τ_{s} is the surface stress.

A number of models exist for the break up due to turbulent shearing for fluid particles. They can be categorized into five scenarios:

1. Turbulent kinetic energy of the particle greater than a critical value
2. Velocity fluctuation around the particle surface greater than a critical value
3. Turbulent kinetic energy of the hitting eddy greater than a critical value
4. Inertial force hitting the eddy greater than the interfacial force of the smallest daughter particle
5. Combination of criteria (3) and (4).

Turbulent kinetic energy of the particle greater than a critical value

With regard to the specific development of this phenomenological model, Coulaloglou and Tavlarides (1977) considered the break up of a drop in turbulent liquid–liquid dispersion. The consideration centers on the drop oscillating and deforming due to local pressure fluctuations. Here, an oscillating deformed drop will break if kinetic energy transmitted from drop–eddy collisions is greater than its surface energy. The break up rate is defined as

$$b(\nu_i) = \left(\frac{1}{\text{break up time}}\right)\left(\frac{\text{fraction of}}{\text{drops breaking}}\right) \tag{4.84}$$

In Eqn (4.84), the fraction of drops breaking up is assumed to be proportional to the fraction of drops that have a turbulent kinetic energy greater than the surface tension. Consider a given droplet with surface energy

$$E_{\mathrm{c}} \approx \sigma d_{\mathrm{p},i}^2 \tag{4.85}$$

The fraction of eddies with a kinetic energy greater than E_{c} will be equivalent to the number fraction of eddies that have velocities greater than a corresponding fluctuating velocity. This fraction of eddies can be represented by

$$\left(\frac{\text{fraction of}}{\text{eddies breaking}}\right) = \left(\frac{\text{fraction of}}{\text{drops breaking}}\right) = \exp\left(-\frac{E_c}{\overline{E}}\right) \tag{4.86}$$

Equation (4.86) assumes that the kinetic energy distribution of the drops is the same as that of the eddies. Since only energies associated with velocity fluctuations of a scale smaller than $d_{p,i}$ will tend to break up a drop of size $d_{p,i}$ (larger eddies will tend to only carry the droplet without breaking it), the mean turbulent kinetic energy is given by

$$\overline{E} \approx \rho^{p} d_{p,i}^{3} u_{t,i}^{2} \tag{4.87}$$

Note that the square of the turbulent velocity is given by Eqn (4.8) in the inertial subrange.

The break up time is determined by assuming that the motion of mass centers of the daughter droplets to be formed via binary breakage is similar to the relative motion of two lumps of fluid in a turbulent flow field as described by Batchelor (1956). At a time equivalent to the break up time, the separation is equal to that for droplet break up. In other words,

$$\left(\frac{1}{\text{break up time}}\right) \approx \frac{(\varepsilon^{c})^{1/3}}{d_{p,i}^{2/3}} \tag{4.88}$$

In mathematical form, the break up rate can thus be expressed as

$$b(v_i) = C_{10}\frac{(\varepsilon^{c})^{1/3}}{d_{p,i}^{2/3}} \exp\left(-C_{11}\frac{\sigma}{\rho^{p}(\varepsilon^{c})^{2/3} d_{p,i}^{5/3}}\right) \tag{4.89}$$

To account for the "damping" effect of droplets on the local turbulent intensities at high volume fraction, Eqn (4.89) can be modified according to

$$b(v_i) = C_{12}\frac{(\varepsilon^{c})^{1/3}}{d_{p,i}^{2/3}} \frac{1}{1+\alpha^{p}} \exp\left(-C_{13}\frac{\sigma(1+\alpha^{p})^{2}}{\rho^{g}(\varepsilon^{c})^{2/3} d_{p,i}^{5/3}}\right) \tag{4.90}$$

Constants C_{12} and C_{13} in Eqn (4.90) can be allotted values of 0.00,481 and 0.08, respectively (Liao and Lucas, 2009).

Nevertheless, Prince and Blanch (1990) pointed out that the model of Coulaloglou and Tavlarides (1977) predicted a break up rate several orders of magnitude lower than experimental results for gas–liquid mixtures. This may be due to the consideration of the density of the dispersed phase in the expression. By considering the density to apply to the continuous phase, at least for gas–liquid flows, the break up criterion is of a similar order to break up due to the turbulent kinetic energy of the hitting eddy being greater than a critical value, a scenario which will be described later.

Velocity fluctuation around the particle surface greater than a critical value

In this theoretical model, Narsimhan et al. (1979) argued that the oscillation and breakage of a drop is a result of the difference in velocity fluctuations between points near the drop surface, which is due to the arrival of turbulent eddies. Whether or not such an event occurs, it is governed by the relative magnitude of the time scales of the eddy arrival process and oscillations of the droplet. Since oscillations of a droplet can be caused only by eddies of a scale smaller than the size of the droplet, very small eddies are thereby sufficient to set the smaller droplets into oscillations. The time scale of oscillations is taken to be inversely proportional to the eddy frequency. Smaller eddies may be expected to create the high frequency oscillations. Thus, the postulation assumes that the time scale of oscillation is smaller than the time scale of the arrival of an eddy. So, once an eddy of sufficiently high energy arrives, breakage of the droplet is induced.

By describing the arrival of eddies via a Poisson process, the probability distribution of the relative velocity between two points separated by some distance is assumed to be normal, which is given by

$$PD(x) = \frac{1}{\sqrt{2\pi\sigma}} \exp\left(\frac{x^2}{2\sigma^2}\right) \tag{4.91}$$

In Eqn (4.91), σ^2 is equivalent to the square of the turbulent velocity which is given by Eqn (4.8) in the inertial subrange. It can be shown that the increase in the surface energy required for break up is minimal if binary breakage occurs. Hence, the minimum increase in the surface energy can be expressed as $(2^{1/3} - 1)\sigma\pi^{1/3}6^{2/3}\nu^{2/3}$, where ν is the drop volume. This energy is realized by the kinetic energy of oscillation of the droplet. Therefore, the critical velocity for a particle of volume $V_{p,i}$ is determined through the energy balance during equal breakage with the assumption that the increase in the surface energy at the case is the minimum:

$$\begin{aligned} \frac{1}{2}(\rho^c\nu_i)u^2_{\text{critical}} &= (2^{1/3} - 1)\sigma\pi^{1/3}6^{2/3}\nu_i^{2/3} \Rightarrow \\ u^2_{\text{critical}} &= 2(2^{1/3} - 1)\frac{\sigma}{\rho^c}\pi^{1/3}6^{2/3}\nu_i^{-1/3} \end{aligned} \tag{4.92}$$

Defining the variable λ to denote the average number of eddies arriving on the surface of the particle, and taking the collision frequency as a constant, multiplying this variable with the probability distribution given in Eqn (4.91) yields the break up rate:

$$\begin{aligned} b(\nu_i) &= \lambda \,\mathrm{erfc}\left(\frac{u_{\text{critical}}}{\sqrt{2}\sigma}\right) \Rightarrow \\ b(\nu_i) &= \lambda \,\mathrm{erfc}\left(\frac{\sqrt{a}\left(\frac{\pi}{6}\right)^{-1/6} d_{p,i}^{-5/6}}{2\varepsilon^c}\right) \end{aligned} \tag{4.93}$$

where $a = 2(2^{1/3} - 1)\pi^{1/3}6^{2/3}\sigma/\rho^c$ and the volume of the particle has been assumed to be that of a spher. According to this model, the probability of equal binary breakage is highest since the required energy considered herein is taken to be the minimum. It should nonetheless be noted that this is inconsistent with experimental observations of Hesketh et al. (1987), which demonstrated that the binary equal-sized breakage has the lowest probability.

Turbulent kinetic energy of the hitting eddy greater than a critical value

The models, as described in previous two sections, have considered that collisions between eddies and drops were the dominant mechanism for break up. By making an analogy with collisions in ideal gases, Prince and Blanch (1990) examined the break up rate through the interaction of bubbles with turbulent eddies. In their model, the break up rate could be derived as the product of collision frequency and break up efficiency. In other words,

$$b(v_i) = w(d_{p,i}, \gamma)P_b(d_{p,i}, \gamma) \tag{4.94}$$

where γ is the eddy length scale. Similar to the coalescence frequency, the collision frequency $w(d_{p,i},\gamma)$ is defined as the effective swept volume which is equal to the product of the collision-sectional area, the relative velocity between bubbles and the number of eddies with size γ:

$$w(d_{p,i}, \gamma) = \int_{n_\gamma} \frac{\pi}{4}(d_{p,i} + \gamma)^2\left(u_{t,i}^2 + u_{t,\gamma}^2\right)^{1/2} dn_\gamma \tag{4.95}$$

In Eqn (4.95), the turbulent velocity for the eddy could be analogously determined by

$$u_{t,\gamma}^2 = C(\varepsilon^c\gamma)^{2/3} \tag{4.96}$$

while the number of eddies n_γ is evaluated by the following relationship derived by Azbel and Athanasios (1983):

$$\begin{gathered} \frac{dn_\gamma(k)}{dk} = \frac{0.1k^2}{\rho^c} \Rightarrow \\ n_\gamma(k) = \int_{k_{min}}^{k_{max}} \frac{0.1k^2}{\rho^c} dk \end{gathered} \tag{4.97}$$

where k is the eddy wave number of eddies, which is related to the eddy size by $2/\gamma$. It should be noted that the upper limit of the integration becomes infinitely large as the eddy size approaches zero. Considering the inertial subrange, a minimum eddy size is prescribed as being of some arbitrary size below which an eddy will not cause any breakage. Prince and Blanch (1990) set the limit to 20% of the particle size. Nevertheless, Lasheras et al. (2002) demonstrated that the upper limit can be rather sensitive for certain flow conditions.

The break up efficiency $P_b(d_{p,i},\gamma)$ is assumed to be equal to the probability that turbulent eddies have sufficient energy to rupture the bubble. Taking the eddy energy to be proportional to the square of its velocity results in a function of the following form for the fraction of eddies for break up efficiency:

$$P_b(d_{p,i},\gamma)=\exp\left(-\frac{u_{c,i}^2}{u_{t,i}^2}\right) \tag{4.98}$$

The square of critical eddy velocity $u_{c,i}^2$ in Eqn (4.98) for break up may be obtained through the critical Weber number, which may be written as

$$u_{c,i}^2=\frac{\sigma \mathrm{We}_{cr}}{d_{p,i}\rho^c} \tag{4.99}$$

Recalling Eqn (4.7) which is the turbulent velocity for the particle of size $d_{p,i}$ and along with Eqns (4.95)–(4.99), the integral in Eqn (4.95) can be determined subsequently for the break up rate.

Tsouris and Tavlarides (1994) also considered the break up rate according to Eqn (4.95). In modeling the eddy–drop collision frequency, they assumed that the turbulence is isotropic, the drop size is in the inertial subrange, and drops can break up only by collisions with smaller eddies or those of the same size. All of these assumptions are similar to those considered in the model developed by Prince and Blanch (1990). The most important difference between the two models is the consideration of the critical energy in the break up efficiency. Rather than the surface energy of the parent drop, Tsouris and Tavlarides (1994) determined the critical energy as the mean value of the energy required for the break up into two equal-sized droplets and a small and big daughter droplet. According to Narsimhan et al. (1979), the critical energy is given by

$$E_{c,i}(d_{p,i})=\pi\left(\frac{d_{p,i}}{2^{1/3}}\right)^2+\frac{\pi\sigma d_{p,\max}^2}{2}+\frac{\pi\sigma d_{p,\min}^2}{2}-\pi\sigma d_{p,i}^2 \tag{4.100}$$

The break up efficiency which yields the probability of an eddy–drop collision resulting in drop break up is then assumed to be given by

$$P_b(d_{p,i},\gamma)=\exp\left(-\frac{E_{c,i}(d_{p,i})}{C_{14}\overline{E}(\gamma)}\right) \tag{4.101}$$

where $\overline{E}(\gamma)$ is the mean kinetic energy for an eddy, which can be calculated as

$$\overline{E}(\gamma)=\frac{1}{2}m_\gamma u_{t,\gamma}^2=\frac{1}{2}\rho^c V_\gamma u_{t,\gamma}^2=\frac{\pi}{12}\rho^c\gamma^3 u_{t,\gamma}^2 \tag{4.102}$$

In Eqn (4.102), the volume of the eddy has been determined through the consideration of a spherical fluid particle. The constant C_{14} is a constant of order unity while the turbulence velocity for the eddy is given in Eqn (4.96). The consideration of the collision frequency is no different from the form in Eqn (4.95). An additional minor modification is the inclusion of a turbulence damping factor into the break up rate which can be estimated as

$$DF(\alpha) = \left[1 + 2.5\alpha\left(\frac{\mu^{\mathrm{p}} + 0.4\mu^{\mathrm{c}}}{\mu^{\mathrm{p}} + \mu^{\mathrm{c}}}\right)\right]^2 \tag{4.103}$$

This model experienced the same difficulty as the model proposed by Prince and Blanch (1990). As suggested by Tsouris and Tavlarides (1994), the smallest and biggest effective eddies are arbitrarily set to one-half of the size of the critical drop size and drop diameter, respectively. The integration between the limits of the eddy wave number is again sensitive to certain flow conditions (Lasheras et al., 2002).

Luo and Svendsen (1996) proposed an alternative theoretical model for the break up rate based on the consideration of kinetic gas theory. The collision frequency of eddies of sizes between γ and $\gamma + \mathrm{d}\gamma$ with particles of size $d_{\mathrm{p},i}$ can be expressed by

$$w(d_{\mathrm{p},i}, \gamma) = \int_{\gamma} \frac{\pi}{4}(d_{\mathrm{p},i} + \gamma)^2 u_{\mathrm{t},\gamma} \dot{n}_\gamma \mathrm{d}\gamma \tag{4.104}$$

According to Tennekes and Lumley (1972), a relationship relating the number density rate of eddies $\dot{n}_\gamma$ and the energy spectrum $E(k)$ can be obtained as follows:

$$\frac{1}{2} m_\gamma u_{\mathrm{t},\gamma}^2 \dot{n}_\gamma \mathrm{d}\gamma = \frac{1}{2}\rho^{\mathrm{c}} \frac{\pi}{6}\gamma^3 u_{\mathrm{t},\gamma}^2 \dot{n}_\gamma \mathrm{d}\gamma = E(k)\rho^{\mathrm{c}}(1-\alpha)(-\mathrm{d}k) \tag{4.105}$$

The functional form of the energy spectrum is well described in the inertial subrange (Tennekes and Lumley, 1972):

$$E(k) = C_{15}(\varepsilon^{\mathrm{c}})^{2/3} k^{-5/3} \tag{4.106}$$

where C_{15} is considered to be a universal constant equal to a value of 1.5 as given by Batchelor (1956) based on turbulence theory. Given that the relationship between the wave number and the size of an eddy is $2/\gamma$, the number density rate of eddies is

$$\dot{n}_\gamma = C_{16}\frac{(1-\alpha)}{\gamma^4} \tag{4.107}$$

where $C_{16} \approx 0.8$. The collision frequency can thus be written as

$$w(d_{\mathrm{p},i}, \gamma) = C_{17}(1-\alpha)\left(\frac{\varepsilon^{\mathrm{c}}}{d_{\mathrm{p},i}^2}\right)^{1/3} \int_{\xi}^{1} \frac{(1+\xi)^2}{\xi^{11/3}} \mathrm{d}\xi \tag{4.108}$$

where $C_{17} = 0.923$ and $\xi = \gamma/d_{\mathrm{p},i}$ is the size ratio between an eddy and a particle.

For the determination of the break up efficiency, the consideration of a suitable expression for probability for particle break up, such as that considered by Luo and Svendsen (1996), is similar to the models proposed by Prince and Blanch (1990) and Tsouris and Tavlarides (1994). Assuming a natural exponential function for the kinetic energy distribution of eddies in turbulence, the conditional break up efficiency can be expressed as

$$P_{\mathrm{b}}\left(d_{\mathrm{p},i}, d_{\mathrm{p},j}, \gamma\right) = 1 - \int_{0}^{\chi_{\mathrm{c}}} \exp(-\chi)d\chi = \exp(-\chi_{\mathrm{c}}) \tag{4.109}$$

where χ_c represents the critical dimensionless energy for break up:

$$\chi_{\mathrm{c}} = \frac{E_{\mathrm{c},i}\left(d_{\mathrm{p},i}\right)}{\overline{E}(\gamma)} \tag{4.110}$$

In Eqn (4.110), the critical energy in this model during break up involves the consideration of when a particle with size $d_{\mathrm{p},i}$ breaks up into two particles with a given break up volume fraction $f_{\mathrm{BV}}(=v_i/v_j)$, which is related to the increase coefficient of surface area:

$$c_{\mathrm{f}} = f_{\mathrm{BV}}^{2/3} + (1 - f_{\mathrm{BV}})^{2/3} - 1 \tag{4.111}$$

The time scale of particle oscillations is assumed to be smaller than the time scale associated with eddy collision. This implies that once an eddy of sufficiently high energy arrives, particle break up occurs; the kinetic energy of collision exceeding the increase of surface energy for break up is given by

$$E_{\mathrm{c},i}\left(d_{\mathrm{p},i}\right) = c_{\mathrm{f}}\pi d_{\mathrm{p},i}^{2}\sigma \tag{4.112}$$

The denominator in Eqn (4.110) is simply the mean kinetic energy for an eddy, which is given by Eqn (4.102). In this model, the determination of the upper and lower limits indirectly includes two unknowns. Owing to the specific consideration in the inertial subrange of isotropic turbulence, it is common practice to replace the lower limit by the minimum size of eddies. This approximation is reasonably acceptable since very small eddies have very low energy levels and very short life spans thereby exerting a negligible effect on the break up of particles. This model also requires the specification of a maximum particle size which then gives the upper limit of the integration.

Martinez-Bazan et al. (1999a) considered a phenomenological model for break up which was derived from the extension of the classical theory of gases. This model assumes that turbulence is fully developed in the scales of interest and local isotropy can thus be applied to describe the underlying turbulence where break up of a bubble can take place. For the bubble to break, the model focuses on the deformation of the bubble and the energy required to deform is provided by the turbulent stresses produced by the surrounding fluid. The surface restoring pressure is given by

$$\tau_s\left(d_{p,i}\right) = 6\frac{\sigma}{d_{p,i}} \tag{4.113}$$

while the average deformation force produced by the turbulent stresses from the velocity fluctuations existing in the liquid is

$$\tau_t\left(d_{p,i}\right) = \frac{1}{2}\rho^c u_{t,i}^2 = \frac{1}{2}\rho^c C(\varepsilon^c)^{2/3} d_{p,i}^{2/3} \tag{4.114}$$

When $\tau_t(d_{p,i}) > \tau_s(d_{p,i})$, the bubble will deform and eventually break. By equating the surface energy with the deformation energy, the minimum size of the particle is given by

$$\begin{aligned} \frac{1}{2}\rho^c C(\varepsilon^c)^{2/3} d_{p,\min}^{2/3} &= 6\frac{\sigma}{d_{p,i}} \Rightarrow \\ d_{p,\min} &= \left(\frac{12\sigma}{C\rho^c d_{p,i}}\right)^{3/2} (\varepsilon^c)^{-1} \end{aligned} \tag{4.115}$$

Martinez-Bazan et al. (1999a) postulated that the larger the difference between the gradient of pressure produced by turbulent fluctuations on the bubble surface and restoring pressure by surface tension, the larger the probability that the bubble will break in a certain time. In other words, the break up rate should decrease to zero as this difference vanishes. By estimating the break up time according to

$$t_b\left(d_{p,i}\right) \propto \frac{d_{p,i}}{\sqrt{u_{t,i}^2 - 12\sigma/\left(\rho^c d_{p,i}\right)}} \tag{4.116}$$

where the denominator denotes the characteristic velocity of the bubble during the break up process, the break up rate is then given by

$$b(v_i) = \frac{1}{t_b\left(d_{p,i}\right)} = C_{18}\frac{\sqrt{u_{t,i}^2 - 12\sigma/\left(\rho^c d_{p,i}\right)}}{d_{p,i}} \tag{4.117}$$

where C_{18} takes on a value of 0.25, which has been experimentally determined (Martinez-Bazan et al., 1999a). This particular model is similar to all of the aforementioned models because it assumes that a bubble or droplet will break up if the turbulent kinetic energy in the continuous phase is larger than a critical value. It should be noted that the model is restricted to homogeneous and isotropic turbulent flows.

Inertial force of the hitting turbulent eddy greater than the interfacial force of the smallest daughter particle

Lehr and Mewes (2001) proposed a model based on a force balance between the inertial force of the arriving eddy and the interfacial force of the smallest daughter particle. This model determines the break up rate in a similar fashion as in the previous section, through the product between the arrival frequency and the corresponding break up efficiency of the eddy. The collision frequency of eddies of sizes between γ and $\gamma + d_{p,i}$ with particles of size $d_{p,i}$ can be described using the same expression as in Eqn (4.104). For the break up efficiency, it is assumed to be dependent on the angle φ at which the eddy collides with the bubble. By taking the force balance between the inertial and interfacial forces which is given by

$$\cos(\varphi)\frac{1}{2}\rho^c u_\gamma^2 = \frac{\sigma}{d_{p,j}} \tag{4.118}$$

and assuming that the relative probability is equal for all steradians, the break up efficiency can be expressed by

$$P_b\left(d_{p,i}, d_{p,j}, \gamma\right) \approx \frac{\sigma}{\rho^c (\varepsilon^c \gamma)^{2/3} d_{p,j}^4} \tag{4.119}$$

The break up rate can thus be written as

$$b(v_i, v_j) = w\left(d_{p,i}, \gamma\right) P_b\left(d_{p,i}, d_{p,j}, \gamma\right) \Rightarrow$$

$$b(v_i, v_j) = C_{19} \int_{\gamma_{min}}^{\gamma_{max}} \frac{\sigma(1-\alpha^p)}{(\varepsilon^c)^{1/3}\rho^c d_{p,j}^4} \frac{\left(d_{p,i} + \gamma\right)^2}{\gamma^{12/3}} d\gamma \tag{4.120}$$

where C_{19} has a constant value of 3.55 (Lehr and Mewes, 2001). Hence, the total break up frequency of fluid particles with size $d_{p,i}$ is given as

$$b(v_i) = \int_0^{0.5} \Omega\left(v_i, v_j\right) df_{BV} \tag{4.121}$$

In Eqn (4.120), the upper limit of integration is taken to be equal to the bubble diameter that includes break up:

$$\gamma_{max} = d_{p,i} \tag{4.122}$$

For the lower limit of integration, by satisfying the force balance in Eqn (4.118), it is set by

$$\gamma_{min} = \max\left(d_{p,j}, \frac{\sigma^{3/2}}{(\rho^c)^{3/2}\varepsilon^c d_{p,j}^{3/2}}\right) \tag{4.123}$$

In Lehr et al. (2002), the break up efficiency is calculated based on the criterion that the kinetic energy of the eddy exceeds a critical energy which is obtained from the force balance equation. By assuming a normal distribution of the mean turbulent velocity of eddies with length scale γ, the break up efficiency is

$$P_{\mathrm{b}}\left(d_{\mathrm{p},i}, d_{\mathrm{p},j}, \gamma\right) \approx \frac{\sigma}{\rho^{\mathrm{c}}\left(\varepsilon^{\mathrm{c}}\gamma\right)^{2/3} d_{\mathrm{p},j}^{4}} \exp\left(-\frac{2\sigma}{\rho^{\mathrm{c}}\left(\varepsilon^{\mathrm{c}}\right)^{2/3}\gamma^{2/3} d_{\mathrm{p},j}}\right) \tag{4.124}$$

Employing the above break up efficiency in place of Eqn (4.120) yields an improved model for the break up rate.

Combination of different criteria for break up efficiency

Wang et al. (2003) proposed an alternative model for the break up rate in order to overcome a number of shortcomings that exist in models such as those proposed by Luo and Svendsen (1996), Lehr and Mewes (2001), and Lehr et al. (2002). In the model of Luo and Svendsen (1996), only the energy constant is considered during the break up process, which implies that the kinetic energy of an eddy is considered to be larger than the increase of surface energy due to break up of a fluid particle. For break up with a small fraction, the capillary pressure is usually very high when the radius of curvature tends to zero. Therefore, an eddy with a larger kinetic energy may not be able to produce sufficient dynamic pressure to overcome the capillary pressure. Also, when an eddy arrives with energy greater than or equal to the minimum energy required for break up, it can cause all fluid particles to break up with a break up fraction smaller than the given break up volume fraction. In the model of Lehr and Mewes (2001) and Lehr et al. (2002), the inertial force of the colliding eddy during break up is usually larger than the interfacial force, and the deformation of the fluid particle is strengthened until break up occurs. This means that the force balance may not be satisfied during the break up process. In order to circumvent these shortcomings, both the energy constraint and the capillary pressure constraint are adopted in order to develop a new model for the break up rate.

Two different break up efficiencies are developed. The energy distribution of eddies with size γ according to Angelidou et al. (1979) is given by the following density function:

$$P_{\mathrm{e}}(E(\gamma)) = \frac{1}{\overline{E}(\gamma)} \exp\left(-\frac{E(\gamma)}{\overline{E}(\gamma)}\right) \tag{4.125}$$

where $\overline{E}(\gamma)$ is the mean kinetic energy already given in Eqn (4.102). When break up occurs, the dynamic pressure should satisfy the following equation:

$$\frac{1}{2}\rho^{\mathrm{c}} u_{\gamma}^{2} \geq \frac{\sigma}{d_{\mathrm{p},j}} \tag{4.126}$$

The left-hand side of Eqn (4.126) represents the capillary pressure where the minimum radius of curvature has been set equal to the size of the daughter fluid particle for simplification. Since $d_{p,j} \geq \sigma V_\gamma / E(\gamma)$, the minimum break up fraction can be determined as

$$f_{BV,min} = \left(\frac{d_{p,j}}{d_{p,i}}\right)^3 = \left(\frac{\pi \gamma^3 \sigma}{6E(\gamma) d_{p,i}}\right)^3 \tag{4.127}$$

Also, the eddy energy $E(\gamma)$ should be larger than or equal to the increase of surface energy for break up:

$$E(\gamma) \geq c_f \pi d_{p,i}^2 \sigma \tag{4.128}$$

$$c_{f,max} \geq \min\left(2^{1/3} - 1, \frac{E(\gamma)}{\pi d_{p,i}^2 \sigma}\right) \tag{4.129}$$

Based on Eqn (4.129), the maximum break up fraction $f_{BV,max}$ is determined. When a fluid particle of size $d_{p,i}$ is bombarded by an eddy with size γ and kinetic energy $E(\gamma)$, the possible break up fraction lies between $f_{BV,min}$ and $f_{BV,max}$. The density function is thus determined by

$$P_b\left(d_{p,i}, d_{p,j}, E(\gamma), \gamma\right) = \frac{1}{f_{Bv,max} - f_{Bv,min}} \tag{4.130}$$

Finally, the break up efficiency is then calculated by using

$$P_e\left(d_{p,i}, d_{p,j}, \gamma\right) = \int_0^\infty \frac{1}{f_{Bv,max} - f_{Bv,min}} \frac{1}{E(\gamma)} \exp\left(-\frac{E(\gamma)}{\bar{E}(\gamma)}\right) dE(\gamma) \tag{4.131}$$

Similarly to Wang et al. (2003), Zhao and Ge (2007) also considered both the energy constraint and the force balance constraint in their model development. However, they introduce a so-called eddy efficiency, which is related to the ratio between the half period of bubble oscillation and the eddy lifetime. The eddy efficiency can be defined as

$$C_{eddy} = \min\left(\frac{(\varepsilon^c)^{1/3}}{2f(n)\gamma^{2/3}}, 1\right) \tag{4.132}$$

where $f(n)$ is the fluctuation frequency of the bubble surface in a developed turbulent flow field given by

$$f(n) = \left[\left(\frac{2\sigma}{\pi^2 d_{p,i}^3}\right)\left(\frac{(n+2)(n+1)n(n-1)}{(n+1)\rho^p + n\rho^c}\right)\right]^{1/2} \tag{4.133}$$

in which the $n=2$ mode has been chosen to describe a shape oscillation beginning from a spherical shape and passing through oblate spherical shapes and prolate spherical shapes, which closely resemble the shapes occurring during actual bubble oscillation. This eddy efficiency can also be viewed as the portion of energy that can be extracted from the eddy and converted to the surface energy during bubble deformation. As the kinetic energy of the hitting eddy exceeds the increase of surface required for break up, it can be shown that

$$E(\Upsilon) \geq \frac{E_{c,i}(d_{p,i})}{C_{eddy}} \tag{4.134}$$

Moreover,

$$\frac{1}{2}\rho^c u_\Upsilon^2 \geq \frac{\sigma}{d_{p,j}} \Rightarrow \frac{1}{2}\rho^c u_\Upsilon^2 \geq \frac{\sigma}{d_{p,i}\min(f_{BV}, 1-f_{BV})^{1/3}} \tag{4.135}$$

Consequently, the break up efficiency of the bubble when bombarded by the eddy should be equal to the break up efficiency of the eddy where a kinetic energy has no less than the minimum energy required to break the bubble. In other words,

$$P_b(d_{p,i}, d_{p,j}=\Upsilon) = P_e(E(\Upsilon) \geq E_{cr}(d_{p,i}\Upsilon)) = \exp(-E_{cr}(d_{p,i}\Upsilon)) \tag{4.136}$$

where $E_{cr}(d_{p,i},\Upsilon)$ is the larger of the minimum energy values obtained from two constraints:

$$E_{cr}(d_{p,i}, \Upsilon) = \max\left(\frac{E_{c,i}(d_{p,i})}{C_{eddy}}, \frac{\sigma V_\Upsilon}{d_{p,i}\min(f_{BV}, 1-f_{BV})^{1/3}}\right) \tag{4.137}$$

4.3.2 Break up Due to Viscous Shear Force

It can be shown that the deformation of a fluid particle due to viscous shear force primarily depends on the capillary number Ca, which is defined as the ratio of viscous stress over the surface tension. The deformation of the fluid particle increases with Ca. When Ca is increased to a critical value, the fluid particle will become unstable and break up. At this instance, it will break into two equal-sized fragments accompanied by a few much smaller satellite fluid particles. However, when Ca is suddenly increased to a value well above the critical value, the fluid particle is rapidly elongated into a long cylindrical fluid thread, which subsequently breaks into a series of fragments due to the growth of wavelike-shaped distortions, designated by capillary instabilities.

The average break up time in simple shear flow has been determined through the experimental results of Grace (1982) and Wieringa et al. (1996) to be

$$t_{b,v}(d_{p,i}) = \frac{\mu^c d_{p,i}}{2\sigma} f(p) \tag{4.138}$$

where the function $f(p)$ is characterized by the flow type with p being the viscosity ratio: $p = \mu^p/\mu^c$. This function $f(p)$ can be correlated either in the form of

$$f(p) = C_1' p^n \tag{4.139}$$

or in the form adopted in Lo and Zhang (2009):

$$f(p) = C_2' + C_3' \log(p^n) + C_4' (\log(p^n))^2 \tag{4.140}$$

The break up rate is subsequently given by

$$b(v_i) = \frac{C_{20}}{t_{b,v}(d_{p,i})} \tag{4.141}$$

where C_{20} is an adjustable constant.

4.3.3 Break up Due to Interfacial Instability and Shearing Off

For the influence of interfacial instability and shearing off on the break up process, the corresponding theory is rather limited when compared to break up due to turbulent fluctuation and viscous flow. Consideration of these mechanisms has been primarily concerned with the flow of cap/slug bubbles in air–water mixtures. Sun et al. (2004) indicated that there are two possible mechanisms that could lead to break up due to interfacial instability, one being due to turbulent fluctuations or random collisions while the other could be due to wake interactions. In this instance, the bubble becomes unstable and disintegrates when the bubble volume exceeds the maximum stable limit. Additionally, Fu and Ishii (2002) considered the premise of shearing mechanism where small bubbles have been seen to originate from the shearing off of cap/slug bubbles due to the penetration of gases in the thin interfacial layer into the liquid film around the rim of a bubble. Nonetheless, it should be recognized that whereas binary break up can be assumed for the development of models for break up rates due to turbulent shearing and viscous shear, such an assumption may not be strictly applicable to the models pertaining to interfacial instability and shearing off on the break up process. Further experimental investigations are still required to determine reliable correlations for which physical systems and under which flow conditions the break up outcomes can occur.

4.3.4 Comments on Daughter Particle Size Distribution

Aside from the consideration of appropriate models for the break up rate, knowing the number of daughter particles produced by a parent particle is also of importance. The fragmentation of parent particles into two daughter particles has been assumed in the majority of investigations. The various shapes of the daughter size distributions based on the phenomenological models developed for the break up process due to turbulent shearing and viscous shear are explored next.

Martinez-Bazan et al. (1999b) proposed a statistical model based on energy principles to describe the daughter particle size distribution. The model assumes that when a parent particle in the inertial subrange breaks, two daughter particles are formed—one with size $d_{p,j}$ and the other with a size in the range of $d_{p,min} \leq d_{p,j} \leq d_{p,max}$. Fundamentally, the model postulates that the probability of the present parent particle of size $d_{p,i}$ breaking up into the daughter particle of any size between the minimal possible size $d_{p,min}$ and its complementary size $d_{p,max}$, is proportional to the difference between the turbulent stresses over a length equal to its size and the confinement forces and, surface pressure forces holding the parent particle together. For the formation of a daughter particle of size $d_{p,j}$, the difference in stresses is given by $\frac{1}{2}\rho^c C_{23}(\varepsilon^c d_{p,j})^{2/3} - 6\sigma/d_{p,i}$ where C_{23} is a constant taken to have a value of 8.2 (Batchelor, 1956). For each daughter particle of size $d_{p,j}$, a complementary daughter particle of size $(d_{p,i}^3 - d_{p,j}^3)^{1/3}$ is formed by the difference stresses given by $\frac{1}{2}\rho^c C_{23}(\varepsilon^c (d_{p,i}^3 - d_{p,j}^3)^{1/3})^{2/3} - 6\sigma/d_{p,i}$. Thus, the probability of forming a daughter particle of size $d_{p,j}$ and its complementary element $(d_{p,i}^3 - d_{p,j}^3)^{1/3}$ is related to the product of the excess stresses corresponding to the two daughter particles:

$$P_1 \propto \left[\frac{1}{2}\rho^c C_{23}\left(\varepsilon^c d_{p,j}\right)^{2/3} - 6\sigma/d_{p,i}\right]\left[\frac{1}{2}\rho^c C_{23}\left(\varepsilon^c\left(d_{p,i}^3 - d_{p,j}^3\right)^{1/3}\right)^{2/3} - 6\sigma/d_{p,i}\right] \tag{4.142}$$

Also, assuming that the size of the daughter particle is to have a stochastic value and be uniformly distributed in the range of $[0, d_{p,i}]$, the distribution probability density of any size in $[0, V_{p,i}]$ would always be $P_2 = 1/V_{p,i}$. The total probability can be written as

$$\begin{aligned} P(v_i, v_j) \propto P_1 P_2 = \frac{1}{v_i}\left[\frac{1}{2}\rho^c C_{23}\left(\varepsilon^c d_{p,j}\right)^{2/3} - 6\sigma/d_{p,i}\right] \\ \times\left[\frac{1}{2}\rho^c C_{23}\left(\varepsilon^c\left(d_{p,i}^3 - d_{p,j}^3\right)^{1/3}\right)^{2/3} - 6\sigma/d_{p,i}\right] \end{aligned} \tag{4.143}$$

Utilizing the normalization condition $\int_0^{V_{p,i}} P(v_i, v_j)\mathrm{d}v_j = 1$, the dimensionless daughter particle size distribution can be expressed by

$$\beta(f_{BV}, 1) = \frac{\left[\frac{1}{2}\rho^c C_{23}\left(\varepsilon^c d_{p,j}\right)^{2/3} - 6\sigma/d_{p,i}\right]\left[\frac{1}{2}\rho^c C_{23}\left(\varepsilon^c\left(d_{p,i}^3 - d_{p,j}^3\right)^{1/3}\right)^{2/3} - 6\sigma/d_{p,i}\right]}{\int_0^1 \left[\frac{1}{2}\rho^c C_{23}\left(\varepsilon^c d_{p,j}\right)^{2/3} - 6\sigma/d_{p,i}\right]\left[\frac{1}{2}\rho^c C_{23}\left(\varepsilon^c\left(d_{p,i}^3 - d_{p,j}^3\right)^{1/3}\right)^{2/3} - 6\sigma/d_{p,i}\right]\mathrm{d}f_{BV}} \tag{4.144}$$

Based on this phenomenological model, this model predicts a symmetric distribution with the highest probability for equal-sized particle break up. This type of "bell-shape" distribution, such as that shown in Figure 4.8, is in accordance with the experimental data of Risso and Fabre (1998), who found equal-sized daughter distribution to be more common for bubble break up than unequal distribution.

Luo and Svendsen (1996) derived an expression for the break up density of a particle with size $d_{p,i}$ breaking into a daughter particle of size $d_{p,j}$ and a complementary daughter particle of size $(d_{p,i}^3 - d_{p,j}^3)^{1/3}$ using energy arguments similar to that of Tsouris and Tavlarides (1994). This present model gives both a partial break up rate, which is the break up rate for a particle of size $d_{p,i}$ undergoing a binary break up resulting in a particle of size $d_{p,j}$ and its complementary particle, and an overall break up rate. The daughter particle size distribution can thus be determined by normalizing the partial break up rate by the overall break up rate:

$$\beta(f_{BV}, 1) = \frac{\displaystyle\int_{\xi_{min}}^{1} \frac{(1+\xi)^2}{\xi^{11/3}} \exp(-\chi_c) d\xi}{\displaystyle\frac{1}{2}\int_0^1 \int_{\xi_{min}}^{1} \frac{(1+\xi)^2}{\xi^{11/3}} \exp(-\chi_c) d\xi df_{BV}} \tag{4.145}$$

The above daughter particle size distribution results and is calculated directly from the model. It has a minimum at equal-sized break up and maximum when the volume fraction $f_{BV} \to 0$ or 1 (Figure 4.9). Such a "U-shape" distribution, as depicted in Figure 4.9, has a nonzero minimum and exhibits a dependency on the parent particle size, which is in accordance with the experimental data of Hesketh et al. (1991). Hesketh and co-workers concluded that an

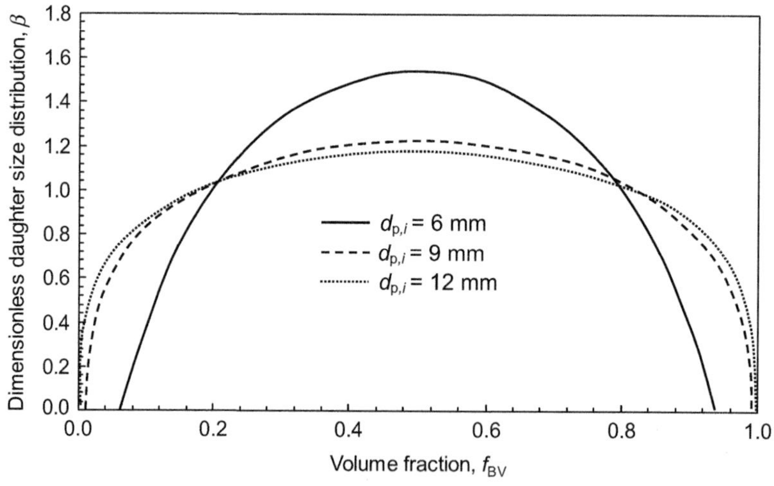

Figure 4.8
Daughter size distribution by Martinez-Bazan et al. (1999b).

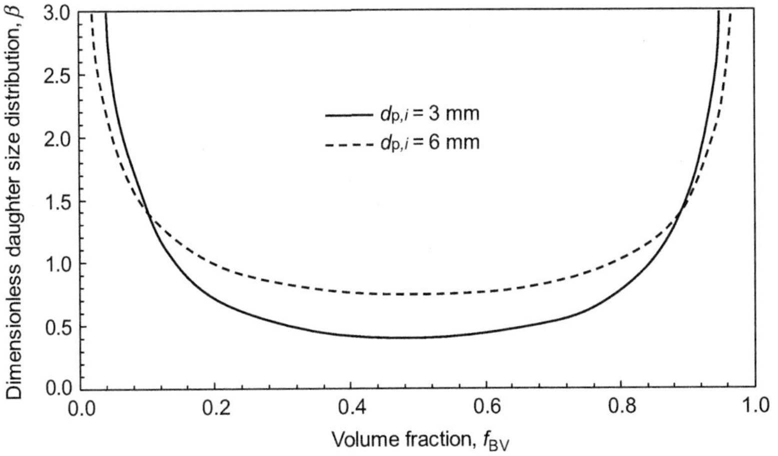

Figure 4.9
Daughter size distribution by Luo and Svendsen (1996).

unequal size distribution is more probable since it has a lower probability than an equal size distribution.

Lehr et al. (2002) derived a different expression for the daughter particle size distribution. Normalizing the partial break up rate by the overall break up rate yields

$$\beta(f_{BV},1)=\frac{\int_{\gamma_{\min}}^{\gamma_{\max}}\frac{1}{d_{p,j}^{4}}\frac{(d_{p,i}+\gamma)^{2}}{\gamma^{12/3}}\exp\left(-\frac{2\sigma}{\rho^{c}(\varepsilon^{c})^{2/3}\gamma^{2/3}d_{p,j}}\right)d\gamma}{\frac{1}{2}\int_{0}^{1}\int_{\gamma_{\min}}^{\gamma_{\max}}\frac{1}{d_{p,j}^{4}}\frac{(d_{p,i}+\gamma)^{2}}{\gamma^{12/3}}\exp\left(-\frac{2\sigma}{\rho^{c}(\varepsilon^{c})^{2/3}\gamma^{2/3}d_{p,j}}\right)d\gamma df_{BV}} \tag{4.146}$$

Prediction of the daughter particle size distribution illustrates that the probability of small and large daughter particles increases rapidly with the parent particle size. With the increase of the parent particle size, the distribution changes from a single modal, "bell-shape", to a bimodal, "M-shape", which indicates that equal-sized break up is more likely for small particles than large particles (Figure 4.10). This is similar to the assumption of Nambiar et al. (1992). It should be noted that the position of the peak does not change much, which is inconsistent with the fact that the minimum break up fraction depends on the parent particle size.

Wang et al. (2003) determined the daughter particle size distribution as

$$\beta(f_{BV},1)=\frac{\int_{\gamma_{\min}}^{\gamma_{\max}}\frac{(d_{p,i}+\gamma)^{2}}{\gamma^{11/3}}\int_{0}^{\infty}\frac{1}{f_{BV,\max}-f_{BV,\min}}\frac{1}{E(\gamma)}\exp\left(-\frac{E(\gamma)}{\overline{E}(\gamma)}\right)dE(\gamma)d\gamma}{\frac{1}{2}\int_{0}^{1}\int_{\gamma_{\min}}^{\gamma_{\max}}\frac{(d_{p,i}+\gamma)^{2}}{\gamma^{11/3}}\int_{0}^{\infty}\frac{1}{f_{BV,\max}-f_{BV,\min}}\frac{1}{E(\gamma)}\exp\left(-\frac{E(\gamma)}{\overline{E}(\gamma)}\right)dE(\gamma)d\gamma df_{BV}} \tag{4.147}$$

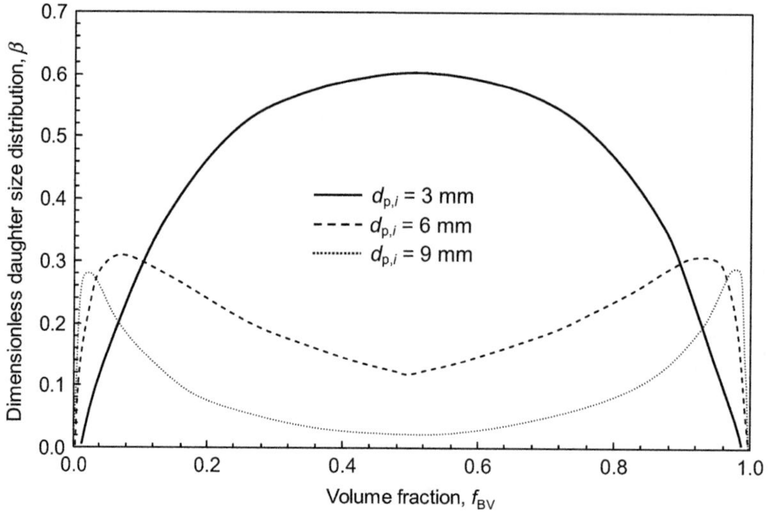

Figure 4.10
Daughter size distribution by Lehr et al. (2002).

According to Wang et al. (2003), the daughter particle size distribution must satisfy four requirements: (1) a local minimum but nonzero at equal break up, (2) distribution function depends on both the parent particle size and dynamics in the continuous phase, in particular the energy dissipation rate, (3) probability density of the daughter particles approaches zero when the break up fraction approaches zero, and (4) function form should not be dependent on experimental conditions or include singularity. This phenomenological model yields the characteristic "M-shape" distribution depicted in Figure 4.11, as was proposed by

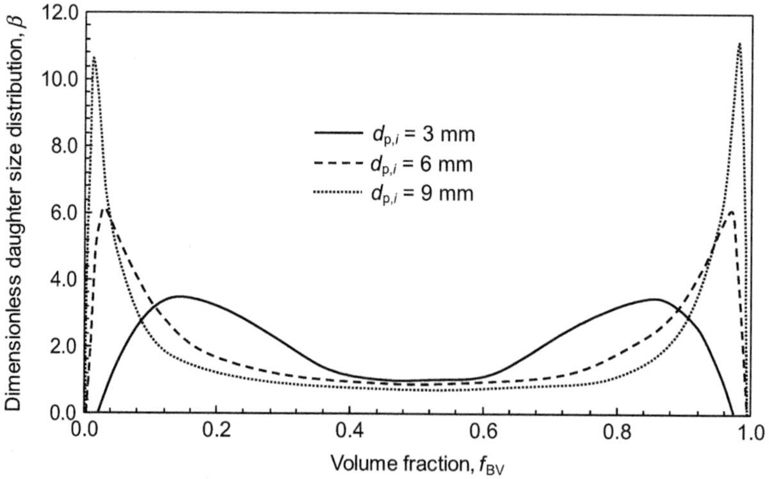

Figure 4.11
Daughter size distribution by Wang et al. (2003).

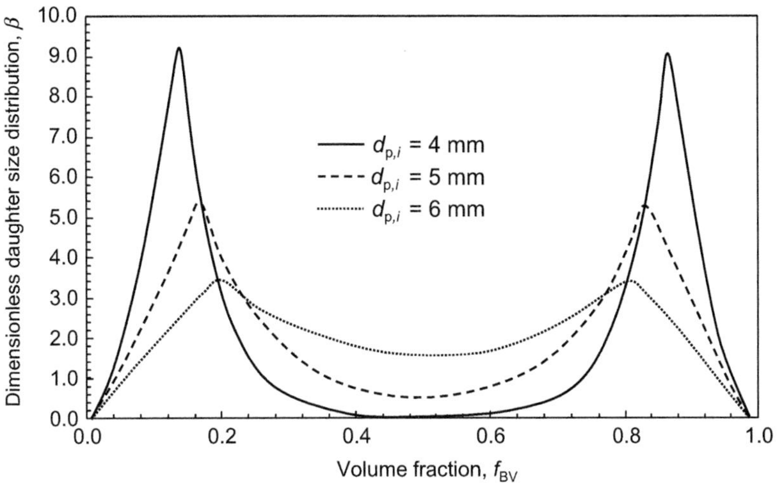

Figure 4.12
Daughter size distribution by Zhao and Ge (2007).

Lehr et al. (2002). Accordingly, based on the formulation by Zhao and Ge (2007), where they assumed one of the daughter particles in binary break up having the size of a turbulent interacting eddy, the density probability of daughter particle with size $d_{p,j}$ should comprise the sum of the break up induced by an eddy with size $\gamma = (6V_{p,j}/\pi)^{1/3}$ and with size $\gamma = (6(V_{p,i} - V_{p,j})/\pi)^{1/3}$. The daughter particle size distribution can be expressed by

$$\beta(f_{\mathrm{BV}}, 1) = \frac{d_{\mathrm{p},i} w\left(d_{\mathrm{p},i}, \gamma\right) P_{\mathrm{b}}\left(d_{\mathrm{p},i}, \gamma\right)}{3 f_{BV}^{2/3} \int_{\gamma_{\min}}^{d_{\mathrm{p},i}} w\left(d_{\mathrm{p},i}, \gamma\right) P_{\mathrm{b}}\left(d_{\mathrm{p},i}, \gamma\right) \mathrm{d}\gamma} + \frac{d_{\mathrm{p},i} w\left(d_{\mathrm{p},i}, d_{\mathrm{p},i} - \gamma\right) P_{\mathrm{b}}\left(d_{\mathrm{p},i}, d_{\mathrm{p},i} - \gamma\right)}{3(1 - f_{\mathrm{BV}})^{2/3} \int_{\gamma_{\min}}^{d_{\mathrm{p},i}} w\left(d_{\mathrm{p},i}, d_{\mathrm{p},i}\right) P_{\mathrm{b}}\left(d_{\mathrm{p},i}, d_{\mathrm{p},i}\right) \mathrm{d}\gamma} \tag{4.148}$$

Figure 4.12 illustrates the daughter size distribution proposed by Zhao and Ge (2007).

The rather deviating experimental data and constraining daughter particle size distributions indicate that further investigations are still required in order to arrive at reliable correlations that can be employed for physical systems and flow conditions wherein the equal and unequal break up outcomes can occur.

4.4 Mechanisms and Kernels of Fluid Particle Coalescence and Break up for One-group, Two-group and Multigroup Formulation

This modeling framework includes the classification of bubbles of different sizes and shapes into different groups. It entails additional transport equations aptly describing the transport

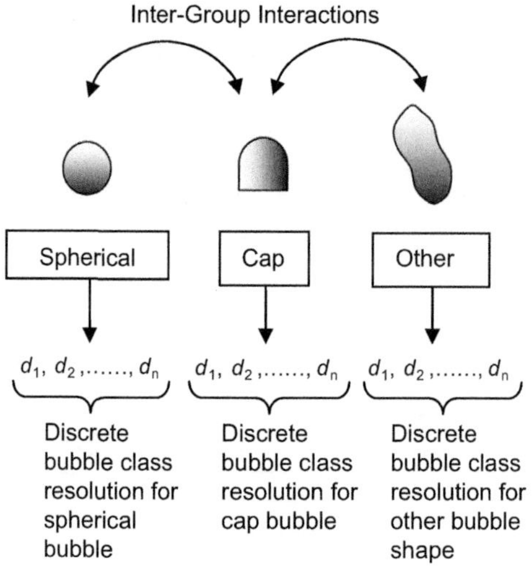

Figure 4.13
An illustration of the multigroup multibubble class model.

phenomena of these distinct groups of bubbles, as proposed by Ishii et al. (2002). Such a model possesses the capability of accounting for the wide spectrum of bubble sizes that may exist in different flow regimes. A schematic representation for the multigroup multibubble class model is illustrated in Figure 4.13. For each bubble shape, a set of discrete bubble class equations is solved. Referring to Figure 4.13, a two-group multibubble class model primarily accommodates the consideration of intragroup interactions for the spherical and cap/slug bubbles. Essentially, two sets of population balance equations are solved for each intragroup. Possible interactions of two-group bubbles that can be treated as sources and sinks to the population balance equations are illustrated in Figure 4.14.

As shown in Figure 4.15, interactions between bubbles can be described by five mechanisms:

- Coalescence due to random collisions driven by liquid turbulence
- Coalescence due to wake entrainment
- Break up due to the impact of turbulent eddies
- Shearing-off of small bubbles from cap/slug bubbles
- Break up of large cap bubbles due to flow instability on the bubble surface.

For intragroup mechanisms of spherical bubbles, the usual coalescence and break up processes due to random collisions and turbulent impact can be adopted; these have been extensively described in previous sections. However, the intragroup mechanisms for cap/slug bubbles require more complex treatment where coalescence of bubbles can be subjected to random collisions and wake entrainment, while bubble break up can be attributed to turbulent impact, shearing off, and surface instability. Referring to Figure 4.15, wake entrainment of

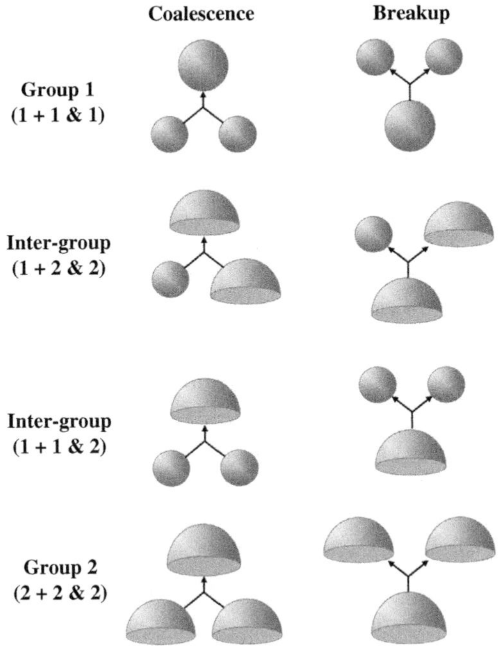

Figure 4.14
Classification of possible interactions of two-group bubbles. *After Hibiki and Ishii (2009).*

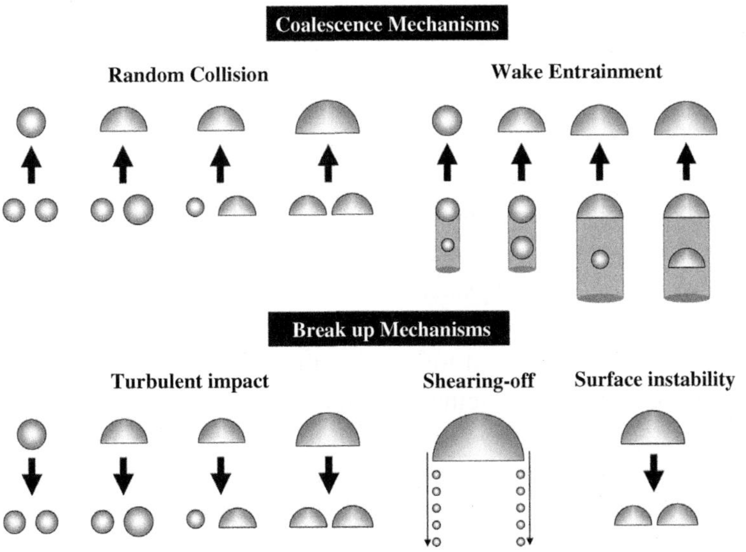

Figure 4.15
A schematic illustration of the mechanisms associated with the two-group bubble interactions. *After Hibiki and Ishii (2009).*

cap/slug bubbles governs the cap/slug bubble number density, which significantly affects the flow structure and intensiveness of intergroup interactions. When bubbles enter the critical wake region of a preceding bubble, they can accelerate and coalesce with this preceding bubble according to the mechanisms based upon the critical distance of a probability approach such as that proposed by Sun et al. (2004). Nevertheless, cap/slug bubble disintegration due to surface instability can be enhanced significantly by the high turbulent intensity and active eddy–bubble interaction in the wake region of the slug bubbles. This occurs when the volume of the resulting bubble from bubble coalescence exceeds the maximum bubble stable limit; it becomes unstable and disintegrates. The mechanism for the shearing-off effect is reflected by the fact that the bubbles shearing off from the cap/slug bubbles act as a primary source for the spherical bubbles and ultimately the interfacial area concentration. More details on the mechanisms for cap/slug bubbles can be found in Hibiki and Ishii (2000b), Fu and Ishii (2002), and Sun et al. (2004).

In addition to the intragroup interactions, it has been demonstrated that the intergroup interactions also contribute significantly to the transport of the interfacial states. Similar mechanisms to the above considerations for intragroup interactions may be derived accordingly for the intergroup interactions.

4.5 Summary

In this chapter, mechanistic models for fluid particle coalescence and break up have been described. With regard to the coalescence processes, possible mechanisms that have been considered for the coalescence among fluid particles are turbulent random motion-induced collisions, wake entrainment, velocity gradient-induced collisions, buoyancy-induced collision, and capture in a turbulent eddy. In addition, three postulated theories, or criteria, are proposed which include (1) the film drainage model where coalescence can occur when the liquid film ruptures before the bubbles or drops separate, (2) the energy model where coalescence will proceed depending upon the impact of colliding fluid particles, and (3) the critical approach velocity model where small approach velocities result in high coalescence probability. The characteristics of physical models for the collision frequency and coalescence efficiency are discussed. With regard to the break up processes, various mechanisms that have been considered for causing the break up among fluid particles in turbulent dispersions are: turbulent fluctuation and collision, viscous shear forces, interfacial stability, and the shearing-off process. The characteristics of physical models for the break up frequency and daughter size distribution are discussed. A modeling framework that includes the classification of bubbles of different sizes and shapes into different groups is proposed. Consideration of a multigroup multibubble class model is presented.

CHAPTER 5

Mechanistic Models for Gas–Particle Liquid–Particle Flows

5.1 Introduction

Motion of individual solid particles traveling in a gas phase or liquid phase is normally characterized by gas–particle or liquid–particle flows. Individual gas bubbles or liquid drops can coalesce into large bubbles or drops through the breaking down of the interfaces between the deforming phases and these entities at the same time can break up due to the shearing by the fluid into two smaller bubbles or drops. Nevertheless, the phenomenon of particle aggregation in gas–liquid or liquid–particle flows can effectively occur when solid particles are brought together into close contact with each other in order that collision between particles can be realized. If the net interparticle force is attractive and strong enough to overcome the hydrodynamic forces, these particles will subsequently stick together resulting in the formation of an aggregate. For the phenomenon of particle breakage, the discussion is directed toward the reduction of the size of the aggregate, which is affected by the fluid shear overcoming the attractive interparticle force holding the individual solid particles together within the aggregate. The distinctive difference between the break up of bubbles or drops and breakage of aggregates is that the former generally assume a binary break up while the latter may undergo binary, ternary or even quaternary breakage.

Many industrial applications of gas–particle or liquid–particle flows involve a wide range of situations and concern the design and control of various multiphase processes. For liquid–particle flows, the removal of fine particles from a liquid suspension by means of coagulation (aggregation and flocculation) is encountered at various stages in water purification and waste treatment. Aggregation of fine particles, which can be achieved through mixing a flocculant, is also necessary for gravity thickening in hydrometallurgical industries like alumina and mineral sand processing operations. Characterization and control of aggregate properties are of considerable interest for the overall effective operation, as aggregates or flocs with certain size, structure, shape, or density may be more suitable for a specific process in comparison to others. For gas–particle flows, dust particle clogging is a major problem especially for the radiator of a construction vehicle operating in a dusty environment and heat exchangers associated with the vehicle cooling system. This has led to the need of frequent vehicle downtime in order to properly maintain and promote longevity of such industrial systems. Dust fouling has also been a major maintenance concern for electronic equipment and micro electromechanical systems. Small dust particles adhering to electrical circuit boards can short out the electrical system. Production of nanoparticles is of growing interest due to their use in antiabrasion coatings, as fillers for advanced composite materials, and as coatings for advanced sensors, catalysts and battery electrodes.

The computational models for solution of flows with aggregation and breakage of solid particles are quite varied in approach. In principle, these models can be divided into two general classes—(1) population balance approach and (2) discrete element method. The first relates the rate of change of the number of aggregates of a certain size, and of number density N_i, to various effects that can lead to generation or elimination of aggregates of size N_i, resulting from aggregate collision and breakage processes. Suitable expressions for the source and sink terms will be described in the next few sections. This approach has been sufficiently refined to provide reasonably accurate prediction of aggregate size distribution in different types of flow fields.

For predictions of micromechanics and microstructure of aggregates or of their interactions with each other and with immersed surfaces in the flow, especially in microfluidic flows, *the adoption of a discrete element method* (*where transport and interactions of each particle are computationally followed*) and an overpopulation balance approach could overcome the difficulty associated with the simulation of adhesive particle flows. A discrete element method has been employed extensively for particles of large sizes, where adhesion effects are negligible and particles are of nanoscale, the particle sizes being compatible with the adhesion length scales, such that molecular dynamics can be directly applied. It is the regime in between these two extremes, involving particles with a diameter much larger than the adhesion length scale but still small enough to exhibit significant particle adhesion, with which the computational models for adhesive particles will deal in this chapter. The challenge is the necessity to include a wide range of forces and torques acting on the particles due to collision events that include the elastic and dissipative normal forces and the resistance from particle sliding, twisting and rolling motions. Most of these collision forces and torques are significantly affected by the adhesive characteristic, which plays a critical role in the aggregate and breakage dynamics.

Various mechanisms and available kernel models for particle aggregation and breakage are first described for the application of the population balance approach. In the discrete element method, a soft-sphere model, in contrast to the hard-sphere model described in Chapter 2, is presented in which the model allows particles to overlapping each other in order to keep track of their deformation during a collision. Here, contact forces and torques for adhesive particles are derived from a point on the overlapping bodies for the soft-sphere model.

5.2 Mechanisms and Kernel Models of Solid Particle Aggregation

In accordance with the macroscopic population balance formulation adopted for the discrete birth and death rates due to fluid particle coalescence in Chapter 4, the population balance equation of the discrete particle number density can also be applied to resolve the birth and death rates due to particle aggregation. First, the birth of solid particles of volume v_i due to aggregation results from the aggregation of all solid particles smaller than v_i. Therefore, the birth rate for aggregates that form an aggregate of volume v_i is

$$S^{+}_{C,i}(\mathbf{r},\mathbf{Y},t) = \int_{v_i}^{v_{i+1}} \mathrm{d}v \left[\frac{1}{2} \sum_{j=0}^{i-1} \int_{v_j}^{v_{j+1}} \underbrace{a\left(\tilde{V}_p, \tilde{\mathbf{r}}; V'_p, \mathbf{r}', \mathbf{Y}\right)}_{\text{aggregation frequency}} f_1\left(\tilde{V}_p, \tilde{\mathbf{r}}, t\right) f_1(V', \mathbf{r}', t)\mathrm{d}v' \right] \tag{5.1}$$

Second, the death of aggregates due to aggregation between two aggregates of volume v_i or between one aggregate of volume v_i can also be determined by

$$S^{-}_{C,i}(\mathbf{r},\mathbf{Y},t) = \int_{v_i}^{v_{i+1}} \mathrm{d}v \left[\sum_{j=0}^{M} \int_{v_j}^{v_{j+1}} \underbrace{a\left(V'_p, \mathbf{r}'; V_p, \mathbf{r}, \mathbf{Y}\right)}_{\text{aggregation frequency}} f_1\left(V'_p, \mathbf{r}', t\right) f_1\left(V_p, \mathbf{r}, t\right)\mathrm{d}v' \right] \tag{5.2}$$

For the specific treatment of particle aggregation in gas–particle or liquid–particle flows, the aggregation frequency is also expressed as the product of an aggregation kernel that describes the frequency at which particles will collide and the collision efficiency. The general expression for the aggregation frequency can be expressed in terms of volumes v_i and v_j as

$$a\left(v_i, v_j\right) = \alpha\left(v_i, v_j\right)\beta\left(v_i, v_j\right) \tag{5.3}$$

5.2.1 Aggregation Due to Interparticle Collision

In general, the process of aggregation comprises two important stages. Particles must initially be brought together into close proximity with each other by a transport mechanism in order that interparticle collision can occur. An aggregate is subsequently formed if the net interparticle force is attractive and strong enough to overcome the hydrodynamic forces to make the particles adhere or fuse. Depending on the size of the particles, there are a number of mechanisms that can induce relative movement among particles and, hence, lead to collisions. Three different mechanisms that can be operative during the aggregation of a given set of particles are: (1) perikinetic coagulation due to Brownian motion, (2) shear-induced orthokinetic coagulation and (3) differential sedimentation aggregation. As aggregates become larger, particle breakage will become important. In some cases, the competition between aggregation and breakage can lead to a steady state particle size distribution (Vigil and Ziff, 1989).

Aggregation Kernel Due to Brownian Motion Induced Collisions

Brownian motion is presumably the random motion of particles suspended in a fluid (gas or liquid) resulting from their collisions by fast-moving atoms or molecules in the gas or liquid. For submicrometer particles, usually smaller than 1 μm in diameter, Brownian motion induced collisions are the controlling mechanism for the process of aggregation. When aggregation occurs, three regimes categorized by free molecular, transition, and continuum, can be appropriately described by the Knudsen number:

$$Kn \equiv \frac{\text{mean free path of gas molecules}}{\text{physical length scale}} = \frac{\lambda}{L} \tag{5.4}$$

More specifically, the different regimes are:

- Free molecular, $Kn >> 1$: the path between the particles is much *larger* than the particle diameter. It is anticipated that the particles in this regime are free to move around.
- Transition, $0.1 < Kn < 1$: this represents the state between the free molecular regime and the continuum regime.
- Continuum, $Kn << 1$: the path between the particles is much *smaller* than the particle diameter. The particles are rather crowded and the movement in this regime is close to being a continuous flow.

By definition, the mean free path of gas molecules λ, which is the average distance the particle travels between collisions with other particles, may be calculated based on the particle diameter d from the following expression:

$$\lambda = \frac{k_B T^c}{(2)^{1/2} \pi R_p^2 p_0} \tag{5.5}$$

where k_B is the Boltzmann constant, which is essentially the ratio between the universal gas constant ($\approx$8.314 J/mol K) and the Avogadro number ($\approx 6.022 \times 10^{-23}$ 1/mol), and p_0 is the fixed ambient pressure. Adopting the particle radius as the physical length scale, the Knudsen number becomes: $Kn = 2\lambda/R_p$. On the basis of Eqn (5.5), the Knudsen number is directly dependent only on the particle diameter R_p. As dictated by the size of the particles, the collision kernel of the free molecular regime and the continuum regime can be determined by:

Free molecular regime

$$\begin{aligned} \beta^{\text{Brownian,F}}(v_i, v_j) &= K_f (v_i^{1/3} + v_j^{1/3})^2 \sqrt{\frac{1}{v_i} + \frac{1}{v_j}} \\ &= K_f' (R_{p,i} + R_{p,j})^2 \left(\frac{1}{R_{p,i}^3} + \frac{1}{R_{p,j}^3} \right)^{1/2} \end{aligned} \tag{5.6}$$

Continuum regime

$$\beta^{\text{Brownian,C}}(v_i, v_j) = K_c\left(v_i^{1/3} + v_j^{1/3}\right)\left(\frac{C(v_i)}{v_i^{1/3}} + \frac{C(v_j)}{v_j^{1/3}}\right)$$
$$= K_c\left(R_{p,i} + R_{p,j}\right)\left(\frac{C(v_i)}{R_{p,i}} + \frac{C(v_j)}{R_{p,j}}\right) \tag{5.7}$$

In Eqn (5.6), $K_f = (3/4\pi)^{1/6}\sqrt{6k_B T^c/\rho^p}$ and $K'_f = \sqrt{6k_B T^c/\rho^p}$ for nonsticking particles, while $K_f = \sqrt{8k_B T^c/\rho^p V^p}$ and $K'_f = \sqrt{8k_B T^c/\rho^p V^p}$ for sticking particles. In Eqn (5.7), $K_c = 2k_B T^c/3\mu^c$. Parameters T^c and μ^c in K_f, K'_f and K_c are defined as the continuous suspending temperature and the dynamic viscosity of the continuous phase. $R_{p,i}$ and $R_{p,j}$ are the collision radii of the *i*- and *j*-sized particles or aggregates. The Cunningham slip correction factor $C(v_l)$ in Eqn (5.7) can be obtained from

$$C(v_l) = 1 + 1.257Kn_l \tag{5.8}$$

Note that the Knudsen number changes accordingly with the particle radius based on the specific size of volume v_l in the above Cunningham slip correction factor. In the transition regime, the coagulation rate is typically determined by the harmonic mean of the continuum and free molecular rates.

$$S_{C,i}^{\text{Brownian}} = \frac{S_{C,i}^{\text{Brownian,C}} S_{C,i}^{\text{Brownian,F}}}{S_{C,i}^{\text{Brownian,C}} + S_{C,i}^{\text{Brownian,F}}} \tag{5.9}$$

Aggregation Kernel Due to Shear-Induced Collisions

For particles with a diameter in the range of 1–50 μm, the process of aggregation is now mostly caused by collisions by orthokinetic coagulation (velocity gradients). The collision kernel for shear-induced collisions is expressed as

$$\beta^{\text{Shear}}(v_i, v_j) = C_1 G\left(R_{p,i} + R_{p,j}\right)^3 \tag{5.10}$$

where C_1 is a numerical constant, which depends on the type of flow, and G is the characteristic shear rate (velocity gradient) of the flow field. In a simple shear laminar flow, the constant C_1 is taken to have a value of 4/3 (Smoluchowski, 1917) and G refers to the laminar shear velocity of the fluid. In the viscous subrange where the particles are smaller than the Kolmogorov length scale, particles are influenced by the local shear within the eddy. Saffman and Turner (1956) derived the numerical constant Ψ with a value of $8\pi/15$, while the characteristic shear rate is related to the local energy dissipation rate and the kinematic viscosity according to $G=(\varepsilon^c/v^c)^{1/2}$.

In the inertial subrange where the particles are larger than the smallest eddy, particles are now dragged by the velocity fluctuation in the flow field. Abrahamson (1975) proposed a collision kernel that is given by

$$\beta^{\text{Shear}}(v_i, v_j) = 2^{3/2}\sqrt{\pi}\sqrt{U_{p,i}^2 + U_{p,j}^2}\left(R_{p,i} + R_{p,j}\right)^3 \tag{5.11}$$

where $U_{p,i}^2$ and $U_{p,j}^2$ are the mean square velocities for *i*- and *j*-sized particles.

Aggregation Kernel Due to Differential Sedimentation Induced Collisions

Differential sedimentation induced collisions involve a physical property of the particles, namely, their density-excess ratio $(\rho^p - \rho^c)/\rho^c$ over that of the fluid. Based on their different densities, collisions and subsequent aggregation may occur when larger or heavier particles overtake the smaller or lighter particles. Collisions caused by differential sedimentation become important for particles or aggregates larger than ~50 μm. The collision kernel for differential sedimentation induced collisions is expressed as

$$\beta^{\text{Differential}}(v_i, v_j) = \frac{2\pi}{9}k_D\left(R_{p,i} + R_{p,j}\right)^2\left|R_{p,i}^2 + R_{p,j}^2\right| \tag{5.12}$$

where k_D is given by $g(\rho^p - \rho^c)/\mu^c$.

Collision Efficiency of Particles

The collision efficiency is generally a function of the hydrodynamic interactions, short-range forces that act on the particles upon collisions, and the structure of aggregates concerning how porous and permeable the existing aggregates are within the flow. In the most practical multiphase system, viscous resistance is reflected in the collision efficiency, which is the ratio of the actual frequency at which particles collide with and stick to each other, and the theoretical aggregation frequency, described by Brownian motion induced and shear-induced collisions (see previous sections).

For impermeable particles, the collision efficiency can be expressed in the form of

$$\alpha(v_i, v_j) = C_2 Fl^{-0.18} \tag{5.13}$$

where C_2 is an adjustable prefactor depending on the expected and observed particle sizes. In Wang et al. (2005) and Cheng et al. (2009), a value of 0.43 is adopted for this prefactor. The flow number for the aggregation of solid particles is given by

$$Fl = \frac{6\pi\mu\left(r_{p,i} + r_{p,j}\right)^3 G}{8H} \tag{5.14}$$

Equation (5.14) can be interpreted as the ratio between shear-induced and van der Waals forces in which *H* signifies the Hamaker constant. In a laminar flow, *G* simply refers to the laminar shear velocity of the fluid. In a turbulent flow, the characteristic shear rate G can be determined according to $(\varepsilon^{c}/\nu^{c})^{1/2}$.

For permeable particles, Kusters et al. (1997) proposed correcting the flow number for the fact that the attractive force between porous particles is smaller than that between solid particles. The only consideration for the attractive interaction between two porous aggregates is between the two nearest primary particles since the other particles are separated by a distance too large for the interaction force to be effective (Firth and Hunter, 1976). Therefore, the van der Waals attraction between two porous aggregates may be estimated by the force between two primary particles residing in each aggregate as

$$F_{\text{agg}} = \frac{R_0 H}{12h^2} \tag{5.15}$$

where R_0 is the radius of the primary particle, H is the Hamaker constant, ranging between 10^{-19} J and 10^{-20} J for liquid–particle systems, and h is the binding distance between the two porous aggregates. The attractive force can be written as

$$F_{\text{sp}} = \frac{R_{\text{p},i} R_{\text{p},j} H}{6h^2 \left(R_{\text{p},i} + R_{\text{p},j}\right)} \tag{5.16}$$

In order to obtain the flow number corresponding to porous aggregates, the flow number given in Eqn (5.14) can be multiplied by the ratio of the attractive forces of solid and porous particles. In other words,

$$Fl = \frac{6\pi\mu \left(R_{\text{p},i} + R_{\text{p},j}\right)^3 G}{8H} \times \frac{2R_{\text{p},i} R_{\text{p},j}}{R_0 \left(R_{\text{p},i} + R_{\text{p},j}\right)} \tag{5.17}$$

As the collision efficiency decreases with increasing flow number, it follows from Eqn (5.13) that the efficiency decreases more rapidly with increasing particle size than the collision efficiency of solid particles.

According to Wang et al. (2005) and Cheng et al. (2009), the collision radius of an aggregate $R_{\text{p},i}$ can be taken to be proportional to the number of primary particles in the aggregate i, in accordance with

$$R_{\text{p},i} = R_0 \left(\frac{i}{k}\right)^{1/D_{\text{f}}} \tag{5.18}$$

where D_{f} is the mass fractal dimension with a value close to 3 and k is the lacunarity usually taken to be unity (Kusters et al., 1997; Flesch et al., 1999). Hence, the aggregate volume of v_i for a fractal aggregate is related to its characteristic size $R_{\text{p},i}$ by

$$v_i = k v_0 \left(\frac{R_{\text{p},i}}{R_0}\right)^{D_{\text{f}}} \tag{5.19}$$

in which v_0 is the volume of the primary particle. It should be noted that the use of the fractal exponent is only strictly justified for an aggregate composed of a large of number of primary particles. This assumption is commonly adopted and extended to all aggregate sizes in order to simplify the calculations.

5.3 Mechanisms and Kernel Models of Solid Particle Breakage

Similar to the particle aggregation in gas–liquid or liquid–liquid flows, the breakage of aggregates is significantly influenced by continuous phase hydrodynamics and interfacial interactions. For the phenomenon of particle breakage, the main consideration is the reduction of the size of aggregates, which is affected by the fluid flow overcoming the attractive interparticle force holding the individual solid particles together within the aggregates. In accordance with the particle aggregation, a practical consideration of the birth and death rates due to particle breakage is adopted. Appropriate source and sink terms are henceforth developed directly on the averaging scales performing analysis of particle breakage.

The macroscopic population balance formulation of the discrete birth and death rates due to particle breakage are delineated. First, the birth of aggregates of volume v_i due to breakage results from the breakage of all aggregates smaller than v_i. The birth rate for aggregates of volume v_i can be obtained by use of:

$$S^{+}_{\mathrm{B},i} = (\mathbf{r},\mathbf{Y},t)\int_{v_i}^{v_{i+1}} \mathrm{d}v \left[\sum_{j=0}^{M} \int_{v_j}^{v_{j+1}} v\left(V'_{\mathrm{p}},\mathbf{r},\mathbf{Y}\right) \underbrace{b\left(V'_{\mathrm{p}},\mathbf{r},\mathbf{Y}\right)}_{\text{breakage kernel}} \times \underbrace{P\left(V_{\mathrm{p}},\mathbf{r}\middle|V'_{\mathrm{p}},\mathbf{r},\mathbf{Y},t\right)}_{\text{fragment distribution function}} f_1\left(V'_{\mathrm{p}},\mathbf{r},t\right)\mathrm{d}v' \right] \tag{5.20}$$

Second, the death of volume v_i due to breakage results from the breakage of the aggregates within this class. In other words,

$$S^{-}_{\mathrm{B},i}(\mathbf{r},\mathbf{Y},t) = \int_{v_i}^{v_{i+1}} \mathrm{d}v \left[\underbrace{b\left(V_{\mathrm{p}},\mathbf{r},\mathbf{Y}\right)}_{\text{breakage kernel}} f_1\left(V_{\mathrm{p}},\mathbf{r},t\right) \right] \tag{5.21}$$

For the breakage of aggregates, the general expressions for the breakage kernel and fragment distribution function in gas–particle or liquid–particle flows can normally be expressed in terms of volumes v_i and v_j as $b(v_i)$ and $P(v_j,v_i)$, respectively.

5.3.1 Breakage Due to Hydrodynamic Stresses

Among the many mechanisms that can occur in practical multiphase systems, the process of breakage or fragmentation of aggregates due to hydrodynamic stresses is mostly encountered. In general, the theory for particle breakage is not as well developed as the theory of particle aggregation. The breakage functions can be factored into two parts. First, the breakage kernel that is the rate coefficient for breakage of a particle needs to be appropriately defined based on relevant hydrodynamic breakage mechanisms. Second, several functional forms of the fragment distribution function (daughter size distribution function) need to be developed in which some physical constraints have to be fulfilled: the number of fragments formed has to be correctly represented (the masses of the fragments have to sum up to the original particle mass).

Breakage Kernel of Particles

A number of breakage kernels have been proposed based on flow-induced stresses for solid particles. The mechanisms that dominate the breakage of a large aggregate depend largely on the strength of the aggregate bonds between the solid particles in regard to the hydrodynamic stresses that act on the aggregate.

For laminar flow, Wang et al. (2005) have proposed that the breakage rate is exponentially distributed dependent on the characteristic shear rate G so that

$$b(v_i) = C_3 G \exp\left(-\frac{G_{c,i}}{G}\right) \tag{5.22}$$

where C_3 is a constant that is prescribed according to $\sqrt{2/\pi}$ and $G_{c,i}$ is the critical shear rate that causes breakage, which can be related to the particles size as

$$G_{c,i} = \frac{B}{\sqrt{R_{p,i}}} \tag{5.23}$$

in which B is a fitting parameter related to the binding strength between monomer particles and the aggregation morphology. Wang et al. (2005) has correlated this parameter by a power law: $B = A'\overline{G}^{y'}$, where $\overline{G}$ is the spatial mean shear rate ranging between 1.86 1/s and 5.59 1/s. The value of A' has been found to be equal to 7.74×10^{-2} and the value of y' is 0.22.

For turbulent flow, the breakage of solid aggregates is highly dependent on the ratio between the particle sizes and the smallest turbulent eddy. When the size of a particle is smaller than the turbulent microscale, breakage is very likely to be caused by shear stresses that originated from the turbulent dynamic velocity difference acting on the opposite sides of the particle. The exponential form as depicted in Eqn (5.22) can be adopted with the characteristic shear rate now given by $G = (\varepsilon^c / \nu^c)^{1/2}$ while critical shear rate can be written as a function of the aggregate strength τ_f as

$$G_{c,i} = \frac{\tau_f}{\mu^c} \tag{5.24}$$

According to Marchisio et al. (2003), the aggregate strength can be estimated by

$$\tau_f = \frac{9}{8} k_c \phi F_{agg} \frac{1}{\pi L_0^2} \tag{5.25}$$

where L_0 is the diameter of the primary particle, and the interparticle force F_{agg} between two primary particles is given by Eqn (5.15). The coordination number in Eqn (5.25) can be approximated by the volume fraction of solid within the aggregates ϕ according to

$$k_c \approx 15\phi^{1.2} \tag{5.26}$$

Alternatively, a semitheoretical expression that has also found application to a wide variety of breakage phenomena in turbulent flow is the power-law breakage kernel:

$$b(v_i) = C_4 (\varepsilon^c)^{\beta'} (\nu^c)^{-\alpha'} D_{p,i}^{\gamma'} \tag{5.27}$$

where C_4 is a constant value between 10^{-4} and 10^{-3}. Classifying the failure modes through manifestation, induction and location of the failure, and the state at which aggregates break when the maximum eigenvalue of the stress tensor is greater than the aggregate strength, the number of aggregate bonds with a strength at or below the aggregate strength is thus not a linear function of the strain rate, and size-dependency is considered in the breakage model as given in Eqn (5.27). Several values of the exponent γ' have been proposed. In particular, the exponent γ' can vary between 0 and 1 according to Luo and Svendsen (1996) or between 1 and 3 according to Peng and Williams (1994). For the exponent β', Peng and Williams (1994) have employed a value of 0.5. Nonetheless, Serra and Casamitjana (1998) have found that the relationship between the breakage kernel and turbulent dissipation rate depends on the volume fraction of the solid. When the volume fraction is low, β' has a value of 0.9. For a higher volume fraction, the exponent β' is higher, which can be explained in that particle–particle collisions are more effective if the volume fraction is higher. A value of greater than 0.5 may be applied for exponent α'; Cheng et al. (2009) have adopted a value of 2 for their study of the precipitation process in a stirred tank.

Fragment Distribution Function of Particles

The specific form of the fragment distribution function to describe the hydrodynamic breakage of the primary solid aggregates depends upon a number of factors, which include particle properties such as strength and morphology and the particular breakage mechanism being experienced by the solid aggregates. Since the fragment distribution function is likely to be system-specific and is notoriously difficult to experimentally determine, several forms of the fragment distribution functions have been proposed. Some of these include uniform distribution of fragments, formation of two equal fragments—symmetric fragmentation, formation of two fragments with fixed mass ratio, erosion of particles, and parabolic distribution of fragments. The distribution functions corresponding to these cases are:

Uniform

$$P(\nu_j, \nu_i) = \begin{cases} \frac{1}{\nu_i} & \text{if } 0 < \nu_j < \nu_i \\ 0 & \text{otherwise} \end{cases} \tag{5.28}$$

Symmetric fragmentation

$$P(\nu_j, \nu_i) = \begin{cases} 2 & \text{if } \nu_j = \frac{1}{2}\nu_i \\ 0 & \text{otherwise} \end{cases} \tag{5.29}$$

Mass ratio 1:4

$$P(\nu_j, \nu_i) = \begin{cases} 1 & \text{if } \nu_j = \frac{1}{5}\nu_i \\ 1 & \text{if } \nu_j = \frac{4}{5}\nu_i \\ 0 & \text{otherwise} \end{cases} \tag{5.30}$$

Erosion

$$P(\nu_j, \nu_i) = \begin{cases} 1 & \text{if } \nu_j = \nu_0 \\ 1 & \text{if } \nu_j = \nu_i - \nu_0 \\ 0 & \text{otherwise} \end{cases} \tag{5.31}$$

Parabolic

$$P(\nu_j, \nu_i) = \frac{1}{2}\frac{C_5}{\nu_i} + \frac{1 - C_5/2}{\nu_i}\left[12\left(\frac{\nu_j}{\nu_i}\right)^2 - 12\left(\frac{\nu_j}{\nu_i}\right) + 3\right] \tag{5.32}$$

In Eqn (5.32), the parabolic distribution collapses to a uniform distribution when $C_5 = 2$. If $0 < C_5 < 2$, a concave parabola is attained which means that more unequal-sized fragments are obtained than equal-sized fragments. If $2 < C_5 < 3$, a convex parabola is however attained. It should be noted that values beyond the range of 0 and 3 are not allowed because of the distribution having a negative value.

Nonetheless, Diemer and Olson (2002) have proposed a generalized fragment distribution function that allows multiple breakage fragments (>2) to be simulated. Considering the self-similar behavior of the daughter or fragment distribution where the similarity z is the ratio of daughter-to-fragment size (Ramkrishna, 2000), a suitable form of $P(v, v')$ is then given by

$$P(v, v') = \frac{\theta(z)}{v'} \tag{5.33}$$

where $z \equiv v/v'$. The generalized form of $\theta(z)$ can be expressed as

$$\theta(z) = \sum_i w_i p_i \frac{z^{q_i - 1}(1 - z)^{r_i - 1}}{\mathbf{B}(q_i, r_i)} \tag{5.34}$$

where w_i is the weighting factor, p_i is the number of fragments produced in breakage, q_i and r_i are the exponents, and $\mathbf{B}(q_i,r_i)$ is the beta function. The average number of fragments is given by

$$\overline{p} = \sum_i w_i p_i \tag{5.35}$$

in which weights w_i are expected to equal unity, in other words, $\sum_i w_i = 1$, and the mass balance imposes the following constraint:

$$\sum_i \left(\frac{p_i q_i}{p_i + q_i} \right) = 1 \tag{5.36}$$

Table 5.1 presents a variety of distribution shapes that can be obtained via the generalized fragment distribution function. In this table, δ is the Dirac delta function, w is the weighting coefficient and ε, v, v_1 and v_2 are user-defined parameters.

5.3.2 Breakage Due to Other Mechanisms

In a flaming process, particulate smoke can be formed in the gas phase as a result of incomplete combustion and high temperature pyrolysis reactions at low oxygen concentrations. In general, particulate smoke consists of the agglomeration of minute soot particles that come together to form complex chains and clusters at a particular flame location. The disappearance of soot particles toward the flame tip is attributed to the completion of carbonization before they are released into the surroundings.

Soot production is inherently a chemically-controlled phenomenon. Nearly all phases including nucleation, condensation, coagulation, surface growth, agglomeration and oxidation chemistry play an important role. Once soot particles are formed through nucleation, which is based on acetylene as the soot precursor, they can grow by three mechanisms: condensation, coagulation and surface growth. Condensation results from the

Table 5.1: Daughter distributions of various fragment distribution functions.

Type	$\theta(z)$	p	Constraints
Equisized (Kostoglou et al., 1997)	$p\delta\left(z-\frac{1}{p}\right)$	p	$p\geq 2$
Attrition (Kostoglou et al., 1997)	$\delta(z-1+\varepsilon)+\delta(z-\varepsilon)$	2	$\varepsilon \ll 1$
Power law (Vigil and Ziff, 1989)	$(\nu+1)z^{\nu-1}$	$\frac{\nu+1}{\nu}$	$0<\nu\leq 1$
Parabolic (Vigil and Ziff, 1989)	$(\nu+1)(\nu+2)z^{\nu-1}(1-z)$	$\frac{\nu+2}{\nu}$	$0<\nu\leq 2$
Parabolic (Hill and Ng, 1995)	$12(h\nu'-2)z(1-z)+2(3-h\nu')$	2	
Austin (Austin et al., 1976)	$w(\nu_1+1)z^{\nu_1-1}+(1-w)(\nu_2+1)z^{\nu_2-1}$	$w\left(1+\frac{1}{\nu_1}\right)+(1-w)\left(1+\frac{1}{\nu_2}\right)$	$\nu_1,\nu_2>0$; $w\geq \nu_1\left(\frac{\nu_2-1}{\nu_2-\nu_1}\right)$
Binary beta (Hsia and Tavlarides, 1983)	$60z^2(1-z)^2$	2	
Binary beta (McCoy and Wang, 1994)	$\frac{2}{P(\nu,\nu)}z^{\nu-1}(1-z)^{\nu-1}$	2	$\nu>0$
Uniform (Vigil and Ziff, 1989)	$(p-1)(1-z)^{p-2}$	p	$p\geq 2$

growth of particles via condensation of two-dimensional polyaromatic hydrocarbons (PAH) on three-dimensional PAH. Coagulation involves the aggregation of two soot particles leading to the formation of a larger spheroid. Since soot particles are of nanometer size, perikinetic coagulation due to Brownian motion that could occur in the free, transition and molecular regimes, as described in the section "Aggregation Kernel Due to Brownian Motion Induced Collisions" (see p. 140), is normally considered as the primary mechanism of the collision among soot particles. Surface growth of particles proceeds in conjunction with coagulation. The particles grow via the addition of carbon atoms on the surface of a soot particle due to chemical reactions with the gas phase through the hydrogen-abstraction-carbon-addition mechanism (Markatou et al., 1993). Older particles will undergo agglomeration where large clusters of particles are subsequently formed. These clusters are now the primary soot particles in the flaming process. Oxidation primarily occurs as a result of attack by oxygen molecules and hydroxyl radicals and such a process serves to contribute to the reduction of the particle size. This represents the *main breakage mechanism* of soot production. It should be noted that during surface growth, oxidants may also move to the particles and react with them to form some surface intermediates, which may then be desorbed and converted back into the gas phase. This represents another breakage mechanism during the soot production process.

5.4 Discrete Element Method—Soft-Sphere Model

The soft-sphere model requires that particle collisions are considered to be of finite duration. Duration of contact is related to the nonfinite particle stiffness, which is usually specified as the particle property. In this model, the force at contact is thus continuously varying as the particles are being deformed. The deformation of the particle is represented by the assumption of a small overlap. Forces at all contacts are determined at one instant and Newton's equations of motion are then solved to obtain new particle locations and velocities. For dense flows, this model is considered to be much more efficient than the hard-sphere model described in Chapter 2. More importantly, it can be applied to any configurations, including static and dynamic situations.

At any instant during the collision of two particles, as shown in Figure 5.1, forces acting between the two particles (assuming spheres for the purpose of illustration) can be decomposed into normal and tangential components. The formation of normal and tangential stresses during impact can be described as a decoupled problem. The normal unit vector is given by

$$\mathbf{n} = \frac{\mathbf{x}_i - \mathbf{x}_j}{|\mathbf{x}_i - \mathbf{x}_j|} \tag{5.37}$$

and the tangential direction is expressed by

$$\mathbf{t} = \frac{\mathbf{V}^{\mathrm{t}}}{|\mathbf{V}^{\mathrm{t}}|} \tag{5.38}$$

where $\mathbf{V}^{\mathrm{t}}$ is the slip velocity of the point of contact and tangential velocity can be obtained from

$$\mathbf{V}^{\mathrm{t}} = \mathbf{V}_{\mathrm{r}} - (\mathbf{V}_{\mathrm{r}} \cdot \mathbf{n})\mathbf{n} \tag{5.39}$$

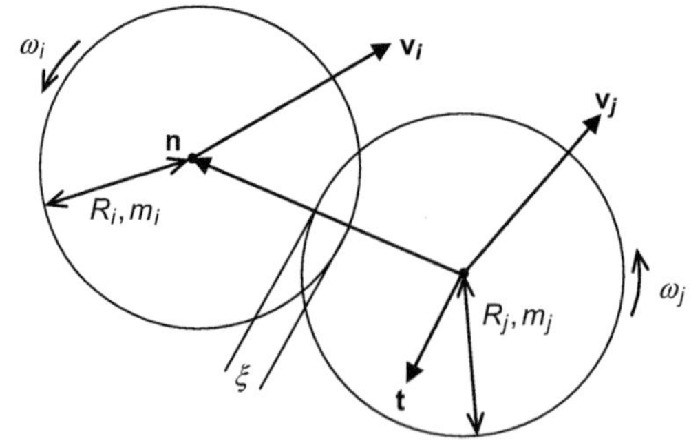

Figure 5.1
Contact between two particles for soft-sphere model.

with the relative velocity of each particle

$$\mathbf{V}_r = \mathbf{V}_i - \mathbf{V}_j - (R_i\boldsymbol{\omega}_i + R_j\boldsymbol{\omega}_j) \times \mathbf{n} \tag{5.40}$$

Particle deformation during contact is characterized by the normal overlap or displacement ξ of the two particles:

$$\xi = -(R_i + R_j) + (\mathbf{x}_i - \mathbf{x}_j) \cdot \mathbf{n} \tag{5.41}$$

while the normal displacement rate $\dot{\xi}$ can be obtained from the relative translational velocities projected in the direction of the normal unit vector, in other words,

$$\dot{\xi} = -(\mathbf{V}_i - \mathbf{V}_j) \cdot \mathbf{n} \tag{5.42}$$

The total collision force and torque fields on particle i can be written as

$$\mathbf{F}^A = \mathbf{F}^n + \mathbf{F}^t \tag{5.43}$$

$$\mathbf{M}^A = R_i\mathbf{n} \times \mathbf{F}^t \tag{5.44}$$

where $\mathbf{F}^n$ and $\mathbf{F}^t$ are the normal and tangential contact forces.

5.4.1 Particle–Particle Interaction Without Adhesion

The forces and torques acting on the particles through the soft-sphere model can be described as those acting along the line normal to the particles' centers, and those relating to the resistances from the sliding, twisting, and rolling of one particle over another (Figure 5.2).

The normal forces that are applicable according to the dependency of the normal force on the degree of overlap and the displacement rate can be categorized by different models, as reviewed in Kruggel-Emden et al. (2007, 2008). They are: continuous potential, linear and nonlinear viscoelastic and hysteretic. The most relevant models are described in the following sections.

Normal Force Due to Continuous Potentials

Continuous potentials between particles are widely used, such as those of the Lennard-Jones potential for molecular dynamics simulations on the atomic or molecular level (Verlet, 1967). Aoki and Akiyama (1995) proposed that the normal force combined the Lennard-Jones potential ($F^n_{\text{potential}}$) and linear dissipation of kinetic energy ($F^n_{\text{dissipation}}$):

$$\mathbf{F}^n = F^n_{\text{potential}}\mathbf{n} + F^n_{\text{dissipation}}\mathbf{n} = -\frac{d\phi}{dX}\mathbf{n} + F^n_{\text{dissipation}}\mathbf{n} \tag{5.45}$$

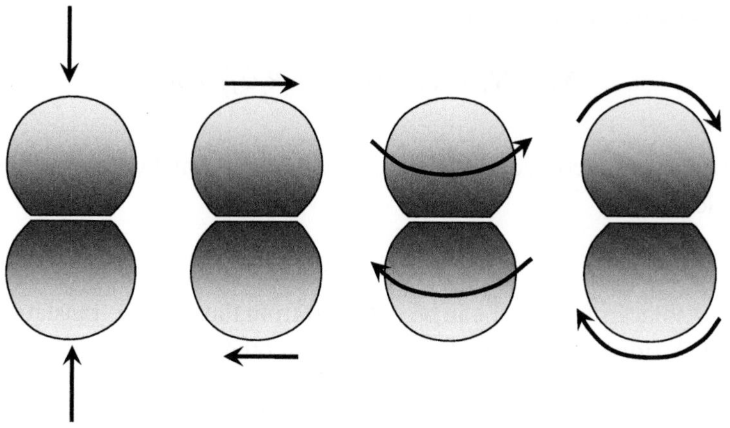

Figure 5.2
Modes of particle–particle interaction.

Given that $\phi = \varepsilon^{\text{n}}\left[\left(\frac{d_{\text{p}}}{X}\right)^{12} - \left(\frac{d_{\text{p}}}{X}\right)^{6} + \frac{1}{4}\right]$ where ε^{n} is the characteristic energy and $X = R_i + R_j - \xi$, being the distance between the two particle centers, the potential force is expressed by

$$F^{\text{n}}_{\text{potential}} = -\frac{\text{d}\phi}{\text{d}X} = -\varepsilon^{\text{n}}\left[\frac{6d_{\text{p}}^{6}}{X^{7}} - \frac{12d_{\text{p}}^{12}}{X^{13}}\right] \tag{5.46}$$

while the dissipation force can be modeled as

$$F^{\text{n}}_{\text{dissipation}} = -m_{\text{p}}\gamma^{\text{n}}\dot{\xi} \tag{5.47}$$

where γ^{n} is the velocity proportional damping coefficient. Both ε^{n} and γ^{n} are adjustable parameters. Alternatively, Langston et al. (1994) derived an equation incorporating the linear stiffness k^{n} and a velocity proportional damping γ^{n} according to

$$\mathbf{F}^{\text{n}} = F^{\text{n}}_{\text{potential}}\mathbf{n} + F^{\text{n}}_{\text{dissipation}}\mathbf{n} = -k(r)\xi\mathbf{n} - \gamma^{\text{n}}\mathbf{V}^{\text{n}} \tag{5.48}$$

where $k(r) = -\frac{\text{d}F^{\text{n}}_{\text{potential}}}{\text{d}X} = \frac{d^2\phi}{\text{d}X^2}$ with $\phi = \frac{d_{\text{p}}mg}{c}\left(\frac{d_{\text{p}}}{X}\right)^{c}$ yielding $k(r) = \left(m_{\text{p}}g(c+1)\frac{d_p^{c+1}}{X^{c+2}}\right)$ being the linear stiffness, and $\gamma^{\text{n}} = \beta^{\text{n}} \cdot 2\sqrt{m_{\text{p}}k^{\text{n}}}$ being the damping proportional to velocity; $\mathbf{V}^{\text{n}} = \dot{\xi}\mathbf{n}$ is the normal deformation velocity. The dimensionless exponent c is prescribed as a value of 36 and the damping ratio β^{n} is used as an adjustable parameter.

Normal Force Due to Linear Viscoelastic Model

In retrospect, the most frequently employed models in the discrete element method are linear (Tsuji et al., 1993). Here, the normal force comprises the modeling of elastic repulsion and viscous dissipation. The damped harmonic oscillator force can be written in a similar form as stipulated in Eqn (5.45) as

$$\mathbf{F}^{\mathrm{n}} = F^{\mathrm{n}}_{\mathrm{elastic}}\mathbf{n} + F^{\mathrm{n}}_{\mathrm{dissipation}}\mathbf{n} = -k^{\mathrm{n}}\xi\mathbf{n} - \overline{\gamma}^{\mathrm{n}}\mathbf{V}^{\mathrm{n}} \tag{5.49}$$

where k^{n} is the linear spring stiffness and $\overline{\gamma}^{\mathrm{n}}$ is the constant of velocity proportional damper. In this model, a mean coefficient of restitution and a mean collision time are defined. According to Schafer et al. (1996), the related spring stiffness and the damping coefficient can be determined via the mean coefficient of restitution e^{n} and mean collision time t^{n}:

$$e^{\mathrm{n}} = \exp\left[\frac{\gamma^{\mathrm{n}}}{2m_{\mathrm{eff}}}\pi\left(\frac{k^{\mathrm{n}}}{m_{\mathrm{eff}}} - \left(\frac{\gamma^{\mathrm{n}}}{2m_{\mathrm{eff}}}\right)^2\right)^{-1/2}\right] \tag{5.50}$$

$$t^{\mathrm{n}} = \pi\left(\frac{k^{\mathrm{n}}}{m_{\mathrm{eff}}} - \left(\frac{\gamma^{\mathrm{n}}}{2m_{\mathrm{eff}}}\right)^2\right)^{-1/2} \tag{5.51}$$

where m_{eff} is the reduced mass of the colliding bodies given by $\frac{1}{m_{\mathrm{eff}}} = \frac{1}{m_i} + \frac{1}{m_j}$.

Normal Force Due to Nonlinear Viscoelastic Model

Nonetheless, limitations of a constant coefficient of restitution and a constant duration of contact of the linear model can be overcome by applying the nonlinear spring damper models. Lee and Herrmann (1993) have proposed a partly nonlinear model:

$$\mathbf{F}^{\mathrm{n}} = F^{\mathrm{n}}_{\mathrm{elastic}}\mathbf{n} + F^{\mathrm{n}}_{\mathrm{dissipation}}\mathbf{n} = -\tilde{k}^{\mathrm{n}}\xi^{3/2}\mathbf{n} - \tilde{\gamma}^{\mathrm{n}}\mathbf{V}^{\mathrm{n}} \tag{5.52}$$

with $\tilde{k}^{\mathrm{n}}$ and $\tilde{\gamma}^{\mathrm{n}}$ being the spring stiffness and velocity damping constant, respectively. Here, the dissipative force still depends only on the linear displacement rate. According to the contact theory of Hertz (1882), the stiffness can be calculated when the physical properties such as Young's modulus (E) and Poisson ratio (σ) are known. Hence,

$$\tilde{k}^{\mathrm{n}} = \frac{4}{3}E_{\mathrm{eff}}\sqrt{R_{\mathrm{eff}}} \tag{5.53}$$

where E_{eff} is the effective Young's modulus determined by $\frac{1}{E_{\mathrm{eff}}} = \frac{1-\sigma_i^2}{E_i} + \frac{1-\sigma_j^2}{E_j}$ while R_{eff} is the reduced radius given by $\frac{1}{R_{\mathrm{eff}}} = \frac{1}{R_i} + \frac{1}{R_j}$. However, Kuwabara and Kono (1987) have proposed a fully nonlinear model:

$$\mathbf{F}^{\mathrm{n}} = F^{\mathrm{n}}_{\mathrm{elastic}}\mathbf{n} + F^{\mathrm{n}}_{\mathrm{dissipation}}\mathbf{n} = -\tilde{k}^{\mathrm{n}}\xi^{3/2}\mathbf{n} - \tilde{\gamma}^{\mathrm{n}}\mathbf{V}^{\mathrm{n}}\xi^{1/2} \tag{5.54}$$

with $\tilde{k}^{\mathrm{n}}$ evaluated according to Eqn (5.53) and the dissipative factor $\tilde{\gamma}^{\mathrm{n}}$ being an adjustable parameter. Comparing to Kuwabara and Kono (1987), Tsuji et al. (1992) have alternatively proposed a slight modification to the dissipative force with a different exponent:

$$\mathbf{F}^{\mathrm{n}} = F^{\mathrm{n}}_{\mathrm{elastic}}\mathbf{n} + F^{\mathrm{n}}_{\mathrm{dissipation}}\mathbf{n} = -\tilde{k}^{\mathrm{n}}\xi^{3/2}\mathbf{n} - \tilde{\gamma}^{\mathrm{n}}\mathbf{V}^{\mathrm{n}}\xi^{1/4} \tag{5.55}$$

Normal Force Due to Hysteretic Model

In order to include the effect of plasticity and to avoid the usage of the velocity dependent damping, hysteretic models that may be linear or nonlinear have been proposed. Notable contributors to the development of these models are: Walton and Braun (1986), Sadd et al. (1993), Thornton (1997), Vu-Quoc and Zhang (1999) and Tomas (2003). In all hysteretic models, the materials in contact suffer permanent deformation. We primarily focus on the models developed by Walton and Braun (1986) and Thornton (1997); the former proposed a partially-latching-linear-spring model, which is given by

$$\mathbf{F}^{\mathrm{n}} = F^{\mathrm{n}}\mathbf{n} \quad \text{where} \qquad F^{\mathrm{n}} = \begin{cases} -k_{\mathrm{l}}\xi & \dot{\xi} \geq 0 \\ -k_{\mathrm{ul}}(\xi - \xi_{\mathrm{f}}) & \dot{\xi} < 0 \end{cases} \tag{5.56}$$

where k_{l} and k_{ul} refer to different spring stiffnesses and ξ_{f} is the final resulting deformation of the impacting particle. Thornton (1997), on the other hand, developed a theoretical model for elastic, perfectly plastic material:

$$\mathbf{F}^{\mathrm{n}} = F^{\mathrm{n}}\mathbf{n} \quad \text{where} \qquad F^{\mathrm{n}} = \begin{cases} -\tilde{k}^{\mathrm{n}}\xi^{3/2} & \dot{\xi} \geq 0 \wedge \xi < \xi_{\mathrm{y}} & \text{elastic} \\ -\left(\tilde{k}^{\mathrm{n}}\xi_{\mathrm{y}}^{3/2} + \pi\tilde{p}_{\mathrm{y}}r_{\mathrm{eff}}(\xi - \xi_{\mathrm{f}})\right) & \dot{\xi} \geq 0 \wedge \xi \geq \xi_{\mathrm{y}} & \text{plastic} \\ -\tilde{k}^{\mathrm{n}}_{\mathrm{ul}}(\xi - \xi_{\mathrm{max}}) & \dot{\xi} < 0 & \text{nonlinear elastic} \end{cases} \tag{5.57}$$

For the force model by Walton and Braun (1986), the simplicity of the model yields an analytic solution for the final particle deformation, which can be written as

$$\xi_{\mathrm{f}} = \frac{v_0\sqrt{m_{\mathrm{eff}}}(k_{\mathrm{ul}} - k_{\mathrm{l}})}{k_{\mathrm{ul}}\sqrt{k_{\mathrm{l}}}} \tag{5.58}$$

where v_0 is the initial velocity, assuming that the particles begin to touch at $t=0$. The coefficient of restitution and duration of contact can be expressed in terms of the different spring stiffness according to

$$e^{\mathrm{n}} = \sqrt{\frac{k_{\mathrm{l}}}{k_{\mathrm{ul}}}} \tag{5.59}$$

$$t^{\mathrm{n}} = \frac{\pi\sqrt{m_{\mathrm{eff}}}\left(\sqrt{k_{\mathrm{l}}} + \sqrt{k_{\mathrm{ul}}}\right)}{2\sqrt{k_{\mathrm{l}}k_{\mathrm{ul}}}} \tag{5.60}$$

where $\frac{1}{m_{\mathrm{eff}}} = \frac{1}{m_i} + \frac{1}{m_j}$. If k_{l} and k_{ul} are kept constant and are of different values, the force model leads to a constant coefficient of restitution and constant duration of contact. The spring stiffness k_{ul} may be modeled as a combination of the spring stiffness k_{l} and actual maximum force $F^{\mathrm{n}}_{\mathrm{max}}$ by a linear relation through the parameter s:

$$k_{\mathrm{ul}} = k_{\mathrm{l}} + sF^{\mathrm{n}}_{\mathrm{max}} \tag{5.61}$$

With $F^{\mathrm{n}}_{\mathrm{max}} = k_{\mathrm{l}}\xi_{\mathrm{max}} = k_{\mathrm{l}}\sqrt{m_{\mathrm{eff}}}\, v_0/\sqrt{k_{\mathrm{l}}}$, the resulting coefficient of restitution and duration of contact depend on v_0:

$$e^{\mathrm{n}} = \sqrt{\frac{k_{\mathrm{l}}}{1 - sv_0\sqrt{m_{\mathrm{eff}}/k_{\mathrm{l}}}}} \tag{5.62}$$

$$t^{\mathrm{n}} = \frac{\pi\sqrt{m_{\mathrm{eff}}}\left(\sqrt{k_{\mathrm{l}}} + \sqrt{k_{\mathrm{l}} - s\sqrt{k_{\mathrm{l}}}\sqrt{m_{\mathrm{eff}}}\, v_0}\right)}{4\sqrt{k_{\mathrm{l}}^2 - sk_{\mathrm{l}}^2\sqrt{m_{\mathrm{eff}}}\, v_0/\sqrt{k_{\mathrm{l}}}}} \tag{5.63}$$

For the force model by Thornton (1997), the spring stiffness $\tilde{k}^{\mathrm{n}}$ in Eqn (5.55) is evaluated according to Eqn (5.51). The displacement at the yield point is determined by

$$\xi_{\mathrm{y}} = R_{\mathrm{eff}}\left(\frac{\pi\tilde{p}_{\mathrm{y}}}{2E_{\mathrm{eff}}}\right) \tag{5.64}$$

where $\frac{1}{E_{\mathrm{eff}}} = \frac{1-\sigma_i^2}{E_i} + \frac{1-\sigma_j^2}{E_j}$, $\frac{1}{R_{\mathrm{eff}}} = \frac{1}{R_i} + \frac{1}{R_j}$ and the contact pressure $\tilde{p}_{\mathrm{y}}$, a strict material property, is often used as an adjustable model parameter. In Eqn (5.55), the stiffness $\tilde{k}^{\mathrm{n}}_{\mathrm{ul}}$ for the unloading cycle, which follows a nonlinear elastic behavior, is calculated based on the product of the reduced radius during unloading and the actual maximum force. Since the reduced radius during unloading is given by

$$R_{\mathrm{eff,ul}} = \frac{R_{\mathrm{eff}}\tilde{k}^{\mathrm{n}}\xi^{3/2}}{F^{\mathrm{n}}_{\mathrm{max}}} \tag{5.65}$$

the unloading stiffness can be obtained as

$$\tilde{k}^{\mathrm{n}}_{\mathrm{ul}} = \frac{4}{3} E_{\mathrm{eff}} \sqrt{R_{\mathrm{eff,ul}}} \tag{5.66}$$

The resulting coefficient of restitution depending only on the nonlinear elastic stiffness $\tilde{k}^{\mathrm{n}}$ and contact pressure $\tilde{p}_{\mathrm{y}}$ is:

$$e^{\mathrm{n}} = \begin{cases} \sqrt{\frac{6\sqrt{3}}{5}} \sqrt{1 - \frac{1}{6}\left(\frac{v_{\mathrm{y}}}{v_0}\right)^2} \left[\frac{\left(\frac{v_{\mathrm{y}}}{v_0}\right)}{\left(\frac{v_{\mathrm{y}}}{v_0}\right) + 2\sqrt{\frac{6}{5} - \frac{1}{5}\left(\frac{v_{\mathrm{y}}}{v_0}\right)^2}} \right]^{1/4} & \dot{\xi} \geq \dot{\xi}_{\mathrm{y}} \\ 1 & \dot{\xi} < \dot{\xi}_{\mathrm{y}} \end{cases} \tag{5.67}$$

with the displacement rate at yield given by

$$\dot{\xi}_{\mathrm{y}} = \left(\frac{\pi}{2E_{\mathrm{eff}}}\right)^2 \sqrt{\frac{8\pi R_{\mathrm{eff}}^3}{15 m_{\mathrm{eff}}}} \tilde{p}_{\mathrm{y}}^{5/2} \tag{5.68}$$

The aforementioned models are limited to a certain deformation mechanism. In Kruggel-Emdeen et al. (2007), extension of models to overcome certain limitations imposed by current models have been discussed. Interested readers can refer to the article for further description on new improved models to evaluate the normal force.

Tangential Force

Tangential force arises when two particles are involved in an oblique collision or when the particles are spinning. More importantly, tangential force prevents the particle media from adapting the state of lowest energy and is therefore responsible for many observable phenomena. It is therefore essential to model the tangential force as realistically as possible. Similarly to the normal direction, there is a wide variety of complex models to describe the force displacement behavior for the tangential direction. Different models, namely linear and nonlinear, can be applied. The former models exhibit a linear dependence on the tangential displacement while the latter models reveal a nonlinear dependence on the tangential displacement or a mixed nonlinear dependence on quantities from both normal and tangential directions.

Based on the proposal by Kruggel-Emden et al. (2008), the tangential force for sliding resistance can be determined by

$$\mathbf{F}^{\mathrm{t}} = -\min\left(\underbrace{\gamma^{\mathrm{t}}|\mathbf{V}^{\mathrm{t}}|}_{\text{viscous friction}} , \underbrace{\mu|\mathbf{F}^{\mathrm{n}}|}_{\text{Coulomb friction}} \right) \mathbf{t} \tag{5.69}$$

where γ^t is a phenomenological chosen parameter influencing $\mathbf{F}^t$, only for tangential velocities close to zero, and μ is the friction coefficient. Equation (5.69) poses certain limitations especially for dense granular structures; it is unable to handle the existence of tangential elasticity, making a reversal of the tangential velocity during contact impossible and leads to a zero tangential force for zero tangential velocities. This therefore results in the collapse of quasistatic granular structures. In order to overcome such difficulties, Cundall and Strack (1979) have proposed evaluating the elongation ζ of the virtual tangential spring, being stretched and shortened while the particles stay in contact by

$$\zeta = \left(\int_{t_0}^{t} \mathbf{V}^t(t')dt' \right) \cdot \mathbf{t} \tag{5.70}$$

Equation (5.70) can adopt positive or negative values. The vector $\boldsymbol{\zeta} = \zeta \cdot \mathbf{t}$ may be oriented in the opposite direction to $\mathbf{t}$ thereby allowing the tangential static force $\mathbf{F}^t_{static} = -k^t\boldsymbol{\zeta}$, where k^t is the spring stiffness in the tangential direction to reverse the slip or tangential velocity $\mathbf{V}^t$. This static tangential force needs to be limited by the Coulomb force to allow for sliding. This leads to

$$\mathbf{F}^t = -\min\left(\underbrace{k^t|\zeta|}_{\text{elastic}}, \quad \underbrace{\mu|\mathbf{F}^n|}_{\text{Coulomb friction}} \right) \frac{\boldsymbol{\zeta}}{|\boldsymbol{\zeta}|} \tag{5.71}$$

Similar to the normal force, realistically behaving tangential force models can be expressed in the form of

$$\mathbf{F}^t = -\underbrace{k^t\boldsymbol{\zeta}}_{\text{elastic}} - \underbrace{\gamma^t\mathbf{V}^t}_{\text{viscous friction}} \tag{5.72}$$

The spring stiffness in the tangential direction k^t may be specified as a constant or determined through nonlinear models. For the latter, under the assumption of constant normal force solution for purely elastic bodies by Mindlin and Deresiewicz (1953), Tsuji et al. (1992) have proposed the spring stiffness to be evaluated based on the shear modulus and Poisson ratio according to

$$k^t = 8G_{\text{eff}}\sqrt{R_{\text{eff}}\xi} \tag{5.73}$$

where G_{eff} is the effective shear modulus determined by $\dfrac{1}{G_{\text{eff}}} = \dfrac{2-\sigma_i}{G_i} + \dfrac{2-\sigma_j}{G_j}$ while, here again, R_{eff} is the reduced radius given by $\dfrac{1}{R_{\text{eff}}} = \dfrac{1}{R_i} + \dfrac{1}{R_j}$. It should be noted that if the tangential force is limited by the Coulomb friction, the tangential displacement is given by

$$\zeta = \frac{\mathbf{F}^{\mathrm{n}}}{k^{\mathrm{t}}} \tag{5.74}$$

Alternatively, Walton and Braun (1986) have derived an additive scheme for the elastic contribution of the tangential force, which is a simplification of the approach introduced by Mindlin and Deresiewicz (1953) for the case of a constant normal force:

$$\mathbf{F}^{\mathrm{t}}_{i+1} = \mathbf{F}^{\mathrm{t}}_i + k^{\mathrm{t}}\Delta\zeta_i \tag{5.75}$$

where $\mathbf{F}^{\mathrm{t}}_{i+1}$ is the tangential force at time t, $\mathbf{F}^{\mathrm{t}}_i$ is the tangential force at the previous step and $\Delta\zeta_i$ is the change of the tangential displacement vector. The tangential spring stiffness is given by

$$k^{\mathrm{t}} = \begin{cases} k^{\mathrm{t}}_0\left(1 - \frac{\mathbf{F}^{\mathrm{t}}_i - \mathbf{F}^{\mathrm{t}^*}_i}{\mu\mathbf{F}^{\mathrm{n}}_i - \mathbf{F}^{\mathrm{t}^*}_i}\right)^{1/3} & \text{for increasing } \mathbf{F}^{\mathrm{t}}_i \\ k^{\mathrm{t}}_0\left(1 - \frac{\mathbf{F}^{\mathrm{t}}_i - \mathbf{F}^{\mathrm{t}^*}_i}{\mu\mathbf{F}^{\mathrm{n}}_i - \mathbf{F}^{\mathrm{t}^*}_i}\right)^{1/3} & \text{for decreasing } \mathbf{F}^{\mathrm{t}}_i \end{cases}$$

$$k^{\mathrm{t}}_0 = k^{\mathrm{n}}\frac{1-\sigma}{1-\sigma/2} \tag{5.76}$$

where k^{t}_0 is the initial tangential stiffness. The quantity $\mathbf{F}^{\mathrm{t}^*}_i$ is initially set to zero and set to the current value of $\mathbf{F}^{\mathrm{t}}_i$ whenever the force reverses direction. In Eqn (5.76), the change in the normal force that inevitably occurs during impact of particles is accounted for by using the instantaneous value of $\mathbf{F}^{\mathrm{n}}_i$ in determining the spring stiffness in the tangential direction.

Sliding, Twisting and Rolling Resistance

Besides sliding resistance, twisting resistance and possibly rolling resistance may need to be accounted for when two particles collide. The former occurs when two colliding particles have different rotation rates while the latter is related to the change of position of the particle–particle contact point due to the particle motion. According to Marshall (2009), the twisting resistance can be expressed by analogy to the friction model used for sliding in the form of

$$\mathbf{M}^{\mathrm{t}} = -k^{\mathrm{q}}\left(\int_{\mathrm{t}_0}^{\mathrm{t}} \boldsymbol{\Omega}^{\mathrm{t}}(t')\mathrm{d}t'\right)\cdot\mathbf{t} - \gamma^{\mathrm{q}}\boldsymbol{\Omega}^{\mathrm{t}} \tag{5.77}$$

where k^{q} is the torsional stiffness and γ^{q} is the torsional friction coefficient. The relative twisting rate $\boldsymbol{\Omega}^{\mathrm{t}}$ is defined by

$$\boldsymbol{\Omega}^{\mathrm{t}} = \left(\boldsymbol{\omega}_i - \boldsymbol{\omega}_j\right)\cdot\mathbf{n} \tag{5.78}$$

The rolling resistance can also be expressed in a similar form to Eqn (5.77) according to

$$\mathbf{M}^{\mathrm{r}} = -k^{\mathrm{q}}\left(\int_{\mathrm{t}_0}^{\mathrm{t}} \mathbf{V}^{\mathrm{r}}(t')\mathrm{d}t'\right) \cdot \mathbf{t}^{\mathrm{r}} - \gamma^{\mathrm{r}}\mathbf{V}^{\mathrm{r}} \cdot \mathbf{t}^{\mathrm{r}} \tag{5.79}$$

where k^{r} is the rolling stiffness and γ^{r} is the rolling friction coefficient. The rolling velocity can be ascertained from

$$\mathbf{V}^{\mathrm{t}} = R_{\mathrm{eff}}\left(\boldsymbol{\omega}_i - \boldsymbol{\omega}_j\right) \times \mathbf{n} - \frac{1}{2}\left(\frac{R_j - R_i}{R_j + R_i}\right)\mathbf{V}_{\mathrm{r}} \tag{5.80}$$

where $\dfrac{1}{R_{\mathrm{eff}}} = \dfrac{1}{R_i} + \dfrac{1}{R_j}$ with the direction of rolling $\mathbf{t}^{\mathrm{r}}$ defined as

$$\mathbf{t}^{\mathrm{r}} = \frac{\mathbf{V}^{\mathrm{r}}}{|\mathbf{V}^{\mathrm{r}}|} \tag{5.81}$$

The twisting and rolling torques are added into Eqn (5.44) to determine the total torque fields on the particle.

5.4.2 Particle–Particle Interaction Due to Adhesion

The phenomenon of adhesion of particles is described in this section. Within the context of the discrete element method, the adhesive forces are assumed to act on length scales much smaller than the particle size. This means that the adhesive forces have no effect until two or more particles collide. For the formation of particle agglomerates, various forces between particles act on a length scale much smaller than the particle size that give rise to adhesion between particles; these forces include van der Waals, liquid bridging due to a thin film about a particle, and gas and capillary adhesion of particles floating on an interface—which are discussed in subsequent sections. Other types of particle–particle interaction at large distances due to the surrounding electric field, magnetic and thermal gradients are also discussed. All of these different types of adhesive forces, which are schematically illustrated in Figure 5.3, are by no means exhaustive but are representative enough to give the reader a sense of the modeling challenges involving the use of an adhesive discrete element method in numerous engineering problems of interest.

van der Waals

The van der Waals force can be considered as a short-range force causing particle adhesion. This generally occurs as two particles move toward each other during a collision; the contact region grows with time along with the formation of adhesive bonds. The van der Waals force between molecules is proportional to h^{-6}, where h is the separation distance between molecular centroids. This separation distance is assumed to remain uniform throughout the contact region and constant in time, such that as the particle centroids move toward each other

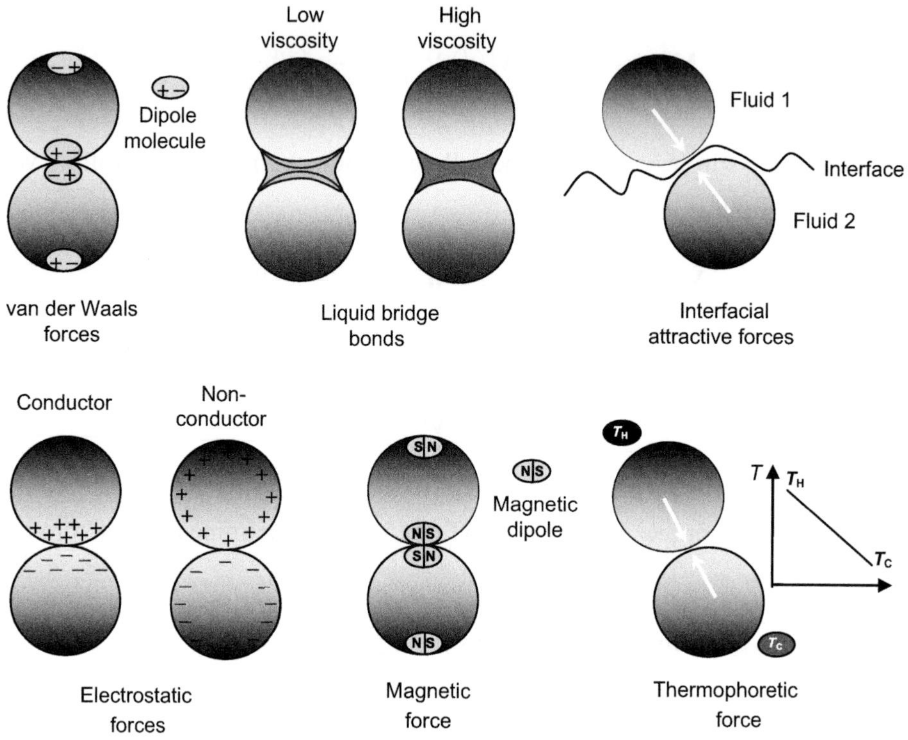

Figure 5.3
Particle adhesion effects. (For the color version of this figure, the reader is referred to the online version of this book.)

the contact region radius increases but the separation distance remains fixed. For smooth spherical particles, the separation distance is determined by the equilibrium between the attractive adhesion and the repulsive van der Waals forces. This distance is primarily determined by the fluid lubrication forces in the squeeze flow between the particle surfaces within the contact region (Davis et al., 1986; Serayssol and Davis, 1986). Since the resulting distance is of the order of 10 nm, the distance with real particles is determined by the length scale of particle surface roughness.

Essentially, there are two classic adhesion contact models that have been developed that are suitable to be embedded in discrete element method simulations of small particles. The first model proposed by Johnson, Kendall and Roberts (1971) is known as the JKR theory and the second model by Derjaguin, Muller and Toporov (1975) is known as the DMT theory; both models are for elastic contact.

The JKR theory is based on the balance among surface, potential and elastic energies. The separation distance between two spheres is given by

$$D = \frac{a^2}{R_{\text{eff}}} \tag{5.82}$$

where a is the contact region radius and $\frac{1}{R_{\mathrm{eff}}} = \frac{1}{R_i} + \frac{1}{R_j}$. In the JKR model, the relationship between the normal elastic-adhesive force and contact region radius takes the form:

$$a^3 = \frac{3R_{\mathrm{eff}}}{4E_{\mathrm{eff}}}\left(F_{\mathrm{ne}} + 6\pi\gamma R_{\mathrm{eff}} + \sqrt{12\pi\gamma R_{\mathrm{eff}}F_{\mathrm{ne}} + \left(6\pi\gamma R_{\mathrm{eff}}\right)^2}\right) \tag{5.83}$$

where γ is the adhesive energy density on the surface. The normal overlap or displacement of the two particles is calculated from the contact region radius by $\xi = a^2/r_{\mathrm{eff}} - 2\sqrt{\pi\gamma a/E_{\mathrm{eff}}}$, where $\frac{1}{E_{\mathrm{eff}}} = \frac{1-\sigma_i^2}{E_i} + \frac{1-\sigma_j^2}{E_j}$. When the surface energy is zero, i.e. $\gamma = 0$, the Hertz equation for contact between two spheres is recovered. By setting $a = 0$, the pull-off force at which the spheres are separated is predicted to be

$$F_{\mathrm{c}} = -3\gamma\pi R_{\mathrm{eff}} \tag{5.84}$$

The DMT theory differs from the JKR theory in that the model assumes there is only compressive Hertzian stress inside the contact area and adhesive stress outside the contact area. The adhesion between the two particles does not alter the profile of the asperity during the contact time. Therefore, a Hertzian profile is assumed. The adhesive contact in the DMT model comprises the superposition of an elastic Hertz repulsive force and a constant adhesion force, which can be written as

$$a^3 = \frac{3R_{\mathrm{eff}}}{4E_{\mathrm{eff}}}\left(F_{\mathrm{ne}} + 4\pi\gamma R_{\mathrm{eff}}\right) \tag{5.85}$$

where the contact radius region satisfies $a = \sqrt{\xi R_{\mathrm{eff}}}$. By setting $a = 0$, the pull-off force at which the spheres are separated is

$$F_{\mathrm{c}} = -4\gamma\pi R_{\mathrm{eff}} \tag{5.86}$$

It should be noted that DMT and JKR have been derived from different assumptions. Tabor (1977) pointed out that each model is valid under different conditions. In particular, JKR is suitable for *soft* materials with a large surface energy while DMT is suitable for *hard* materials with a low surface energy. The transition between these two regimes can be characterized by the Tabor parameter, which can defined as

$$\mu_{\mathrm{Ta}} = \left(\frac{4R_{\mathrm{eff}}\gamma^2}{E_{\mathrm{eff}}^2 D^3}\right)^{1/3} \tag{5.87}$$

which indicates a measure of the ratio of elastic deformation to the range of surface force. Liu et al. (2009) discussed the applicability of different adhesion models for use in a discrete element method of adhesive particle flows. They have found that for most practical applications, DMT may apply for $\mu_{Ta} < 0.1$ while JKR applies for $\mu_{Ta} > 3.0$. In between these two limits, it would appear that JKR could give good predictions of contact size and compliance even in conditions well outside the expected JKR range. For example, Greenwood (1997) has reported that the representative radii of contact can be well predicted by the JKR theory where μ_{Ta} is as small as 0.3.

Liquid Bridging

The prevalence of this adhesive force results in the formation of a *liquid bridge* stretching between two particles as they collide, especially in a humid environment. This liquid bridge thereby introduces a capillary force that pulls the particles toward each other for adhesion to occur. Also, the liquid film introduces an enhanced frictional force between the particles due to the higher viscosity of the liquid filling the contact region when compared to the surrounding gas. The total liquid bridge therefore consists of the sum of the capillary and enhanced frictional forces. To account for particle collision with liquid bridging, appropriate models for the forces are required as well as the criterion for rupture of the liquid bridges at a critical separation distance.

According to Maugis (1987), the capillary force for equal-sized spherical particles with an effective radius R_{eff} can be expressed as

$$F_{cap} = 4\pi R_{eff} G_f \cos\theta \tag{5.88}$$

where θ is the static contact angle and $\dfrac{1}{R_{eff}} = \dfrac{1}{R_i} + \dfrac{1}{R_j}$. Among the many different expressions, Pitois et al. (2000) have shown that Eqn (5.88) compares rather well with experimental data. The coefficient G_f in Eqn (5.88) is defined by

$$G_f = 1 - \frac{1}{\sqrt{1 + V_L/\pi R_{eff} h^2}} \tag{5.89}$$

such that G_f approaches unity for cases where the particles are touching. In Eqn (5.89), h is the minimum separation distance between spherical particles and V_L is the liquid bridge volume. As the separation distance increases, the capillary force monotonically decays. Nonetheless, it should be noted that Eqn (5.89) can also be reasonably applied for unequal particle size over the full range of separation distances with the exception being at close contact and very close to the rupture distance.

In order to determine the critical particle separation distance at which the liquid bridge ruptures, Pitois et al. (2000) proposed an expression that accounts for the effects of particle motion on rupture as

$$h_{\text{rupture}} = (1 + 0.5\theta)\left(1 + \sqrt{Ca}\right)\left(\frac{V_{\text{L}}}{4R_{\text{eff}}^2}\right) \tag{5.90}$$

where Ca is the capillary number defined as $\mu^c|\mathbf{v_r} \cdot \mathbf{n}|/\sigma$, which has been written in terms of the normal component of the particle relative velocity. For $Ca = 0$, Eqn (5.54) reduces to the static rupture criterion (Lian et al., 1993). The factor $1 + \sqrt{Ca}$ has been introduced based on the best fit to experimental data for Ca values between $0.001 \leq Ca \leq 0.1$.

The mechanics of thin liquid films are well described by the well-known Reynolds equation. Based on lubrication theory, the enhanced frictional force acting on spherical particles according to Matthewson (1988) has the form:

$$F_{\text{visc}} = 6\pi\mu^c R_{\text{eff}}^2 G_{\text{f}}^2 \frac{1}{h}\frac{\mathrm{d}h}{\mathrm{d}t} \tag{5.91}$$

where μ^c is the viscosity of the continuous phase. This expression can be related to the liquid bridge volume via the coefficient G_{f} in Eqn (5.89).

Interfacial Attractive Force

For particles that float on an interface between a liquid and a gas or between two immiscible liquids, an attractive force is exhibited pulling these particles toward each other due to the deformation of the interface in the region between them. Such interfacial deformation leads to the presence of a capillary force, which can be evaluated by the solution of the Young-Laplace equation for the interface shape around each particle. Nicolson (1949) demonstrated that the force calculation can be simplified for situations where the following criterion is satisfied. In other words,

$$\frac{Bo(1-\chi)}{\chi} << 1 \tag{5.92}$$

where Bo is the particle Bond number and χ is the density ratio. When Eqn (5.92) is satisfied, the Young-Laplace equation can be linearized and analytically solved for each particle; then, superimpose the individual particle solutions to obtain an approximate solution for interface deformation. Based on the linear superposition approximation theory, the capillary force on the ith particle caused by jth particle can be expressed as

$$\mathbf{F}_{\text{int}} = -2\pi\sigma Q_{\text{p}}^2 q K_1(qX)\mathbf{n} \tag{5.93}$$

where $K_1(X)$ is the modified Bessel function of the second kind of order 1 and X is the distance between the centroids of two particles, i.e. $X = R_i + R_j - \xi$. In Eqn (5.93), q and Q_{p} are respectively defined by

$$q = \sqrt{\frac{g\rho^{c}}{\sigma}} \tag{5.94}$$

$$Q_{p} = \frac{-\tan(\theta - \beta)}{qK_{1}\left(0.5qd_{p}\sin\beta\right)} \tag{5.95}$$

where β is the cone angle that is determined by the balance of forces normal to the interface.

For small Bond number, $qd_{p} = \sqrt{Bo} << 1$ so that $K_{1}(0.5qd_{p}\sin\beta) \approx 2/qd\sin\beta$. Equation (5.95) can thus be simplified as

$$Q_{p} = -\frac{d_{p}}{2}\sin\beta\tan(\theta - \beta) \tag{5.96}$$

Also, for cases with a significant interparticle attractive force, it can be concluded that $qX << 1$. Employing an asymptotic expansion for the Bessel function with small argument yields

$$K_{1}(qX) \approx 1/qX \tag{5.97}$$

Substituting Eqns (5.96) and (5.97) into Eqn (5.93) gives the approximation of the interaction forces between interfacial particles:

$$\mathbf{F}_{\text{int}} = -\pi\sigma\frac{d_{p}^{2}}{2}\sin^{2}\theta\tan^{2}(\theta - \beta)\frac{\mathbf{n}}{r_{ij}} \tag{5.98}$$

Other Types of Field—Particle Interaction

In addition to the different types of surrounding fluid forces that may act on particles as described in Chapter 2, particle interactions in some cases may be subjected to external electrical, magnetic and temperature fields. Owing to the perturbations in these fields, the effects of field—particle interaction can significantly influence particles to be attracted toward each other thereby providing the premise for particle collision and eventual adhesion between particles.

First, the problem of particle transport and adhesion under the action of electric fields can be modeled in the context of the discrete element method by considering the electric field forces acting on the particles. For each particle with a charge Q and an effective dipole moment $\mathbf{p}$, the particle is subjected to both Coulomb force $\mathbf{F}_{c}$, dielectrophoretic force $\mathbf{F}_{p}$ and torque $\mathbf{M}_{p}$ (Li et al., 2011):

$$\mathbf{F}_{c} = Q\mathbf{E} \tag{5.99}$$

$$\mathbf{F}_{\mathrm{p}} = \mathbf{p} \cdot \nabla \mathbf{E} \tag{5.100}$$

$$\mathbf{M}_{\mathrm{p}} = \mathbf{p} \times \mathbf{E} \tag{5.101}$$

In Eqns (5.100) and (5.101), the particle dipole moment $\mathbf{p}$ comprises the sum of the permanent dipole $\mathbf{p}_{\mathrm{o}}$ and the induced dipole moment $\mathbf{p}_{\mathrm{i}}$:

$$\mathbf{p} = \mathbf{p}_{\mathrm{o}} + \mathbf{p}_{\mathrm{i}} \tag{5.102}$$

Suitable values for the permanent dipole can usually be obtained based on particle materials while the induced dipole moment is determined depending on the particular electric fields acting on the particles.

In a direct current electric field, the induced dipole moment is aligned parallel to the electric field vector $\mathbf{E}$ by

$$\mathbf{p}_{\mathrm{i}} = \frac{\pi}{2} \varepsilon^{\mathrm{f}} K d_{\mathrm{p}}^{3} \mathbf{E} \tag{5.103}$$

where K is the Clausius–Mossotti function, which depends on the fluid and particle dielectric permittivities, ε^{f} and ε^{p}, for a dielectric material and on the fluid and particle conductivities, λ^{f} and λ^{p}, for a conducting material such that

$$K = \frac{\varepsilon^{\mathrm{p}} - \varepsilon^{\mathrm{f}}}{\varepsilon^{\mathrm{p}} + 2\varepsilon^{\mathrm{f}}} \quad \text{or} \quad K = \frac{\lambda^{\mathrm{p}} - \lambda^{\mathrm{f}}}{\lambda^{\mathrm{p}} + 2\lambda^{\mathrm{f}}} \tag{5.104}$$

varying within the range $-0.5 \leq K \leq 1$.

In an alternating current field, the electric field vector $\mathbf{E}$ is of a more complex nature, which is given by

$$\mathbf{E}(\mathbf{x}, t) = \mathrm{Re}\{\overline{\mathbf{E}}(\mathbf{x}) \exp(j\omega t)\} \tag{5.105}$$

where $\overline{\mathbf{E}}(\mathbf{x})$ is complex-valued, $j \equiv \sqrt{-1}$, ω is the oscillation frequency, t is the time and Re{ } is the real part. In this particular case, the time-average dielectrophoretic force and torques become

$$\overline{\mathbf{F}}_{\mathrm{p}} = \frac{\pi}{4} \varepsilon^{\mathrm{f}} d_{\mathrm{p}}^{3} \, \mathrm{Re}\{K(\omega)\} \nabla E_{\mathrm{rms}}^{2} \tag{5.106}$$

$$\overline{\mathbf{M}}_{\mathrm{p}} = -\frac{\pi}{2} \varepsilon^{\mathrm{f}} d_{\mathrm{p}}^{3} \, \mathrm{Im}\{K(\omega)\} E_{\mathrm{o}}^{2} \widehat{\mathbf{z}} \tag{5.107}$$

where $\widehat{\mathbf{z}}$ is a unit vector orthogonal to $\mathbf{E}$ such that $\overline{\mathbf{E}} = E_{\mathrm{o}}(\widehat{\mathbf{x}} - j\widehat{\mathbf{y}})$ and E_{rms} is the root-mean-square electric field value. In Eqns (5.106) and (5.107), the Clausius–Mossotti function for particles with ohmic (or dielectric) losses takes the form

$$K(\omega) = \frac{\varepsilon^{\rm p} - \varepsilon^{\rm f} - \dfrac{j(\lambda^{\rm p} - \lambda^{\rm f})}{\omega}}{\varepsilon^{\rm p} + 2\varepsilon^{\rm f} - \dfrac{j(\lambda^{\rm p} - 2\lambda^{\rm f})}{\omega}} \tag{5.108}$$

Defining the Maxwell–Wagner relaxation time scale as

$$\tau_{\rm MW} = \frac{\varepsilon^{\rm p} + 2\varepsilon^{\rm f}}{\lambda^{\rm p} + 2\lambda^{\rm f}} \tag{5.109}$$

the following limiting scales can be observed:

$$\mathrm{Re}\{K(\omega)\} \rightarrow \begin{cases} \dfrac{\lambda^{\rm p} - \lambda^{\rm f}}{\lambda^{\rm p} + 2\lambda^{\rm f}} & \text{for} \quad \omega\tau_{\rm MW} << 1 \\ \dfrac{\varepsilon^{\rm p} - \varepsilon^{\rm f}}{\varepsilon^{\rm p} + 2\varepsilon^{\rm f}} & \text{for} \quad \omega\tau_{\rm MW} >> 1 \end{cases} \tag{5.110}$$

It can be seen that the Clausius–Mossotti function varies with the oscillation frequency of the electric field. In some cases, it can even change sign as the electric field increases. The different directions of the dielectrophoretic force acting on particles with positive and negative Clausius–Mossotti function values play a significant role in the distribution of particles in the surrounding fluid.

Second, particle transport and adhesion under the action of magnetic fields can also be modeled in a similar manner to the electric field forces acting on the particles. For each particle, the force and torques on the ith particle due to magnetism (Kargulewicz et al., 2012) are:

$$\mathbf{F}_i^{\rm m} = \nabla(\mathbf{m}_i \cdot \mathbf{B}(\mathbf{x}_i)) \tag{5.111}$$

$$\mathbf{M}_i^{\rm m} = \mathbf{m}_i \times \mathbf{B}(\mathbf{x}_i) \tag{5.112}$$

where $\mathbf{m}_i$ is the magnetic dipole moment of the ith particle. The resulting field $\mathbf{B}(\mathbf{x}_i)$ in Eqns (5.111) and (5.112) is given by

$$\mathbf{B}(\mathbf{x}_i) = \mu_0\left(\sum_j^{n_{\rm m}} \mathbf{H}(\mathbf{x}_i, \mathbf{x}_j) + \mathbf{H}^{\rm ext}\right) \tag{5.113}$$

where μ_0 is the magnetic permeability in a vacuum ($= 4\pi \times 10^{-7}$ N/A^2). The total secondary magnetic field generated by other magnetized particles $\sum_j^{n_{\rm m}} \mathbf{H}(\mathbf{x}_i, \mathbf{x}_j)$ is:

$$\sum_j^{n_m} \mathbf{H}(\mathbf{x}_i, \mathbf{x}_j) = \frac{1}{4\pi R_{ij}^3}\left(3\hat{\mathbf{e}}_{ij}(\mathbf{m}_j \cdot \hat{\mathbf{e}}_{ij}) - \mathbf{m}_j\right) \tag{5.114}$$

with $\mathbf{x}_{ij} = \mathbf{x}_i - \mathbf{x}_j$, $R_{ij} = |\mathbf{x}_{ij}|$ and $\widehat{\mathbf{e}}_{ij} = \mathbf{x}_{ij}/R_{ij}$, while $\mathbf{H}^{\text{ext}}$ is the externally applied primary magnetic field. It should be noted that Eqn (5.114) has been arrived at after the consideration of ferromagnetic materials where the phenomena of saturation and hysteresis occur. For sufficiently large external magnetic field, a saturated magnetization occurs.

Third, the phenomenon of thermophoresis can be described by which a particle suspended in a fluid moves from hot to cold regions due to a force induced by an external temperature gradient (Tyndall, 1870). The thermophoretic force is strongly dependent on the Knudsen number (Kn). In the free-molecule regime ($Kn >> 1$) an expression for the thermophoretic force proposed by Waldman (1966) takes the form:

$$F_{\text{th}} = -\frac{8}{15}\frac{\lambda^{\text{tr}} d_{\text{p}}^2}{\overline{c}} \nabla T \tag{5.115}$$

where T is the absolute temperature and λ^{tr} is the gas thermal conductivity that represents only the translational part of the thermal energy. The mean speed $\overline{c}$ of gas molecules can be written as

$$\overline{c} = \sqrt{\frac{8k_{\text{B}}T}{\pi m^{g}}} \tag{5.116}$$

where m^g is the mass of gas molecule and k_{B} is the Boltzmann constant.

For immediate Knudsen numbers, a general form according to Li et al. (2011) can be expressed by

$$F_{\text{th}} = -\frac{\sqrt{\pi}}{8} h(Kn, \kappa_{21}) \; \frac{\lambda^{\text{tr}} d_{\text{p}}^2}{\overline{c}} \nabla T \tag{5.117}$$

Talbot et al. (1980) presented an empirical formula for the dimensionless thermophoretic force $h(Kn,\kappa_{21})$, which was based on the consideration of the slip regime or near-continuum regime. Yamamoto and Ishihara (1988) and Loyalka (1992) applied the linearized Boltzmann equation to derive suitable expressions for $h(Kn,\kappa_{21})$ which are dependent on the Knudsen number (Kn) and the gas–particle thermal conductivity ratio (κ_{21}). More details on the formulation of the different expressions for the dimensionless thermophoretic force are left to interested readers.

5.5 Summary

Mechanistic models for solid particle aggregation and breakage have been described in the first half of this chapter. With regard to the aggregation processes of a given set of particles, they are: perikinetic coagulation due to Brownian motion, shear-induced orthokinetic coagulation, and differential sedimentation aggregation. The characteristics of physical models for the aggregation kernel and collision frequency are discussed. With regard to breakage, fragmentation of aggregates due to hydrodynamic stresses is considered. It should be noted that the theory for particle breakage is not as well developed as the theory of particle aggregation. The characteristics of physical models for the breakage kernel and fragment distribution (daughter size distribution) are discussed. A distinctive difference between the break up of bubbles or drops and breakage of aggregates is that the former assumes a binary break up while the latter may experience binary, ternary or even quaternary breakage.

The discrete element method based on the soft-sphere model has been described in the second half of this chapter. A wide range of forces and torques acting on the particles due to collision events, which include the normal forces, elastic and nonelastic, and the resistance from particle sliding, twisting and rolling motions, is discussed. Collision forces and torques playing a critical role in determining the aggregate and breakage dynamics of particles are also discussed, which include adhesion driven by van der Waals, liquid bridging due to a thin film about a particle and a gas and capillary adhesion of particles floating on an interface as well as other types of particle–particle interaction at large distances due to the surrounding electric field, and magnetic and thermal gradients.

CHAPTER 6

Solution Methods and Turbulence Modeling

6.1 Introduction

Depending on which frame is being considered to obtain the required computational solutions for multiphase flows, there are essentially two reference frames for which solution methods are specifically developed. The first reference frame is the Eulerian reference frame—fluid passes through fixed differential control volumes of which the characteristics of the continuous phase are obtained by solving transport equations in *partial differential* form in a given coordinate system.

In the computational multiphase fluid dynamics framework, the process of obtaining computational solutions generally comprises two stages. In the *first stage*, the partial differential equations based on conservation laws including the consideration of boundary and initial conditions are required to be converted into a system of discrete algebraic equations. This stage is normally known as the discretization stage. In the *second stage*, appropriate numerical methods are implemented to provide approximate or computational solutions to the governing equations of the multifluid model. Among the many available discretization techniques, the *finite volume method* is described in this chapter in order to demonstrate the step-by-step process of converting the governing equations into a system of algebraic equations.

In the generic population balance framework, the equation governing the particle number density function undergoes the same two stages analogous to the governing equations in the computational multiphase fluid dynamics framework since it is also solved in the same Eulerian reference frame. Nevertheless, the complex phenomenological nature of particle dynamics demands a variety of different numerical approaches to solve the population balance equation and determine the particle size distribution in space as well as in time.

The second reference frame is the Lagrangian reference frame, which involves the tracking of individual particles (or parcels of particles) as they move through the computational domain. In this instance, the reference frame moves with the particles, and the instantaneous position of a particle can be considered as a function of the location where from it originated and the time elapsed. The need to determine the motion of particles in the Lagrangian reference frame requires numerical procedures that can aptly solve equations of the *ordinary differential* form. Consideration of appropriate solution methods that are generally employed across a wide range of length and time scales for Lagrangian procedures such as molecular dynamics, Brownian dynamics and the discrete element method, is described herein.

Since almost all multiphase flows of engineering significance are turbulent in nature, one practical approach to resolve turbulent multiphase flows is to employ time or mass-weighted averaging of the equations governing the mass, momentum and energy. The Favre-averaged (mass-weighted averaging) approach is introduced, which leads to the well-known Reynolds-Averaged Navier–Stokes (RANS) equations. Since no single turbulence model can be readily employed to span the wide range of turbulence states and none is expected to be universally valid for all types of multiphase flows, useful turbulence models can be applied to resolve turbulent multiphase flows, which are introduced and described in this chapter. With the advancement of quicker and robust numerical algorithms as well as increasingly powerful digital computers, the ability to calculate turbulent multiphase flows at sufficiently high enough spatial and temporal resolution, subject to the availability of computational resources, allows the concept of Large Eddy Simulation (LES) to be employed. Such an approach involves the resolution of the structure of the turbulent flow as distinct transport of large and small-scale motions. On this basis, the large-scale motion is directly simulated on a scale the underlying computational mesh will allow, while the small-scale motion is accordingly modeled. Since the large-scale motion is generally much more energetic and by far the more effective transporter of the conserved properties than are the small-scale ones, such an approach treats the large eddies precisely but approximates the small eddies so long as the smaller scalar turbulence physically exhibits isotropic turbulence behavior in reality. Typical LES models applicable for turbulent multiphase flows are described in this chapter.

6.2 Solution Methods for Eulerian Models

In the computational multiphase fluid dynamics framework, all solution methods for Eulerian models are required to utilize some form of discretization.

The oldest discretization approach is the *finite difference method*. Through consideration of Taylor series expansions, appropriate finite difference expressions can be formulated to aptly approximate the partial derivatives of the governing equations, which yield an algebraic equation for the flow solution at each point of the grid system. This method is commonly applied to *structured* grids since it requires a mesh having a high degree of regularity. The grid spacing between the nodal points need not be uniform; there are limits on the amount of grid stretching or distortion that can be imposed to maintain accuracy. These finite difference structured grids must conform to the constraints of general coordinate systems such as Cartesian grids comprised of four- or six-sided computational domains in two or three dimensions. However, the use of an intermediate coordinate mapping, such as the body-fitted coordinate system, allows this major geometrical constraint to be relaxed in order to accommodate complex shapes.

Other discretization methods that are also available include the *finite element method* and *spectral method*. On one hand, the finite element method employs simple piecewise polynomial functions on local elements to describe the variations of the unknown flow variables. The concept of weighted residuals is introduced to measure the errors associated with the approximate functions, which are minimized with successive solution iterations, up to the limits of accuracy achievable by the level of geometric discretization. A set of nonlinear algebraic equations for the unknown terms of the approximating functions is solved, yielding a flow solution. This method has the ability to accommodate *unstructured* grids for arbitrary geometries. However, it generally requires greater computational resources and computer processing power. On the other hand, the unknowns of the governing equations are replaced with a truncated series via the spectral method. The main difference between the spectral method and finite difference and finite element methods is that global approximation is adopted either by means of a truncated Fourier series or a series of Chebyshev polynomials for the entire flow domain instead of local approximations. The discrepancy between the exact solution and the approximation is dealt with by using a weighted residuals concept similar to the finite element method.

The *finite volume method* is the most commonly used discretization approach in today's computational multiphase fluid dynamics. Like the finite element method, it has the capacity of handling arbitrary geometries with ease. It can be applied to structured and unstructured grids. It also bears many similarities to the finite difference method; the method is therefore simply applied. Because of the many associated advantages this method possesses, it will be further described and elaborated on in Section 6.4.

In multiphase flow problems, a range of explicit or implicit procedures such as the *fully explicit*, *partially implicit* or *fully implicit methods* may need to be adopted to better handle the interfacial source terms that appear in the transport equations governing the conservation of mass, momentum and energy in order to promote convergence of the numerical solution. For cases where the coupling between the phases is very *tight*, an implicit treatment is required. The fully implicit Partial Elimination Algorithm and the semi-implicit simultaneous solution of nonlinearly coupled equations (SINCE) will be described for the effective treatment of the interphase coupling terms. Otherwise, the explicit method can be employed when the coupling is relatively *weak*.

In the Eulerian framework, the algebraic equations can be solved according to the full or partial simultaneous solution or sequential solution of the balance equations. The approach based on the full or partial simultaneous solution necessitates the treatment of the interphase coupling in a more implicit fashion, which in turn reduces the likelihood of the numerical solutions diverging. This approach, which solves the algebraic equations via coupled solvers, in general, requires relatively larger computational resources. In contrast, the approach based on the sequential solution consists of the numerical solution of the algebraic equations being attained through the application of commonly available iterative solvers except that the interphase terms appearing in the balance equations require special treatment. It may be necessary to implement the more implicit procedure of the interphase slip algorithm (IPSA) such as the interphase slip algorithm-coupled (IPSA-C)—an alternative solution algorithm to the sequential iterative approach—when the interphase coupling is very tight.

As will be discussed in the next section, the systematic arrangement of discrete number of points throughout the flow field, normally called a *mesh*, is a significant consideration in computational multiphase fluid dynamics. Application on different types of meshes that can be generated for a given problem is a serious matter. This is because generation of elements to fill the domain can result in either great success or utter failure in attaining the numerical solutions of the multiphase flow problems. *Grid generation* remains an active area of ongoing research and development and it suffices to say that materials that will be presented henceforth will serve only to scratch the surface of this important research area. Primarily focusing on the application of meshes targeted for multiphase flow problems, the materials in the next section should serve, at the very least, to provide the reader with some basic ideas of grid generation and practical guidelines in the prospect of handling more sophisticated multiphase flow configurations that are found in practice.

6.3 Mesh Systems

By definition, a *structured mesh* is a mesh containing cells having either a regular-shaped element with four-nodal corner points in two dimensions or a hexahedral-shaped element with eight-nodal corner points in three dimensions, while an *unstructured mesh* can be described as a mesh overlay with cells in the form of either a triangle-shape element in two dimensions or a tetrahedron-shape element in three dimensions.

In handling simple multiphase flow problems, the use of a structured mesh brings about a number of benefits in multiphase computations. Being relatively easy to construct, it provides the foundation whereby novel ideas or concepts to resolve complex features of multiphase flows can be investigated in a more efficient way. This then allows a more rigorous and thorough assessment of any proposals of new models or enhancements to the numerical algorithms that can be performed without specifically dealing with the complexities associated with grid generation. Nevertheless, most practical multiphase problems involve complex geometries that do not exactly fit in Cartesian coordinates where a structure mesh could be immediately constructed. Because of this, the use of an unstructured mesh is becoming more prevalent and widespread in many computational multiphase fluid dynamics applications. Nowadays, the majority of computer codes

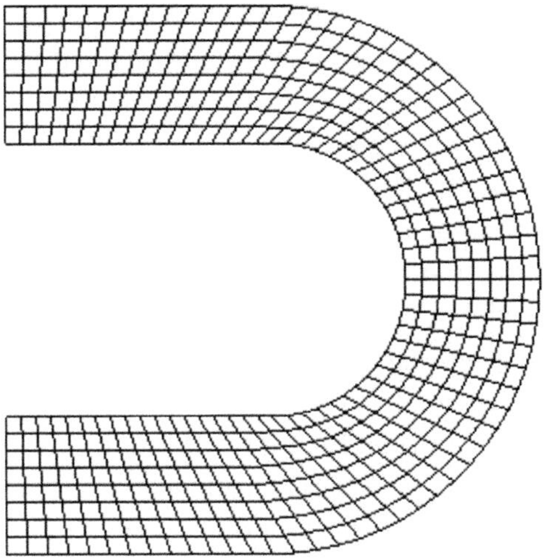

Figure 6.1
Structured mesh for the 180° bend geometry.

commercially available are based on the generation of unstructured elements because such an approach provides unlimited geometric flexibility and allows the most efficient use of computing resources for complex flows.

For the purpose of illustration, consider the problem of a multiphase flow inside a 180° bend. Figures 6.1 and 6.2 demonstrate two typical grid designs that can be achieved through the use of structured and unstructured mesh elements. It can be clearly observed that there is an order of regularity to the arrangement of the elements filling the interior region of the 180° bend

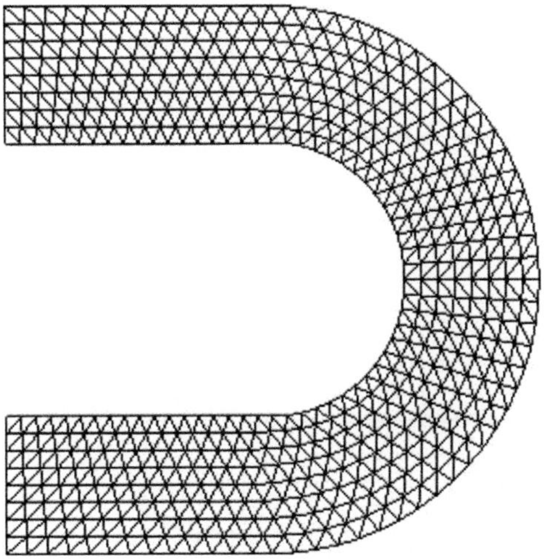

Figure 6.2
Unstructured mesh for the 180° bend geometry.

geometry in the overlay structured mesh while the opposite is true for the overlay unstructured mesh. For more complex geometries, unstructured meshing has the edge over structured meshing by providing maximum flexibility in matching the elements with highly curved boundaries and inserting the required elements to resolve multiphase flow regions where they matter most in areas of significant high gradients. Nevertheless, it is well known that in order to achieve comparable accuracies, unstructured meshing generally results in an overlay mesh comprising more triangle- or tetrahedron-shaped elements in comparison to structured meshing of four-nodal or hexahedral-shape elements for the same geometry. Also, the use of triangle- or tetrahedron-shaped elements should be avoided for near-wall flows to resolve the viscous boundary layers. In these regions, prismatic or hexahedral cells are preferred because of their regular shape. Figure 6.3 illustrates a *hybrid* mesh that combines different element types in matching appropriate elements with boundary surfaces and allocating various element types to other parts of the complex flow regions. For the particular example of the 180° bend geometry, grid quality can be enhanced through the placement of quadrilateral elements in resolving the viscous boundary layers near the walls, while triangular elements are generated for the rest of the flow domain. This normally leads to both accurate solutions and better convergence for the numerical solution methods. Furthermore, there is increasing interest toward the development of a mesh containing *polyhedral* element in resolving a range of practical multiphase flow problems. A polyhedral mesh may be created by cell agglomeration that combines tetrahedral and polyhedral elements. Figure 6.4 depicts an example of a polyhedral mesh for the 180° bend geometry, which results in a considerable reduction of the overall element count. Cell agglomeration has the capacity of improving the original mesh by converting particular regions with highly-skewed tetrahedral elements to those with polyhedral elements, thereby improving mesh quality. The use of a polyhedral

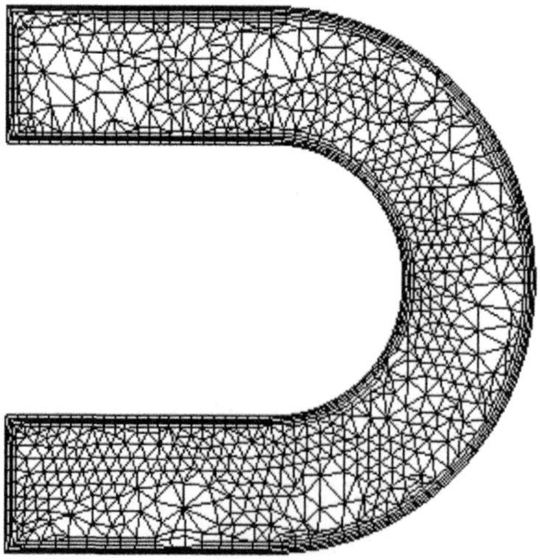

Figure 6.3
Hybrid mesh for the 180° bend geometry.

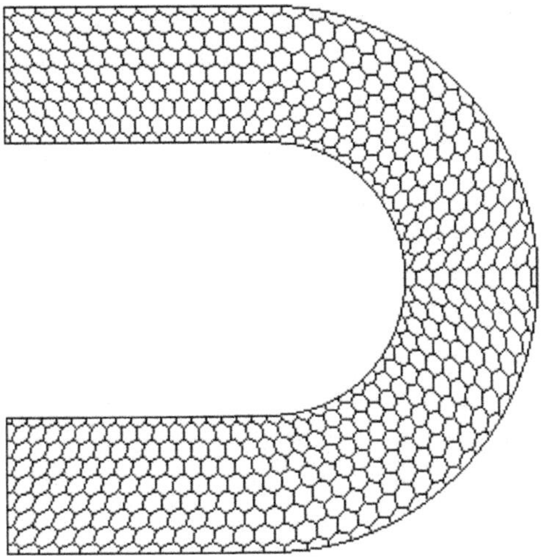

Figure 6.4
Polyhedral mesh for the 180° bend geometry.

mesh also leads to quicker convergence of the numerical solution. A clear potential benefit of applying a polyhedral mesh is that it allows the flexibility of an unstructured mesh to be applied to a complex geometry without the computational overheads associated with a large amount of tetrahedral elements. Polyhedral meshing has been shown thus far to have considerable advantages over tetrahedral meshing with regard to the attained accuracy and efficiency of numerical computations.

Grid quality of a generated mesh depends on the consideration of the cell shape: *aspect ratio*, *skewness*, *warp angle* or included angle of adjacent faces. Figure 6.5 illustrates a quadrilateral cell having mesh spacing of $\Delta\eta$ and $\Delta\xi$ and an angle of θ between the grid lines of the cell. Accordingly, we can define the grid *aspect ratio* of the cell as $AR = \Delta y/\Delta x$. Large aspect

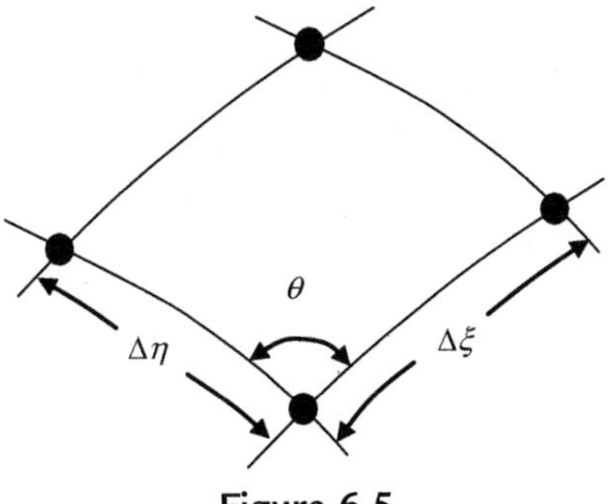

Figure 6.5
A quadrilateral cell having mesh spacing of $\Delta\xi$ and $\Delta\eta$ and an angle of θ between the grid lines of the cell.

ratios should always be avoided in important regions inside the interior flow domain as they can degrade the solution accuracy and may result in possible poor iterative convergence (or divergence) depending on the flow solver during the numerical computations. Whenever possible, it is recommended that AR is maintained within the range of $0.2 < AR < 5$ within the interior region. For near-wall boundaries the condition for AR can, however, be relaxed. If the fluid flow is in the y direction, the need to appropriately choose small Δx mesh spacing in the x direction will generally yield $AR > 5$. In such cases, the approximated first and second order gradients are now only biased in the y direction mimicking more of a one-dimensional flow behavior along this direction. This behavior is also exemplified where $AR < 0.2$ if the fluid flow is in the x direction. Such consideration can assist in possibly alleviating convergence difficulties and enhancing the solution accuracy especially in appropriately resolving the wall boundary layers where the rapid solution change exists along the perpendicular direction of the fluid flow.

For grid *distortion* or *skewness*, which relates to the angle θ between the grid lines as indicated in Figure 6.5, it is desirable that the grid lines should be optimized in such a way that the angle θ is approximately 90° (orthogonal). If the angle $\theta < 45°$ or $\theta > 135°$, the mesh contains these highly-skewed cells and often exhibits a deterioration of the computational results or leads to numerical instabilities. For some complicated geometries, there is a high probability that the generated mesh may contain cells that are just bordering the *skewness* angle limits. The convergence behavior of such a mesh may be hampered due to the significant influence of additional terms in the discretized form of the transformed equations. It is necessary to avoid nonorthogonal cells near the geometry walls. The angle between the grid lines and the boundary of the computational domain should be maintained as close as possible to 90°. If an unstructured mesh is adopted, special care needs to be taken to ensure that the *warp angles* measuring between the surfaces normal to the triangular parts of the faces are not greater than 175° as indicated by the angle β in Figure 6.6. Cells with large

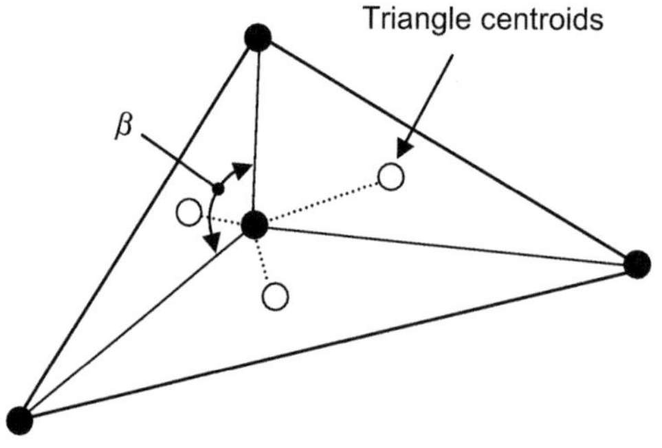

Figure 6.6
A triangular cell having an angle of β between the surfaces normal to the triangular parts of the faces connected to two adjacent triangles.

deviations from the coplanar faces can lead to serious convergence problems and deterioration in the computational results. In many grid generation packages, the problem can be overcome by a grid smoothing algorithm to improve the element *warp angles*.

6.4 Numerical Discretization

6.4.1 Finite Volume Method

The finite volume method centers on the concept of *control volume integration*. For ease of illustration, the generic transport equation as already described in Chapter 2 is simplified into its integral forms of a steady state process. In vector form, Eqn (2.159) for the multifluid model reduces to

$$\underbrace{\int_A \left(\alpha^k \rho^k \mathbf{U}^k \phi^k\right) \cdot \mathbf{n}\, dA}_{\text{advection}} = \underbrace{\int_A \left(\alpha^k \Gamma^k_{\phi^k} \nabla \phi^k\right) \cdot \mathbf{n}\, dA}_{\text{diffusion}} + \underbrace{\int_V S^k_{\phi^k}\, dV}_{\text{source term}} \tag{6.1}$$

In essence, the finite volume method discretizes the integral forms of the conservation equations directly in the physical space. If we consider the physical domain to be subdivided into a number of finite contiguous control volumes, the resulting statements express the exact conservation of property ϕ^k from each of the control volumes. In a control volume, the bounding surface areas of the element are, in general, directly linked to the discretization of the advection and diffusion terms in Eqn (6.1). The discretized form of the advection term for the multifluid model is thus given by

$$\int_A \left(\alpha^k \rho^k \mathbf{U}^k \phi^k\right) \cdot \mathbf{n}\, dA \approx \sum_f \left(\alpha^k \rho^k \mathbf{U}^k \cdot \mathbf{n} \phi^k\right)_f A_f \tag{6.2}$$

where the summation is over the number of faces and A_f is the area of the face of the control volume. Similarly, the discretized form of the diffusion term from which the surface fluxes are determined at the control volume faces is:

$$\int_A \left(\alpha^k \Gamma^k_{\phi^k} \nabla \phi^k\right) \cdot \mathbf{n}\, dA \approx \sum_f \left(\alpha^k \Gamma^k_{\phi^k} \nabla \phi^k \cdot \mathbf{n}\right)_f A_f \tag{6.3}$$

while the source term can be approximated by

$$\int_V S^k_{\phi^k}\, dV \approx S^k_{\phi^k} \Delta V \tag{6.4}$$

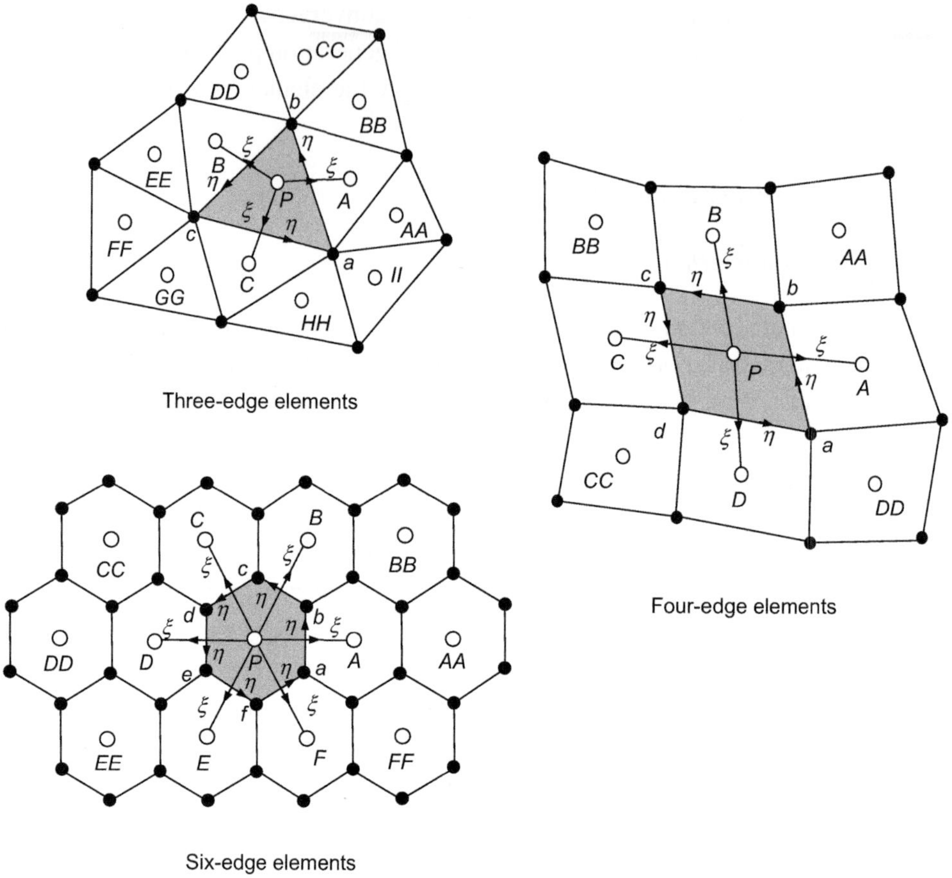

Figure 6.7
Different mesh arrangements consisting of three-, four- or six-edge elements.

The advection and diffusion terms can be formulated based on different mesh arrangements consisting of three-, four- or six-edge elements as shown in Figure 6.7. Initially focusing on the three-edge element of point *P*, which represents the centroid of the control volume in Figure 6.8, this point can be taken to be connected with the surrounding control volumes at the respective centroids indicated by points *A*, *B* and *C*. In general, the gradient of a generic property ϕ^k in Eqn (6.3) can be expressed in terms of the Cartesian coordinates as

$$\nabla\phi^k = \frac{\partial\phi^k}{\partial x}\mathbf{i} + \frac{\partial\phi^k}{\partial y}\mathbf{j} \tag{6.5}$$

The representation of the gradients in terms of the Cartesian coordinates is not particularly useful since cell centroid values of ϕ^k are unknown along these directions.

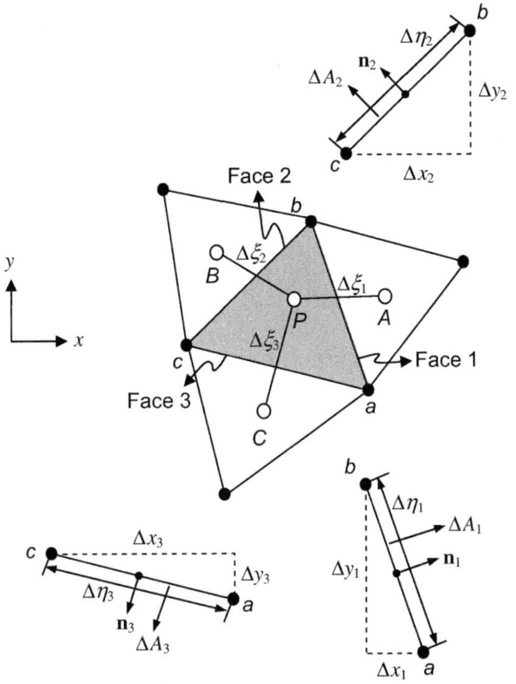

Figure 6.8
Schematic illustration of shaded three-edge element of point *P*.

Consider instead the coordinate system $\xi - \eta$ on the element faces. In order to express the gradients in Eqn (6.5) with respect to the curvilinear coordinates ξ and η, applying the chain rule,

$$\frac{\partial\phi^k}{\partial\xi} = \frac{\partial x}{\partial\xi}\frac{\partial\phi^k}{\partial x} + \frac{\partial y}{\partial\xi}\frac{\partial\phi^k}{\partial y} \tag{6.6}$$

$$\frac{\partial\phi^k}{\partial\eta} = \frac{\partial x}{\partial\eta}\frac{\partial\phi^k}{\partial x} + \frac{\partial y}{\partial\eta}\frac{\partial\phi^k}{\partial y} \tag{6.7}$$

Equations (6.6) and (6.7) can be solved to obtain $\partial\phi^k/\partial x$ and $\partial\phi^k/\partial y$. The solution becomes

$$\frac{\partial\phi^k}{\partial x} = \frac{1}{J}\left(\frac{\partial y}{\partial\eta}\frac{\partial\phi^k}{\partial\xi} - \frac{\partial y}{\partial\xi}\frac{\partial\phi^k}{\partial\eta}\right) \tag{6.8}$$

$$\frac{\partial\phi^k}{\partial y} = \frac{1}{J}\left(-\frac{\partial x}{\partial\eta}\frac{\partial\phi^k}{\partial\xi} + \frac{\partial x}{\partial\xi}\frac{\partial\phi^k}{\partial\eta}\right) \tag{6.9}$$

In Eqns (6.8) and (6.9), J denotes the Jacobian, which is given by

$$J = \left(\frac{\partial x}{\partial \xi}\frac{\partial y}{\partial \eta} - \frac{\partial x}{\partial \eta}\frac{\partial y}{\partial \xi}\right) \tag{6.10}$$

Substituting Eqns (6.8) and (6.9) into Eqn (6.5) yields

$$\nabla\phi^k = \frac{1}{J}\left(\frac{\partial y}{\partial \eta}\frac{\partial \phi^k}{\partial \xi} - \frac{\partial y}{\partial \xi}\frac{\partial \phi^k}{\partial \eta}\right)\mathbf{i} + \frac{1}{J}\left(-\frac{\partial x}{\partial \eta}\frac{\partial \phi^k}{\partial \xi} + \frac{\partial x}{\partial \xi}\frac{\partial \phi^k}{\partial \eta}\right)\mathbf{j} \tag{6.11}$$

Given that the vertices a, b and c are denoted by the coordinates (x_a, y_a), (x_b, y_b) and (x_c, y_c) and lines ab, ac and bc are segments joining the vertices a and b, a and c, and b and c of the three-edge element, the outward normal vector $\mathbf{n}_1$ for face 1 is defined by

$$\mathbf{n}_1 = \frac{\Delta y_1}{A_1}\mathbf{i} - \frac{\Delta x_1}{A_1}\mathbf{j} \tag{6.12}$$

The elemental area A_1 in Eqn (6.12) is $A_1 = \sqrt{(\Delta x_1)^2 + (\Delta y_1)^2}$, where $\Delta x_1 = x_b - x_a$ and $\Delta y_1 = y_b - y_a$. For faces 2 and 3, the outward normal vectors are accordingly defined by

$$\mathbf{n}_2 = \frac{\Delta y_2}{A_2}\mathbf{i} - \frac{\Delta x_2}{A_2}\mathbf{j} \tag{6.13}$$

$$\mathbf{n}_3 = \frac{\Delta y_3}{A_3}\mathbf{i} - \frac{\Delta x_3}{A_3}\mathbf{j} \tag{6.14}$$

with $A_2 = \sqrt{(\Delta x_2)^2 + (\Delta y_2)^2}$, where $\Delta x_2 = x_b - x_c$, $\Delta y_2 = y_b - y_c$ and $A_3 = \sqrt{(\Delta x_3)^2 + (\Delta y_3)^2}$, where $\Delta x_3 = x_a - x_c$ and $\Delta y_3 = y_a - y_c$. Using Eqn (6.11) and the unit vectors given in Eqns (6.12–6.14), the diffusion term can be formulated as

$$\begin{aligned}
\sum_f \left(\alpha^k \Gamma_{\phi^k} \nabla\phi^k \cdot \mathbf{n}\right)_f A_f &= \sum_{i=1}^{N_a=3} \left(\alpha^k \Gamma_{\phi^k} \nabla\phi^k \cdot \mathbf{n}\right)_i A_i \\
&= \left[\frac{\alpha^k \Gamma_{\phi^k}}{J}\left(\frac{\partial y}{\partial \eta}\Delta y + \frac{\partial x}{\partial \eta}\Delta x\right)\frac{\partial \phi^k}{\partial \xi} - \frac{\alpha^k \Gamma_{\phi^k}}{J}\left(\frac{\partial y}{\partial \xi}\Delta y + \frac{\partial x}{\partial \xi}\Delta x\right)\frac{\partial \phi^k}{\partial \eta}\right]_1 \\
&+ \left[\frac{\alpha^k \Gamma_{\phi^k}}{J}\left(\frac{\partial y}{\partial \eta}\Delta y + \frac{\partial x}{\partial \eta}\Delta x\right)\frac{\partial \phi^k}{\partial \xi} - \frac{\alpha^k \Gamma_{\phi^k}}{J}\left(\frac{\partial y}{\partial \xi}\Delta y + \frac{\partial x}{\partial \xi}\Delta x\right)\frac{\partial \phi^k}{\partial \eta}\right]_2 \\
&+ \left[\frac{\alpha^k \Gamma_{\phi^k}}{J}\left(\frac{\partial y}{\partial \eta}\Delta y + \frac{\partial x}{\partial \eta}\Delta x\right)\frac{\partial \phi^k}{\partial \xi} - \frac{\alpha^k \Gamma_{\phi^k}}{J}\left(\frac{\partial y}{\partial \xi}\Delta y + \frac{\partial x}{\partial \xi}\Delta x\right)\frac{\partial \phi^k}{\partial \eta}\right]_3
\end{aligned} \tag{6.15}$$

where J is the Jacobian in the form given in Eqn (6.10). For face 1, the geometrical quantities in Eqn (6.15) may be determined by

$$\left.\frac{\partial x}{\partial \xi}\right|_1 \approx \frac{x_A - x_P}{\Delta \xi_1} \quad \left.\frac{\partial y}{\partial \xi}\right|_1 \approx \frac{y_A - y_P}{\Delta \xi_1}$$

$$\left.\frac{\partial x}{\partial \eta}\right|_1 \approx \frac{x_b - x_a}{\Delta \eta_1} \quad \left.\frac{\partial y}{\partial \eta}\right|_1 \approx \frac{y_b - y_a}{\Delta \eta_1}$$

where $\Delta \xi_1$ is the distance between the points A and P and $\Delta \eta_1$ is the distance between the vertices a and b. Similar evaluations of the geometrical quantities based on the above procedure for face 1 may be carried for the rest of faces 2 and 3 of the triangular element.

Focusing next on the rectangular element of point P, which represents the centroid of the control volume in Figure 6.9, this point can now be taken to be connected with the surrounding control volumes at the respective centroids indicated by points A, B, C and D. Given that the vertices a, b, c and d are denoted by the coordinates (x_a, y_a), (x_b, y_b), (x_c, y_c) and (x_d, y_d) and lines ab, bc, cd and ad are segments joining the vertices a and b, b and

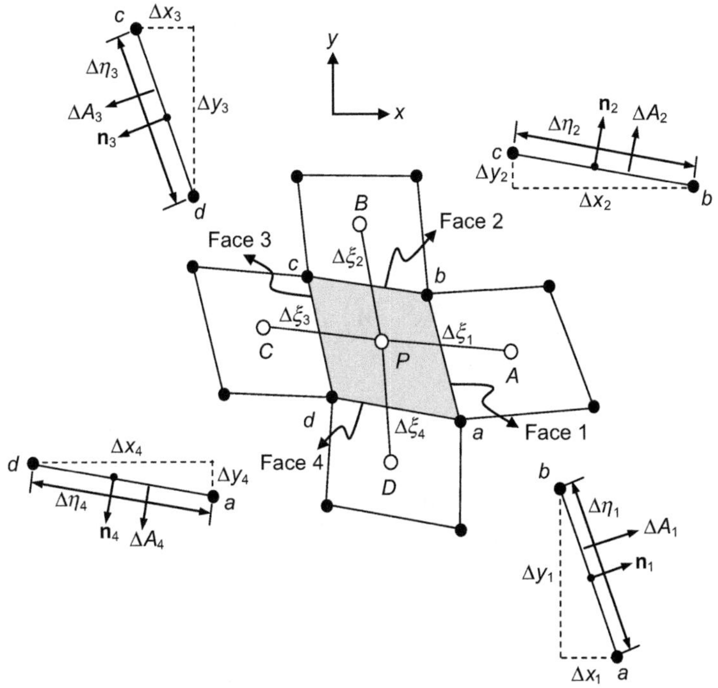

Figure 6.9
Schematic illustration of shaded four-edge element of point P.

c, c and d and a and d of the four-edge element, the outward normal vectors for faces 1, 2, 3 and 4 are defined by

$$\mathbf{n}_1 = \frac{\Delta y_1}{A_1}\mathbf{i} - \frac{\Delta x_1}{A_1}\mathbf{j} \tag{6.16}$$

$$\mathbf{n}_2 = \frac{\Delta y_2}{A_2}\mathbf{i} - \frac{\Delta x_2}{A_2}\mathbf{j} \tag{6.17}$$

$$\mathbf{n}_3 = \frac{\Delta y_3}{A_3}\mathbf{i} - \frac{\Delta x_3}{A_3}\mathbf{j} \tag{6.18}$$

$$\mathbf{n}_4 = \frac{\Delta y_4}{A_4}\mathbf{i} - \frac{\Delta x_4}{A_4}\mathbf{j} \tag{6.19}$$

with

$$A_1 = \sqrt{(\Delta x_1)^2 + (\Delta y_1)^2} \qquad A_2 = \sqrt{(\Delta x_2)^2 + (\Delta y_2)^2}$$

$$A_3 = \sqrt{(\Delta x_3)^2 + (\Delta y_3)^2} \qquad A_4 = \sqrt{(\Delta x_4)^2 + (\Delta y_4)^2}$$

where

$$\Delta x_1 = x_b - x_a, \quad \Delta y_1 = y_b - y_a \quad \Delta x_2 = x_c - x_b, \quad \Delta y_2 = y_c - y_b$$

$$\Delta x_3 = x_d - x_c, \quad \Delta y_3 = y_d - y_c \quad \Delta x_4 = x_a - x_d, \quad \Delta y_4 = y_a - y_d$$

Using Eqn (6.11) and the unit vectors given in Eqns (6.16–6.19), the diffusion term can henceforth be formulated as

$$\begin{aligned}
\sum_f \left(\alpha^k \Gamma_{\phi^k} \nabla \phi^k \cdot \mathbf{n}\right)_f A_f &= \sum_{i=1}^{N_a=4} \left(\alpha^k \Gamma_{\phi^k} \nabla \phi^k \cdot \mathbf{n}\right)_i A_i \\
&= \left[\frac{\alpha^k \Gamma_{\phi^k}}{J}\left(\frac{\partial y}{\partial \eta}\Delta y + \frac{\partial x}{\partial \eta}\Delta x\right)\frac{\partial \phi^k}{\partial \xi} - \frac{\alpha^k \Gamma_{\phi^k}}{J}\left(\frac{\partial y}{\partial \xi}\Delta y + \frac{\partial x}{\partial \xi}\Delta x\right)\frac{\partial \phi^k}{\partial \eta}\right]_1 \\
&+ \left[\frac{\alpha^k \Gamma_{\phi^k}}{J}\left(\frac{\partial y}{\partial \eta}\Delta y + \frac{\partial x}{\partial \eta}\Delta x\right)\frac{\partial \phi^k}{\partial \xi} - \frac{\alpha^k \Gamma_{\phi^k}}{J}\left(\frac{\partial y}{\partial \xi}\Delta y + \frac{\partial x}{\partial \xi}\Delta x\right)\frac{\partial \phi^k}{\partial \eta}\right]_2 \\
&+ \left[\frac{\alpha^k \Gamma_{\phi^k}}{J}\left(\frac{\partial y}{\partial \eta}\Delta y + \frac{\partial x}{\partial \eta}\Delta x\right)\frac{\partial \phi^k}{\partial \xi} - \frac{\alpha^k \Gamma_{\phi^k}}{J}\left(\frac{\partial y}{\partial \xi}\Delta y + \frac{\partial x}{\partial \xi}\Delta x\right)\frac{\partial \phi^k}{\partial \eta}\right]_3 \\
&+ \left[\frac{\alpha^k \Gamma_{\phi^k}}{J}\left(\frac{\partial y}{\partial \eta}\Delta y + \frac{\partial x}{\partial \eta}\Delta x\right)\frac{\partial \phi^k}{\partial \xi} - \frac{\alpha^k \Gamma_{\phi^k}}{J}\left(\frac{\partial y}{\partial \xi}\Delta y + \frac{\partial x}{\partial \xi}\Delta x\right)\frac{\partial \phi^k}{\partial \eta}\right]_4
\end{aligned} \tag{6.20}$$

For the special case of an orthogonal mesh, Eqn (6.20) reduces to

$$\sum_f \left(\alpha^k \Gamma_{\phi^k} \nabla \phi^k \cdot \mathbf{n}\right)_f A_f = \sum_{i=1}^{N_a=4} \left(\alpha^k \Gamma_{\phi^k} \nabla \phi^k \cdot \mathbf{n}\right)_i A_i$$

$$= \left[\alpha^k \Gamma_{\phi^k} \Delta y \frac{\partial \phi^k}{\partial \xi}\right]_1 + \left[\alpha^k \Gamma_{\phi^k} \Delta x \frac{\partial \phi^k}{\partial \xi}\right]_2 + \left[\alpha^k \Gamma_{\phi^k} \Delta y \frac{\partial \phi^k}{\partial \xi}\right]_3$$

$$+ \left[\alpha^k \Gamma_{\phi^k} \Delta x \frac{\partial \phi^k}{\partial \xi}\right]_4 \tag{6.21}$$

From the above equation, it can be seen that the diffusion transport normal to the elemental faces can be written purely in terms of values of property ϕ^k at the centroids sharing the faces. In developing suitable expressions for the diffusion terms on three- and four-edge elements, the consideration of six-edge element primarily entails the formulation of additional geometrical relationships for the additional faces of the element. Although the assumption of two-dimensionality has been invoked for the purpose of illustration, the above development holds for three-dimensional situations.

For the advection term, the normal velocity of the multifluid model for a three-edge element at face 1 can be evaluated in terms of the Cartesian velocity components as

$$\left(\mathbf{U}^k \cdot \mathbf{n}_1\right) = u_1^k \frac{\Delta y_1}{A_1} - v_1^k \frac{\Delta x_1}{A_1} \tag{6.22}$$

In accordance with Eqn (6.22), similar evaluations of the normal velocities can also be performed for the remaining faces of the triangular element. The normal kth phase velocities for faces 2 and 3 are thus given by

$$\left(\mathbf{U}^k \cdot \mathbf{n}_2\right) = u_2^k \frac{\Delta y_2}{A_2} - v_2^k \frac{\Delta x_2}{A_2} \tag{6.23}$$

$$\left(\mathbf{U}^k \cdot \mathbf{n}_3\right) = u_3^k \frac{\Delta y_3}{A_3} - v_3^k \frac{\Delta x_3}{A_3} \tag{6.24}$$

Using Eqns (6.22–6.24), the advection term can be approximated by

$$\sum_f \left(\alpha^k \rho^k \mathbf{U}^k \cdot \mathbf{n} \phi^k\right)_f A_f = \sum_{i=1}^{N_a=3} \alpha_i^k \rho_i^k \left(\mathbf{U}^k \cdot \mathbf{n}_i\right) A_i \phi_i^k$$

$$= \alpha_1^k \rho_1^k \left(\mathbf{U}^k \cdot \mathbf{n}_1\right) A_1 \phi_1^k + \alpha_2^k \rho_2^k \left(\mathbf{U}^k \cdot \mathbf{n}_2\right) A_2 \phi_2^k + \alpha_3^k \rho_3^k \left(\mathbf{U}^k \cdot \mathbf{n}_3\right) A_3 \phi_3^k \tag{6.25}$$

For a four-edge element, the advection term is simply given by

$$\sum_f \left(\alpha^k \rho^k \mathbf{U}^k \cdot \mathbf{n}\phi^k\right)_f A_f = \sum_{i=1}^{N_a=4} \alpha_i^k \rho_i^k (\mathbf{U}^k \cdot \mathbf{n}_i) A_i \phi_i^k$$
$$= \alpha_1^k \rho_1^k (\mathbf{U}^k \cdot \mathbf{n}_1) A_1 \phi_1^k + \alpha_2^k \rho_2^k (\mathbf{U}^k \cdot \mathbf{n}_2) A_2 \phi_2^k + \alpha_3^k \rho_3^k (\mathbf{U}^k \cdot \mathbf{n}_3) A_3 \phi_3^k + \alpha_4^k \rho_4^k (\mathbf{U}^k \cdot \mathbf{n}_4) A_4 \phi_4^k \qquad (6.26)$$

6.4.2 Basic Approximation of the Diffusion Term

The flux gradients of property ϕ^k at the control volume faces are normally approximated from the discrete quantities of the surrounding elements or control volumes. For the three-edge element at face 1, as depicted in Figure 6.8, the two flux gradients can be determined by

$$\left.\frac{\partial \phi^k}{\partial \xi}\right|_1 \approx \frac{\phi_A^k - \phi_P^k}{\Delta \xi_1} \quad \left.\frac{\partial \phi^k}{\partial \eta}\right|_1 \approx \frac{\phi_b^k - \phi_a^k}{\Delta \eta_1} \qquad (6.27)$$

where $\Delta\xi_1$ is the distance between the points A and P and $\Delta\eta_1$ is the distance between the vertices a and b. To determine the values of ϕ_a, which is located at the vertex a of the three-edge element, one possible approach is to adopt a simple averaging over the neighboring points that leads to

$$\phi_a^k = \frac{\phi_P^k + \phi_A^k + \phi_{AA}^k + \phi_{II}^k + \phi_{HH}^k + \phi_C^k}{6} \qquad (6.28)$$

In a similar manner, ϕ_b and ϕ_c can be ascertained by

$$\phi_b^k = \frac{\phi_P^k + \phi_A^k + \phi_{BB}^k + \phi_{CC}^k + \phi_{DD}^k + \phi_B^k}{6}$$
$$\phi_c^k = \frac{\phi_P^k + \phi_B^k + \phi_{EE}^k + \phi_{FF}^k + \phi_{GG}^k + \phi_C^k}{6} \qquad (6.29)$$

Equation (6.15) can be rewritten in an algebraic form in terms of discrete quantities of adjoining elements as

$$\sum_f \left(\Gamma_{\phi^k} \nabla \phi^k \cdot \mathbf{n}\right)_f A_f = D_1^k(\phi_A^k - \phi_P^k) + D_2^k(\phi_B^k - \phi_P^k) + D_3^k(\phi_C^k - \phi_P^k) + S_{\text{non}}^k \qquad (6.30)$$

In the above equation, the diffusive flux parameters D_1^k, D_2^k and D_3^k and added contribution term S_{non}^k due to nonorthogonality are, respectively,

$$D_1^k = \left[\frac{1}{\Delta\xi}\frac{\alpha^k\Gamma_{\phi^k}}{J}\left(\frac{\partial y}{\partial\eta}\Delta y + \frac{\partial x}{\partial\eta}\Delta x\right)\right]_1 \quad D_2^k = \left[\frac{1}{\Delta\xi}\frac{\alpha^k\Gamma_{\phi^k}}{J}\left(\frac{\partial y}{\partial\eta}\Delta y + \frac{\partial x}{\partial\eta}\Delta x\right)\right]_2$$

$$D_3^k = \left[\frac{1}{\Delta\xi}\frac{\alpha^k\Gamma_{\phi^k}}{J}\left(\frac{\partial y}{\partial\eta}\Delta y + \frac{\partial x}{\partial\eta}\Delta x\right)\right]_3$$

$$\begin{aligned} S_{\text{non}}^k = &-\left[\frac{1}{\Delta\eta}\frac{\alpha^k\Gamma_{\phi^k}}{J}\left(\frac{\partial y}{\partial\xi}\Delta y + \frac{\partial x}{\partial\xi}\Delta x\right)\right]_1 (\phi_b^k - \phi_a^k) \\ &-\left[\frac{1}{\Delta\eta}\frac{\alpha^k\Gamma_{\phi^k}}{J}\left(\frac{\partial y}{\partial\xi}\Delta y + \frac{\partial x}{\partial\xi}\Delta x\right)\right]_2 (\phi_c^k - \phi_b^k) \\ &-\left[\frac{1}{\Delta\eta}\frac{\alpha^k\Gamma_{\phi^k}}{J}\left(\frac{\partial y}{\partial\xi}\Delta y + \frac{\partial x}{\partial\xi}\Delta x\right)\right]_3 (\phi_a^k - \phi_c^k) \end{aligned}$$

Note that appropriate interpolation methods are employed to facilitate the evaluation of the diffusion coefficients ($\Gamma_{\phi^k}^k$) as well as the volume fractions (α^k) at the elemental faces.

For the four-edge element at face 1, as illustrated in Figure 6.9, the two flux gradients can also be determined in the same fashion as for the three-edge element as in Eqn (6.27). To determine the values of ϕ_a^k, which is located at the vertex a of the four-edge element, a simple averaging over the neighboring points for the four-edge element now leads to

$$\phi_a^k = \frac{\phi_P^k + \phi_A^k + \phi_{DD}^k + \phi_D^k}{4} \tag{6.31}$$

In a similar manner, ϕ_b^k, ϕ_c^k and ϕ_d^k can be ascertained by

$$\begin{aligned} \phi_b^k &= \frac{\phi_P^k + \phi_A^k + \phi_{AA}^k + \phi_B^k}{4} \\ \phi_c^k &= \frac{\phi_P^k + \phi_B^k + \phi_{BB}^k + \phi_C^k}{4} \\ \phi_d^k &= \frac{\phi_P^k + \phi_C^k + \phi_{CC}^k + \phi_D^k}{4} \end{aligned} \tag{6.32}$$

Equation (6.20) can be rewritten in an algebraic form in terms of discrete quantities of adjoining elements as

$$\begin{aligned} \sum_f \left(\Gamma_{\phi^k}\nabla\phi^k\cdot\mathbf{n}\right)_f A_f = &D_1^k(\phi_A^k - \phi_P^k) + D_2^k(\phi_B^k - \phi_P^k) \\ &+ D_3^k(\phi_C^k - \phi_P^k) + D_4^k(\phi_D^k - \phi_P^k) + S_{\text{non}}^k \end{aligned} \tag{6.33}$$

with

$$D_1^k = \left[\frac{1}{\Delta\xi} \frac{\alpha^k \Gamma_{\phi^k}}{J} \left(\frac{\partial y}{\partial \eta} \Delta y + \frac{\partial x}{\partial \eta} \Delta x \right) \right]_1 \quad D_2^k = \left[\frac{1}{\Delta\xi} \frac{\alpha^k \Gamma_{\phi^k}}{J} \left(\frac{\partial y}{\partial \eta} \Delta y + \frac{\partial x}{\partial \eta} \Delta x \right) \right]_2$$

$$D_3^k = \left[\frac{1}{\Delta\xi} \frac{\alpha^k \Gamma_{\phi^k}}{J} \left(\frac{\partial y}{\partial \eta} \Delta y + \frac{\partial x}{\partial \eta} \Delta x \right) \right]_3 \quad D_4^k = \left[\frac{1}{\Delta\xi} \frac{\alpha^k \Gamma_{\phi^k}}{J} \left(\frac{\partial y}{\partial \eta} \Delta y + \frac{\partial x}{\partial \eta} \Delta x \right) \right]_4$$

$$\begin{aligned} S_{\text{non}}^k = & - \left[\frac{1}{\Delta\eta} \frac{\alpha^k \Gamma_{\phi^k}}{J} \left(\frac{\partial y}{\partial \xi} \Delta y + \frac{\partial x}{\partial \xi} \Delta x \right) \right]_1 (\phi_b^k - \phi_a^k) \\ & - \left[\frac{1}{\Delta\eta} \frac{\alpha^k \Gamma_{\phi^k}}{J} \left(\frac{\partial y}{\partial \xi} \Delta y + \frac{\partial x}{\partial \xi} \Delta x \right) \right]_2 (\phi_c^k - \phi_b^k) \\ & - \left[\frac{1}{\Delta\eta} \frac{\alpha^k \Gamma_{\phi^k}}{J} \left(\frac{\partial y}{\partial \xi} \Delta y + \frac{\partial x}{\partial \xi} \Delta x \right) \right]_3 (\phi_d^k - \phi_c^k) \\ & - \left[\frac{1}{\Delta\eta} \frac{\alpha^k \Gamma_{\phi^k}}{J} \left(\frac{\partial y}{\partial \xi} \Delta y + \frac{\partial x}{\partial \xi} \Delta x \right) \right]_4 (\phi_a^k - \phi_d^k) \end{aligned}$$

Here again, appropriate interpolation methods are employed to facilitate the evaluation of the diffusion coefficients ($\Gamma_{\phi^k}^k$) as well as the volume fractions (α^k) at the elemental faces.

6.4.3 Basic Approximation of Advection Term

The principal problem in the discretization of the advection term is the calculation of property ϕ^k at the elemental faces and the determination of convective fluxes across these boundaries, which generally require a special treatment. Based on the assumption that the convective fluxes can be appropriately resolved at the elemental faces, the different interpolation approaches to evaluate the property ϕ^k at these faces are discussed in this section. Normally, it is desirable to express the property ϕ^k across the elemental surface in terms of the quantity at the central point of the element in question and in connection with the quantities at the neighboring points of the surrounding elements. Although the judicious choice of interpolation approaches appears rather straightforward, it has been found that the stability of the numerical solution is strongly dependent on the flow direction of the fluid.

A linear interpolation could be realized between the central and neighboring nodes with the assumption of a piecewise linear gradient profile. With reference to the meshes illustrated in Figures 6.8 and 6.9, the face value of a generic property ϕ_1^k can be determined as

$$\phi_1^k = \frac{1}{2}\left(\phi_P^k + \phi_A^k\right) \tag{6.34}$$

Similar considerations can also be realized at the other elemental faces. The above formulae are second order accurate and such an interpolation procedure is commonly recognized as the *central differencing scheme*. In spite of its second order accuracy, it does not exhibit any bias on the flow direction. It has been well documented in the literature (Patankar, 1980; Versteeg and Malalasekera, 2007) that the inadequacy of this scheme in a strongly convective flow is its inability to identify the flow direction. The above formulae usually result in large "undershoots" and "overshoots" in some flow problems, eventually causing the numerical calculations to diverge. In some circumstances, this may yield nonphysical solutions. One possibility is to significantly increase the mesh resolution for the computational domain with very small grid spacing until stability is achieved during the numerical calculations. Nevertheless, such an approach is deemed not viable, and ineffective for practical flow calculations.

In order to overcome the problem associated with central differencing, much emphasis has been placed on developing an array of interpolation schemes that can accommodate some recognition of the flow direction. Through central differencing approximation, the face value of ϕ_1^k is always assumed to be weighted by the influence of the available variables at the neighboring points; the downstream value of ϕ_A^k is always required during the evaluation of ϕ_1^k, which is usually not known *a priori*. By exerting an unequal weighting influence, a numerical solution can nonetheless be designed to recognize the direction of the flow in order to appropriately determine the interface values. This is essentially the hallmark of the *upwind* or *donor-cell* concept. The convection flux parameter, which can be taken to be equivalent to the mass flow rate to the surface element, can be defined by

$$F_f = \rho_f(\mathbf{U}\cdot\mathbf{n})_f A_f \tag{6.35}$$

where the subscript f denotes the face of the control volume. If $F_1 > 0$, the face value of ϕ_1 according to the *donor-cell* concept can be approximated according to its upstream neighboring counterpart as

$$\phi_1^k = \phi_p^k \tag{6.36}$$

Similarly, if $F_1 < 0$, ϕ_1 is conversely evaluated by

$$\phi_1^k = \phi_A^k \tag{6.37}$$

This scheme, known as the *upwind scheme*, promotes numerical stability, satisfies transportiveness (flow direction), boundness (diagonally dominant matrix coefficients ensuring numerical convergence) and conservativeness (fluxes that are represented in a consistent manner). Nevertheless, this scheme is only first-order accurate and yields *false numerical diffusion.*

In order to improve the solution accuracy, Spalding (1972) developed a scheme that combines the central and upwind differencing schemes by employing piecewise formulae based on the local Peclet number Pe. The local Peclet number is a nondimensional number that measures the relative strengths of the convective and diffusive fluxes. It can be evaluated at any face of a control volume as

$$Pe_f = \frac{F_f}{D_f} \tag{6.38}$$

where D_f is the diffusion flux parameter. As formulated for different mesh arrangements in the previous section, the diffusion flux parameter for the multifluid model is given by D_i^k. This so-called *hybrid differencing scheme* retains a second-order accuracy for small Peclet numbers due to central differencing but reverts to the first-order upwind differencing for large Peclet numbers. The face value of ϕ_1 according to the hybrid differencing formulae is given by

$$\phi_1^k = \begin{cases} \frac{1}{2}(\phi_P^k + \phi_A^k) & \text{if } |Pe_1| \le 2 \\ \phi_P^k & \text{if } Pe_1 > 2 \\ \phi_A^k & \text{if } Pe_1 < -2 \end{cases} \tag{6.39}$$

Similar to upwind differencing, this scheme is highly stable, satisfies transportiveness and produces physically realistic solutions. In spite of the exploitation of the favorable properties of the upwind and central differencing schemes, the accuracy of this scheme is still only first order. In most cases of real practical flows, the majority of the local Peclet numbers will be greater than 2 due to the large flow velocities that exist within the flow system.

Another popular scheme that is considered to yield better results than the hybrid scheme is the *power-law differencing scheme* of Patankar (1980). Here, the upwind differencing becomes effective only when the local Peclet number is greater than 10. This particular scheme also possesses similar properties to the hybrid scheme. The face value of ϕ_1 according to the power-law differencing formulae can be determined as

$$\phi_1^k = \begin{cases} (1-\chi_1)\phi_P^k + \chi_1\phi_A^k & \text{if } |Pe_1| \le 10 \\ \phi_P^k & \text{if } Pe_1 > 10 \\ \phi_A^k & \text{if } Pe_1 < -10 \end{cases} \tag{6.40}$$

where $\chi_1 = (1 - 0.1Pe_1)^5/Pe_1$.

The inherent first-order accuracy in all of the above schemes, especially due to the consideration of the upwind concept, makes them prone to unwanted numerical diffusion errors. In order to reduce these numerical errors, higher order approximations such as the *second order upwind differencing scheme* and *third-order Quadratic Upstream Interpolation for Convective Kinetics (QUICK) scheme* of Lenoard (1979) have been proposed.

One other possible way of evaluating the face value of the transport quantity is via the least-squares gradient reconstruction technique (Barth and Jespersen, 1989). Taking $\Delta\mathbf{r}$ to be the distance vector from point P to the face of the control volume, the face value of the transported quantity can be evaluated by means of

$$\phi_f^k = \phi_P^k + \left(\nabla\phi^k\right)_P \cdot \Delta\mathbf{r} \tag{6.41}$$

In this approach, a gradient at each control-volume center is constructed using all available nearest neighboring surrounding information. It can be shown that Eqn (6.41) is of a second order approximation since the neglected terms in the Taylor series expansion of the transport quantity are proportional to the square of the distance $\Delta\mathbf{r}$. Defining $\Delta\mathbf{r}$ to be equivalent to $\Delta x\mathbf{i} + \Delta y\mathbf{j}$, $\nabla\phi$ at point P could be conveniently expressed in terms of the Cartesian gradients. Equation (6.41) becomes

$$\phi_f^k = \phi_P^k + \left.\frac{\partial\phi^k}{\partial x}\right|_P \Delta x_f + \left.\frac{\partial\phi^k}{\partial y}\right|_P \Delta y_f \tag{6.42}$$

In reference to Figure 6.10, a set of three equations can be ascertained as

$$\begin{aligned}
\phi_1^k &= \phi_P^k + \left.\frac{\partial\phi^k}{\partial x}\right|_P (x_1 - x_P) + \left.\frac{\partial\phi^k}{\partial y}\right|_P (y_1 - y_P) \\
\phi_2^k &= \phi_P^k + \left.\frac{\partial\phi^k}{\partial x}\right|_P (x_2 - x_P) + \left.\frac{\partial\phi^k}{\partial y}\right|_P (y_2 - y_P) \\
\phi_3^k &= \phi_P^k + \left.\frac{\partial\phi^k}{\partial x}\right|_P (x_3 - x_P) + \left.\frac{\partial\phi^k}{\partial y}\right|_P (y_3 - y_P)
\end{aligned} \tag{6.43}$$

which can be assembled into a matrix form:

$$\underbrace{\begin{bmatrix} x_1 - x_P & y_1 - y_P \\ x_2 - x_P & y_2 - y_P \\ x_3 - x_P & y_3 - y_P \end{bmatrix}}_{\mathbf{A}} \underbrace{\begin{bmatrix} \left.\dfrac{\partial\phi^k}{\partial x}\right|_P \\ \left.\dfrac{\partial\phi^k}{\partial y}\right|_P \end{bmatrix}}_{\mathbf{X}} = \underbrace{\begin{bmatrix} \phi_1^k - \phi_P^k \\ \phi_2^k - \phi_P^k \\ \phi_3^k - \phi_P^k \end{bmatrix}}_{\mathbf{B}} \tag{6.44}$$

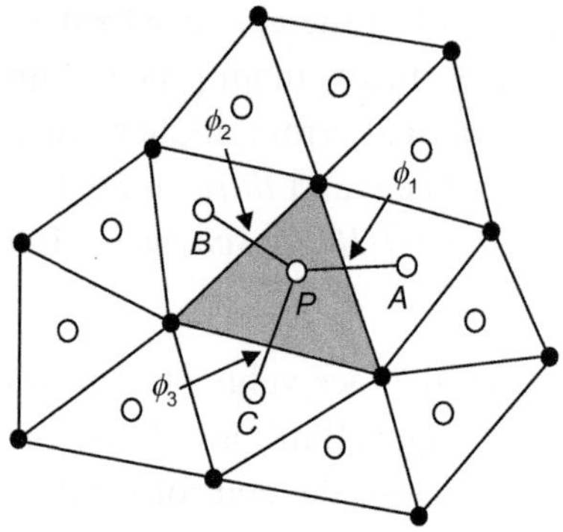

Figure 6.10
A control volume and the surrounding face quantities in a three-edge element.

Equation (6.44) is an overdetermined linear equation system. Nonetheless, by multiplying both sides of the equation with the transpose $\mathbf{A}^{\mathrm{T}}$:

$$\mathbf{A}^{\mathrm{T}} = \begin{bmatrix} x_1 - x_P & x_2 - x_P & x_3 - x_P \\ y_1 - y_P & y_2 - y_P & y_3 - y_P \end{bmatrix}$$

we have

$$\underbrace{\begin{pmatrix} a_{11} & a_{12} \\ a_{21} & a_{22} \end{pmatrix}}_{\mathbf{A}^{\mathrm{T}}\mathbf{A}} \underbrace{\begin{bmatrix} \left.\dfrac{\partial \phi^k}{\partial x}\right|_P \\ \left.\dfrac{\partial \phi^k}{\partial y}\right|_P \end{bmatrix}}_{\mathbf{X}} = \underbrace{\begin{bmatrix} b_1 \\ b_2 \end{bmatrix}}_{\mathbf{A}^{\mathrm{T}}\mathbf{B}} \tag{6.45}$$

where

$$\begin{aligned} a_{11} &= (x_1 - x_P)^2 + (x_2 - x_P)^2 + (x_3 - x_P)^2 \\ a_{12} = a_{21} &= (x_1 - x_P)(y_1 - y_P) + (x_2 - x_P)(y_2 - y_P) + (x_3 - x_P)(y_3 - y_P) \\ a_{22} &= (y_1 - x_P)^2 + (y_2 - x_P)^2 + (y_3 - x_P)^2 \\ b_1 &= (x_1 - x_P)\left(\phi_1^k - \phi_P^k\right) + (x_2 - x_P)\left(\phi_2^k - \phi_P^k\right) + (x_3 - x_P)\left(\phi_3^k - \phi_P^k\right) \\ b_2 &= (y_1 - x_P)\left(\phi_1^k - \phi_P^k\right) + (y_2 - x_P)\left(\phi_2^k - \phi_P^k\right) + (y_3 - x_P)\left(\phi_3^k - \phi_P^k\right) \end{aligned}$$

It can be seen that the matrix $\mathbf{A}^{\mathrm{T}}\mathbf{A}$ in Eqn (6.45) is simply a 2×2 matrix, which can be easily inverted to solve for $\mathbf{X}$. Since matrix $\mathbf{A}$ depends on geometry, this calculation needs to be performed only once for each node. The required Cartesian gradients at point P are obtained from

$$\mathbf{X} = \left(\mathbf{A}^{\mathrm{T}}\mathbf{A}\right)^{-1}\mathbf{A}^{\mathrm{T}}\mathbf{B} \tag{6.46}$$

Equation (6.45) can also be directly applied to any elemental shapes in an unstructured mesh. This entails merely expanding the matrices **A**, $\mathbf{A}^{\mathrm{T}}$ and **B** according to

$$\mathbf{A} = \begin{bmatrix} x_1 - x_P & y_1 - y_P \\ x_2 - x_P & y_2 - y_P \\ x_3 - x_P & y_3 - y_P \\ \vdots & \vdots \\ x_{N_a} - x_P & y_{N_a} - y_P \end{bmatrix} \quad \mathbf{A}^{\mathrm{T}} = \begin{bmatrix} x_1 - x_P & x_2 - x_P & x_3 - x_P & \cdots & x_{N_a} - x_P \\ y_1 - y_P & y_2 - y_P & y_3 - y_P & \cdots & y_{N_a} - y_P \end{bmatrix}$$

$$\mathbf{B} = \begin{bmatrix} \phi_1 - \phi_P \\ \phi_2 - \phi_P \\ \phi_3 - \phi_P \\ \vdots \\ \phi_{N_a} - \phi_P \end{bmatrix}$$

6.4.4 Basic Approximation of Time-Advancing Solutions

In order to illustrate the approximate form of the unsteady transport equation of property ϕ^k, Eqn (2.159) needs to be further augmented with an integration over a finite time step Δt. By changing the order of integration in the time derivative terms, we obtain

$$\int_V \left(\int_t^{t+\Delta t} \frac{\partial(\alpha^k \rho^k \phi^k)}{\partial t} \mathrm{d}t \right) \mathrm{d}V + \int_t^{t+\Delta t} \left(\int_A (\alpha^k \rho^k \mathbf{U}^k \phi^k) \cdot \mathbf{n}\, \mathrm{d}A \right) \mathrm{d}t$$
$$= \int_t^{t+\Delta t} \left(\int_A (\alpha^k \Gamma^k_{\phi^k} \nabla \phi^k) \cdot \mathbf{n}\, \mathrm{d}A \right) \mathrm{d}t + \int_t^{t+\Delta t} \int_V S^k_{\phi^k} \mathrm{d}V \mathrm{d}t \tag{6.47}$$

The previously described discretization methods are equally applicable for the treatment of the advection, diffusion and source terms in the above equations. Using the finite volume approximations as exemplified in Eqns (6.2–6.4) and approximating the time derivative of the above equation similar to the source term,

$$\left(\int_t^{t+\Delta t} \frac{\partial(\alpha^k \rho^k \phi^k)}{\partial t} \mathrm{d}t \right) \Delta V + \int_t^{t+\Delta t} \left(\sum_f (\alpha^k \rho^k \mathbf{U}^k \cdot \mathbf{n} \phi^k)_f \Delta A_f \right) \mathrm{d}t$$
$$= \int_t^{t+\Delta t} \left(\sum_f (\alpha^k \Gamma^k_{\phi^k} \nabla \phi^k \cdot \mathbf{n})_f \Delta A_f \right) \mathrm{d}t + \int_t^{t+\Delta t} S_{\phi^k} \Delta V\, \mathrm{d}t \tag{6.48}$$

To numerically solve the above equation, suitable methods necessary for time integration are required. In the majority of cases, the first-order accurate time derivative is approximated by

$$\int_{t}^{t+\Delta t} \frac{\partial\left(\alpha^k\rho^k\phi^k\right)}{\partial t}\mathrm{d}t = \frac{\left(\alpha^k\rho^k\phi^k\right)^{n+1} - \left(\alpha^k\rho^k\phi^k\right)^{n}}{\Delta t} \tag{6.49}$$

where Δt is the incremental time step and the superscripts n and $n+1$ denote the previous and current time levels, respectively. Introducing a weighting parameter θ between 0 and 1 to generalize the integration of the advection, diffusion and source terms over the time step Δt and utilizing the approximations in Eqns (6.48) and (6.49), Eqn (6.47) can be written as

$$\begin{aligned}
&\left(\frac{\left(\alpha^k\rho^k\phi^k\right)^{n+1} - \left(\alpha^k\rho^k\phi^k\right)^{n}}{\Delta t}\right)\Delta V + (1-\theta)\left(\sum_f \left(\alpha^k\rho^k\mathbf{U}^k\cdot\mathbf{n}\phi^k\right)_f\Delta A_f\right)^n \\
&+ \theta\left(\sum_f \left(\alpha^k\rho^k\mathbf{U}^k\cdot\mathbf{n}\phi^k\right)_f\Delta A_f\right)^{n+1} \\
&= (1-\theta)\left(\sum_f \left(\alpha^k\Gamma^k_{\phi^k}\nabla\phi^k\cdot\mathbf{n}\right)_f\Delta A_f\right)^n + \theta\left(\sum_f \left(\alpha^k\Gamma^k_{\phi^k}\nabla\phi^k\cdot\mathbf{n}\right)_f\Delta A_f\right)^{n+1} \\
&+ (1-\theta)S^n_{\phi^k}\Delta V + \theta S^{n+1}_{\phi^k}\Delta V
\end{aligned} \tag{6.50}$$

Note that the algebraic expressions of the advection and diffusion terms for particular mesh systems can be found in previous sections. Depending on the values of θ, different time-marching methods can be realized for Eqn (6.50).

For an *explicit* method, θ is set to zero and all values in the advection, diffusion and source terms are known at the previous time level n. Assuming that the values of density and volume fraction are known at $n+1$, property $(\phi^k)^{n+1}$ is immediately evaluated. Nevertheless, if θ is set to unity, this method, which is commonly referred to as the *fully implicit* procedure, results in the need of calculating all values in the time derivative, advection, diffusion and source terms at the current time level $n+1$; an iterative procedure is generally required. The explicit and fully implicit approaches are nonetheless methods of only first order in time. Similar to the first order in space, these methods may also cause unwanted numerical diffusion in time. In order to reduce these numerical errors, the use of second-order approximations such as the *explicit Adams–Bashforth*, *semi-implicit Crank–Nicolson* and *fully implicit* methods could be employed.

As an extension to the first-order explicit method, the *second-order explicit Adams–Bashforth* method now requires the values at time level n as well as at time level $n-1$. The unsteady transport equations can be formulated according to

$$\left(\frac{\left(\alpha^k\rho^k\phi^k\right)^{n+1}-\left(\alpha^k\rho^k\phi^k\right)^{n}}{\Delta t}\right)\Delta V=\left(\frac{3}{2}\frac{\partial\left(\alpha^k\rho^k\phi^k\right)}{\partial t}\bigg|^{n}-\frac{1}{2}\frac{\partial\left(\alpha^k\rho^k\phi^k\right)}{\partial t}\bigg|^{n-1}\right)\Delta V$$
$$=-\frac{3}{2}\left(\sum_f\left(\alpha^k\rho^k\mathbf{U}^k\cdot\mathbf{n}\phi^k\right)_f\Delta A_f\right)^{n}$$
$$+\frac{1}{2}\left(\sum_f\left(\alpha^k\rho^k\mathbf{U}^k\cdot\mathbf{n}\phi^k\right)_f\Delta A_f\right)^{n-1}$$
$$+\frac{3}{2}\left(\sum_f\left(\alpha^k\Gamma_{\phi^k}\nabla\phi^k\cdot\mathbf{n}\right)_f\Delta A_f\right)^{n}$$
$$-\frac{1}{2}\left(\sum_f\left(\alpha^k\Gamma_{\phi^k}\nabla\phi^k\cdot\mathbf{n}\right)_f\Delta A_f\right)^{n-1}$$
$$+\left(\frac{3}{2}S^{n}_{\phi^k}-\frac{1}{2}S^{n+1}_{\phi^k}\right)\Delta V \tag{6.51}$$

In the *second-order semi-implicit Crank–Nicolson* method, this special type of differencing in time requires the solution to be attained by averaging the values between time levels n and $n+1$. According to Eqns (6.50) and (6.51), the weighting parameter is prescribed midway between time levels n and $n+1$, i.e. $\theta=\frac{1}{2}$, yielding

$$\left(\frac{\left(\alpha^k\rho^k\phi^k\right)^{n+1}-\left(\alpha^k\rho^k\phi^k\right)^{n}}{\Delta t}\right)\Delta V+\frac{1}{2}\left(\sum_f\left(\alpha^k\rho^k\mathbf{U}^k\cdot\mathbf{n}\phi^k\right)_f\Delta A_f\right)^{n}$$
$$+\frac{1}{2}\left(\sum_f\left(\alpha^k\rho^k\mathbf{U}^k\cdot\mathbf{n}\phi^k\right)_f\Delta A_f\right)^{n+1}$$
$$=\frac{1}{2}\left(\sum_f\left(\alpha^k\Gamma^k_{\phi^k}\nabla\phi^k\cdot\mathbf{n}\right)_f\Delta A_f\right)^{n}+\frac{1}{2}\left(\sum_f\left(\alpha^k\Gamma^k_{\phi^k}\nabla\phi^k\cdot\mathbf{n}\right)_f\Delta A_f\right)^{n+1}$$
$$+\frac{1}{2}\left(S^{n}_{\phi^k}+S^{n+1}_{\phi^k}\right)\Delta V \tag{6.52}$$

In contrast to the first-order fully implicit method, the *second-order fully implicit* method approximates the time derivatives of ϕ^m and ϕ^k according to

$$\int_t^{t+\Delta t}\frac{\partial\left(\alpha^k\rho^k\phi^k\right)}{\partial t}dt=\frac{3\left(\alpha^k\rho^k\phi^k\right)^{n+1}-4\left(\alpha^k\rho^k\phi^k\right)^{n}-\left(\alpha^k\rho^k\phi^k\right)^{n-1}}{2\Delta t} \tag{6.53}$$

As in Adams—Bashforth, this method also requires the values at time level $n-1$. Using the above equations in place of the first-order approximations and setting the weighting parameter $\theta = 1$, the unsteady transport equations become

$$\left(\frac{3\left(\alpha^k\rho^k\phi^k\right)^{n+1} - 4\left(\alpha^k\rho^k\phi^k\right)^{n} - \left(\alpha^k\rho^k\phi^k\right)^{n-1}}{2\Delta t}\right)\Delta V + \left(\sum_f \left(\alpha^k\rho^k\mathbf{U}^k\cdot\mathbf{n}\phi^k\right)_f\Delta A_f\right)^{n+1}$$
$$= \left(\sum_f \left(\alpha^k\Gamma^k_{\phi^k}\nabla\phi^k\cdot\mathbf{n}\right)_f\Delta A_f\right)^{n+1} + S^{n+1}_{\phi^k}\Delta V \tag{6.54}$$

Another higher order explicit method that could also be possibly considered is the *third- or fourth-order Runge Kutta* method. The interested reader should refer to the article by Rai and Moin (1991) for more details on the numerical implementation of this particular time-marching method. In practice, this method allows a larger time step to be adopted for the same order of accuracy to be achieved when compared to the Adams—Bashforth method and, thus, marginally compensates for the increased amount of computations that are experienced during the numerical calculations. It should, however, be noted that explicit methods based on the Adams—Bashforth and third- or fourth-order Runge Kutta formulations are subjected to time step restriction during the numerical calculations. They are normally required to satisfy the Courant—Friedrichs—Levy number condition in order to ensure numerical stability. In contrast, implicit methods are not subject to stability restrictions. This is because the solution domain of influence includes the whole system through the coupling between adjacent points of the control volume at the new time step.

6.4.5 Algebraic Form of Discretized Equations

The diffusion fluxes can be approximated and rewritten through all faces of the elemental volume for multifluid model as

$$\sum_f \alpha^k_f D^k_f\left(\phi^k_N - \phi^k_P\right) + S^k_{\mathrm{DC,non}} \tag{6.55}$$

where the subscript N denotes the neighboring nodes. From the above, the diffusion flux parameter D^k_f along with the added contribution $S^k_{\mathrm{DC,non}}$ due to nonorthogonality needs to be determined for the particular mesh systems of different element shapes. It should be noted that $S^k_{\mathrm{DC,non}}$ drops out for an orthogonal mesh. Similarly, the convective fluxes for the multifluid model can be expressed as

$$\sum_f \alpha^k_f F^k_f\left[\phi^k_U + \frac{1}{2}\psi\left(r^k\right)\left(\phi^k_D - \phi^k_U\right)\right] \tag{6.56}$$

where r^k is determined by

$$r^k = \frac{\phi_P^k - \phi_D^k}{\phi_U^k - \phi_P^k} \tag{6.57}$$

and the subscripts U and D denote the upstream and downstream values of property ϕ^k around the node P. For numerical stability, the source term is treated according to

$$S_{\phi^k}^k \Delta V = \left(S_u^k - S_P^k \phi_P^k\right) \Delta V_P \tag{6.58}$$

When Eqs (6.55), (6.56) and (6.58) are substituted into the unsteady flow Eqn (6.50), assuming a first-order fully implicit procedure,

$$\left(\frac{\left(\alpha_P^k \rho_P^k \phi_P^k\right)^{n+1} - \left(\alpha_P^k \rho_P^k \phi_P^k\right)^n}{\Delta t} \right) \Delta V_P + \left(\sum_f \alpha_f^k F_f^k \left[\phi_U^k + \frac{1}{2} \psi(r^k) \left(\phi_D^k - \phi_U^k\right) \right] \right)^{n+1}$$

$$= \left(\sum_f \alpha_f^k D_f^k \left(\phi_N^m - \phi_P^m\right) \right)^{n+1} + \left(S_{\mathrm{DC,non}}^k\right)^{n+1} + \left(S_u^k - S_P^k \phi_P^k\right)^{n+1} \Delta V_P \tag{6.59}$$

Dropping the superscript $n+1$, which by default denotes the current time level, Eqn (6.59) can be rearranged as

$$a_P^k \phi_P^k = \sum_{nb} a_{nb}^k \phi_{nb}^k + S_{\mathrm{DC,off}}^k + S_{\mathrm{DC,non}}^k + S_u^k \Delta V_P + \frac{\left(\alpha_P^k \rho_P^k \phi_P^k\right)^n \Delta V_P}{\Delta t} \tag{6.60}$$

where

$$a_P^k = \sum_{nb} a_{nb}^k + S_P^k \Delta V_P + \sum_f \alpha_f^k F_f^k + \frac{\left(\alpha_P^k \rho_P^k\right) \Delta V_P}{\Delta t} \tag{6.61}$$

From the above, a_P^k is the diagonal matrix coefficient of ϕ_P^k, $\sum_f \alpha_f^k F_f^k$ is the mass imbalance over all faces of the control volume, and S_P^k is the coefficient that is extracted from the treatment of the source term in order to further increase its diagonal dominance. The coefficients of any neighboring nodes for any surrounding control volumes a_{nb}^k can be expressed by

$$a_{nb}^k = \alpha_f^k D_f^k + \alpha_f^k \max\left(-F_f^k, 0\right) \tag{6.62}$$

To guarantee diagonal dominance, the well-known deferred correction approach is adopted to appropriately treat the off-diagonal contributions arising from the use of higher order approximations, which is represented in the source term $S_{\mathrm{DC,off}}^k$:

$$S_{\mathrm{DC,off}}^k = \sum_f \alpha_f^k F_f^k \left[\frac{1}{2} \psi(r^k) \left(\phi_D^k - \phi_U^k\right) \right] \tag{6.63}$$

The deferred correction approach is also employed for flow problems where the mesh is nonorthogonal, which is accommodated through the source term $S^k_{\text{DC,non}}$. In Eqn (6.63), the generalized upwind-biased expression conforms to the basic approximations of the first-order upwind differencing scheme, second-order central differencing scheme, second-order upwind differencing scheme and QUICK scheme:

$\psi(r^k) = 0$	—	first-order upwind differencing scheme
$\psi(r^k) = 1$	—	second-order central differencing scheme
$\psi(r^k) = r^k$	—	second-order upwind differencing scheme
$\psi(r^k) = 0.25r^k + 0.75$	—	QUICK scheme

For an unstructured mesh arrangement, the recommendation by Darwish and Moukalled (2003) may be adopted to evaluate r^k:

$$r^k = \left[\frac{\left(2\nabla\phi_P^k \cdot \Delta\mathbf{r}_{PA}\right)}{\phi_D^k - \phi_U^k} - 1\right] \tag{6.64}$$

where $\Delta\mathbf{r}_{PA}$ represents the distance vector between the central point P and surrounding point A. The gradient $\nabla\phi_P^k$ in Eqn (6.64) may be realized according to the least-squares gradient reconstruction technique as described in the previous section.

One important characteristic feature of multiphase flows is the transfer of mass, momentum and energy between the phases. When the coupling is relatively weak, the interfacial source terms, especially in the governing equations of the multifluid model, may be safely lumped together and treated as general source terms in the governing equations. However, when the coupling is very tight, the interfacial transport is such that a condition necessitates a proper treatment in order to promote numerical stability and avoid the divergence of the numerical solution. One effective way of overcoming this problem is the linearization of the interfacial source terms. In a number of flow cases, these couplings can be very tight, which results in the slow convergence or divergence of the numerical solution if the interphase interaction terms are not properly treated. In order to increase the diagonal dominance, some degree of implicitness is required in the treatment of these terms.

Consider the algebraic equation of the generic dependent variable ϕ^k, which can be expressed in a general form as

$$a_P^k\phi_P^k = \sum_{nb} a_{nb}^k\phi_{nb}^k + S'^k + \sum_{l=1}^{N_p}\left(\left(\dot{M}_{lk}\right)_P\phi_P^l - \left(\dot{M}_{kl}\right)_P\phi_P^k\right) + \sum_{l=1}^{N_p}(C_{kl})_P\left(\phi_P^l - \phi_P^k\right) \tag{6.65}$$

where $(\dot{M}_{lk})_P \equiv (\dot{m}_{lk})_P \Delta V_P$, $(\dot{M}_{kl})_P \equiv (\dot{m}_{kl})_P \Delta V_P$ and $(C_{kl})_P \equiv (c_{kl})_P \Delta V_P$. The diagonal coefficient a_P^k is given by Eqn (6.61), while the source term S'^k is comprised of the off-diagonal, nonorthogonal and unlinearized contributions of the transport equation as well as values at time level n. In the same equation, the term $((\dot{M}_{lk})_P \phi_P^l - (\dot{M}_{kl})_P \phi_P^k)$ arises only if the interfacial mass transfer takes place and the term $(C_{kl})_P(\phi_P^l - \phi_P^k)$ describes the interfacial transfer between phase k and phase l, which accounts for the interfacial drag force in the momentum equation or the interfacial heat source in the energy equation. It should be noted that the interphase coefficient c_{kl} in the term C_{kl} has the following properties: $c_{kk} = c_{ll} = 0$ and $c_{kl} = c_{lk}$. Hence, the sum over all phases of all the interfacial transfer terms is effectively zero. Karema and Lo (1999) have provided a detailed discussion on the treatment of these two terms. Different interfacial coupling algorithms that can be employed are reviewed below.

Based upon the *fully explicit* treatment, this method simply substitutes existing values of the dependent variables in the interfacial source term. Consequently, only the constant part of linearization is utilized, which results in

$$a_P^k \phi_P^k = \sum_{nb} a_{nb}^k \phi_{nb}^k + S'^k + \sum_{l=1}^{N_p} (\dot{M}_{lk})_P \phi_P^{l*} + \sum_{l=1}^{N_p} (C_{kl})_P (\phi_P^{l*} - \phi_P^{k*}) \tag{6.66}$$

where the diagonal coefficient a_P^k is now expressed as

$$a_P^k = \sum_{nb} a_{nb}^k + \sum_{l=1}^{N_p} (\dot{M}_{kl})_P + S_P^k \Delta V_P + \sum_f \alpha_f^k F_f^k + \frac{(\alpha_P^k \rho_P^k) \Delta V_P}{\Delta t} \tag{6.67}$$

In Eqn (6.66), ϕ^{l^*} and ϕ^{k^*} represent existing values of the dependent variables. In the *partially implicit* treatment, the interfacial source term is linearized in the most possible natural way by treating the term containing the interface transfer coefficients multiplied by the current phase variable as the first-order term and the remaining portion as the constant term. The transfer of all terms depending on the variable ϕ^{k^*} to the left-hand side of Eqn (6.66) yields

$$a_P^k \phi_P^k = \sum_{nb} a_{nb}^k \phi_{nb}^k + S'^k + \sum_{l=1}^{N_p} (\dot{M}_{lk})_P \phi_P^{l*} + \sum_{l=1}^{N_p} (C_{kl})_P \phi_P^{l*} \tag{6.68}$$

with

$$a_P^k = \sum_{nb} a_{nb}^k + \sum_{l=1}^{N_p} (\dot{M}_{kl})_P + \sum_{l=1}^{N_p} (C_{kl})_P + S_P^k \Delta V_P + \sum_f \alpha_f^k F_f^k + \frac{(\alpha_P^k \rho_P^k) \Delta V_P}{\Delta t} \tag{6.69}$$

It is clear that the implementation of the partially implicit treatment is straightforward and gives marginally better performance in comparison to the fully explicit treatment without significantly increasing the computational costs.

There is, nonetheless, a tendency for the coefficient of the first-order term (interface transfer coefficient) to become very large when the values of the dependent variable in different phases are close to each other. A situation therefore arises where a large number multiplies a very small difference to the values of the dependent variable, creating a condition where it is prone to the divergence of the numerical solution. This particularly pertains to the interfacial momentum transfer in a bubbly air–water flow where the coupling between the phases is very tight due to its rather short characteristic time scale. In order to provide a necessary remedy to the problem, the interfacial transport in tight coupling condition demands a more implicit treatment.

For general multiphase conditions, it is required that the SINCE (simultaneous solution of non-linearly coupled equations) developed by Lo (1989, 1990) is employed to implicitly treat the interaction of multiple phases within the interfacial source terms. The generic transport equation can be written as

$$a_P^k\phi_P^k = \sum_{nb} a_{nb}^k\phi_{nb}^k + S'^k + \sum_{l=1}^{N_p}\left(\dot{M}_{lk}\right)_P\phi_P^{l*} + \sum_{l=1}^{N_p}(C_{kl})_P\left(\phi_P^l - \phi_P^k\right) \tag{6.70}$$

with the diagonal coefficient a_P^k given by Eqn (6.67). By transferring the coefficients multiplying ϕ_P^k to the left-hand side, Eqn (6.70) can be arranged to the following system of linear equations:

$$D_P^1\phi_P^1 = (C_{12})_P\phi_P^2 + (C_{13})_{PP}\phi_P^3 + \cdots + \left(C_{1N_p}\right)_P\phi_P^{N_p} + \sum_{nb} a_{nb}^1\phi_{nb}^{1*} + S'^1 + \sum_{l=1}^{N_p}\left(\dot{M}_{l1}\right)_P\phi_P^{l*}$$

$$\vdots$$

$$D_P^{N_p}\phi_P^{N_p} = \left(C_{N_p1}\right)_P\phi_P^1 + \left(C_{N_p2}\right)_P\phi_P^2 + \cdots + \left(C_{N_p(N_p-1)}\right)_P\phi_P^{N_p-1} + \sum_{nb} a_{nb}^{N_p}\phi_{nb}^{N_p*} + S'^{N_p}$$

$$+\sum_{l=1}^{N_p}\left(\dot{M}_{lN_p}\right)_P\phi_P^{l*} \tag{6.71}$$

where

$$D_P^k = a_P^k + \sum_{l=1}^{N_p}(C_{kl})_P$$

Equation (6.71) is readily interpreted in a matrix form as

$$\underbrace{\begin{bmatrix} D_P^1 & \cdots & -\left(C_{1N_p}\right)_P \\ \vdots & & \vdots \\ -\left(C_{N_p1}\right)_P & \cdots & D_P^{N_p} \end{bmatrix}}_{\mathbf{A}} \underbrace{\begin{bmatrix} \phi_{nb}^{1*} \\ \vdots \\ \phi_{nb}^{N_p*} \end{bmatrix}}_{\mathbf{X}} = \underbrace{\begin{bmatrix} \sum_{nb} a_{nb}^1\phi_{nb}^{1*} + S'^1 + \sum_{l=1}^{N_p}\left(\dot{M}_{l1}\right)_P\phi_P^{l*} \\ \vdots \\ \sum_{nb} a_{nb}^{N_p}\phi_{nb}^{N_p*} + S'^{N_p} + \sum_{l=1}^{N_p}\left(\dot{M}_{lN_p}\right)_P\phi_P^{l*} \end{bmatrix}}_{\mathbf{B}} \tag{6.72}$$

The coefficient matrix **A** and the source matrix **B** are calculated based upon the existing values. Solving the matrix Eqn (6.72) element by element, a new intermediate estimate ϕ_P^{k*} for the dependent variable in all phases can be obtained. This process can be performed as

$$\mathbf{X} = (\mathbf{A})^{-1}\mathbf{B} \tag{6.73}$$

These new intermediate estimates from Eqn (6.73) are subsequently substituted for the interfacial coupling term on the right-hand side of Eqn (6.70), which result in an explicit evaluation for ϕ_P^k:

$$D_P^k \phi_P^k = \sum_{nb} a_{nb}^k \phi_{nb}^k + S'^k + \sum_{l=1}^{N_p} (\dot{M}_{lk})_P \phi_P^{l*} + \sum_{l=1}^{N_p} (C_{kl})_P \phi_P^{l*} \tag{6.74}$$

Only the contribution due to the interfacial drag force in the momentum equation or the interfacial heat source in the energy equation is treated semi-implicitly in SINCE. The interfacial mass transfer contribution is handled through the *partially implicit* treatment.

In cases where the consideration of multifluid is exercised to better resolve the complex topologies of multiphase flows, the algorithm based on SINCE provides the only option in order to appropriately treat the tight coupling of the interfacial source terms of short characteristic time scales. It should be noted that SINCE is computationally more involved when compared to the simple approaches based on the fully explicit and partially implicit treatments. Nevertheless, if the time step during the numerical simulation could be maintained below the characteristic time scale of the interfacial transfer processes, it is possible that the partially implicit treatment of the interfacial source terms could be applied to avoid the unnecessarily large number of numerical operations associated with the use of more implicit schemes. Otherwise, SINCE becomes indispensable for achieving a better convergence of the numerical solutions of multiphase flows.

6.5 Numerical Solvers

6.5.1 Iterative Calculations for the Segregated Approach

For the finite volume method, the resultant matrix of the algebraic equations as described in the previous section is typically sparse, which means that most of the elements are zero and the nonzero terms are close to the diagonal. For two- or three-dimensional calculations of multiphase flows, iterative calculations are normally adopted. Based on the repeated applications of an algorithm leading to its eventual convergence after a number of repetitions, they are generally considered to be much more economical since only nonzero terms of the algebraic equations are required to be stored in the core memory. For calculations of multiphase flows that are typically nonlinear problems, they are used out of necessity. In order to assemble the complete solution procedure for an implicit-type algorithm, the sequence of iterative calculations leading to the final converged solution satisfying all the governing equations can be summarized as:

1. Initialize all field values by an initial guess.
2. Solve the algebraic momentum equations to obtain u_P^{k*} and v_P^{k*} based on the guessed pressure $p*$.
3. Calculate the face velocities employing the Rhie–Chow interpolation method (Rhie and Chow, 1983).
4. Solve the pressure correction equation to obtain p' through the IPSA or IPSA-C algorithm.
5. Update pressure according by $p = p^* + p'$ through the available solution of the pressure correction field.
6. Correct the velocities at element centers and element faces.
7. Solve additional equations for any property ϕ^k governing the flow process, if necessary, such as enthalpy.
8. Using the corrected velocities, mass fluxes and pressure as the prevailing fields for the new iteration cycle, return to step (2).

The sequence of steps (2)–(7) is repeated until convergence is achieved. One criterion to terminate the solution procedure is the *mass residual*. Another way of determining whether convergence has been achieved is through the sum of absolute *imbalances* (*residuals*) of the discretized equations at all computational nodes. The *imbalance* of property ϕ^k at any computational grid point can be calculated as $R = \left|\sum_{nb} \frac{a_{nb}^k \phi_{nb}^{k*}}{a_P^k} + B^{\phi^k} - \phi^k\right|$, where B^{ϕ^k} represents the accumulated sources/sinks for the property ϕ^k. The sum is given by $SR_{\phi^k} = \sum_{l=1}^{M} R_l^n$, where M is the total number of grid points and n is the iteration counter. The majority of computer codes impose their own convergence criteria, which are generally applicable to a wide range of flow conditions. Interested readers may wish to investigate the specified tolerance levels employed by a number of computer codes commercially available. Appropriate settings of convergence criteria are still usually determined from practical experience and the application of particular methods.

The calculation of the face velocities via the Rhie–Chow interpolation formula for step (2) is described in the following. Based on the velocity vector $\mathbf{U}^k = u^k\mathbf{i} + v^k\mathbf{j}$, the Rhie–Chow interpolation formula is applied for the evaluation of the normal component of velocity, i.e. $(\mathbf{U}^k \cdot \mathbf{n})_f$. For face 1 of the meshes shown in Figures 6.8 and 6.9, the normal velocity can be written as

$$(\mathbf{U}^k \cdot \mathbf{n})_1 A_1 = u_1^k \Delta y_1 - v_1^k \Delta x_1 \tag{6.75}$$

The velocity components u^k and v^k can be determined in accordance with

$$u^k = \overline{u^k} + \overline{(D^{k,u})}\frac{\partial p}{\partial x} - \overline{(D^{k,u})\frac{\partial p}{\partial x}} \tag{6.76}$$

$$v^k = \overline{v^k} + \overline{(D^{k,v})}\frac{\partial p}{\partial y} - \overline{(D^{k,v})\frac{\partial p}{\partial y}} \tag{6.77}$$

where in terms of nodes P and A:

$$u_1^k = \frac{1}{2}\left[u_P^k + u_A^k\right] + \frac{1}{2}\overline{\left(\alpha_P^k \frac{\Delta V_P}{a_P^{k,u}}\right)_1}\left[\left.\frac{\partial p}{\partial x}\right|_P + \left.\frac{\partial p}{\partial x}\right|_A\right] - \overline{\left(\alpha_P^k \frac{\Delta V_P}{a_P^{k,u}}\right)_1}$$
$$\times\left[\left(\frac{1}{J}\frac{\partial y}{\partial \eta}\right)_1 \frac{(p_A - p_P)}{\Delta \xi_1} - \left(\frac{1}{J}\frac{\partial y}{\partial \xi}\right)_1 \frac{(p_b - p_a)}{\Delta \eta_1}\right] \tag{6.78}$$

$$v_1^k = \frac{1}{2}\left[v_P^k + v_A^k\right] + \frac{1}{2}\overline{\left(\frac{\alpha_P^k \Delta V_P}{a_P^{k,v}}\right)_1}\left[\left.\frac{\partial p}{\partial y}\right|_P + \left.\frac{\partial p}{\partial y}\right|_A\right] - \overline{\left(\frac{\alpha_P^k \Delta V_P}{a_P^{k,v}}\right)_1}$$
$$\times\left[-\left(\frac{1}{J}\frac{\partial x}{\partial \eta}\right)_1 \frac{(p_A - p_P)}{\Delta \xi_1} + \left(\frac{1}{J}\frac{\partial x}{\partial \xi}\right)_1 \frac{(p_b - p_a)}{\Delta \eta_1}\right] \tag{6.79}$$

where the Cartesian pressure gradients in Eqns (6.78) and (6.79) at nodes P and A can be evaluated using an appropriate gradient reconstruction method (see Section 6.4.3). Analogous Rhie–Chow interpolation expressions can also be applied to determine the necessary normal velocities at the other faces of the elemental volume.

Provided that the pressure field is sufficiently smooth, the Rhie–Chow interpolation expressions as derived from the above for different mesh systems may suffer from large errors, especially in the case of rapidly changing source terms – for example, the prevalence of a strong buoyancy force on different sides of a sharp interface in multiphase flows. In such a situation, the *standard* Rhie–Chow interpolation formula that involves only the evaluation of the velocities and pressure gradients by linear interpolation from cell centers may not be sufficiently accurate, and in turn could result in significant mass imbalance in cells adjacent to the interface. Admittedly, the momentum source terms on a staggered mesh or mesh-oriented arrangement are primitively defined at their staggered or mesh-oriented locations, and, hence, they are automatically balanced by the first-order pressure gradients. In order to overcome the problem associated with collocated arrangement, one possible solution is to somehow mimic the staggered or mesh-oriented situation by ensuring that the source term is primitively defined at cell faces, and the Rhie–Chow interpolation expressions are accordingly modified to include the effect of this term into the formula in order to appropriately evaluate the face velocities. This so-called *improved* Rhie–Chow interpolation method is illustrated as follows.

The essential idea of the improved Rhie–Chow interpolation method is that the source term is either partly or completely immersed in the pressure gradient. Considering the u-momentum equation, the source term having rapid changes is initially separated according to

$$B_P^u = B_P^{u,\text{imp}} + B_P^{\prime u,\text{imp}} \tag{6.80}$$

where $B_P^{u,\text{imp}}$ and $B_P^{\prime u,\text{imp}}$ are part of the improved treatment and the remaining part of a customary source term. All the extracted source terms, which are defined at element centers, are later interpolated to element faces by linear interpolation. Thus, the evaluation of the velocity components u_1^k and v_1^k through the *improved* Rhie–Chow interpolation method becomes

$$\begin{aligned} u_1^k = {} & \frac{1}{2}\left[u_P^k + u_A^k\right] + \frac{1}{2}\overline{\left(\alpha_P^k \frac{\Delta V_P}{a_P^{k,u}}\right)_1}\left[\left.\frac{\partial p}{\partial x}\right|_P + \left.\frac{\partial p}{\partial x}\right|_A\right] - \overline{\left(\alpha_P^k \frac{\Delta V_P}{a_P^{k,u}}\right)_1}\left[\left(\frac{1}{J}\frac{\partial y}{\partial \eta}\right)_1 \frac{(p_A - p_P)}{\Delta \xi_1}\right. \\ & \left. - \left(\frac{1}{J}\frac{\partial y}{\partial \xi}\right)_1 \frac{(p_b - p_a)}{\Delta \eta_1}\right] - \overline{\left(\frac{1}{a_P^{k,u}}\right)_1}\left(\overline{\left(B_P^{\prime u,\text{imp}}\right)_1} - B_1^{u,\text{imp}}\right) \end{aligned} \tag{6.81}$$

$$\begin{aligned} v_1^k = {} & \frac{1}{2}\left[v_P^k + v_A^k\right] + \frac{1}{2}\overline{\left(\frac{\alpha_P^k \Delta V_P}{a_P^{k,v}}\right)_1}\left[\left.\frac{\partial p}{\partial y}\right|_P + \left.\frac{\partial p}{\partial y}\right|_A\right] - \overline{\left(\frac{\alpha_P^k \Delta V_P}{a_P^{k,v}}\right)_1}\left[-\left(\frac{1}{J}\frac{\partial x}{\partial \eta}\right)_1 \frac{(p_A - p_P)}{\Delta \xi_1}\right. \\ & \left. + \left(\frac{1}{J}\frac{\partial x}{\partial \xi}\right)_1 \frac{(p_b - p_a)}{\Delta \eta_1}\right] - \overline{\left(\frac{1}{a_P^{k,v}}\right)_1}\left(\overline{\left(B_P^{\prime v,\text{imp}}\right)_1} - B_1^{v,\text{imp}}\right) \end{aligned} \tag{6.82}$$

where

$$B_P^{\prime u,\text{imp}} = \frac{1}{3}\left(B_1^{u,\text{imp}} + B_2^{u,\text{imp}} + B_3^{u,\text{imp}}\right) \tag{6.83}$$

$$B_P^{\prime v,\text{imp}} = \frac{1}{3}\left(B_1^{v,\text{imp}} + B_2^{v,\text{imp}} + B_3^{v,\text{imp}}\right) \tag{6.84}$$

and

$$\begin{aligned} B_1^{u,\text{imp}} &= \frac{1}{2}\left(B_P^{u,\text{imp}} + B_1^{u,\text{imp}}\right) \\ B_1^{v,\text{imp}} &= \frac{1}{2}\left(B_P^{v,\text{imp}} + B_1^{v,\text{imp}}\right) \\ B_2^{u,\text{imp}} &= \frac{1}{2}\left(B_P^{u,\text{imp}} + B_2^{u,\text{imp}}\right) \\ B_2^{v,\text{imp}} &= \frac{1}{2}\left(B_P^{v,\text{imp}} + B_2^{v,\text{imp}}\right) \\ B_3^{u,\text{imp}} &= \frac{1}{2}\left(B_P^{u,\text{imp}} + B_3^{u,\text{imp}}\right) \\ B_3^{v,\text{imp}} &= \frac{1}{2}\left(B_P^{v,\text{imp}} + B_3^{v,\text{imp}}\right) \end{aligned} \tag{6.85}$$

Analogous Rhie–Chow interpolation expressions (as derived from above) for different mesh systems are henceforth applied to determine the necessary normal velocities at the other faces of the elemental volume.

6.5.2 Application of IPSA or IPSA-C for the Segregated Approach

Originally proposed by Patankar and Spalding (1972), the solution method based on SIMPLE (Semi-Implicit for Method Pressure-Linkage Equations) is well suited to solve the discretized macroscopic balance equations of mass, momentum and energy for single phase flows. This pressure correction technique is basically an iterative approach that caters for implicit-type algorithms of steady or unsteady solutions and centers on the basic philosophy of effectively coupling between the pressure and the velocity. The pressure is linked to the velocity via the construction of a pressure field to guarantee conservation of mass and the equation for the mass conservation becomes a *kinematic* constraint on the velocity field rather than a *dynamic* equation. In the consideration of solution methods for multiphase flows, it can be demonstrated that the IPSA and its variant IPSA-C are mere extensions of the well-known SIMPLE algorithm.

Interphase slip algorithm

The SIMPLE algorithm, such as that adopted for single phase flows, is essentially a "guess-and-correct" procedure for the calculation of pressure through the solution of a pressure correction equation. It is, therefore, an iterative procedure. In particular, computation of multiphase flows through the IPSA method—an extension to the SIMPLE algorithm—can be realized in the context of a computational multiphase fluid dynamics framework through the use of the shared pressure approximation; all phase pressures in the momentum equation are assumed to be equal, meaning that $p^k = p$. As a result, the pressure gradient term of all phases appears as a product of common pressure gradient ∇p and the respective volume fraction α^k, and accordingly forms a pressure shared by volume fraction.

To illustrate the IPSA method, the discretized equation governing mass conservation for an unstructured mesh of three-edge elements can be expressed in terms of the normal face velocities as

$$\alpha_1^k \rho_1^k (\mathbf{U}^k \cdot \mathbf{n}_1) A_1 + \alpha_2^k \rho_2^k (\mathbf{U}^k \cdot \mathbf{n}_2) A_2 + \alpha_3^k \rho_3^k (\mathbf{U}^k \cdot \mathbf{n}_3) A_3 = -\frac{\left(\alpha_P^k \rho_P^k - \left(\alpha_P^k \rho_P^k\right)^n\right) \Delta V_P}{\Delta t} \tag{6.86}$$

or an unstructured mesh of four-edge elements by

$$\begin{aligned} &\alpha_1^k \rho_1^k (\mathbf{U}^k \cdot \mathbf{n}_1) A_1 + \alpha_2^k \rho_2^k (\mathbf{U}^k \cdot \mathbf{n}_2) A_2 + \alpha_3^k \rho_3^k (\mathbf{U}^k \cdot \mathbf{n}_3) A_3 + \alpha_4^k \rho_4^k (\mathbf{U}^k \cdot \mathbf{n}_4) A_4 \\ &= -\frac{\left(\alpha_P^k \rho_P^k - \left(\alpha_P^k \rho_P^k\right)^n\right) \Delta V_P}{\Delta t} \end{aligned} \tag{6.87}$$

For the purpose of illustration, the discretized phase momentum equations of the multifluid model can be written for the guessed velocities u_P^k and v_P^{k*} with a guessed pressure field p^* as:

$$u_P^{k*} = \sum_{nb} \frac{a_{nb}^{k,u} u_{nb}^{k*}}{a_P^{k,u}} - \alpha_P^k \frac{\Delta V_P}{a_P^{k,u}} \left.\frac{\partial p^*}{\partial x}\right|_P + B_P^{u^k} \tag{6.88}$$

$$v_P^{k*} = \sum_{nb} \frac{a_{nb}^{k,v} v_{nb}^{k*}}{a_P^{k,v}} - \alpha_P^k \frac{\Delta V_P}{a_P^{k,v}} \left.\frac{\partial p^*}{\partial y}\right|_P + B_P^{v^k} \tag{6.89}$$

where $B_P^{u^k}$ and $B_P^{v^k}$ are the remaining kth phase source terms after the pressure gradient source terms have been taken out. The corrected velocities u_P^k and v_P^k with the correct pressure field p may be determined according to

$$u_P^k = \sum_{nb} \frac{a_{nb}^{k,u} u_{nb}^k}{a_P^{k,u}} - \alpha_P^k \frac{\Delta V_P}{a_P^{k,u}} \left.\frac{\partial p}{\partial x}\right|_P + B_P^{u^k} \tag{6.90}$$

$$v_P^k = \sum_{nb} \frac{a_{nb}^{k,v} v_{nb}^k}{a_P^{k,v}} - \alpha_P^k \frac{\Delta V_P}{a_P^{k,v}} \left.\frac{\partial p}{\partial y}\right|_P + B_P^{v^k} \tag{6.91}$$

Similar to the single phase SIMPLE algorithm, the interdependency of elements is limited by discarding the neighboring terms of the element under consideration. The velocity corrections for the kth phase Cartesian velocity components become

$$u_P^k - u_P^{k*} = -D_P^{k,u} \left.\frac{\partial p'}{\partial x}\right|_P \tag{6.92}$$

$$v_P^k - v_P^{k*} = -D_P^{k,v} \left.\frac{\partial p'}{\partial y}\right|_P \tag{6.93}$$

where $D^{k,u} = \alpha_P^k \dfrac{\Delta V_P}{a_P^{k,u}}$, $D^{k,v} = \alpha_P^k \dfrac{\Delta V_P}{a_P^{k,v}}$ and $p' = (p - p^*)$.

Using suitable normal vectors and since the velocity vector is defined as $\mathbf{U}^k = u^k\mathbf{i} + v^k\mathbf{j}$, the pressure correction for an unstructured mesh of three-edge elements via Eqn (6.86) and using Eqns (6.92) and (6.93) can be obtained as

$$\begin{aligned}
&\left[\frac{1}{\Delta\xi}\frac{\alpha^k\rho^k D^{k,u}}{J}\left(\frac{\partial y}{\partial\eta}\Delta y\right) + \frac{1}{\Delta\xi}\frac{\alpha^k\rho^k D^{k,v}}{J}\left(\frac{\partial x}{\partial\eta}\Delta x\right)\right]_1 (p'_A - p'_P) + \\
&\left[\frac{1}{\Delta\xi}\frac{\alpha^k\rho^k D^{k,u}}{J}\left(\frac{\partial y}{\partial\eta}\Delta y\right) + \frac{1}{\Delta\xi}\frac{\alpha^k\rho^k D^{k,v}}{J}\left(\frac{\partial x}{\partial\eta}\Delta x\right)\right]_2 (p'_B - p'_P) + \\
&\left[\frac{1}{\Delta\xi}\frac{\alpha^k\rho^k D^{k,u}}{J}\left(\frac{\partial y}{\partial\eta}\Delta y\right) + \frac{1}{\Delta\xi}\frac{\alpha^k\rho^k D^{k,v}}{J}\left(\frac{\partial x}{\partial\eta}\Delta x\right)\right]_3 (p'_C - p'_P) + S_{\text{non}}^{p'} = b'
\end{aligned} \tag{6.94}$$

where

$$S_{\text{non}}^{p'} = -\left[\frac{1}{\Delta\eta}\frac{\alpha^k\rho^k D^{k,u}}{J}\left(\frac{\partial y}{\partial\xi}\Delta y\right) + \frac{1}{\Delta\eta}\frac{\alpha^k\rho^k D^{k,v}}{J}\left(\frac{\partial x}{\partial\xi}\Delta x\right)\right]_1 (p'_b - p'_a)$$

$$-\left[\frac{1}{\Delta\eta}\frac{\alpha^k\rho^k D^{k,u}}{J}\left(\frac{\partial y}{\partial\xi}\Delta y\right) + \frac{1}{\Delta\eta}\frac{\alpha^k\rho^k D^{k,v}}{J}\left(\frac{\partial x}{\partial\xi}\Delta x\right)\right]_2 (p'_c - p'_b)$$

$$-\left[\frac{1}{\Delta\eta}\frac{\alpha^k\rho^k D^{k,u}}{J}\left(\frac{\partial y}{\partial\xi}\Delta y\right) + \frac{1}{\Delta\eta}\frac{\alpha^k\rho^k D^{k,v}}{J}\left(\frac{\partial x}{\partial\xi}\Delta x\right)\right]_3 (p'_a - p'_c)$$

$$b' = -\left(\alpha_1^k\rho_1^k(\mathbf{U}^k\cdot\mathbf{n}_1)^* A_1 + \alpha_2^k\rho_2^k(\mathbf{U}^k\cdot\mathbf{n}_2)^* A_2 + \alpha_3^k\rho_3^k(\mathbf{U}^k\cdot\mathbf{n}_3)^* A_3 + \frac{(\alpha_P^k\rho_P^k - (\alpha_P^k\rho_P^m)^n)\Delta V_P}{\Delta t}\right)$$

or for an unstructured mesh of four-edge elements via Eqn (6.87) and using Eqns (6.92) and (6.93) through

$$\begin{aligned}
&\left[\frac{1}{\Delta\xi}\frac{\alpha^k\rho^k D^{k,u}}{J}\left(\frac{\partial y}{\partial\eta}\Delta y\right) + \frac{1}{\Delta\xi}\frac{\alpha^k\rho^k D^{k,v}}{J}\left(\frac{\partial x}{\partial\eta}\Delta x\right)\right]_1 (p'_A - p'_P) + \\
&\left[\frac{1}{\Delta\xi}\frac{\alpha^k\rho^k D^{k,u}}{J}\left(\frac{\partial y}{\partial\eta}\Delta y\right) + \frac{1}{\Delta\xi}\frac{\alpha^k\rho^k D^{k,v}}{J}\left(\frac{\partial x}{\partial\eta}\Delta x\right)\right]_2 (p'_B - p'_P) + \\
&\left[\frac{1}{\Delta\xi}\frac{\alpha^k\rho^k D^{k,u}}{J}\left(\frac{\partial y}{\partial\eta}\Delta y\right) + \frac{1}{\Delta\xi}\frac{\alpha^k\rho^k D^{k,v}}{J}\left(\frac{\partial x}{\partial\eta}\Delta x\right)\right]_3 (p'_C - p'_P) + \\
&\left[\frac{1}{\Delta\xi}\frac{\alpha^k\rho^k D^{k,u}}{J}\left(\frac{\partial y}{\partial\eta}\Delta y\right) + \frac{1}{\Delta\xi}\frac{\alpha^k\rho^k D^{k,v}}{J}\left(\frac{\partial x}{\partial\eta}\Delta x\right)\right]_4 (p'_D - p'_P) + S_{\text{non}}^{p'} = b'
\end{aligned} \tag{6.95}$$

where

$$S_{\text{non}}^{p'} = -\left[\frac{1}{\Delta\eta}\frac{\alpha^k\rho^k D^{k,u}}{J}\left(\frac{\partial y}{\partial\xi}\Delta y\right) + \frac{1}{\Delta\eta}\frac{\alpha^k\rho^k D^{k,v}}{J}\left(\frac{\partial x}{\partial\xi}\Delta x\right)\right]_1 (p'_b - p'_a)$$

$$-\left[\frac{1}{\Delta\eta}\frac{\alpha^k\rho^k D^{k,u}}{J}\left(\frac{\partial y}{\partial\xi}\Delta y\right) + \frac{1}{\Delta\eta}\frac{\alpha^k\rho^k D^{k,v}}{J}\left(\frac{\partial x}{\partial\xi}\Delta x\right)\right]_2 (p'_c - p'_b)$$

$$-\left[\frac{1}{\Delta\eta}\frac{\alpha^k\rho^k D^{k,u}}{J}\left(\frac{\partial y}{\partial\xi}\Delta y\right) + \frac{1}{\Delta\eta}\frac{\alpha^k\rho^k D^{k,v}}{J}\left(\frac{\partial x}{\partial\xi}\Delta x\right)\right]_3 (p'_d - p'_c)$$

$$-\left[\frac{1}{\Delta\eta}\frac{\alpha^k\rho^k D^{k,u}}{J}\left(\frac{\partial y}{\partial\xi}\Delta y\right) + \frac{1}{\Delta\eta}\frac{\alpha^k\rho^k D^{k,v}}{J}\left(\frac{\partial x}{\partial\xi}\Delta x\right)\right]_4 (p'_a - p'_d)$$

$$b' = -\left(\alpha_1^k\rho_1^k\left(\mathbf{U}^k\cdot\mathbf{n}_1\right)^*A_1 + \alpha_2^k\rho_2^k\left(\mathbf{U}^k\cdot\mathbf{n}_2\right)^*A_2 + \alpha_3^k\rho_3^k\left(\mathbf{U}^k\cdot\mathbf{n}_3\right)^*A_3 + \alpha_4^k\rho_4^k\left(\mathbf{U}^k\cdot\mathbf{n}_4\right)^*A_4 + \frac{\left(\alpha_P^k\rho_P^k - \left(\alpha_P^k\rho_P^m\right)^n\right)\Delta V_P}{\Delta t}\right)$$

Collecting terms, the desired pressure correction for three- or four-edge elements can be written in a general form as

$$a_P^k p'_P = \sum_{nb} a_{nb}^k p'_{nb} + S_{\text{non}}^{p\prime} - \sum_{l=1}^{N_p}\frac{R_P^l}{\rho_P^l} \tag{6.96}$$

For an unstructured mesh of three-edge elements, the neighboring coefficients a_{nb}^k comprise

$$a_1^k = \sum_{l=1}^{N_P}\frac{\alpha_1^l\rho_1^l}{\rho_P^l}\left[\frac{1}{\Delta\xi}\frac{D^{l,u}}{J}\left(\frac{\partial y}{\partial\eta}\Delta y\right) + \frac{1}{\Delta\xi}\frac{D^{l,v}}{J}\left(\frac{\partial x}{\partial\eta}\Delta x\right)\right]_1$$

$$a_2^k = \sum_{l=1}^{N_P}\frac{\alpha_2^l\rho_2^l}{\rho_P^l}\left[\frac{1}{\Delta\xi}\frac{D^{l,u}}{J}\left(\frac{\partial y}{\partial\eta}\Delta y\right) + \frac{1}{\Delta\xi}\frac{D^{l,v}}{J}\left(\frac{\partial x}{\partial\eta}\Delta x\right)\right]_2$$

$$a_3^k = \sum_{l=1}^{N_P}\frac{\alpha_3^l\rho_3^l}{\rho_P^l}\left[\frac{1}{\Delta\xi}\frac{D^{l,u}}{J}\left(\frac{\partial y}{\partial\eta}\Delta y\right) + \frac{1}{\Delta\xi}\frac{D^{l,v}}{J}\left(\frac{\partial x}{\partial\eta}\Delta x\right)\right]_3$$

with

$$\begin{aligned} S_{\text{non}}^{p'} = &-\sum_{l=1}^{N_P}\frac{\alpha_1^l\rho_1^l}{\rho_P^l}\left[\frac{1}{\Delta\eta}\frac{D^{l,u}}{J}\left(\frac{\partial y}{\partial\xi}\Delta y\right) + \frac{1}{\Delta\eta}\frac{D^{l,v}}{J}\left(\frac{\partial x}{\partial\xi}\Delta x\right)\right]_1\left(p'_b - p'_a\right) \\ &-\sum_{l=1}^{N_P}\frac{\alpha_2^l\rho_2^l}{\rho_P^l}\left[\frac{1}{\Delta\eta}\frac{D^{l,u}}{J}\left(\frac{\partial y}{\partial\xi}\Delta y\right) + \frac{1}{\Delta\eta}\frac{D^{l,v}}{J}\left(\frac{\partial x}{\partial\xi}\Delta x\right)\right]_2\left(p'_c - p'_b\right) \\ &-\sum_{l=1}^{N_P}\frac{\alpha_3^l\rho_3^l}{\rho_P^l}\left[\frac{1}{\Delta\eta}\frac{D^{l,u}}{J}\left(\frac{\partial y}{\partial\xi}\Delta y\right) + \frac{1}{\Delta\eta}\frac{D^{l,v}}{J}\left(\frac{\partial x}{\partial\xi}\Delta x\right)\right]_3\left(p'_a - p'_c\right) \end{aligned}$$

while the last term that denotes the phase mass residual is given by

$$R_P^l = \alpha_1^l\rho_1^l\left(\mathbf{U}^l\cdot\mathbf{n}_1\right)^*A_1 + \alpha_2^l\rho_2^l\left(\mathbf{U}^l\cdot\mathbf{n}_2\right)^*A_2 + \alpha_3^l\rho_3^l\left(\mathbf{U}^l\cdot\mathbf{n}_3\right)^*A_3 + \frac{\left(\left(\alpha_P^l\rho_P^l\right) - \left(\alpha_P^l\rho_P^l\right)^n\right)\Delta V_P}{\Delta t}$$

For an unstructured mesh of four-edge elements, the neighboring coefficients a^k_{nb} are

$$a^k_1 = \sum_{l=1}^{N_P} \frac{\alpha^l_1 \rho^l_1}{\rho^l_P}\left[\frac{1}{\Delta\xi}\frac{D^{l,u}}{J}\left(\frac{\partial y}{\partial\eta}\Delta y\right) + \frac{1}{\Delta\xi}\frac{D^{l,v}}{J}\left(\frac{\partial x}{\partial\eta}\Delta x\right)\right]_1$$

$$a^k_2 = \sum_{l=1}^{N_P} \frac{\alpha^l_2 \rho^l_2}{\rho^l_P}\left[\frac{1}{\Delta\xi}\frac{D^{l,u}}{J}\left(\frac{\partial y}{\partial\eta}\Delta y\right) + \frac{1}{\Delta\xi}\frac{D^{l,v}}{J}\left(\frac{\partial x}{\partial\eta}\Delta x\right)\right]_2$$

$$a^k_3 = \sum_{l=1}^{N_P} \frac{\alpha^l_3 \rho^l_3}{\rho^l_P}\left[\frac{1}{\Delta\xi}\frac{D^{l,u}}{J}\left(\frac{\partial y}{\partial\eta}\Delta y\right) + \frac{1}{\Delta\xi}\frac{D^{l,v}}{J}\left(\frac{\partial x}{\partial\eta}\Delta x\right)\right]_3$$

$$a^k_4 = \sum_{l=1}^{N_P} \frac{\alpha^l_4 \rho^l_4}{\rho^l_P}\left[\frac{1}{\Delta\xi}\frac{D^{l,u}}{J}\left(\frac{\partial y}{\partial\eta}\Delta y\right) + \frac{1}{\Delta\xi}\frac{D^{l,v}}{J}\left(\frac{\partial x}{\partial\eta}\Delta x\right)\right]_4$$

with

$$\begin{aligned}
S^{p'}_{\text{non}} = &-\sum_{l=1}^{N_P} \frac{\alpha^l_1 \rho^l_1}{\rho^l_P}\left[\frac{1}{\Delta\eta}\frac{D^{l,u}}{J}\left(\frac{\partial y}{\partial\xi}\Delta y\right) + \frac{1}{\Delta\eta}\frac{D^{l,v}}{J}\left(\frac{\partial x}{\partial\xi}\Delta x\right)\right]_1 (p'_b - p'_a)\\
&-\sum_{l=1}^{N_P} \frac{\alpha^l_2 \rho^l_2}{\rho^l_P}\left[\frac{1}{\Delta\eta}\frac{D^{l,u}}{J}\left(\frac{\partial y}{\partial\xi}\Delta y\right) + \frac{1}{\Delta\eta}\frac{D^{l,v}}{J}\left(\frac{\partial x}{\partial\xi}\Delta x\right)\right]_2 (p'_c - p'_b)\\
&-\sum_{l=1}^{N_P} \frac{\alpha^l_3 \rho^l_3}{\rho^l_P}\left[\frac{1}{\Delta\eta}\frac{D^{l,u}}{J}\left(\frac{\partial y}{\partial\xi}\Delta y\right) + \frac{1}{\Delta\eta}\frac{D^{l,v}}{J}\left(\frac{\partial x}{\partial\xi}\Delta x\right)\right]_3 (p'_d - p'_c)\\
&-\sum_{l=1}^{N_P} \frac{\alpha^l_4 \rho^l_4}{\rho^l_P}\left[\frac{1}{\Delta\eta}\frac{D^{l,u}}{J}\left(\frac{\partial y}{\partial\xi}\Delta y\right) + \frac{1}{\Delta\eta}\frac{D^{l,v}}{J}\left(\frac{\partial x}{\partial\xi}\Delta x\right)\right]_4 (p'_a - p'_d)
\end{aligned}$$

while the phase mass residual is:

$$\begin{aligned}
R^l_P = &\alpha^l_1\rho^l_1(\mathbf{U}^l\cdot\mathbf{n}_1)^* A_1 + \alpha^l_2\rho^l_2(\mathbf{U}^l\cdot\mathbf{n}_2)^* A_2 + \alpha^l_3\rho^l_3(\mathbf{U}^l\cdot\mathbf{n}_3)^* A_3 + \alpha^l_4\rho^l_4(\mathbf{U}^l\cdot\mathbf{n}_3)^* A_4\\
&+ \frac{\left((\alpha^l_P\rho^l_P) - (\alpha^l_P\rho^l_P)^n\right)\Delta V_P}{\Delta t}
\end{aligned}$$

It should be noted that Eqn (6.96) has been normalized by the phase density ρ^l_P in order to avoid bias toward the heavier fluid. The diagonal coefficient a^k_P for both meshes is effectively the sum of the neighboring coefficients. For convenience, the contribution due to mesh nonorthogonality may be neglected especially for a nearly orthogonal mesh, and in any case, the pressure corrections vanish for a converged solution.

Interphase slip algorithm-coupled

The primary aim for the development of the IPSA-C method is to alleviate the problem associated with the strong coupling between two or multiple phases. As described in the previous section for the IPSA method, the interfacial source terms have been treated in accordance with the *partially implicit* algorithm. In the IPSA-C method, the performance of the pressure correction step is improved by the semi-implicit inclusion of the interfacial source terms following the idea of the SINCE method.

To illustrate the IPSA-C method, reconsider again the discretized phase momentum equations of the multifluid model for the guessed velocities u_P^{k*} and v_P^{k*} with a guessed pressure field p^*. In contrast to the IPSA method, the equations can now be written in the form as in the SINCE algorithm in which the momentum transport by the change of phase and the momentum transport by the interfacial force are treated in *partially* and *fully implicits*. They are:

$$u_P^{k*} = \sum_{nb} \frac{a_{nb}^{k,u} u_{nb}^{k*}}{a_P^{k,u}} - \alpha_P^k \frac{\Delta V_P}{a_P^{k,u}} \left.\frac{\partial p^*}{\partial x}\right|_P + \sum_{l=1}^{N_p} \frac{(C_{kl})_P}{a_P^{k,u}} \left(u_P^{l*} - u_P^{k*}\right) + B_P^{\prime u^k} \tag{6.97}$$

$$v_P^{k*} = \sum_{nb} \frac{a_{nb}^{k,v} v_{nb}^{k*}}{a_P^{k,v}} - \alpha_P^k \frac{\Delta V_P}{a_P^{k,v}} \left.\frac{\partial p^*}{\partial y}\right|_P + \sum_{l=1}^{N_p} \frac{(C_{kl})_P}{a_P^{k,v}} \left(v_P^{l*} - v_P^{k*}\right) + B_P^{\prime v^k} \tag{6.98}$$

where $B_P^{\prime u^k}$ and $B_P^{\prime v^k}$ are the remaining kth phase source terms after the pressure gradient source and interfacial force terms have been taken out. The corrected velocities u_P^k and v_P^k with the correct pressure field p may be determined according to

$$u_P^{k} = \sum_{nb} \frac{a_{nb}^{k,u} u_{nb}^{k}}{a_P^{k,u}} - \alpha_P^k \frac{\Delta V_P}{a_P^{k,u}} \left.\frac{\partial p}{\partial x}\right|_P + \sum_{l=1}^{N_p} \frac{(C_{kl})_P}{a_P^{k,u}} \left(u_P^{l} - u_P^{k}\right) + B_P^{\prime u^k} \tag{6.99}$$

$$v_P^{k} = \sum_{nb} \frac{a_{nb}^{k,v} v_{nb}^{k}}{a_P^{k,v}} - \alpha_P^k \frac{\Delta V_P}{a_P^{k,v}} \left.\frac{\partial p}{\partial y}\right|_P + \sum_{l=1}^{N_p} \frac{(C_{kl})_P}{a_P^{k,v}} \left(v_P^{l} - v_P^{k}\right) + B_P^{\prime v^k} \tag{6.100}$$

Following the similar treatment in the IPSA methodology, the velocity corrections may be written as

$$u_P^{\prime k} = -\alpha_P^k \frac{\Delta V_P}{a_P^{k,u}} \left.\frac{\partial p'}{\partial x}\right|_P + \sum_{l=1}^{N_p} \frac{(C_{kl})_P}{a_P^{k,u}} \left(u_P^{\prime l} - u_P^{\prime k}\right) \tag{6.101}$$

$$v_P^{\prime k} = -\alpha_P^k \frac{\Delta V_P}{a_P^{k,v}} \left.\frac{\partial p'}{\partial y}\right|_P + \sum_{l=1}^{N_p} \frac{(C_{kl})_P}{a_P^{k,v}} \left(v_P^{\prime l} - v_P^{\prime k}\right) \tag{6.102}$$

where $u_P^{\prime k} = u_P^k - u_P^{k*}$, $u_P^{\prime l} = u_P^l - u_P^{l*}$, $v_P^{\prime k} = v_P^k - v_P^{k*}$, $v_P^{\prime l} = v_P^l - v_P^{l*}$ and $p' = (p - p^*)$.

To solve the velocity correction $u_P'^k$, Eqn (6.101) is adapted to an equivalent structure of the SINCE algorithm. This can be achieved by defining the following phase pressure correction parameters:

$$\Psi_P'^k = \frac{u_P'^k}{-\left.\dfrac{\partial p'}{\partial x}\right|_P} \qquad \Psi_P'^l = \frac{u_P'^l}{-\left.\dfrac{\partial p'}{\partial x}\right|_P} \tag{6.103}$$

With the above parameters, Eqn (6.103) can be expressed in accordance with the SINCE method as

$$\Psi_P'^k = \sum_{l=1}^{N_p} \frac{(C_{kl})_P}{a_P^{k,u}} \left(\Psi_P'^l - \Psi_P'^k\right) + \alpha_P^k \frac{\Delta V_P}{a_P^{k,u}} \tag{6.104}$$

Equation (6.104) can be arranged to the following system of linear equations:

$$\begin{aligned} D_P^1 \Psi_P'^1 &= \frac{(C_{12})_P}{a_P^{1,u}} \Psi_P'^2 + \frac{(C_{13})_P}{a_P^{1,u}} \Psi_P'^3 + \cdots + \frac{\left(C_{1N_p}\right)_P}{a_P^{1,u}} \Psi_P'^{N_p} + \alpha_P^1 \frac{\Delta V_P}{a_P^{1,u}} \\ &\vdots \\ D_P^{N_p} \Psi_P'^{N_p} &= \left(C_{N_p 1}\right)_P \Psi_P'^1 + \left(C_{N_p 2}\right)_P \Psi_P'^2 + \cdots + \left(C_{N_p (N_p - 1)}\right)_P \Psi_P'^{N_p - 1} + \alpha_P^{N_p} \frac{\Delta V_P}{a_P^{N_p,u}} \end{aligned} \tag{6.105}$$

where

$$D_P^k = 1 + \sum_{l=1}^{N_p} \frac{(C_{kl})_P}{a_P^{k,u}}$$

which can be interpreted as the matrix equation

$$\underbrace{\begin{bmatrix} D_P^1 & \cdots & -\dfrac{\left(C_{1N_p}\right)_P}{a_P^{1,u}} \\ \vdots & & \vdots \\ -\dfrac{\left(C_{N_p 1}\right)_P}{a_P^{N_p,u}} & \cdots & D_P^{N_p} \end{bmatrix}}_{\mathbf{A}} \underbrace{\begin{bmatrix} \Psi_P'^{1*} \\ \vdots \\ \Psi_P'^{N_p*} \end{bmatrix}}_{\mathbf{X}} = \underbrace{\begin{bmatrix} \alpha_P^1 \dfrac{\Delta V_P}{a_P^{1,u}} \\ \vdots \\ \alpha_P^{N_p} \dfrac{\Delta V_P}{a_P^{N_p,u}} \end{bmatrix}}_{\mathbf{B}} \tag{6.106}$$

In Eqn (6.106), the coefficient matrix **A** and the source matrix **B** are calculated based upon the existing values. Solving the matrix Eqn (6.106), the phase pressure correction parameters $\Psi_P'^{k*}$ can be calculated as

$$\mathbf{X} = (\mathbf{A})^{-1}\mathbf{B} \tag{6.107}$$

which are then used as intermediate estimates to determine the interfacial coupling terms. By first substituting the parameters from Eqn (6.107) into Eqn (6.105) to yield

$$D_P^k \Psi_P^k = \sum_{l=1}^{N_p} \frac{(C_{kl})_P}{a_P^{k,u}} \Psi_P^{k*} + \alpha_P^k \frac{\Delta V_P}{a_P^{k,u}} \tag{6.108}$$

the velocity–pressure interdependency for the velocity correction $u_P'^k$ is thus recovered through Eqn (6.103) according to

$$u_P'^k = -\Psi_P^k \left.\frac{\partial p'}{\partial x}\right|_P = -\underbrace{\left(\frac{\sum_{l=1}^{N_p} \frac{(C_{kl})_P}{a_P^{k,u}} \Psi_P^{k*} + \alpha_P^k \frac{\Delta V_P}{a_P^{k,u}}}{1 + \sum_{l=1}^{N_p} \frac{(C_{kl})_P}{a_P^{k,u}}} \right)}_{H_P^{k,u}} \left.\frac{\partial p'}{\partial x}\right|_P \tag{6.109}$$

Following the similar procedure as described above, the velocity correction $v_P'^k$ is

$$v_P'^k = -\underbrace{\left(\frac{\sum_{l=1}^{N_p} \frac{(C_{kl})_P}{a_P^{k,v}} \Psi_P^{k*} + \alpha_P^k \frac{\Delta V_P}{a_P^{k,v}}}{1 + \sum_{l=1}^{N_p} \frac{(C_{kl})_P}{a_P^{k,v}}} \right)}_{H_P^{k,v}} \left.\frac{\partial p'}{\partial y}\right|_P \tag{6.110}$$

In the forms presented in Eqns (6.109) and (6.110), the IPSA-C method as described above is equally applicable for the evaluation of the velocity correction $u_P'^k$ and $v_P'^k$ in complex meshes due to the general manner in which the phase pressure correction parameters are normalized. In an unstructured mesh, the pressure correction gradients are required to be evaluated using an appropriate gradient reconstruction method in order to evaluate $H^{l,u}$ and $H^{l,v}$. From this point onward, the IPSA-C method proceeds in the same manner as the pressure correction step of the IPSA method. The velocity–pressure interdependency expressions are substituted into the joint equation governing mass conservation with the same element face approximations as described in the previous section. This leads to the pressure correction equations with coefficients $D^{l,u}$ and $D^{l,v}$ replaced by $H^{l,u}$ and $H^{l,v}$.

6.5.3 Comments on Matrix Solvers

An iterative solver is based on performing a series of repeated operations where the error in the approximate solution is reduced by each application of the operations. Through these repeated operations, a solution to the set of algebraic multiphase equations is achieved, which is the essence of the segregated approach such as that described by the IPSA or IPSA-C method.

In general, this system of equations can usually be written in matrix form as:

$$\mathbf{AX} = \mathbf{B} \tag{6.111}$$

where $\mathbf{X}$ is the unknown nodal variables of the transport property and matrix $\mathbf{A}$ contains the nonzero entries:

$$\mathbf{A} = \begin{bmatrix} A_{11} & A_{12} & A_{13} & \cdots & A_{1n} \\ A_{21} & A_{22} & A_{23} & \cdots & A_{2n} \\ A_{31} & A_{32} & A_{33} & \cdots & A_{3n} \\ \vdots & \vdots & \vdots & \ddots & \vdots \\ A_{n-21} & A_{n-22} & A_{n-23} & \cdots & A_{n-2n} \\ A_{n-11} & A_{n-12} & A_{n-13} & \cdots & A_{n-1n} \\ A_{n1} & A_{n2} & A_{n3} & \cdots & A_{nn} \end{bmatrix} \tag{6.112}$$

with $\mathbf{B}$ comprised of known values of $\mathbf{X}$, for example, which are given by the boundary conditions or source/sink terms. When matrix $\mathbf{A}$ is dense—few nonzero entries—the direct solution of Eqn (6.111) is obtained using Gauss elimination. This preferred technique is carried out in two stages. It is first factorized into lower triangular $\mathbf{L}$ and upper triangular $\mathbf{U}$ factors. The factor form of Eqn (6.111) is then solved according to

$$\mathbf{UX} = \mathbf{L}^{-1}\mathbf{B} \tag{6.113}$$

Because of the structure of $\mathbf{U}$, Eqn (6.113) only involves a *back substitution*, once $\mathbf{L}^{-1}\mathbf{B}$ is formed. Generally speaking, the inversion of matrix $\mathbf{A}$ through Gauss elimination can be rather computationally expensive.

For iterative techniques, Eqn (6.111) can be rewritten as

$$(\mathbf{N} - \mathbf{P})\mathbf{X} = \mathbf{B} \tag{6.114}$$

where $\mathbf{N}$ is taken to be close to $\mathbf{A}$ but computationally more efficient to factorize, e.g. $\mathbf{N}$ may contain the diagonal entries while $\mathbf{P}$ comprises the off-diagonal entries. Equation (6.113) can be reexpressed as

$$\mathbf{X}^k = \mathbf{N}^{-1}\mathbf{P}\mathbf{X}^{k-1} + \mathbf{N}^{-1}\mathbf{B} \tag{6.115}$$

where $\mathbf{X}^{k-1}$ are the known variables at iteration step $k-1$ while $\mathbf{X}^k$ are the unknown variables at iteration step k. The simplest iterative method for solving Eqn (6.115) is the *Jacobi method.* With $\mathbf{N}$ representing the diagonal entries, i.e. $\mathbf{DI}$, where $\mathbf{I}$ is the identity matrix, and $\mathbf{P}$ is given by $\mathbf{P} = \mathbf{L} + \mathbf{U}$, the method entails solving for

$$\phi_i^k = \left(B_i - \sum_{j,j \neq i}^{N_c} A_{ij}\phi_j^{k-1} \right) \Big/ A_{ii} \tag{6.116}$$

A more immediate improvement to the *Jacobi method* is provided by the *Gauss–Siedel method*. Through each iteration step, values from the lower triangular matrix $\mathbf{L}$ are immediately employed as soon as they are made available. In this case, $\mathbf{N} = \mathbf{DI} - \mathbf{L}$ and $\mathbf{P} = \mathbf{U}$, and Eqn (6.115) becomes

$$\phi_i^k = \left(B_i - \sum_{j=1}^{i-1} A_{ij}\phi_j^k - \sum_{j,j\neq i}^{N_c} A_{ij}\phi_j^{k-1} \right) \Big/ A_{ii} \tag{6.117}$$

Most engineering applications of multiphase flows require the feasibility of constructing high quality meshes to resolve the physical flow structures. Such flow problems usually amount to substantial grid points to achieve adequate resolution for the whole physical domain. If very fine meshes are required, the *Jacobi method* and *Gauss–Siedel method* suffer from slow convergence on very fine meshes. The number of iterations for these methods is linearly proportional to the number of grid points in one coordinate direction. This behavior is related to the fact that information has to travel back and forth across the domain several times. For a discrete grid increment Δ_m, the high frequency errors can be represented by the smallest value $\lambda_{\min} = 2\Delta_m$. These iterative methods are rather efficient in removing the high frequency errors in a few iterations in a coarse mesh. It is the removal of the low frequency errors, and equivalently of the residual, in a sufficiently fine mesh that causes the slow convergence, especially on a fixed grid. To overcome the convergence problem, a sequence of grids may be constructed in order that low frequency errors on a fine mesh can become high frequency errors on a coarser mesh. These high frequency errors are now essentially lost or hidden in the coarse mesh and the solution procedure begins to dampen at a more rapid rate than would have taken place in the fine mesh because of the larger Δ_m. By progressively moving the intermediate results to coarser meshes, the low frequency errors are essentially damped; when these results are transferred back to the fine mesh, the low frequency errors would indeed be much smaller than they would have been for an equal number of iterations that are to be performed on the fine mesh itself. Multigrid methods seek, therefore, to exploit the high frequency smoothing of iterative methods in the following way.

Consider Eqn (6.111) to be solved on a sequence of grids $m = 1, \ldots, M$, which can be written for the finest grid as

$$\mathbf{A}^M\mathbf{X}^M = \mathbf{B}^M \tag{6.118}$$

The residual (or defect) $\mathbf{R}^M$ for the solution of this grid satisfies

$$\mathbf{R}^M = \mathbf{B}^M - \mathbf{A}^M\mathbf{Y}^M \tag{6.119}$$

where $\mathbf{Y}^M$ represents the intermediate solution of Eqn (6.118). We can define the error $\mathbf{e}^M$ as the difference between the true solution $\mathbf{X}^M$ and the intermediate solution $\mathbf{Y}^M$:

$$\mathbf{e}^M = \mathbf{X}^M - \mathbf{Y}^M \tag{6.120}$$

A relationship can thus be established between the error and residual as:

$$\mathbf{A}^M\mathbf{e}^M = \mathbf{R}^M \tag{6.121}$$

On the next coarser grid, $M-1$, we work with the form presented in Eqn (6.121) instead of solving for the solution of $\mathbf{y}^M$. In other words,

$$\mathbf{A}^{M-1}\mathbf{e}^{M-1} = \mathbf{R}^{M-1} \tag{6.122}$$

Given the values of the residual $\mathbf{R}^M$ on the fine mesh, a suitable averaging procedure to find the residual $\mathbf{R}^{M-1}$ is realized on the coarse grid. The entries in matrix $\mathbf{A}^{M-1}$ may be recomputed on the coarser grid or evaluated from the fine grid entries of matrix $\mathbf{A}^M$ using some form of averaging or interpolation technique. An adequate number of iterations are performed to obtain the converged solution of the error $\mathbf{e}^{M-1}$. The process of transferring variables from a fine grid to a coarse grid is commonly known as *restriction.* It may be carried out into a number of increasingly coarse levels as illustrated by the typical V-cycle or W-cycle with five different grid levels shown in Figures 6.11 and 6.12. After the converged solution is attained, we reverse the process along the direction from the coarse to the fine grid, which is known as *prolongation.* A convenient linear interpolation procedure can be used to generate values for the prolonged error $\mathbf{e}'^M$ at intermediate points in the fine grid. At the finest grid level, the intermediate fine grid solution can be subsequently corrected according to

$$\mathbf{Y}^{\text{Improved}} = \mathbf{Y}^M + \mathbf{e}'^M \tag{6.123}$$

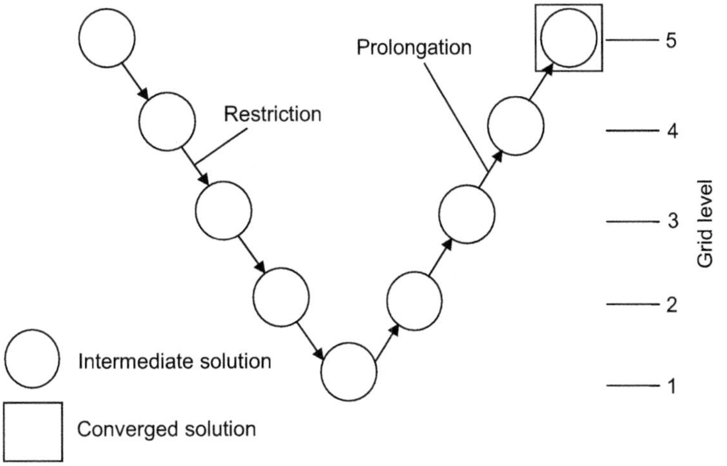

Figure 6.11
Schematic representation of a multigrid method using a V-cycle.

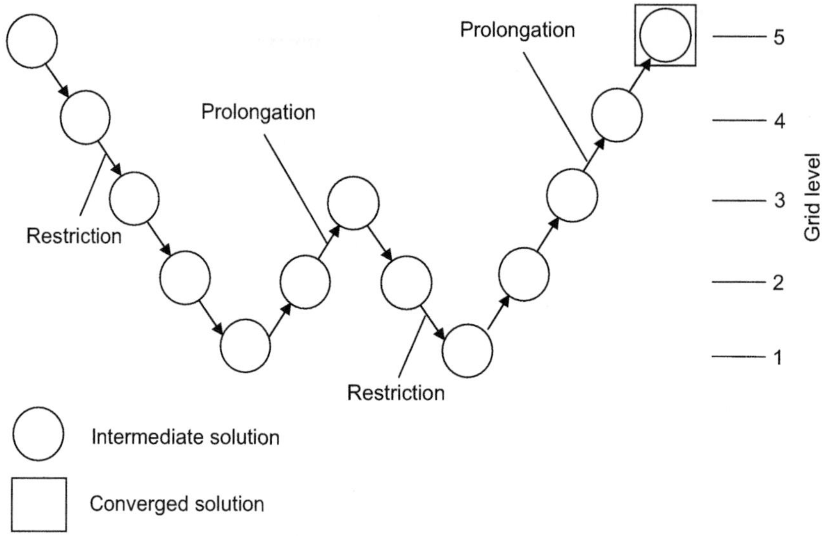

Figure 6.12
Schematic representation of a multigrid method using a W-cycle.

There are also other strategies that may be used for cycling between the coarse and fine grids. Efficiency may be improved by the decision to switch from one grid to another on the depending rate of convergence through the combination of V-W cycles or other possible combinations.

A multigrid method can also be characterized as being either *geometric* or *algebraic*. A geometric multigrid, also known as the Full Approximation Scheme multigrid, involves a hierarchy of meshes (cycling between fine and coarse grids) and the discretized equations are evaluated on every level. Within each level, simple point-by-point iterative methods such as *Jacobi* and *Gauss–Siedel* could be employed for the coarse level equations to determine the immediate values for ϕ^k. In an algebraic multigrid, the coarse level equations are generated without any geometry or rediscretization on the coarse levels—a feature that makes this method amenable for the use on unstructured meshes. Once linearization is performed on the system of equations, the nonlinear properties are not experienced by the solver until at the fine level where the operator is finally updated. The multigrid approach is more a strategy than a particular method and the interested reader should consult Wesseling (1995), Timmermann (2000) and Thomas et al. (2003) for the latest trends and developments of multigrid methods.

For a general transport equation such as those formulated for unstructured meshes, the *Generalized Minimal Residual* (GMRES) method can be applied to solve the nonsymmetric matrix **A**. It is essentially a projection method based on the projection of matrix **A** onto the Krylov subspace. Such a projection will be missing much information, but depending on how well the subspace is chosen, a good approximation to the expected result can be achieved.

The aim of the GMRES method is to seek the vector **X** within the Krylov subspace such that the Euclidean norm or L^2 norm of the residual **R(Y)** is minimized:

$$\mathbf{R}(\mathbf{y}) = \|\mathbf{B} - \mathbf{AX}\| = \|\mathbf{B} - \mathbf{A}(\mathbf{X} + \mathbf{V}^k\mathbf{Y})\| \tag{6.124}$$

where $\mathbf{V}^k$ is the projection matrix whose columns represent the orthonormal basis set of vectors, then spans the Krylov subspace and **y** is a *k*-dimensional vector within the Krylov subspace. Starting with an initial guess of the solution $\mathbf{X}^0$, the residual vector is calculated:

$$\mathbf{R}^0 = \mathbf{B} - \mathbf{AX}^0 \tag{6.125}$$

and the normalized residual vector comprising the first column of the subspace matrix $\mathbf{V}^k$ is determined:

$$\mathbf{V}^1 = \frac{\mathbf{R}^0}{\|\mathbf{R}^0\|} \tag{6.126}$$

During the iteration stage, the Hessenburg matrix and the vectors that consist of $\mathbf{V}^k$ are created. The next vector in the Krylov subspace is realized as

$$\mathbf{W}^j = \mathbf{AV}^j \tag{6.127}$$

In order to form the orthogonal basis of the Krylov subspace, $\mathbf{W}^j$ is modified by first ascertaining:

$$\mathbf{H}^{i,j} = \mathbf{W}^j \cdot \mathbf{V}^i \tag{6.128}$$

which is an entry to the Hessenburg matrix $\overline{\mathbf{H}}^k$ and subsequently computed as

$$\mathbf{W}^j = \mathbf{AV}^j - \mathbf{H}^{i,j}\mathbf{V}^i \tag{6.129}$$

Next, the Euclidean norm or L^2 norm of $\mathbf{W}^j$ is determined:

$$\mathbf{H}^{j+1,j} = \|\mathbf{W}^j\| \tag{6.130}$$

as well as the next column that defines the subspace:

$$\mathbf{V}^{j+1} = \frac{\mathbf{W}^j}{\mathbf{H}^{j+1,j}} \tag{6.131}$$

This ends the iteration stage. The approximate solution can be formed according to

$$\mathbf{X}^k = \mathbf{X}^0 + \mathbf{V}^k\mathbf{Y}^k \tag{6.132}$$

where $\mathbf{Y}^k$ minimizes the Euclidean norm or L^2 norm of the residual: $\|\mathbf{B} - \mathbf{AX}^k\|$ where $\mathbf{X}^k$ are the possible solutions within the Krylov space defined by $\mathbf{V}^k$. Here, the method requires the solution of a least-squares problem. Fortunately, it can be solved on the reduced order defined by the Krylov space, assuming that an approximate solution vector $\mathbf{X}^k$ exists within a Krylov subspace defined by $\mathbf{V}^k$ according to

$$\begin{aligned} \mathbf{B} - \mathbf{AX} &= \mathbf{B} - \mathbf{A}\left(\mathbf{X}^0 + \mathbf{V}^k\mathbf{Y}\right) \\ &= \mathbf{R}^0 - \mathbf{AV}^k\mathbf{Y} \end{aligned} \tag{6.133}$$

Defining the truncated Hessenburg matrix $\mathbf{H}^k$ whose last row has been removed, the transformed version of $\mathbf{A}$ in the Krylov space is: $(\mathbf{V}^k)^T\mathbf{AV}^k = \mathbf{H}^k$ since $(\mathbf{V}^k)^T = (\mathbf{V}^k)^{-1}$. Also,

$$\mathbf{AV}^k = \mathbf{V}^{k+1}\overline{\mathbf{H}}^k \tag{6.134}$$

Substituting Eqn (6.134) into Eqn (6.133) yields

$$\begin{aligned} \mathbf{B} - \mathbf{AX} &= \mathbf{R}^0 - \mathbf{AV}^k\mathbf{Y} \\ &= \|\mathbf{R}^0\| - \mathbf{V}^{k+1}\overline{\mathbf{H}}^k\mathbf{Y} \\ &= \beta\mathbf{V}^1 - \mathbf{V}^{k+1}\overline{\mathbf{H}}^k\mathbf{Y} \\ &= \mathbf{V}^{k+1}\left(\beta\mathbf{e}_1 - \overline{\mathbf{H}}^k\mathbf{Y}\right) \end{aligned} \tag{6.135}$$

where the vector $\mathbf{e}_1$ is the first column of the $(k+1) \times (k+1)$ identity matrix. The least-squares system of Eqn (6.135):

$$\mathbf{Y}^k = \left\|\beta\mathbf{e}_1 - \overline{\mathbf{H}}^k\mathbf{Y}\right\| \tag{6.136}$$

is reduced to upper triangular and the resulting system is solved via a Gaussian elimination with the last row removed to determine $\mathbf{Y}^k$. Hence, the resulting nonzero entry remaining in the last row on the right-hand side is the residual. This allows the new estimate for $\mathbf{X}^k$ to be obtained according to Eqn (6.132). The number of operations in the calculation of $\mathbf{X}^k$ generally increases linearly with the number of iterations, which may become rather expensive in terms of both computation time and storage requirements for large systems of equations. To overcome the problem, the method can be restarted every m steps, where m is an integer parameter. The approximate solution can now be formed according to

$$\mathbf{X}^m = \mathbf{X}^0 + \mathbf{V}^m\mathbf{Y}^m \tag{6.137}$$

where $\mathbf{Y}^m$ minimizes $\|\beta\mathbf{e}_1 - \overline{\mathbf{H}}^m\mathbf{Y}\|$, where $\mathbf{X}^m$ are the possible solutions within the Krylov space defined by $\mathbf{V}^m$. The residual is computed according to $\mathbf{R}^m = \mathbf{B} - \mathbf{AX}^m$ and if additional steps are required, $\mathbf{X}^0$ and $\mathbf{V}^k$ are reset to $\mathbf{X}^m$ and $\mathbf{V}^1 = \mathbf{R}^m/\|\mathbf{R}^m\|$, and iteration proceeds by determining the Hessenburg matrix and the vectors that comprise $\mathbf{V}^k$. More details of the method can be found in Saad and Schultz (1985). Alternatively, the *Biconjugate Gradient Stabilized* developed by Van der Host (1992) can be applied to overcome the computation time and storage requirements for the calculation of nonsymmetric systems. This method has been found to be robust and considerably more efficient when compared to GMRES. More details of the method are left to interested readers, which can be found in Van der Host (1992).

6.5.4 Coupled Equation System

For multiphase flows, it may be necessary to exploit the simultaneous solution of the governing equations describing the multifluid model due to the coupling between phases and dominant source terms. In order to possibly overcome numerical instabilities that may result from the sequential solution of governing equations, the *coupled* solution approach, a more robust alternative to the segregated approach, is required to be adopted in simultaneously solving the velocity, pressure and volume fraction equations.

Consider for simplicity the full coupling of the system of equations of a two-phase flow problem. The linear equations for (u^1, u^2, v^1, v^2, p) of the nodal point in a single control volume can generally be expressed in a matrix form of $\mathbf{AX} = \mathbf{B}$ as

$$\underbrace{\begin{bmatrix} A_{u^1u^1} & A_{u^1u^2} & A_{u^1v^1} & A_{u^1v^2} & A_{u^1p} \\ A_{u^2u^1} & A_{u^2u^2} & A_{u^2v^1} & A_{u^2v^2} & A_{u^2p} \\ A_{v^1u^1} & A_{v^1u^2} & A_{v^1v^1} & A_{v^1v^2} & A_{v^1p} \\ A_{v^2u^1} & A_{v^2u^2} & A_{v^2v^1} & A_{v^2v^2} & A_{v^2p} \\ A_{pu^1} & A_{pu^2} & A_{pv^1} & A_{pv^2} & A_{pp} \end{bmatrix}}_{A'_{ij}} \underbrace{\begin{bmatrix} u_P^1 \\ u_P^2 \\ v_P^1 \\ v_P^2 \\ p_P \end{bmatrix}}_{\phi'_i} = \underbrace{\begin{bmatrix} S_{u^1} \\ S_{u^2} \\ S_{v^1} \\ S_{v^2} \\ S_p \end{bmatrix}}_{B'_i} \tag{6.138}$$

In order to close the system of equations such as depicted in Eqn (6.138), an equation for the shared pressure p is realized through the improved Rhie–Chow interpolation expressions of the face velocities into the discretized joint equation governing the total mass balance. Since the joint equation is now dependent upon the pressure field, a coupled formulation is realized. Alternatively, it has been deemed necessary to solve the volume fraction simultaneously in conjunction with the system of momentum and pressure equations for strong coupling between phases and dominant source terms. The linear equations for $(u^1, u^2, v^1, v^2, p, \alpha^1)$ of the nodal point in a single control volume can now be expressed in a matrix form as

$$\underbrace{\begin{bmatrix} A_{u^1u^1} & A_{u^1u^2} & A_{u^1v^1} & A_{u^1v^2} & A_{u^1p} & A_{u^1\alpha^1} \\ A_{u^2u^1} & A_{u^2u^2} & A_{u^2v^1} & A_{u^2v^2} & A_{u^2p} & A_{u^2\alpha^1} \\ A_{v^1u^1} & A_{v^1u^2} & A_{v^1v^1} & A_{v^1v^2} & A_{v^1p} & A_{v^1\alpha^1} \\ A_{v^2u^1} & A_{v^2u^2} & A_{v^2v^1} & A_{v^2v^2} & A_{v^2p} & A_{v^2\alpha^1} \\ A_{pu^1} & A_{pu^2} & A_{pv^1} & A_{pv^2} & A_{pp} & A_{p\alpha^1} \\ A_{\alpha^1u^1} & A_{\alpha^1u^2} & A_{\alpha^1v^1} & A_{\alpha^1v^2} & A_{\alpha^1P} & A_{\alpha^1\alpha^1} \end{bmatrix}}_{A'_{ij}} \underbrace{\begin{bmatrix} u_P^1 \\ u_P^2 \\ v_P^1 \\ v_P^2 \\ p_P \\ \alpha_P^1 \end{bmatrix}}_{\phi'_i} = \underbrace{\begin{bmatrix} S_{u^1} \\ S_{u^2} \\ S_{v^1} \\ S_{v^2} \\ S_p \\ S_{\alpha^1} \end{bmatrix}}_{B'_i} \tag{6.139}$$

It should be noted that the volume fraction α^2 can be obtained from the algebraic constraint of the local volume fraction.

The matrix elemental entries A'_{ij} as depicted in Eqn (6.38) or Eqn (6.139) can be assembled into a large matrix including the unknown values of velocity, pressure, volume fraction and other transport property including their sources/sinks in a general form:

$$\begin{bmatrix} A'_{11} & & \cdots & \cdots & & A'_{1n} \\ & A'_{22} & & & & \\ \vdots & & \cdots & & & \vdots \\ \vdots & & & A'_{ii} & & \vdots \\ & & & & \cdots & \\ A'_{n1} & & \cdots & \cdots & & A'_{nn} \end{bmatrix} \begin{bmatrix} \phi'_1 \\ \phi'_2 \\ \vdots \\ \phi'_i \\ \vdots \\ \phi'_{N_c} \end{bmatrix} = \begin{bmatrix} B'_1 \\ B'_2 \\ \vdots \\ B'_i \\ \vdots \\ B'_{N_c} \end{bmatrix} \tag{6.140}$$

where N_c is the total number of unknown nodal variables for the entire physical domain in question. The system as shown in Eqn (6.140) may be solved using Gaussian elimination. A more sophisticated approach is to consider the unknowns of the neighboring nodal points in a more implicit manner that entails the construction of a larger matrix **A**. This matrix may then be solved via the algebraic multigrid strategy in conjunction with the simple point-by-point *Jacobi method* or *Gauss–Siedel method*. It can immediately be seen that the linear system of Eqn (6.139) is amenable to be solved beyond a two-phase flow problem. The advantages of a coupled treatment over a segregated approach are: *robustness*, *efficiency*, *generality* and *simplicity*. Nevertheless, the principal drawback is the high storage requirements for all of the nonzero matrix entries.

6.6 Solution Methods for the Population Balance Equation

Owing to the complex phenomenological nature of particle dynamics, analytical solutions only exist in very few cases for the population balance equation of which aggregation/coalescence and breakage/break up kernels are substantially simplified (Scott, 1968; McCoy and Madras, 2003). Because of the enormous practical interest, numerical approaches have been developed to solve the population balance equation. The most common methods are the Monte Carlo method, method of moments and class method. Theoretically speaking, the Monte Carlo method, which solves the population balance equation based on a statistical ensemble approach (Domilovskii et al., 1979; Liffman, 1992; Debry et al., 2003; Maisels et al., 2004), is attractive in contrast to other methods. The main advantage of the method is the flexibility and accuracy to track particle changes in multidimensional systems. Nonetheless, as the accuracy of the Monte Carlo method is directly proportional to the number of simulation particles, extensive computational time is invariably required. In addition, the incorporation of the method in conjunction with computational multiphase fluid dynamics is not a straightforward process. Other more useful solution methods, such as the class method and method of moments to determine the particle size distribution and ease of coupling with computational multiphase fluid dynamics, are discussed and presented.

6.6.1 Class Method

The class method directly solves, for example, the particle number density, which has received considerable attention due to its rather straightforward implementation within the framework of computational multiphase fluid dynamics. Particularly in the method of discrete classes, the continuous size range of particles can be realized through the discretization of the particle size distribution into a series number of classes of discrete sizes. For each class, the equation of the number density of particles is solved to accommodate the population changes caused by intra/intergroup aggregation/coalescence and breakage/break up of particles. The particle size distribution is thereby approximated as:

$$f_1\left(V_p, \mathbf{r}, t\right) \approx \sum_{i=1}^{N} N_i(\mathbf{r}, t)\delta\left(V_p - v_i\right) \tag{6.141}$$

where N_i represents the number density of the ith class and consists of all particles per unit volume with a pivot size v_i. In this method, the sizes of class methods are continuously fixed and aligned in the state space. An illustration of the class method in approximating the particle size distribution is shown in Figure 6.13.

Based on Eqn (6.141), Eqn (3.15), which is expressed in terms of the particle volume V_p being the internal coordinate and in the absence of particle growth, becomes

$$\frac{\partial N_i(\mathbf{r}, t)}{\partial t} + \nabla_{\mathbf{r}} \cdot \left(\mathbf{v}_{p,i}(\mathbf{r}, \mathbf{Y}, t)N_i(\mathbf{r}, t)\right) = S_{N_i}(\mathbf{r}, \mathbf{Y}, t) \tag{6.142}$$

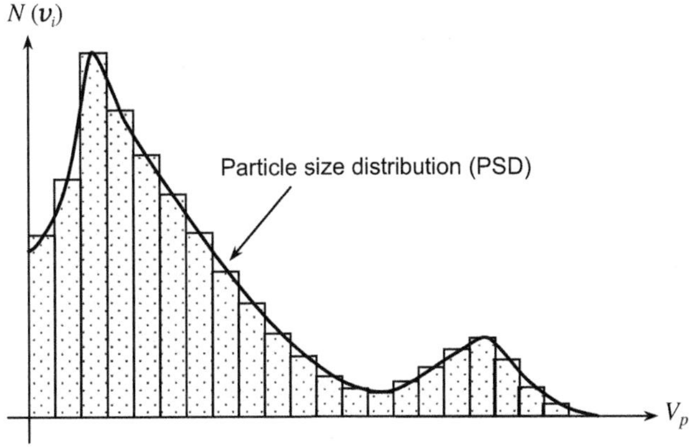

Figure 6.13
An illustration of the class method.

where $\mathbf{v}_{p,i}(\mathbf{r}, \mathbf{Y}, t)$ denotes the local particle velocity weighted by the discrete number density. In the absence of particle growth, Eqn (6.142) reduces to

$$\frac{\partial N_i(\mathbf{r},t)}{\partial t} + \nabla \cdot \left(\mathbf{v}_{p,i}(\mathbf{r},\mathbf{Y},t)N_i(\mathbf{r},t)\right) = S_{N_i}(\mathbf{r},\mathbf{Y},t) \tag{6.143}$$

which is of the same form as derived in Eqn (3.51). To feasibly solve Eqn (6.143), the multiple-size-group (MUSIG) model, which was first introduced by Lo (1996), is widely adopted. In essence, the MUSIG model approximates the continuous particle size distribution by M number of size fractions; the mass conservation of each of the size fractions are balanced by the interfraction mass transfer due to the mechanisms of particle–particle phenomena due to aggregation/coalescence and breakage/break up processes. The overall evolution of the particle size distribution can now be explicitly resolved via source/sink terms within the transport equations. The discrete number density can be alternatively expressed as the size fraction of the dispersed phase in order to be consistent with the variables used in the multifluid model. As initially considered in Eqn (3.54), Eqn (6.143) of the discrete number density can be reexpressed in terms of size fraction as

$$\frac{\partial\left(\rho_i^p \alpha^p f_i\right)}{\partial t} + \nabla \cdot \left(\rho_i^p \alpha^p \mathbf{v}_{p,i} f_i\right) = S'_{N_i} \tag{6.144}$$

Considering that the large range of particles sizes may be required within the multiphase flow, the number of conservation equations to be solved in the context of computational multiphase fluid dynamics for each particle size may become insurmountable and impractical. In order to feasibly obtain solutions to an already complex problem, some simplifications are introduced to the class method. The simplified class method can be categorized according to the well-known *homogeneous* and *inhomogeneous* MUSIG models. A schematic illustration of the homogeneous and inhomogeneous MUSIG models can be seen in Figure 6.14.

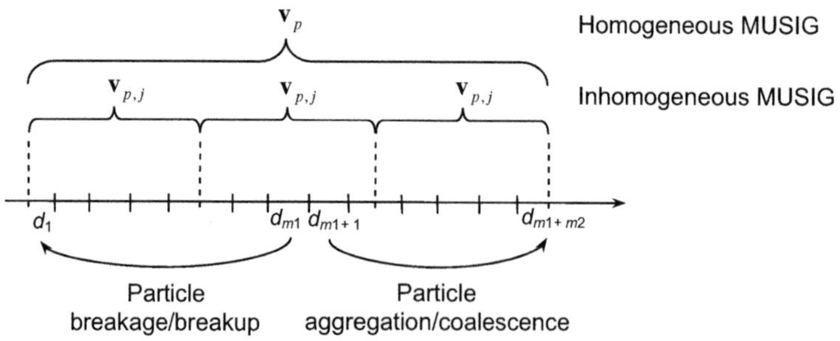

Figure 6.14
An illustration of the MUSIG models. Note that d is the particle diameter, which can be determined through the discrete volume v_i.

In principle, the homogeneous MUSIG model assumes that all particles travel with a common velocity $\mathbf{v}_p$ (Lo, 1996; Pochorecki et al., 2001; Olmos et al., 2001; Yeoh and Tu, 2005; Cheung et al., 2007b). With the assumption of a common velocity and since $\rho_i^p = \rho^p$, Eqn (6.144) is reduced to

$$\frac{\partial(\rho^p\alpha^p f_i)}{\partial t} + \nabla\cdot(\rho^p\alpha^p\mathbf{v}^p f_i) = S'_{N_i} \tag{6.145}$$

The above model works well for a range of particles sizes relatively and moderately large. For multiphase flows where substantial, large particles are found to coexist with small particles, the inhomogeneous MUSIG model developed by Krepper et al. (2005), which consists of subdividing the dispersed phase into N number of velocity fields, may be applied to improve the characterization of the drag between the large and small particles. This extension represents a robust and practical feature for multiphase flows modeling. Useful information on the implementation and application of the inhomogeneous MUSIG model can be found in Shi et al. (2004) and Krepper et al. (2007). Sacrifices are being made to enhance computational efficiency, but the extra computational effort will rapidly diminish due to the foreseeable advancement of computer technology; the inhomogeneous MUSIG model should therefore suffice as the preferred approach in tackling more complex multiphase flows. For this particular model, Eqn (6.144) can be reexpressed in terms of the size fraction of the ith particle size class, where $i \in [1, M_j]$, and velocity group j, where $j \in [1, N]$, according to

$$\frac{\partial\left(\rho_j^p\alpha_j^p f_i\right)}{\partial t} + \nabla\cdot\left(\rho_j^p\alpha_j^p\mathbf{v}_{p,j}f_i\right) = S''_{N_i} \tag{6.146}$$

with additional relations and constraints, are given by:

$$\alpha^p = \sum_{j=1}^{N}\alpha_j^p = \sum_{j=1}^{N}\sum_{i=1}^{M_j}\alpha_{ij}^p \quad \alpha_j^p = \sum_{i=1}^{M_j}\alpha_i^p \quad \sum_{i=1}^{M_j}f_i = 1 \tag{6.147}$$

The treatment of the net source/sink terms of S'_{N_i} in Eqn (6.145) and S''_{N_i} in Eqn (6.146) is described in the following. Source/sink rates due to aggregation/coalescence and breakage/break up are those of birth and death rates of particles. In other words,

$$S'_{N_i} = S''_{N_i} = \underbrace{m_{p,i}S^+_{C,i}}_{B_{C,i}} - \underbrace{m_{p,i}S^-_{C,i}}_{D_{C,i}} + \underbrace{m_{p,i}S^+_{B,i}}_{B_{B,i}} - \underbrace{m_{p,i}S^-_{B,i}}_{D_{B,i}} \tag{6.148}$$

On the basis of the discrete approximation given in Eqn (6.141), the birth and death rates (recalling Eqns (4.1) and (4.2) and Eqns (4.81) and (4.82) in Chapter 4 and Eqns (5.1) and (5.2) and Eqns (5.20) and (5.21) in Chapter 5) can be formulated according to

$$B_{C,i} = m_{p,i}\frac{1}{2}\sum_k \sum_l N_k N_l a(v_k, v_l) \tag{6.149a}$$

$$D_{C,i} = m_{p,i}\sum_k N_i N_k a(v_i, v_k) \tag{6.149b}$$

$$B_{B,i} = m_{p,i}\sum_k N_j b(v_k) P(v_k, v_i) \tag{6.149c}$$

$$D_{B,i} = m_{p,i}\sum_k N_i b(v_k) \tag{6.149d}$$

In order to be consistent with Eqns (6.145) and (6.146), the above birth and death rates are required to be expressed in terms of the size fraction. For the homogeneous MUSIG model, they can be written as:

$$B_{C,i} = \rho^P (\alpha^P)^2 \frac{1}{2}\sum_k \sum_l f_k f_l \frac{v_k + v_l}{v_k v_l} a(v_k, v_l) \tag{6.150a}$$

$$D_{C,i} = \rho^P (\alpha^P)^2 \sum_k f_i f_k \frac{1}{v_k} a(v_i, v_k) \tag{6.150b}$$

$$B_{B,i} = \rho^P \alpha^P \sum_k f_j b(v_k) P(v_k, v_i) \tag{6.150c}$$

$$D_{B,i} = \rho^P \alpha^P \sum_k f_i b(v_k) \tag{6.150d}$$

while for the inhomogeneous MUSIG model, they are:

$$B_{C,i} = \rho_j^P \left(\alpha_j^P\right)^2 \frac{1}{2}\sum_k \sum_l f_k f_l \frac{v_k + v_l}{v_k v_l} a(v_k, v_l) \tag{6.151a}$$

$$D_{C,i} = \rho_j^P \left(\alpha_j^P\right)^2 \sum_k f_i f_k \frac{1}{v_k} a(v_i, v_k) \tag{6.151b}$$

$$B_{B,i} = \rho_j^P \alpha_j^P \sum_k f_k b(v_k) P(v_k, v_i) \tag{6.151c}$$

$$D_{B,i} = \rho_j^P \alpha_j^P \sum_k f_i b(v_k) \tag{6.151d}$$

For the discretized contribution of the birth rate due to aggregation/coalescence, it may be necessary to introduce the aggregation/coalescence matrix as the fraction of volume due to aggregation/coalescence between the kth particle classes at which it goes into the ith particle classes. Defining the aggregation/coalescence matrix as

$$\eta_{kli} = \begin{cases} 1 & \text{if } v_k + v_l > v_i \\ 0 & \text{otherwise} \end{cases}$$

the birth rate due to aggregation/coalescence is accordingly modified by multiplying the above matrix η_{kli} into Eqns (6.150a) and (6.151a), respectively.

Solutions of the class method have been found to be independent of the resolution of the internal coordinate (for this case, volume v_i) if a sufficiently large number of classes is adopted. It should be noted that as the number of transport equations depends on the number of groups being adopted, the homogeneous and inhomogeneous MUSIG models require substantial computational times and resources to achieve stable and accurate numerical predictions.

6.6.2 Standard Method of Moments

Method of moments, first introduced by Hulburt and Katz (1964), has been considered as one of the many promising approaches in viably attaining practical solutions to the population balance equation. The basic idea behind the method of moments centers on the transformation of the problem into lower-order of moments of the particle size distribution. The primary advantage of method of moments is its numerical economy that substantially condenses the flow problem by only tracking the evolution of a small number of moments (Frenklach, 2002). This becomes rather critical in modeling complex flow problems when the range of particles sizes is very large and the particle dynamics is strongly coupled with already time-consuming calculations of turbulence multiphase flows. Another significant aspect of the method of moments is that it does not suffer from any truncation errors in the approximation of the particle size distribution.

Recalling Eqn (3.18), where the mth moments of the number density function are defined in terms of volume: $m_k(\mathbf{r},t) = \int_0^\infty V_p^k f_1(V_p,\mathbf{r},t)\mathrm{d}V_p$, the transport equation for the moments in the absence of particle growth is given in the form of

$$\frac{\partial m_k(\mathbf{r},t)}{\partial t} + \nabla \cdot \left(\mathbf{v}_{p,k}(\mathbf{r},\mathbf{Y},t)m_k(\mathbf{r},t)\right) = S_{m_k}(\mathbf{r},\mathbf{Y},t) \tag{6.152}$$

The moment transform of the source/sink term S_{m_k} is:

$$\begin{aligned} S_{m_k}(\mathbf{r},\mathbf{Y},t) &= \int_0^\infty V_p^k S_{f_1}\left(V_p,\mathbf{r},\mathbf{Y},t\right)\mathrm{d}V_p \\ &= \int_0^\infty V_p^k\left(S_C^+\left(V_p,\mathbf{r},\mathbf{Y},t\right) - S_C^+\left(V_p,\mathbf{r},\mathbf{Y},t\right)\right. \\ &\quad \left. + S_B^+\left(V_p,\mathbf{r},\mathbf{Y},t\right) - S_B^+\left(V_p,\mathbf{r},\mathbf{Y},t\right)\right)\mathrm{d}V_p \end{aligned} \tag{6.153}$$

In most practical applications, the properties are fully determined by the mere consideration of a small number of low-order moments. Hence, the numerical economy of the method of moments results from replacing the infinite number of differential equations to be solved, Eqn (6.153), with a small number of equations for the corresponding moments. One of the simplest ways of accomplishing this is to presume the functional form of the particle size distribution function. According to John et al. (2007), some of the a priori functions employed for the particle size distribution reconstructions are Gaussian (or normal), half-normal, log-normal, β (beta) and γ (gamma). Among these, the Gaussian function is probably the most common one, which is widely applied. A characteristic property of this function is its symmetrical distribution about the mean. However, this property directly leads to the presence of particles with negative sizes. The corresponding amount is sometimes negligible and can be truncated. However, this may not necessarily always be the case in some flows and can lead to nonphysical properties (Baldyga and Bourne, 1999). The half-normal function is a special form of the Gaussian function with a mean of zero and a distribution containing only the positive (physical) particle sizes; the negative half of the curve is neglected. For instance, this particular function could represent the number distribution of particles in an incomplete crystallization process, where the bulk of the particles are the newly nucleated particles, which are highly accumulated near the zero size, considered as the theoretical size of a newly nucleated, molecule-like particle. Nevertheless, a symmetric distribution for a reconstruction may not be desirable in some cases. One of the classical functions for such cases is the log-normal function, which can display a long tail in the direction of the larger particle sizes. This function is widely employed in powder, spray and aerosol technologies. Another function often used in particle technology is the γ-function, which can accommodate various shapes depending on the skewness and kurtosis values deduced from the corresponding moment data (Heinz, 2003). As an alternative, the β-function can be employed, which is capable of taking more various shapes than the γ-function. This β-function can easily exhibit skew on either side of the distribution peak, unlike the log-normal function and γ-function. Therefore, the β-function is widely used in engineering applications (for example, particle and combustion technologies).

All of the aforementioned functions are described in the following. For this purpose, a few general distribution properties like the mean size ($\overline{x}$), coefficient of variation (c_v) and standard deviation (σ) must first be defined, which can be obtained in terms of the three low-order moments (m_0, m_1 and m_2):

$$\overline{x} = \frac{m_1}{m_0} \tag{6.154}$$

$$c_v = \sqrt{\frac{m_0 m_2}{m_1} - 1} \tag{6.155}$$

$$\sigma = c_v \overline{x} \tag{6.156}$$

Incidentally, if the size x is taken to be the particle volume V_p, Eqn (6.154) is identical to Eqn (3.20), which is characterized by the average particle volume $\overline{v}$, and m_0 and m_2 correspond to the particle number density N and interfacial area concentration a_i, respectively. The various distributions with all the moments necessary to explicitly determine the particle size distribution function are:

Gaussian (normal) function

$$f_1(x,\mathbf{r},t)\approx f_1(x) = \frac{1}{\sigma\sqrt{2\pi}}\exp\left(-\frac{(x-\overline{x})^2}{2\sigma^2}\right) \tag{6.157}$$

Half-normal function

$$f_1(x,\mathbf{r},t)\approx f_1(x) = \frac{2\phi}{\pi}\exp\left(-\frac{x^2\phi^2}{\pi}\right) \quad \text{where } \phi = \sqrt{\frac{\pi-2}{2\sigma^2}} \tag{6.158}$$

Log-normal function

$$f_1(x,\mathbf{r},t)\approx f_1(x) = \frac{1}{x\ln\sigma_g\sqrt{2\pi}}\exp\left(-\frac{\ln^2(x/\overline{x}_g)}{2\ln^2\sigma^2}\right) \quad \text{where}$$

$$\overline{x}_g = \frac{\overline{x}}{\exp(0.5\ln^2\sigma_g)} \quad \text{and} \quad \sigma_g = \exp\sqrt{\ln(c_v^2+1)} \tag{6.159}$$

γ-function

$$f_1(x,\mathbf{r},t)\approx f_1(x) = \frac{\mu^\mu x^{(\mu-1)}}{\Gamma(\mu)\overline{x}^\mu}\exp\left(-\frac{\mu x}{\overline{x}}\right) \quad \text{where}$$

$$\mu = \frac{\overline{x}^2}{\sigma^2} \quad \text{and} \quad \Gamma(\mu) = \int_0^\infty z^{(\mu-1)}e^{-z}\,dz \tag{6.160}$$

β-function

$$f_1(x,\mathbf{r},t)\approx f_1(x) = \frac{x^{(\alpha-1)}(1-x)^{(\beta-1)}}{B(\alpha,\ \beta)} \quad \text{where}$$

$$\alpha = \overline{x}\left(\frac{\overline{x}(1-\overline{x})}{\sigma^2}-1\right), \quad \beta = (1-\overline{x})\left(\frac{\overline{x}(1-\overline{x})}{\sigma^2}-1\right) \quad \text{and} \tag{6.161}$$

$$B(\alpha,\beta) = \frac{\Gamma(\alpha)\Gamma(\beta)}{\Gamma(\alpha+\beta)}$$

The reconstruction of a distribution using a few low-order moments through the consideration of a priori assumed function shapes is a very efficient method since the solution can be immediately obtained without any time-consuming computational overheads. The obtained accuracy can be very high when the correct functional shape is employed. But this method requires a lot of information concerning the expected distribution beforehand. Therefore, this method is only accurate, and should always be retained, when the properties of the final particle size distribution are well known (for example, when considering a small variation compared to a known process). On the other hand, when considering an application with unknown particle size distribution, this method cannot be retained in general. One possible alternative is the application of the method of moments with interpolative closure (Frenklach, 2002).

Method of moments with interpolative closure is described in the following. According to Frenklach (1985), the mth moments can be redefined in terms of the discrete number density as $m_k(\mathbf{r},t) = \sum_{i=1}^{\infty} v_i^k N_i(\mathbf{r},t)$ or $m_k = \sum_{i=1}^{\infty} v_i^k N_i$. This simply entails applying the approximation as stipulated by Eqn (6.141). The transport equations for the moments are solved according to Eqn (6.152) with the moment transform of the aggregation/coalescence and breakge/break up of term S_{m_k} represented in terms of the discrete number density, noting that $\rho^p = \rho_i^p = \rho_j^p$, as

$$S_{m_k} = B_{C,k} - D_{C,k} + B_{B,k} - D_{B,k} \tag{6.162}$$

where

$$B_{C,k} = (\rho^p)^k \frac{1}{2} \sum_i \sum_j \left(v_{p,i} + v_{p,j}\right)^k N_i N_j a\left(v_i, v_j\right) \tag{6.163a}$$

$$D_{C,k} = (\rho^p)^k \sum_i \sum_j v_i^k N_i N_j a(v_i, v_k) \tag{6.163b}$$

$$B_{B,k} = (\rho^p)^k \sum_i \sum_j v_i^k N_j b\left(v_j\right) P\left(v_j, v_i\right) \tag{6.163c}$$

$$D_{B,k} = (\rho^p)^k \sum_i \sum_j v_i^k N_i b\left(v_j\right) \tag{6.163d}$$

As will be described later, the difficulty of method of moments with interpolative closure lies in the determination of the fractional-order moments. Consider the aggregation kernel of solid particles in the continuum regime with unity collision efficiency. Equations (6.163a) and (6.163b) become

$$B_{C,k} = (\rho^p)^k \frac{1}{2} \sum_i \sum_j \left(v_{p,i} + v_{p,j}\right)^k N_i N_j \times K_c\left(v_i^{1/3} + v_j^{1/3}\right)\left(\frac{C(v_i)}{v_i^{1/3}} + \frac{C(v_j)}{v_j^{1/3}}\right) \tag{6.164a}$$

$$D_{C,k} = (\rho^p)^k \sum_i \sum_j v_i^k N_i N_j \times K_c\left(v_i^{1/3} + v_j^{1/3}\right)\left(\frac{C(v_i)}{v_i^{1/3}} + \frac{C(v_j)}{v_j^{1/3}}\right) \tag{6.164b}$$

where $K_c = 2k_B T^c / 3\mu^c$ with the Cunningham slip factor given by $C(v_{i \text{ or } j}) = 1 + 1.257\ Kn_{i \text{ or } j}$. The zeroth moment, meaning that $k = 0$, is taken for ease of illustration; Eqns (6.164a) and (6.164b), can be rewritten after some mathematical manipulation as

$$B_{C,0} = K_c\left(m_0^2 + m_{1/3}m_{-1/3} + K_c'\left[m_0 m_{-1/3} + m_{1/3}m_{-2/3}\right]\right) \tag{6.165a}$$

$$D_{C,0} = 2K_c\left(m_0^2 + m_{1/3}m_{-1/3} + K_c'\left[m_0 m_{-1/3} + m_{1/3}m_{-2/3}\right]\right) \tag{6.165b}$$

where $K_c' = 2.514\lambda(\pi\rho^p/6)^{1/3}$. To obtain closure to Eqns (6.165a) and (6.165b), the fractional-order moments have to be evaluated. According to Frenklach (2002), this can be accomplished by interpolation of whole-order moments. By separating the interpolation for positive- and negative-order moments, the Lagrange interpolation among logarithms of whole-order moments can be obtained with sufficient high accuracy. Also consider the aggregation kernel of solid particles in the free molecular regime with unity collision efficiency. Equations (6.163a) and (6.163b) can now be expressed as

$$B_{C,k} = (\rho^p)^k \frac{1}{2} \sum_i \sum_j (v_i + v_j)^k N_i N_j \times K_f (v_i + v_j)^{1/2} v_i^{-1/2} v_j^{-1/2} \left(v_i^{1/3} + v_j^{1/3}\right)^2 \tag{6.166a}$$

$$D_{C,k} = (\rho^p)^k \sum_i \sum_j v_i^k N_i N_j \times K_f (v_i + v_j)^{1/2} v_i^{-1/2} v_j^{-1/2} \left(v_i^{1/3} + v_j^{1/3}\right)^2 \tag{6.166b}$$

where $K_f = (3/4\pi)^{1/6}\sqrt{6k_B T^c/\rho^p}$. The principal difficulty of Eqns (6.166a) and (6.166b) lies in the nonadditive form of summations appearing in the equations. Here again, the zeroth moment, meaning that $k = 0$, is taken for ease of illustration; Eqns (6.166a) and (6.166b) reduce to

$$B_{C,0} = \frac{1}{2} \sum_i \sum_j N_i N_j \times K_f (v_i + v_j)^{1/2} v_i^{-1/2} v_j^{-1/2} \left(v_i^{1/3} + v_j^{1/3}\right)^2 \tag{6.167a}$$

$$D_{C,0} = \sum_i \sum_j N_i N_j \times K_f (v_i + v_j)^{1/2} v_i^{-1/2} v_j^{-1/2} \left(v_i^{1/3} + v_j^{1/3}\right)^2 \tag{6.167b}$$

Because of $(v_i + v_j)^{1/2}$, it is thus impossible to directly perform the summation. By defining $(v_i + v_j)^l$ for $l = 0, 1, 2, \ldots$, the summation can now be expressed in exact terms using fractional-order moments. For example if $l = 1$,

$$B_{C,0}^1 = K_f\left(m_{7/6}m_{-1/2} + m_{1/2}m_{1/6} + 2m_{5/6}m_{-1/6}\right) \tag{6.168a}$$

$$D_{C,0}^1 = 2K_f\left(m_{7/6}m_{-1/2} + m_{1/2}m_{1/6} + 2m_{5/6}m_{-1/6}\right) \tag{6.168b}$$

These fractional-order moments appearing in Eqns (6.168a) and (6.168b) can now be evaluated through the Lagrange interpolation among logarithms of whole-order moments. According to Frenklach (2002), the Lagrange quadratic interpolation for $l = 1/2$ may be performed for the birth terms among $B_{C,0}^0$, $B_{C,0}^1$ and $B_{C,0}^2$ and for the death terms among $D_{C,0}^0$, $D_{C,0}^1$ and $D_{C,0}^2$ or the consideration of additional whole-order terms for higher-order Lagrange interpolation functions. More details on the interpolation procedure can be found in Frenklach (2002).

In general, the mechanistic models developed for the coalescence and break up of bubbles and drops are considerably more sophisticated than the mechanistic models proposed for the aggregation and breakage of solid particles. Quadrature method of moments has been introduced lately, which allows circumventing some of the difficulties being experienced in the method of moments with interpolative closure for modeling complex multiphase flows. The method of moments employing numerical quadrature is discussed in the next section.

6.6.3 Numerical Quadrature

Another different approach for computing the moment is to employ the numerical quadrature scheme, for example the quadrature method of moment as suggested by McGraw (1997). In the quadrature method of moment, instead of space transformation, Gaussian quadrature closure is adopted to approximate the particle size distribution according to a finite set of Dirac delta functions. Taking the particle mass, M, as the internal coordinate, the particle size distribution takes the form:

$$f_1\left(V_p, \mathbf{r}, t\right) \approx \sum_{i=1}^N N_i(\mathbf{r}, t)\delta\left(V_p - \nu_i\right) \tag{6.169}$$

where N_i represents the weight of the ith class and consists of all particles per unit volume with abscissa ν_i. An illustration of the quadrature method of moment in approximating the particle size distribution is given in Figure 6.15.

Although the numerical quadrature approach suffers from truncation errors, it totally eliminates the problem of fractional-order moments. The closure of the method is brought down to solving $2N$ unknowns, N_i and ν_i. A number of approaches in the specific evaluation of the quadrature abscissas and weights have been proposed. McGraw (1997) first introduced the product-difference algorithm formulated by Gordon (1968) for solving monovariate problems. As pointed out by Dorao and Jakobsen (2006a), and by Dorao et al. (2008), the product-difference algorithm is, however, a numerical ill-conditioned method for computing the Gaussian quadrature rule (Lambin and Gaspard, 1982). Comprehensive derivation of the product-difference algorithm can be found in Bove (2005). The computation of the quadrature rule is generally unstable and sensitive to small errors especially for a large

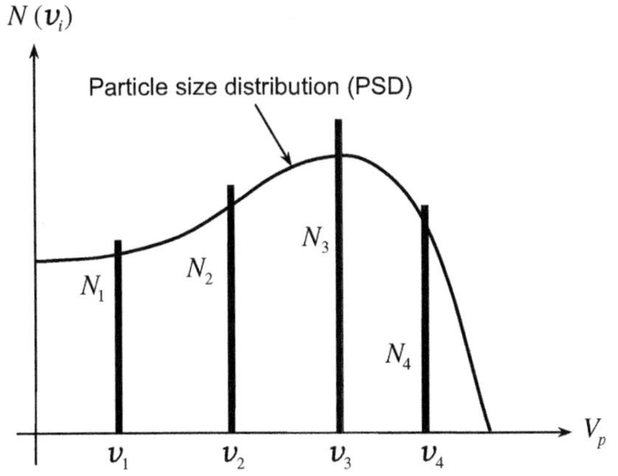

Figure 6.15
An illustration of the quadrature method of moment.

number of moments. McGraw and Wright (2003) derived the Jacobian Matrix Transformation for a multicomponent population, which avoids the instability induced by the product-difference algorithm. Grosch et al. (2007) proposed a generalized framework for various approaches for the quadrature method of moment and evaluated different formulations in terms of numerical quadrature and dynamics simulation. Several studies have also been carried out validating the method against different gas–solid particle problems (Barrett and Webb, 1998; Marchisio et al., 2003a, b, c). Encouraging results obtained thus far clearly demonstrated its usefulness in solving monovariate problems and its potential fusing within computational multiphase fluid dynamics simulations. One of the main limitations of the quadrature method of moment is that each moment is advected with the same phase velocity, which may not be applicable for certain types of multiphase flows such as gas–liquid or liquid–liquid flows.

In order to handle multidimensional flow problems, Marchisio and Fox (2005) extended the quadrature method of moment to the direct quadrature method of moment where the quadrature abscissas and weights are now formulated as transport equations. The main idea of the method is to keep track of the primitive variables appearing in the quadrature approximation, instead of moments of the particle size distribution. This results in the evaluation of the abscissas and weights, which can be obtained using matrix operations. Substituting Eqn (6.169) into Eqn (3.15) and after some mathematical manipulations, the transport equations for weights and abscissas in the absence of particle growth are given by

$$\frac{\partial N_i(\mathbf{r},t)}{\partial t} + \nabla \cdot \left(\mathbf{v}_{p,i}(\mathbf{r},\mathbf{Y},t)N_i(\mathbf{r},t)\right) = a_i \tag{6.170}$$

$$\frac{\partial \zeta_i(\mathbf{r},t)}{\partial t} + \nabla \cdot \left(\mathbf{v}_{p,i}(\mathbf{r},\mathbf{Y},t)\zeta_i(\mathbf{r},t)\right) = b_i \tag{6.171}$$

where $\zeta_i = N_i\nu_i$ is the weighted abscissas and the terms a_i and b_i are related to the birth and death rate of population, which forms $2N$ linear equations where the unknowns can be evaluated via matrix inversion according to

$$A\alpha = d \tag{6.172}$$

The $2N \times 2N$ coefficient matrix $A = [A_1 \quad A_2]$ in the above linear equation takes the form:

$$A_1 = \begin{bmatrix} 1 & \cdots & 1 \\ 0 & \cdots & 0 \\ -\nu_1^2 & \cdots & -\nu_N^2 \\ \vdots & \ddots & \vdots \\ 2(1-N)\nu_1^{2N-1} & \cdots & 2(1-N)\nu_N^{2N-1} \end{bmatrix} \tag{6.173}$$

$$A_2 = \begin{bmatrix} 0 & \cdots & 0 \\ 1 & \cdots & 1 \\ 2\nu_1 & \cdots & 2\nu_N \\ \vdots & \ddots & \vdots \\ (2N-1)\nu_1^{2N-2} & \cdots & (2N-1)\nu_N^{2N-2} \end{bmatrix} \tag{6.174}$$

where the $2N$ vector of unknowns α essentially comprises the terms a_i and b_i in Eqns (6.170) and (6.171):

$$\alpha = [a_1 \cdots a_N \quad b_1 \cdots b_N]^T = \begin{bmatrix} a \\ b \end{bmatrix} \tag{6.175}$$

In Eqn (6.172), d is defined by:

$$d = [S_{m_0} \cdots S_{m_{2N-1}}]^T \tag{6.176}$$

Applying moment transformation, the sources and sinks of the zeroth, first, second, etc., in Eqn (6.176) are identical to S_{m_k} in Eqn (6.162), which can be closed through specification of constitutive relations.

In order to be consistent with the variables used in the multifluid model, the weights and abscissas can be related to the size fraction of the dispersed phase and an effective size, which comprises the product between the volume fraction of the dispersed phase and abscissas. The transport equations for the weights and abscissas can be alternatively written as

$$\frac{\partial(\rho_i^P \alpha_i^P)}{\partial t} + \nabla \cdot (\rho_i^P \alpha_i^P \mathbf{v}_{p,i}) = b_i \tag{6.177}$$

$$\frac{\partial(\rho_i^P \alpha_i^P \nu_i)}{\partial t} + \nabla \cdot (\rho_i^P \alpha_i^P \mathbf{v}_{p,i}\nu_i) = 2\nu_i b_i - \rho_i^P \nu_i^2 a_i - \alpha_i^P \nu_i \frac{\partial \rho_i^P}{\partial t} \tag{6.178}$$

In accordance with the homogeneous MUSIG model, the particles can be assumed to travel with a common velocity ($\mathbf{v}_p$). Since $\rho_i^p = \rho^p$, Eqns (6.177) and (6.178) for the homogeneous direct quadrature method of moment can be expressed in terms of the size fraction of f_i by

$$\frac{\partial(\rho^p \alpha^p f_i)}{\partial t} + \nabla \cdot \left(\rho^p \alpha^p \mathbf{v}_{p,i} f_i\right) = b_i \qquad (6.179)$$

$$\frac{\partial(\rho^p \alpha^p \psi_i)}{\partial t} + \nabla \cdot \left(\rho^p \alpha^p \mathbf{v}_{p,i} \psi_i\right) = 2\nu_i b_i - \rho^p \nu_i^2 a_i \qquad (6.180)$$

where $\psi_i = f_i \nu_i$.

Also in accordance with the inhomogeneous MUSIG model, the inhomogeneous direct quadrature method of moment is thus given by

$$\frac{\partial\left(\rho_j^p \alpha_j^p f_i\right)}{\partial t} + \nabla \cdot \left(\rho_j^p \alpha_j^p \mathbf{v}_{p,i} f_i\right) = b_i \qquad (6.181)$$

$$\frac{\partial\left(\rho_j^p \alpha_j^p \psi_i\right)}{\partial t} + \nabla \cdot \left(\rho_j^p \alpha_j^p \mathbf{v}_{p,j} \psi_i\right) = 2\nu_i b_i - \rho_j^p \nu_i^2 a_i \qquad (6.182)$$

with additional relations and constraints given in Eqn (6.147).

It should be noted that an attractive feature of the direct quadrature method of moment is that the method permits the weights and abscissas to be varied within the state space according to the evolution of the particle size distribution. In summary, the method of moment represents a rather sound mathematical approach and is an elegant tool for solving the population balance equation with limited computational burden. Such an approach is no doubt an emerging technique for solving the population balance equation.

The moment transform of the coalescence and break up of the term S_{m_k} can be expressed as

$$S_{m_k} = B_{C,k} - D_{C,k} + B_{B,k} - D_{B,k} \qquad (6.183)$$

where

$$B_{C,k} = \frac{1}{2} \int_0^\infty \int_0^\infty \left(U + V_p'\right)^k f\left(V_p', t\right) f(U, t) a\left(U, V_p'\right) \mathrm{d}U \mathrm{d}V_p' \qquad (6.184a)$$

$$D_{C,k} = \int_0^\infty \int_0^\infty V_p^k a\left(V_p, V_p'\right) f(V_p, t) f\left(V_p', t\right) \mathrm{d}V_p' \mathrm{d}V_p \qquad (6.184b)$$

$$B_{B,k} = \int_0^\infty \int_0^\infty V_p^k \gamma\left(V_p'\right) b\left(V_p', V_p\right) P\left(V_p', V_p\right) f\left(V_p', t\right) \mathrm{d}V_p' \mathrm{d}V_p \tag{6.184c}$$

$$D_{B,k} = \int_0^\infty V_p^k b\left(V_p, V_p'\right) f\left(V_p, t\right) \mathrm{d}V_p \tag{6.184d}$$

where the terms B and D represent the birth and death rates of the aggregation/coalescence and breakage/break up of particles. In Eqn (6.184a), note that the birth term has been derived by using $U = V_p - V_p'$ and substituting $\mathrm{d}V_p = \mathrm{d}U$. On the basis of the approximation given in Eqn (6.169), the birth and death rates can be written as

$$B_{C,k} = \frac{1}{2} \sum_i \sum_j N_i N_j \left(v_i + v_j\right)^k a\left(v_i, v_j\right) \tag{6.185a}$$

$$D_{C,k} = \sum_i \sum_j v_i^k N_i N_j a\left(v_i, v_j\right) \tag{6.185b}$$

$$B_{B,k} = \sum_i \sum_j v_i^k N_j b\left(v_j\right) P\left(v_j, v_i\right) \tag{6.185c}$$

$$D_k^B = \sum_i \sum_j v_i^k N_i b\left(v_j\right) \tag{6.185d}$$

For the homogeneous direct quadrature method of moment, they can be written as:

$$B_{C,k} = \left(\alpha^p\right)^2 \frac{1}{2} \sum_i \sum_j f_i f_j \frac{\left(v_i + v_j\right)^k}{v_i v_j} a\left(v_i, v_j\right) \tag{6.186a}$$

$$D_{C,k} = \left(\alpha^p\right)^2 \sum_i \sum_j f_i f_j \frac{v_i^{k-1}}{v_j} a\left(v_i, v_j\right) \tag{6.186b}$$

$$B_{B,k} = \alpha^p \sum_i \sum_j f_j \frac{v_i^k}{v_j} b\left(v_j\right) P\left(v_j, v_i\right) \tag{6.186c}$$

$$D_{B,k} = \alpha^p \sum_i \sum_j f_i v_i^{k-1} b\left(v_j\right) \tag{6.186d}$$

while for the inhomogeneous direct quadrature method of moment, they are:

$$B_{C,k} = \left(\alpha_j^p\right)^2 \frac{1}{2} \sum_i \sum_j f_i f_j \frac{\left(v_i + v_j\right)^k}{v_i v_j} a\left(v_i, v_j\right) \tag{6.187a}$$

$$D_{C,k} = \left(\alpha_j^p\right)^2 \sum_i \sum_j f_i f_j \frac{v_i^{k-1}}{v_j} a(v_i, v_j) \tag{6.187b}$$

$$B_{B,k} = \alpha_j^p \sum_i \sum_j f_j \frac{v_i^k}{v_j} b(v_j) P(v_j, v_i) \tag{6.187c}$$

$$D_{B,k} = \alpha_j^p \sum_i \sum_j f_i v_i^{k-1} b(v_j) \tag{6.187d}$$

The discrete volumes in Eqns (6.186a–6.186d) and (6.187a–6.187d) can be evaluated from ψ values once they are determined through Eqn (6.180).

6.6.4 Other Population Balance Methods

In addition to the class method, standard method of moments and method of moments employing numerical quadrature, the least-squares method has been proposed as an alternative to solve the population balance equation as demonstrated by Zhu et al. (2009) for the prediction of fluid particle size distribution in gas–liquid flow. The least-squares method is a well established numerical method for solving a wide range of mathematical problems. The basic idea is to minimize the integral of the square of the residual of a norm-equivalent functional. Consider the system of a population balance equation for the number density function, which can be written in the form:

$$\begin{aligned} \mathscr{L}f_1 &= h_1 \\ \mathscr{B}f_1 &= g_1 \end{aligned} \tag{6.188}$$

where $\mathscr{L}$ is the first-order differential operator, $\mathscr{B}$ is the boundary algebraic operator and h_1 and g_1 are the source terms. Assuming that the system is well-posed, the norm-equivalent functional for the system is given by

$$\mathscr{I}(f_1) = \frac{1}{2}\|\mathscr{L}f_1 - h_1\| + \frac{1}{2}\|\mathscr{B}f_1 - g_1\| \tag{6.189}$$

Introducing $f_1 = f_1 + \varepsilon_1 v_1$ into Eqn (6.189) and taking the partial derivative with respect to ε_1, the minimization statement becomes

$$\lim_{\varepsilon_1 \to 0} \frac{\partial}{\partial \varepsilon_1} \mathscr{I}(f_1 + \varepsilon_1 v_1) = 0 \tag{6.190}$$

where v_1 is a trial function and ε_1 is a small perturbation. Consequently, the necessary condition can be written as

$$A_1(f_1, v_1) = F_1(v_1) \tag{6.191}$$

with

$$\begin{aligned} A_1(f_1, v_1) &= \langle \mathscr{L}f_1, \mathscr{L}v_1 \rangle + \langle \mathscr{B}f_1, \mathscr{B}v_1 \rangle \\ F_1(v_1) &= \langle g_1, \mathscr{L}v_1 \rangle + \langle h_1, \mathscr{B}v_1 \rangle \end{aligned} \tag{6.192}$$

For an accurate prediction of the number density function, Zhu et al. (2009) have adopted the spectral method to discretize Eqn (6.191). Here, a solution is approximated with linear combination of continuous functions (high-order polynomials), generally nonzero throughout the whole fluid domain. The least-squares spectral element method directly solves the number density function instead of solving the moments, such as in approaches employing the standard method or moments and method of moments employing numerical quadrature. Moments are subsequently determined from postcalculations. Whereas a class method requires at least 12–15 classes to achieve good accuracy, which may be deemed to be impractical with the fluid flow calculations, the least-squares spectral element method is highly accurate and compatible with the computational efficiencies associated with the method of moments. Interested readers are encouraged to refer to Dorao and Jakobsen (2006b, 2007a, b) for further details of the methodology.

For complex configurations such as heterogeneous reactors and three-dimensional laminar premixed flame, the most commonly used statistical model is the method of moments with interpolative closure. The main benefit of this particular model is its simplicity. Additionally, the interpolation is well-behaved unless unreasonably high-order interpolation is employed. Mueller et al. (2009a) have shown that the method of moments with interpolative closure can successfully predict the primary particle size due to the distribution of larger particles but not for the persistent nucleation of smaller particles. It is the latter that yields the inherent bimodality nature of the particle size distribution. This can be partly resolved through the consideration of the inherent multimodality of the direct quadrature method of moments. In order to capture the entire particle size distribution, which includes not only the persistent nucleation of smaller particles but also the formation of large fractal aggregates, a hybrid method of moments is proposed by Mueller et al. (2009b), which combines these two methods. In order to capture the influence of smaller particles, a delta function is employed in addition to the method of moments with interpolative closure. Regarding the first delta function in the direct quadrature method of moments, the delta function in the hybrid method of moment is fixed in two-dimensional state space at the nucleated size. Although in the direct quadrature method of moments the first peak is not completely fixed, it does not move much, and no significant error is thus anticipated with fixing its location. More details on the description of the approach are found in Mueller et al. (2009b).

6.7 Solution Methods for Lagrangian Models

In principle, the concept of Lagrangian tracking simply entails following the motion of individual particles. Therefore, Lagrangian particle methods include not only the discrete element method but also molecular dynamics and Brownian dynamics. Figure 6.16 provides a rough schematic representation that indicates the characteristic time and length scales of the various Lagrangian particle methods. The particles being considered, in general, represent a wide range of discrete physical elements including atoms, molecules, nuclei, cells, aerosol or colloidal particles, and granules. Molecular dynamics, Brownian dynamics and the discrete element method share a common characteristic where the discrete elements are allowed to interact for a period of time under prescribed interaction laws, and the solution methods are employed to resolve the motion of each individual element by solving the linear momentum equation, and sometimes the angular momentum equation, subject to forces and torques arising both from particle interaction with each other and those imposed on the particles by the surrounding fluid (Li et al., 2011). The different Lagrangian particle methods primarily differ by the different particle interaction laws as well as by the imposition of random forcing in some models to mimic collisions or interaction with molecules of the surrounding fluid. By carefully identifying the particle interaction laws, the physical particle behavior at a certain range of time and length scales, and for specific types of discrete elements, can be efficiently captured.

6.7.1 Molecular Dynamics

Molecular dynamics is the most detailed molecular simulation method that has been developed to simulate problems where the particles represent individual atoms of approximately the size of 0.1 nm (10^{-10} m). These atoms interact with each other by a

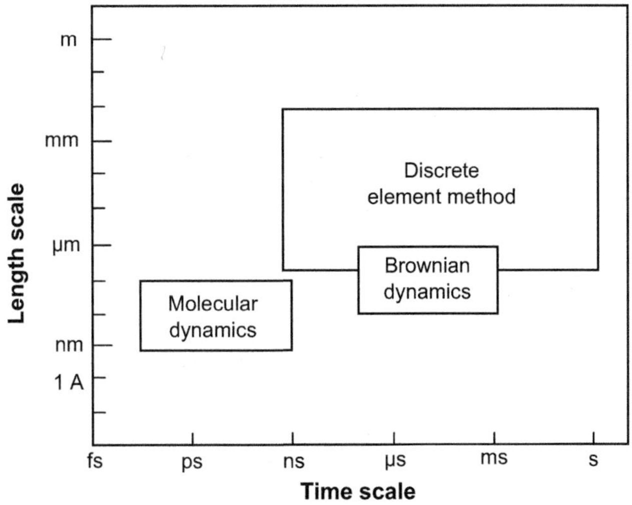

Figure 6.16
An illustration of time and length scales for particle flow simulation methods: molecular dynamics, Brownian dynamics and discrete element method.

combination of nonbonded potentials and repulsive interactions between atoms in a molecule. Because of the need to capture thermal motion of atoms, as well as the stiffness imposed by the chemical bonds with other atoms in the molecule, the time step for molecular dynamics is very small, of the order of 1 fs (10^{-15} s). The most widely used potential for nonbonded interactions is the Lennard-Jones potential:

$$U(R) = 4\varepsilon_{LJ}\left[\left(\frac{R_p}{R}\right)^{12} - \left(\frac{R_p}{R}\right)^{6}\right] \tag{6.193}$$

where R is the distance between two atom centers, ε_{LJ} is the interaction energy parameter (potential well depth) and R_p is the characteristic atom radius (where $U(R_p) = 0$). The first term in Eqn (6.193) represents the short-range, repulsive overlap force while the second term is the longer range, attractive van der Waals force. The Lennard-Jones potential is considered to be a pair-wise potential because it deals with only two particles at a time. In problems involving collisions of three or more atoms, it is sometimes necessary to examine many-body potentials (Tersoff, 1989). The force acting between particles is given by the gradient of the potential energy: $\mathbf{F} = -\nabla U$, which is oriented along the line passing through the particle centers. Based on the linear momentum equation (*Newton's second law*), the acceleration is thus given by

$$\mathbf{a}_p = \frac{D\mathbf{V}_p}{Dt} = -\frac{\nabla U}{m_p} \tag{6.194}$$

As the most computationally intensive component of the computation is the force evaluation, any integration method requiring more than one such calculation per time step is very wasteful unless it can deliver a proportionate increase in the size of the time step while maintaining the same accuracy. The well-known Runge–Kutta method is thus not suitable as it is unable to enlarge the time step due to the prevalence of strongly repulsive force at short distances in the Lennard-Jones potential. In molecular dynamics, the Verlet algorithm is commonly adopted to solve the equation of motion since the solution method requires only the evaluation of the forces at each time step.

The Verlet algorithm (Verlet, 1967) uses information from the current and previous time steps to advance the atomic positions. To derive the algorithm, the positions can be constructed by considering the two Taylor expansions from t to $t + \Delta t$ and $t - \Delta t$. In other words,

$$(\mathbf{x}_p)^{n+1} = (\mathbf{x}_p)^n + (\mathbf{V}_p)^n\Delta t - \frac{1}{2}\frac{(\nabla U)^n}{m_p}\Delta t^2 \tag{6.195}$$

$$(\mathbf{x}_p)^{n-1} = (\mathbf{x}_p)^n - (\mathbf{V}_p)^n\Delta t - \frac{1}{2}\frac{(\nabla U)^n}{m_p}\Delta t^2 \tag{6.196}$$

Adding the above equations yields the equation for advancing the positions in accordance with

$$\left(\mathbf{x}_p\right)^{n+1} = 2\left(\mathbf{x}_p\right)^n - \left(\mathbf{x}_p\right)^{n-1} - \frac{(\nabla U)^n}{m_p}\Delta t^2 \tag{6.197}$$

It can be seen that the algorithm uses positions and accelerations at time t and the positions from time $t - \Delta t$ to calculate new positions at time $t + \Delta t$. The Verlet algorithm does not involve any explicit evaluation of the velocities. The advantages of the algorithm are: (1) it is straightforward and (2) the storage requirements are modest. The disadvantage is that the algorithm only possesses moderate precision.

Some modifications to the Verlet algorithm such as leap-frog and velocity-Verlet algorithms have been proposed in order to improve the solution method in molecular dynamics, especially in the handling of the velocities. For the leap-frog algorithm, the velocities are first calculated at time $t + ½\Delta t$, which are then used to calculate the positions at time $t + \Delta t$. In this way, the velocities leap over the positions, and then the positions leap over the velocities. The algorithm is given by

$$\left(\mathbf{V}_p\right)^{n+1/2} = \left(\mathbf{V}_p\right)^{n-1/2} - \frac{(\nabla U)^n}{m_p}\Delta t \tag{6.198}$$

$$\left(\mathbf{x}_p\right)^{n+1} = \left(\mathbf{x}_p\right)^n + \left(\mathbf{V}_p\right)^{n+1/2}\Delta t \tag{6.199}$$

The advantage of this algorithm is that the velocities are explicitly determined. Nevertheless, the disadvantage is that they are not calculated at the same time as the positions. The velocities at time t can be approximated by averaging the velocities at times $t - ½\Delta t$ and $t + ½\Delta t$:

$$\left(\mathbf{V}_p\right)^n = \frac{1}{2}\left[\left(\mathbf{V}_p\right)^{n+1/2} + \left(\mathbf{V}_p\right)^{n-1/2}\right] \tag{6.200}$$

Alternatively, the velocity-Verlet algorithm evaluates the velocities according to

$$\left(\mathbf{V}_p\right)^{n+1} = \left(\mathbf{V}_p\right)^n + \frac{1}{2}\left[-\frac{(\nabla U)^{n+1}}{m_p} - \frac{(\nabla U)^n}{m_p}\right]\Delta t \tag{6.201}$$

while the positions are calculated through Eqn (6.199). Both leap-frog and velocity-Verlet algorithms are also simple to use and require less memory. The acceleration at time $t + \Delta t$ can be derived from the interaction potential using the new positions at time $t + \Delta t$.

The time and length scale restrictions of molecular dynamics limit the solution method to computations with total duration of about 1 ns (10^{-9} s) and length scales about 10 nm (10^{-8} m). For more complex particle systems, Brownian dynamics has been designed to introduce various simplifications in order to extend the applicability of molecular dynamics to larger systems and longer time periods. It should be noted that atom-level molecular dynamics can be considered to derive an approximate mesoscale interaction model especially between two nanoparticles, which can then be implemented into a discrete element method approach for performing simulation of multiparticle systems.

6.7.2 Brownian Dynamics

Brownian dynamics is a mesoscopic method that has been introduced as a simplification of the Langevin dynamics so as to provide faster calculation of the motion of aerosol or colloidal particles subject to Brownian motion. The Langevin equation consists of three types of forces that act on the particles: (1) a dissipative hydrodynamic drag force, (2) a random Brownian force and (3) nonhydrodynamic forces, which include any external forces and excluded volume interactions. At the core of Brownian dynamics simulation is the assumption that particles are surrounded by a fluid continuum such that the dissipative drag force and random Brownian force are not between individual particles but instead they act between a given particle and a surrounding fluid. By contrast, the different types of molecules of the flow field in molecular dynamics are represented by various types of particles used in the simulation so that there is no background fluid continuum.

The Langevin equation takes the form:

$$m_p \frac{D\mathbf{V}_p(t)}{Dt} = -B\mathbf{V}_p(t) + \sum_i \mathbf{F}_i^{nh}(t) + \mathbf{F}^B(t) \tag{6.202}$$

where B is the drag coefficient given by the Stokes drag law on a sphere for creeping flow and neglecting hydrodynamic interactions (free-draining). Solving Eqn (6.202) yields

$$\mathbf{V}_p(t) = \mathbf{V}_p(t_n)\exp(-B(t-t_n)) + \exp(-B(t-t_n)) \int_{t_n}^{t} \exp(-B(t'-t_n)) \times \frac{1}{m_p}\left(\sum_i \mathbf{F}_i^{nh}(t') + \mathbf{F}^B(t')\right) dt' \tag{6.203}$$

The nonhydrodynamic or particle interaction force $\mathbf{F}_i^{nh}$ may be determined using a conservative potential such as the Lennard-Jones potential given in Eqn (6.193). In Eqn (6.203), the integral over this force can be obtained by expanding it in a power series up to second order. Performing the integration, Eqn (6.203) becomes

$$\mathbf{V}_p(t) = \mathbf{V}_p(t_n)\exp(-B(t-t_n)) + \frac{1}{m_pB}\sum_i \mathbf{F}_i^{nh}(t_n)(1-\exp[-B(t-t_n)])$$
$$+ \frac{1}{m_pB^2}\sum_i \dot{\mathbf{F}}_i^{nh}(t_n)[B(t-t_n)-(1-\exp[-B(t-t_n)])] \tag{6.204}$$
$$+ \frac{1}{m_p}\exp(-B(t-t_n))\int_{t_n}^{t}\exp(-B(t'-t_n))\mathbf{F}^B(t')\mathrm{d}t'$$

Integrating the velocity over a time step $\Delta t = t_{n+1} - t_n$, using partial integration for the term involving $\mathbf{F}^B(t')$, the particle displacements are given as

$$\mathbf{x}_p(t_n+\Delta t) = \mathbf{x}_p(t_n) + \int_{t_n}^{t_n+\Delta t}\mathbf{V}_p(t)\mathrm{d}t$$
$$= \mathbf{x}_p(t_n) + \mathbf{V}_p(t_n)\Delta t\frac{1-\exp(-B\Delta t)}{m_pB} + \frac{1}{m_p}\sum_i \mathbf{F}_i^{nh}(t_n)(\Delta t)^2$$
$$\times\frac{1}{(B\Delta t)^2}[B\Delta t-(1-\exp[-B\Delta t])] + \frac{1}{m_p}\sum_i \dot{\mathbf{F}}_i^{nh}(t_n)(\Delta t)^3$$
$$\times\frac{1}{(B\Delta t)^3}\left(\frac{1}{2}(B\Delta t)^2 - [B\Delta t-(1-\exp[-B\Delta t])]\right) + X_n(\Delta t) \tag{6.205}$$

where

$$X_n(\Delta t) = \frac{1}{m_pB}\int_{t_n}^{t_n+\Delta t}(1-\exp[-B(t_n+\Delta t-t)])\mathbf{F}^B(t)\mathrm{d}t \tag{6.206}$$

Eliminating $\mathbf{V}_p(t_n)$ from Eqn (6.205) by considering the particle displacement at $t_n - \Delta t$:

$$\mathbf{x}_p(t_n+\Delta t) = \mathbf{x}_p(t_n)[1+\exp(-B\Delta t)] - \mathbf{x}_p(t_n-\Delta t)\exp(-B\Delta t)$$
$$+\frac{1}{m_p}\sum_i \mathbf{F}_i^{nh}(t_n)(\Delta t)^2\frac{1}{(B\Delta t)^2}[1-\exp(-B\Delta t)]$$
$$+\frac{1}{m_p}\sum_i \dot{\mathbf{F}}_i^{nh}(t_n)(\Delta t)^3\frac{1}{(B\Delta t)^3}\left(\frac{1}{2}(B\Delta t)^2[1+\exp(-B\Delta t)] \right. \tag{6.207}$$
$$\left. - [1-\exp(-B\Delta t)]\right) + X_n(\Delta t) + \exp(-B\Delta t)X_n(-\Delta t)$$

More details on the derivation of the particle displacements can be found in van Gunsteren and Berendsen (1982). It can be seen that in the limit of zero friction ($B \approx 0$), Eqn (6.207) reduces to

$$\mathbf{x}_p(t_n + \Delta t) = 2\mathbf{x}_p(t_n) - \mathbf{x}_p(t_n - \Delta t) + \frac{1}{m_p}\sum_i \mathbf{F}_i^{nh}(t_n)(\Delta t)^2 \tag{6.208}$$

which is the displacements of the molecular dynamics of Verlet (see Eqn (6.197)).

As an additional simplification of the Langevin equation, the inertia can be neglected due to the small mass of these particles, meaning that there is no bulk motion of the fluid. Because of this approximation, the total force is always taken to be zero. In other words,

$$-B\mathbf{V}_p(t) + \sum_i \mathbf{F}_i^{nh}(t) + \mathbf{F}^B(t) \simeq 0 \tag{6.209}$$

In the absence of particle collisions, Brownian dynamics assumes that the hydrodynamic drag is invariably balanced by the Brownian motion of the particles. It should be noted that this force depends on the set of all particle positions. The Brownian force is, however, taken from a random distribution. In order for the dynamics to satisfy the fluctuation-dissipation theorem, the expectation values for the Brownian force are:

$$\left\langle F^B(t) \right\rangle = 0 \tag{6.210}$$

$$\left\langle F^B(t)F^B(t') \right\rangle = 2Bk_B T\delta(t - t') \tag{6.211}$$

where k_B is the Boltzmann constant, T is the absolute temperature and $\delta(\cdot)$ is the Dirac delta function. The use of the fluid continuum approximation together with the neglect of particle inertia enables Brownian dynamics to be employed at much higher time steps in comparison to the consideration of molecular dynamics (see Figure 6.16). A simple explicit Euler time integration solution method is widely used to solve for the trajectory of the particles:

$$\begin{aligned} \mathbf{x}_p(t_n + \Delta t) &= \mathbf{x}_p(t_n) + \mathbf{V}_p(t_n)\Delta t \Rightarrow \\ \mathbf{x}_p(t_n + \Delta t) &= \mathbf{x}_p(t_n) + \frac{1}{B}\left(\sum_i \mathbf{F}_i^{nh}(t_n) + \mathbf{F}^B(t_n)\right)\Delta t \end{aligned} \tag{6.212}$$

6.7.3 Discrete Element Method

In many aspects, the discrete element method is similar to molecular dynamics and Brownian dynamics. While molecular and Brownian dynamics consider the motion of particles representing individual atoms without any background fluid continuum and the motion of colloidal particles through a simple consideration of Stokes drag law for a sphere for creeping flow and neglecting other hydrodynamic interactions, the discrete element method solves the motion of individual particles of a larger length scale using a continuum approximation of the surrounding fluid. Typical applications of the discrete element method consider particles that are normally large enough such that the Brownian motion can be taken to be negligible. Nevertheless, the Brownian force acting on small particles can be readily added to the discrete element method computations if required. The most significant difference between the discrete element method and molecular dynamics is the consideration of the interparticle contact forces in the model. Whereas it is typical for molecular dynamics to employ simple conservative potentials such as the Lennard-Jones potential to sufficiently describe the interactions of very small particles, such potentials are not adequate to represent the interparticle forces present during the collision of larger-size particles in the discrete element method. It is therefore common to assume that the particle size in discrete element method computations is substantially larger than the characteristic length of the van der Waals length scale, which can be estimated to be $\sim$10 nm (10^{-8} m). For particle sizes on the order of, or much less than, this characteristic length, the van der Waals length scale potentials are a reasonable approximation. For particle sizes much greater than this characteristic length, it is necessary to model instead the detailed mechanics by which the particles will interact during collision. Particle contact forces that include normal impact, sliding and twisting friction between particles, and the effect of adhesion and material damping on impedance of particle rolling, are considerably more complex for the discrete element method. It should be noted that particle adhesion in the discrete element method has no effect until two particles collide or are in near-contact. Unlike in molecular and Brownian dynamics, which are intended for application to small particles, adhesion forces act over the length scales on the order of the particle size.

There are essentially two different approaches that are adopted for the discrete element method. One is the hard-sphere model that has been described in Chapter 2. This model assumes that interactions between particles are pair-wide additive and instantaneous; particle interactions are thus solved using the particle impulse equations. Another is the description of a soft-sphere model in Chapter 5. This time-driven model evolves the contact history using the particle momentum equation through the finite time interval wherein the particles will be in contact. Computationally, the hard-sphere model is more efficient than the soft-sphere model, since the particle collision time scale does not need to be resolved. Nevertheless, it should be pointed out that the hard-sphere model is unable to handle problems dealing with collisions of three or more particles or problems in which particles are in contact for a prolonged time period.

The solution methods for the discrete element method are generally different from those adopted in molecular and Brownian dynamics. As larger time steps can be applied, a suite of effective methods for determining the particle linear velocities and displacements have been proposed. The simplest method is the first-order accurate explicit method by Euler:

$$\mathbf{V}_p(t_n + \Delta t) = \mathbf{V}_p(t_n) + \frac{\sum \mathbf{F}(t_n)}{m_p}\Delta t \quad \mathbf{x}_p(t_n + \Delta t) = \mathbf{x}_p(t_n) + \mathbf{V}_p(t_n)\Delta t \tag{6.213}$$

Another first-order accurate explicit method is that of Newmark:

$$\begin{aligned} \mathbf{V}_p(t_n + \Delta t) &= \mathbf{V}_p(t_n) + \frac{\sum \mathbf{F}(t_n)}{m_p}\Delta t \\ \mathbf{x}_p(t_n + \Delta t) &= \mathbf{x}_p(t_n) + \mathbf{V}_p(t_n)\Delta t + \frac{1}{2}\sum \mathbf{F}(t_n)\Delta t^2 \end{aligned} \tag{6.214}$$

On the other hand, the second-order accurate Adams–Bashforth, Leapfrog and predictor-corrector methods have also been considered in conjunction with the evaluation of the particle velocities and displacements. For the Adams–Bashforth method,

$$\begin{aligned} \mathbf{V}_p(t_n + \Delta t) &= \mathbf{V}_p(t_n) + \frac{1}{m_p}\left(\frac{3}{2}\sum \mathbf{F}(t_n) - \frac{1}{2}\sum \mathbf{F}(t_n - \Delta t)\right)\Delta t \\ \mathbf{x}_p(t_n + \Delta t) &= \mathbf{x}_p(t_n) + \left(\frac{3}{2}\mathbf{V}_p(t_n) - \frac{1}{2}\mathbf{V}_p(t_n - \Delta t)\right)\Delta t \end{aligned} \tag{6.215}$$

Leapfrog method,

$$\begin{aligned} \mathbf{V}_p\left(t_n + \frac{1}{2}\Delta t\right) &= \mathbf{V}_p\left(t_n - \frac{1}{2}\Delta t\right) + \frac{\sum \mathbf{F}(t_n)}{m_p}\Delta t \\ \mathbf{x}_p(t_n + \Delta t) &= \mathbf{x}_p(t_n) + \mathbf{V}_p\left(t_n + \frac{1}{2}\Delta t\right)\Delta t \end{aligned} \tag{6.216}$$

and predictor-corrector method,

$$\begin{aligned} \mathbf{V}_p(t_n + \Delta t) &= \mathbf{V}_p(t_n) + \frac{1}{m_p}\left(\frac{3}{2}\sum \mathbf{F}(t_n) - \frac{1}{2}\sum \mathbf{F}(t_n - \Delta t)\right)\Delta t \\ \mathbf{x}_p(t_n + \Delta t) &= \mathbf{x}_p(t_n) + \left(\frac{3}{2}\mathbf{V}_p(t_n) - \frac{1}{2}\mathbf{V}_p(t_n - \Delta t)\right)\Delta t \\ \mathbf{V}_p(t_n + \Delta t) &= \mathbf{V}_p(t_n) + \frac{1}{2m_p}\left(\sum \mathbf{F}(t_n + \Delta t) + \sum \mathbf{F}(t_n)\right)\Delta t \\ \mathbf{x}_p(t_n + \Delta t) &= \mathbf{x}_p(t_n) + \frac{1}{2}(\mathbf{V}_p(t_n + \Delta t) + \mathbf{V}_p(t_n))\Delta t \end{aligned} \tag{6.217}$$

Note that the velocities at time t_n in the Leapfrog method can be approximated by averaging the velocities at times $t - ½\Delta t$ and $t + ½\Delta t$ (see Eqn (6.200)).

Alternatively, the Runge–Kutta scheme may also be applied to determine the particle velocities and displacements. The second-order accurate modified Runge–Kutta method (Tokoro et al., 2005) can be written as

$$\begin{aligned}
\mathbf{V}_p\left(t+\frac{1}{2}\Delta t\right) &= \mathbf{V}_p(t_n) + \frac{\sum \mathbf{F}(t_n)}{m_p}\frac{\Delta t}{2} \\
\mathbf{x}_p\left(t+\frac{1}{2}\Delta t\right) &= \mathbf{x}_p(t_n) + \mathbf{V}_p(t_n)\frac{\Delta t}{2} \\
\mathbf{V}_p(t_n+\Delta t) &= \mathbf{V}_p(t_n) + \frac{\sum \mathbf{F}(t_n+\frac{1}{2}\Delta t)}{m_p}\Delta t \\
\mathbf{x}_p(t_n+\Delta t) &= \mathbf{x}_p(t_n) + \left(\mathbf{V}_p(t_n+\Delta t)+\mathbf{V}_p(t_n)\right)\frac{\Delta t}{2}
\end{aligned} \tag{6.218}$$

while the fourth-order accurate Runge–Kutta method (Fletcher, 1991) is given by

$$\begin{aligned}
&\frac{D\mathbf{V}_p(t)}{Dt} = \frac{\sum \mathbf{F}(t)}{m_p} = f(t,\mathbf{V}_p(t)) && \frac{D\mathbf{x}_p(t)}{Dt} = \mathbf{V}_p(t) = f(t,\mathbf{x}_p(t)) \\
&k_1 = \Delta t f(t_n,\mathbf{V}_p(t_n)) && k_1' = \Delta t f(t_n,\mathbf{x}_p(t_n)) \\
&k_2 = \Delta t f\left(t+\frac{1}{2}\Delta t,\mathbf{V}_p(t_n)+\frac{1}{2}k_1\right) && k_2' = \Delta t f\left(t+\frac{1}{2}\Delta t,\mathbf{x}_p(t_n)+\frac{1}{2}k_1'\right) \\
&k_3 = \Delta t f\left(t+\frac{1}{2}\Delta t,\mathbf{V}_p(t_n)+\frac{1}{2}k_2\right) && k_3' = \Delta t f\left(t+\frac{1}{2}\Delta t,\mathbf{x}_p(t_n)+\frac{1}{2}k_2'\right) \\
&k_4 = \Delta t f(t+\Delta t,\mathbf{V}_p(t_n)+k_3) && k_4' = \Delta t f(t+\Delta t,\mathbf{x}_p(t_n)+k_3') \\
&\mathbf{V}_p(t_n+\Delta t) = \mathbf{V}_p(t_n) + \frac{\Delta t}{6}(k_1+2k_2+2k_3+k_4) \\
&\mathbf{x}_p(t_n+\Delta t) = \mathbf{x}_p(t_n) + \frac{\Delta t}{6}(k_1'+2k_2'+2k_3'+k_4')
\end{aligned} \tag{6.219}$$

All the aforementioned proposed solution methods can also be similarly employed to determine particle angular velocities and displacements for cases where particles may undergo rotatation in the physical fluid domain.

6.8 Turbulence Modeling for Multiphase Flows

6.8.1 Reynolds-Averaged Equations and Closure

Turbulent flow generally behaves in a random and chaotic manner. It is intrinsically unstable and unsteady, often preceded by a transition phenomenon from the primary laminar state. Deterministically, turbulence can be regarded as having coherence structures; the stability of dynamical flow systems can be studied by perturbating the fluid flow. Statistically, the concept of the cascading of turbulent kinetic energy from large-scale to small-scale structures can be employed, which leads then to turbulence being modeled by the consideration of averaged flow quantities. It is worth noting that the smallest scales by which turbulence dissipates into heat are known as the Kolmogorov scales. In actual flow systems, turbulence is neither organized nor random, but, rather, it tends to oscillate between these two modes in an arbitrary manner. This, therefore, makes the computational prediction of turbulent flow not only challenging but also not as straightforward in practice.

In a range of applications for single-phase flows, the eddy viscosity model has been shown to perform reasonably well to engineering accuracy. First introduced by Boussinesq (1903), this concept suggests that the Reynolds stress can be correlated with the mean rates of deformation. In the context of multifluid modeling, it has been demonstrated by Lopez de Bertodano et al. (1994a, b) that the eddy viscosity and eddy diffusivity hypotheses in describing the turbulence in single-phase flows can also be extended to describe the Reynolds stress and Reynolds flux terms in multiphase flows. The Reynolds stress for different phases can be expressed as

$$-\boldsymbol{\tau}^{k''} = \mu_T^k\left(\nabla \mathbf{U}^k + \left(\nabla \mathbf{U}^k\right)^T\right) - \frac{2}{3}\mu_T^k \nabla \cdot \mathbf{U}^k \delta - \frac{2}{3}\rho^k k^k \delta \tag{6.220}$$

where k^k is the turbulent kinetic energy and μ_T^k is the additional viscosity called the turbulent or eddy viscosity, which is usually taken to be a function of the flow rather than of the fluid and requires to be prescribed. Analogous to the eddy viscosity hypothesis, the eddy diffusivity hypothesis for the Reynolds flux term can be taken to be proportional to the gradient of the transported quantity. For the total enthalpy, the Reynolds flux term can be modeled as

$$-\mathbf{q}_H^{k''} = \Gamma_T^k \nabla H^k \tag{6.221}$$

The term Γ_T^k in the above expression is the eddy diffusivity for total enthalpy of the fluid. Since the turbulent transport of momentum and heat can be attributed through the same mechanisms—eddy mixing—the value of the eddy diffusivity in Eqn (6.221) can be taken to be close to that of the eddy viscosity μ_T^k. According to the definition of the turbulent Prandtl number, i.e. the ratio between momentum diffusivity (viscosity) and thermal diffusivity, the number is thus given by

$$Pr_T^k = \frac{\mu_T^k}{\Gamma_T^k} \tag{6.222}$$

To satisfy dimensional requirements, at least two scaling parameters are required to relate the Reynolds stress to the rate of deformation. One feasible choice is the use of the turbulent kinetic energy k^k and the other is the rate of dissipation of turbulent energy ε^k. The local turbulent viscosity in Eqn (6.222) can be obtained either from dimensional analysis or from analogy to the laminar viscosity as $\mu_T^k \propto \rho^k v_T l$. Based on the characteristic velocity v_T defined as $\sqrt{k^k}$ and the characteristic length l as $(k^k)^{3/2}/\varepsilon^k$, the turbulent viscosity can be calculated according to

$$\mu_T^k = C_\mu \rho^k \frac{(k^k)^2}{\varepsilon^k} \tag{6.223}$$

where C_μ is an empirical constant and values of k^k and ε^k are solved according to their respective transport equations.

On the basis of the eddy viscosity and diffusivity hypotheses, the RANS equations of the multifluid model are:

Mass conservation:

$$\frac{\partial(\alpha^k \rho^k)}{\partial t} + \nabla \cdot (\alpha^k \rho^k \mathbf{U}^k) = \sum_{l=1}^{2} (\dot{m}_{lk} - \dot{m}_{kl}) \tag{6.224}$$

Momentum conservation:

$$\begin{aligned} \frac{\partial(\alpha^k \rho^k \mathbf{U}^k)}{\partial t} + \nabla \cdot (\alpha^k \rho^k \mathbf{U}^k \otimes \mathbf{U}^k) = & -\alpha^k \nabla p + \Big(\nabla \cdot \alpha^k \Big[(\mu^k + \mu_T^k)\big(\nabla \mathbf{U}^k + (\nabla \mathbf{U}^k)^T\big) \\ & - \frac{2}{3}(\mu^k + \mu_T^k)\, \nabla \cdot \mathbf{U}^k \delta - \frac{2}{3}\rho^k k^k \delta \Big] \Big) + \alpha^k \rho^k \mathbf{g} \\ & + \sum_{l=1}^{2} (\dot{m}_{lk} \mathbf{U}^l - \dot{m}_{kl} \mathbf{U}^k) + (p_{\text{int}}^k - p) \nabla \alpha^k + \mathbf{F}_D^{k,\text{drag}} \\ & + \mathbf{F}_D^{k,\text{non-drag}} \end{aligned} \tag{6.225}$$

Energy conservation:

$$\begin{aligned} \frac{\partial(\alpha^k \rho^k H^k)}{\partial t} + \nabla \cdot (\alpha^k \rho^k \mathbf{U}^k H^k) = & \nabla \cdot (\alpha^k \lambda^k \nabla T^k) + \nabla \cdot \left(\alpha^k \frac{\mu_T^k}{Pr_T^k} \nabla H^k \right) \\ & + \sum_{l=1}^{2} (\dot{m}_{lk} H^l - \dot{m}_{kl} H^k) + Q_H^{\text{int}} \end{aligned} \tag{6.226}$$

Two-equation k-ε model

The *standard k-ε model* developed by Launder and Spalding (1974) has been regarded as the "industrial standard" model for most engineering applications of single-phase turbulent fluid flow problems. Derivation of the transport equations for turbulent kinetic energy and dissipation of turbulent kinetic energy generally requires substantial mathematical manipulation. Following similar steps undertaken by Tennekes and Lumley (1976) and Versteeg and Malasekera (1995) in formulating the equations for single-phase turbulent fluid flow, and based on the proposal by Lopez de Bertodano et al. (1994a) and Lahey and Drew (2001) where at high Reynolds numbers, the system of turbulent scalar equations consists of merely straightforward generalizations of the single-phase counterpart via accounting for the fraction of time in which the continuous or dispersed phase occupies a particular given point in space through the local volume fraction α^k, the transport equations for the *two-equation k-ε model* can be written as

$$\frac{\partial}{\partial t}\left(\alpha^k\rho^k k^k\right)+\nabla\cdot\left(\alpha^k\rho^k\mathbf{U}^k k^k\right)$$
$$=\nabla\cdot\left(\alpha^k\frac{\mu_T^k}{\sigma_k}\nabla k^k\right)+\alpha^k\left(P^k+G^k-\rho^k\varepsilon^k\right)+S_{k^k}^{\text{int}} \tag{6.227}$$

$$\frac{\partial}{\partial t}\left(\alpha^k\rho^k\varepsilon^k\right)+\nabla\cdot\left(\alpha^k\rho^k\mathbf{U}^k\varepsilon^k\right)$$
$$=\nabla\cdot\left(\alpha^k\frac{\mu_T^k}{\sigma_\varepsilon}\nabla\varepsilon^k\right)+\alpha^k\frac{\varepsilon^k}{k^k}\left(C_{\varepsilon 1}P^k+C_3\left\|G^k\right\|-C_{\varepsilon 2}\rho^k\varepsilon^k\right)+S_{\varepsilon^k}^{\text{int}} \tag{6.228}$$

where $S_{k^k}^{\text{int}}$ and $S_{\varepsilon^k}^{\text{int}}$ are the additional source or sink terms to accommodate the production and dissipation of turbulence due to the interaction between the continuous and disperse phases. For example, large particulates are known to enhance turbulence due to the production of a turbulent wake behind the particulates while small particulates suppress the turbulence in the flowing fluid. The shear production P^k is

$$P^k=\mu_T^k\nabla\mathbf{U}^k\cdot\left(\nabla\mathbf{U}^k+\left(\nabla\mathbf{U}^k\right)^T\right)-\frac{2}{3}\nabla\cdot\mathbf{U}^k\left(\mu_T^k\nabla\cdot\mathbf{U}^k-\rho^k\mathbf{U}^k\right) \tag{6.229}$$

and G^k is the production due to the gravity, which is valid for weakly compressible flows and can be written as

$$G^k=-\frac{\mu_T^k}{\rho^k\sigma_{\rho^k}}\mathbf{g}\cdot\nabla\rho^k \tag{6.230}$$

where C_3 and σ_{ρ^m} are normally assigned values of unity and $\left\|G^k\right\|$ in Eqn (6.228) is the imposed condition whereby it always remains positive, i.e. max(G^k,0). The constants for the *two-equation k-ε model* are taken to have the same constants that are adopted in the single-phase *standard k-ε model* (see Launder and Spalding, 1974):

$$C_\mu = 0.09 \quad \sigma_k = 1.0 \quad \sigma_\varepsilon = 1.3 \quad C_{\varepsilon 1} = 1.44 \quad C_{\varepsilon 2} = 1.92$$

Comments

1. No single turbulence model can be readily employed to span all turbulent states as none is expected to be universally valid for all types of multiphase flows. The *two-equation k-ε model*, which is a consequence of the eddy viscosity and diffusive hypotheses, assumes that the turbulent stresses are linearly related to the rate of strain by a turbulent viscosity. The model treats C_μ, characteristic velocity v_T and characteristic length l as scalars; the turbulent viscosity is thus a scalar accordingly. This absence of direction dependence implies that the principal strain directions are always aligned to the principal stress directions. Therefore, they behave in an *isotropic* manner. In some flow cases, secondary flows that may exist within the geometry are driven by strongly *anisotropic* turbulence. As such, the assumption of *isotropic* normal Reynolds stresses is unrealistic for such kinds of flows and a more complicated approach by evaluating each of the Reynolds stress components is required.

2. As proposed by Launder (1989) and Rodi (1993) for single-phase turbulent flow, the *Reynolds stress model* directly determines the turbulent stresses by solving a transport equation for each of the stress components. Such equations include the turbulent transport, generation, dissipation and redistribution of Reynolds stresses in the turbulent flow. An additional equation for the dissipation ε is solved to provide a length scale determining quantity. There is no doubt that the *Reynolds stress model* has a greater potential to represent the turbulent flow phenomena more correctly. This type of model can aptly handle complex strain and, in principle, can cope with nonequilibrium flows. Nevertheless, the convergence of the *Reynolds stress model* is, in general, more difficult to be obtained and is very sensitive to the initial conditions of the turbulent stresses. Similar to the *two-equation k-ε model*, the multiphase version of the *Reynolds stress model* is merely a straightforward generalization of the respective single-phase counterpart; more details of the multiphase version can be found in our book Yeoh and Tu (2009).

Shear stress transport (SST) model

With the aim of combining the favorable features of the *two-equation k-ε model* with the *two-equation k-ω model* whereby the inner region of the boundary layer is adequately resolved by the latter while the former is employed to obtain solutions in the outer part of the boundary layer, the *Shear Stress Transport* (SST) variation of *Menter's model* (1993,

1996) has been developed. This model is increasingly being employed and works exceptionally well in handling nonequilibrium boundary layer regions such as areas of flow separation.

The *two-equation k-ω model* developed by Wilcox (1998) represents a useful model for the near-wall treatment of low Reynolds number flow computations in single-phase fluid flow. Principally, the model is based on the transport equations of the turbulent kinetic energy k and the turbulent frequency ω—considered as the ratio of ε to k, i.e. $\omega = \varepsilon/k$. In order to formulate the *SST model*, the *two-equation k-ε model* is required to be transformed into a form consistent with the k-ω formulation. Here, a blending function F_1 is introduced whereby the *two-equation k-ω model* is multiplied by this function F_1 and the transformed *k-ε model* by a function $1 - F_1$. At the boundary layer edge, and outside the boundary layer, the *two-equation k-ε model* is recovered when $F_1 = 0$. The equations of the *SST model* for the multifluid model are:

$$\mu_T^k = \frac{\rho^k a_1 k^k}{\max(a_1 \omega^k, S^k F_2)} \qquad S^k = \sqrt{2S^2}$$

$$S = \left(\frac{\partial u^k}{\partial x}\right) + \left(\frac{\partial v^k}{\partial y}\right) + \left(\frac{\partial w^k}{\partial z}\right) + \frac{1}{2}\left(\frac{\partial u^k}{\partial y} + \frac{\partial v^k}{\partial x}\right) + \frac{1}{2}\left(\frac{\partial u^k}{\partial z} + \frac{\partial w^k}{\partial x}\right) \tag{6.231}$$

$$+ \frac{1}{2}\left(\frac{\partial v^k}{\partial x} + \frac{\partial u^k}{\partial y}\right) + \frac{1}{2}\left(\frac{\partial v^k}{\partial z} + \frac{\partial w^k}{\partial y}\right) + \frac{1}{2}\left(\frac{\partial w^k}{\partial x} + \frac{\partial u^k}{\partial z}\right) + \frac{1}{2}\left(\frac{\partial w^k}{\partial y} + \frac{\partial v^k}{\partial z}\right)$$

$$\frac{\partial}{\partial t}(\alpha^k \rho^k k^k) + \nabla \cdot (\alpha^k \rho^k \mathbf{U}^k k^k)$$
$$= \nabla \cdot \left(\alpha^k \left[\mu^k + \frac{\mu_T^k}{\sigma_{k3}}\right] \nabla k^k\right) + \alpha^k (P^k + G^k - \rho^k \beta' k^k \omega^k) + S_{k^k}^{\text{int}} \tag{6.232}$$

$$\frac{\partial}{\partial t}(\alpha^k \rho^k \omega^k) + \nabla \cdot (\alpha^k \rho^k \mathbf{U}^k \omega^k) = \nabla \cdot \left(\alpha^k \left[\mu^k + \frac{\mu_T^k}{\sigma_{\omega 3}}\right] \nabla \omega^k\right)$$
$$+ 2\alpha^k \rho^k (1 - F_1) \frac{1}{\sigma_{\omega 2} \omega^k} \nabla k^k \nabla \omega^k + \alpha^k \alpha_3 \frac{\omega^k}{k^k} (P^k + C_3 \|G^k\|)$$
$$- \alpha^k \rho^k \beta_3 (\omega^k)^2 + S_{\omega^k}^{\text{int}} \tag{6.233}$$

where P^k and G^k are the same productions terms given in Eqns (6.232) and (6.233) and

$$\sigma_{k3} = F_1 \sigma_{k1} + (1 - F_1)\sigma_{k2} \quad \sigma_{\omega 3} = F_1 \sigma_{\omega 1} + (1 - F_1)\sigma_{\omega 2}$$
$$\alpha_3 = F_1 \alpha_1 + (1 - F_1)\alpha_2 \qquad \beta_3 = F_1 \beta_1 + (1 - F_1)\beta_2$$

The function F_1 is given by

$$F_1 = \tanh(\Phi_1^4) \tag{6.234}$$

with

$$\Phi_1 = \min\left[\max\left(\frac{\sqrt{k^k}}{0.09\omega^k d_n}, \frac{500\mu^k}{\rho^k \omega^k d_n^2}\right), \frac{4\rho^k k^k}{D_\omega^+ \sigma_{\omega 2} d_n^2}\right] \tag{6.235}$$

while the function F_2 can be expressed as

$$F_2 = \tanh(\Phi_2^2) \tag{6.236}$$

with

$$\Phi_2 = \max\left(\frac{2\sqrt{k^k}}{0.09\omega^k d_n}, \frac{500\mu^k}{\rho^k \omega^k d_n^2}\right) \tag{6.237}$$

In Eqn (6.235), the variable D_ω^+ appearing is determined according to

$$D_\omega^+ = \max\left(2\rho^k \frac{1}{\sigma_{\omega 2}\omega^k} \nabla k^k \nabla \omega^k, 10^{-10}\right) \tag{6.238}$$

The distance closest to a fixed wall, which is represented by d_n, is used to calculate the blending functions in order to appropriately switch between the k-ω and k-ε models. The model constants in the transport equations of the *SST model* are given by:

$$\beta' = 0.09 \quad \sigma_{k1} = 1.176 \quad \sigma_{\omega 1} = 2.0 \quad \alpha_1 = 5.0/9.0 \quad \beta_1 = 0.075$$

$$\alpha_1 = 0.31 \quad \sigma_{k2} = 1.0 \quad \sigma_{\omega 2} = 1.168 \quad \alpha_2 = 0.44 \quad \beta_2 = 0.0828$$

Comments

1. Our investigations in Cheung et al. (2007a, b) have shown that the *two-equation SST model* principally derived for single-phase flow could be extended to handle isothermal gas–liquid flow. The model has resulted in significant improvement of the prediction of the liquid or gas velocities as well as the local volume fraction close to a wall.

2. Transport equations for the turbulent kinetic energy and dissipation of turbulent kinetic energy of the *two-equation k-ε* and *SST models* provide the necessary means of determining the turbulent viscosities for the continuous phase as well as the dispersed phase. Nevertheless, owing to the complexity that arises in resolving the effects of small entities such as finite solid particles, liquid drops, or gaseous bubbles (dispersed phase) on the turbulence structure of the carrier medium (continuous phase) in different

types of multiphase flows, it is common practice to determine the turbulent viscosity of the dispersed phase through a simpler prescriptive approach with the primary aim to reduce the computational effort and obviate the necessity of solving further transport equations. For gas—liquid flow, Sato et al. (1981) proposed a model whereby the bubble induced turbulence on the liquid flow is added to the shear-induced turbulence. Denoting the continuous phase by c and the dispersed phase p, the turbulent viscosity from Eqn (6.223) can now be replaced by

$$\mu_T^c = C_\mu \rho^c \frac{(k^c)^2}{\varepsilon^c} + \mu_T^b \tag{6.239}$$

where μ_T^b denotes the extra bubble induced turbulence term:

$$\mu_T^b = C_{\mu b} \rho^c \alpha^p d_b |\mathbf{U}^p - \mathbf{U}^c| \tag{6.240}$$

where $C_{\mu b}$ has a value of 0.6 and d_b is the bubble diameter. The kinematic viscosity of the dispersed phase $\nu_T^p = \mu_T^p / \rho^p$ can be determined from the kinematic viscosity of the continuous phase $\nu_T^c = \mu_T^c / \rho^c$ as

$$\mu_T^p = \frac{\mu_T^c}{\sigma} \frac{\rho^p}{\rho^c} \tag{6.241}$$

where σ represents the turbulent Prandtl number, which has a value of unity. Based on Eqn (6.241), transport equations for the turbulent kinetic energy and dissipation of turbulent kinetic energy are only employed for the continuous phase. Note that the model of Sato et al. (1981) could also be applied, for example, to particles in a gas or a liquid. For gas-particle flow, Tu and Fletcher (1995) have nonetheless adopted the approach proposed by Aldeniji-Fashola and Chen (1990) where the effect of turbulence is modified by a weight factor K^d introduced into Eqn (6.241) according to

$$\mu_T^p = K^p \mu_T^c \frac{\rho^p}{\rho^c} \tag{6.242}$$

and the continuous phase viscosity is simply the shear-induced turbulence:

$$\mu_T^c = C_\mu \rho^c \frac{(k^c)^2}{\varepsilon^c} \tag{6.243}$$

The weight factor K^p in Eqn (6.242) accounts for the transfer of turbulence energy to the dispersed phase due to the inertia of the solid particles.

Near-wall treatment

Near-wall modeling represents an integral part in bridging the low Reynolds flows that exist in the vicinity of a wall with the high Reynolds flows in the bulk fluid beyond the fully-developed turbulence boundary layer. Appropriate near-wall models need to be employed in order to predict and resolve the wall-bounded turbulent flows with sufficient accuracy. The *SST model*, as discussed in the previous section, allows the possibility of fully resolving the flow by extending all the way to the wall boundary. An advantage of employing such a model is that no additional assumptions are required concerning the variation of the variables near the wall. However, the downside is the requirement of a very fine near-wall resolution. For the *two-equation k-ω model*, a wall distance $y^+ \sim 2$ at all the wall nodes is required to sufficiently resolve the fluid flow adjacent to the wall. Such a prerequisite is generally difficult to achieve especially for large full-scale multiphase flow systems.

One feasible approach to overcome the difficulty of modeling the near-wall region is through the prescription of *wall functions* for the *two-equation k-ε model* and *Reynolds Stress model*. By adopting this approach, the difficult near-wall region is not explicitly resolved but is bridged using suitable prescribed functions (Launder and Spalding, 1974). In order to construct these functions, the region close to the wall can usually be characterized by considering the dimensionless velocity U^+ and wall distance y^+ with respect to the local conditions at the wall. The dimensionless wall distance y^+ is defined as $\rho u_\tau(d-y)/\mu$, where very near the wall, $y=d$, while the dimensionless velocity U^+ can be expressed in the form U/u_τ, where U is taken to represent some averaged velocity passing parallel to the wall, and u_τ is the wall friction phase velocity, which is defined with respect to the wall shear stress τ_w as $\sqrt{\tau_w/\rho}$. For the multifluid model, the dimensionless velocity U^+ and wall distance y^+ are similarly employed for each phase or fluid, which are defined in terms of their respective distinct fluid parameters.

For a wall distance of $y^+ < 5$, the boundary layer is predominantly governed by viscous forces that produce the no-slip condition; this region is subsequently referred to as the *viscous sublayer.* By assuming that the shear stress is approximately *constant* and equivalent to the wall shear stress τ_w, a linear relationship between the averaged velocity and the distance from the wall can be obtained yielding

$$U^+ = y^+ \quad \text{for} \quad y^+ < y_0^+ \tag{6.244}$$

With increasing wall distance y^+, turbulent diffusion effects dominate outside the *viscous sublayer.* A logarithmic relationship is employed:

$$U^+ = \frac{1}{\kappa}\ln\left(Ey^+\right) \quad \text{for} \quad y^+ > y_0^+ \tag{6.245}$$

The above relationship is often called the *log-law* and the layer where the wall distance y^+ lies between the range of $30 < y^+ < 500$ is known as the *log-law layer.* The values for κ (~0.4) and E (~9.8) are universal constants valid for all turbulent flows past smooth walls at high Reynolds numbers. For rough surfaces, the constant E is usually reduced. The law of the wall can be modified by scaling the normal wall distance d_n on the equivalent roughness height, h_0 (i.e. y^+ is replaced by d_n/h_0), and appropriate values must be selected from data or literature. The cross-over point y_0^+ can be determined by computing the intersection between the viscous sublayer and the logarithmic region based on the upper root of

$$y_0^+ = \frac{1}{\kappa}\ln\left(Ey_0^+\right) \tag{6.246}$$

A similar universal, nondimensional function can also be constructed to model the effect of the phenomenon of heat transfer. According to Reynolds' analogy, the treatment follows the same law-of-the-wall for the velocity of which the law-of-the-wall for enthalpy comprises: (1) Linear law for the thermal conduction in the sublayer where conduction is important, and (2) Logarithmic law for the turbulent region where effects of turbulence dominate over conduction. The enthalpy in the wall layer is assumed to be:

$$\begin{aligned} H^+ &= Pry^+ && \text{for } y^+ < y_H^+ \\ H^+ &= \frac{Pr_T}{\kappa}\ln\left(F_H y^+\right) && \text{for } y^+ > y_H^+ \end{aligned} \tag{6.247}$$

where F_H is determined by using the empirical formula of Jayatilleke (1969):

$$F_H = E\exp\left\{9.0\kappa\left[\left(\frac{Pr}{Pr_T}\right)^{0.75} - 1\right]\left[1 + 0.28\exp\left(-0.007\frac{Pr}{Pr_T}\right)\right]\right\} \tag{6.248}$$

By definition, the dimensionless enthalpy H^+ is given by:

$$H^+ = \frac{(H_w - H)\rho C_\mu^{0.25}k^{0.5}}{J_H} \tag{6.249}$$

where H_w is the value of enthalpy at the wall, ρ and k are the density and turbulent kinetic energy representing either of the mixture or phase quantities, and the diffusion flux J_H is equivalent to the normal gradient of the enthalpy $(\partial H/\partial n)_{\text{wall}}$ perpendicular to the wall. The thickness of the thermal conduction layer is usually different from the thickness of the viscous sublayer and changes from fluid to fluid. As demonstrated in Eqn (6.246), the crossover point y_H^+ can also be similarly computed through the intersection between the thermal conduction layer and the logarithmic region based on the upper root of

$$Pry_H^+ = Pr_T\frac{1}{\kappa}\ln\left(F_H y_H^+\right) \tag{6.250}$$

The universal profiles derived from the above have been based on an attached two-dimensional Couette flow configuration with *small pressure gradients, local equilibrium of turbulence* (production rate of turbulent kinetic energy equals its destruction rate) and *a constant near-wall stress layer.* For some applications, applying such wall functions may lead to significant inaccuracies in the modeling of wall-bounded turbulent flows. In single-phase flow applications, *nonequilibrium wall functions* and *enhanced wall treatment* that combine a two-layer model with enhanced wall functions are applied to circumvent the limitations imposed through the standard wall functions.

Based on the development by Kim and Choudhury (1995), the key elements of the *nonequilibrium wall functions* are that the log-law is now taken to be sensitized to pressure gradient effects and the two-layer-based concept is adopted to calculate the cell-averaged turbulence kinetic energy production and destruction in wall-adjacent cells. On the basis of the latter aspect, the turbulence kinetic energy budget for the wall-adjacent cells is sensitized to the proportions of the *viscous sublayer* as well as the *fully turbulent layer*, which can significantly vary from cell to cell in highly nonequilibrium flows. This effectively relaxes the *local equilibrium of turbulence* that is adopted by the standard wall functions. In the *enhanced wall treatment*, a single wall law is formulated for the entire wall region. A blending function is introduced to allow a smooth transition between the linear and logarithmic laws. This turbulence law always guarantees the correct asymptotic behavior for large and small values of the wall distance y^+ and provides reasonable representation of the velocity profiles in cases where y^+ lies insides the wall buffer region ($3 < y^+ < 10$). More details of this approach are referred to in Kader (1993). *Nonequilibrium wall functions* and *enhanced wall treatment* are recommended for complex flows that may involve flow separation, flow reattachment and flow impingement.

In multiphase flow analysis, discussions continue to remain open for the application of standard, nonequilibrium and enhanced wall functions in multifluid modeling. Rigorous assessments on the appropriate use of these wall treatments still need to be performed against different types of multiphase flows.

6.8.2 Large Eddy Simulation

The basic idea behind LES is that the large-scale motions are directly solved and the small-scale motions are represented in terms of subgrid scale models. We begin the derivation by assuming the flow domain to be composed of control volumes containing the different fluids occupying a volumetric proportion equal to V^k/V, where V^k is the volume of the kth phase and V is the volume containing all the phases. The consideration of this volume requires the characteristic length scale to be larger than the dispersed phase, which is denoted by the particulate size and/or spacing.

The Favre-averaging approach, as utilized in the RANS framework, is also adopted in the LES framework in order to alleviate the complication of modeling additional correlation terms containing averages of fluctuating quantities. The phase-weighted average for the variable ϕ can be defined by

$$\widetilde{\phi(\mathbf{x}', t)} = \frac{\overline{\mathscr{F}^k(\mathbf{x}', t)\phi(\mathbf{x}', t)}}{\overline{\mathscr{F}^k(\mathbf{x}', t)}} = \frac{\int_\Delta \mathscr{F}^k(\mathbf{x}', t)\phi(\mathbf{x}', t)G(|\mathbf{x} - \mathbf{x}|)d\mathbf{x}'}{\int_\Delta \mathscr{F}^k(\mathbf{x}', t)G(|\mathbf{x} - \mathbf{x}'|)d\mathbf{x}'} \tag{6.251}$$

and the mass-weighted average of the variable ψ in accordance with

$$\widetilde{\psi(\mathbf{x}', t)} = \frac{\overline{\rho^k(\mathbf{x}', t)\psi(\mathbf{x}', t)}}{\overline{\rho^k(\mathbf{x}', t)}} = \frac{\int_\Delta \rho^k(\mathbf{x}', t)\psi(\mathbf{x}', t)G(|\mathbf{x} - \mathbf{x}'|)d\mathbf{x}'}{\int_\Delta \rho^k(\mathbf{x}', t)G(|\mathbf{x} - \mathbf{x}'|)d\mathbf{x}'} \tag{6.252}$$

In the above two equations, $G(|\mathbf{x} - \mathbf{x}'|)$ represents an appropriate spatial filter function and the most common localized filter functions are represented by

Top Hat:

$$G(|\mathbf{x} - \mathbf{x}'|) = \begin{cases} \dfrac{1}{\Delta} & \text{for } |\mathbf{x} - \mathbf{x}'| < \dfrac{\Delta}{2} \\ 0 & \text{otherwise} \end{cases} \tag{6.253}$$

Gaussian:

$$G(|\mathbf{x} - \mathbf{x}'|) = \sqrt{\frac{6}{\pi\Delta^2}} \exp\left(\frac{-6(\mathbf{x} - \mathbf{x}')^2}{\Delta^2}\right) \tag{6.254}$$

Spectral Cutoff:

$$G(|\mathbf{x} - \mathbf{x}'|) = \frac{\sin(k_c(\mathbf{x} - \mathbf{x}'))}{\pi(\mathbf{x} - \mathbf{x}')}, \quad k_c = \frac{\pi}{\Delta} \tag{6.255}$$

The instantaneous variables of ϕ and ψ may now be written as

$$\phi(\mathbf{x}', t) = \widetilde{\phi(\mathbf{x}', t)} + \phi''(\mathbf{x}', t) \tag{6.256}$$

$$\psi(\mathbf{x}', t) = \widetilde{\psi(\mathbf{x}', t)} + \psi''(\mathbf{x}', t) \tag{6.257}$$

where $\widetilde{\phi(\mathbf{x}', t)}$ and $\widetilde{\psi(\mathbf{x}', t)}$ represent the filtered or resolvable components (essentially local averages of the complete field) and $\phi''(\mathbf{x}', t)$ and $\psi''(\mathbf{x}', t)$ are the subgrid scale components that account for unresolved spatial variations at a length smaller than the filter width Δ.

Within the multifluid approach, it is also customary in the context of LES to identify each kth phase by defining a quantity reflecting the averaged volumetric fraction of that phase inside the volume V. The volume fraction can thus be expressed as

$$\alpha^k(\mathbf{x}', t) = \overline{\mathscr{F}^k(\mathbf{x}', t)} = \int_\Delta \mathscr{F}^k(\mathbf{x}', t)G(|\mathbf{x} - \mathbf{x}'|)d\mathbf{x}' \tag{6.258}$$

Dropping the bars that by default denote the Favre-averaging and volume-averaging processes, the filtered equations of the conservation of mass, momentum and energy are

$$\frac{\partial\left(\alpha^k \rho^k\right)}{\partial t}+\nabla \cdot\left(\alpha^k \rho^k \mathbf{U}^k\right)=\mathbf{M}^k \tag{6.259}$$

$$\begin{aligned}
&\frac{\partial\left(\alpha^k \rho^k \mathbf{U}^k\right)}{\partial t}+\nabla \cdot\left(\alpha^k \rho^k \mathbf{U}^k \otimes \mathbf{U}^k\right) \\
&\quad=-\alpha^k \nabla p^k+\left(\nabla \cdot \alpha^k\left[\mu^k\left(\nabla \mathbf{U}^k+\left(\nabla \mathbf{U}^k\right)^T\right)-\frac{2}{3} \mu^k \nabla \cdot \mathbf{U}^k \delta\right]\right) \\
&\qquad-\nabla \cdot\left(\alpha^k \boldsymbol{\tau}^{k''}\right)+\alpha^k \rho^k \mathbf{g}+\mathbf{F}^k
\end{aligned} \tag{6.260}$$

$$\frac{\partial\left(\alpha^k \rho^k H^k\right)}{\partial t}+\nabla \cdot\left(\alpha^k \rho^k \mathbf{U}^k H^k\right)=\nabla \cdot\left(\alpha^k \lambda^k \nabla T^k\right)-\nabla \cdot\left(\alpha^k \mathbf{q}_H^{k''}\right)+\mathbf{E}^k \tag{6.261}$$

where $\mathbf{M}^k$, $\mathbf{F}^k$ and $\mathbf{E}^k$ are the filtered interfacial mass, momentum and energy balance source terms respectively. The unresolved subgrid stress tensor $\boldsymbol{\tau}^{k''}$ in Eqn (6.260) can be modeled according to the Boussinesq hypothesis as

$$\begin{aligned}
\boldsymbol{\tau}^{k''} &= \overline{\rho^k}\widetilde{\mathbf{U}^k \otimes \mathbf{U}^k}-\overline{\rho^k}\tilde{\mathbf{U}}^k \otimes \tilde{\mathbf{U}}^k=-2 \mu_{T_{\mathrm{SGS}}}^k \tilde{\mathbf{S}}^k+\frac{1}{3} \boldsymbol{\tau}^{k''} \delta \\
\tilde{\mathbf{S}}^k &= \frac{1}{2}\left(\nabla \tilde{\mathbf{U}}^k+\left(\nabla \tilde{\mathbf{U}}^k\right)^T\right)-\frac{1}{3} \nabla \cdot \tilde{\mathbf{U}}^k \delta
\end{aligned} \tag{6.262}$$

where $\mu_{T_{\mathrm{SGS}}}^k$ is the subgrid scale eddy viscosity for the kth phase and $\tilde{\mathbf{S}}$ is the strain rate of the large scale or resolved field. Also, the unresolved subgrid enthalpy flux $\mathbf{q}_H^{k''}$ in Eqn (6.261) is modeled in a manner similar to the subgrid turbulence stresses by the standard gradient diffusion hypothesis as

$$\mathbf{q}_H^{k''}=\overline{\rho^k}\left(\widetilde{\mathbf{U}^k H}-\widetilde{\mathbf{U}^k} \tilde{H}\right)=-\frac{\mu_{T_{\mathrm{SGS}}}^k}{Pr_{T_{\mathrm{SGS}}}^k} \nabla \tilde{H} \tag{6.263}$$

where $Pr_{T_{\mathrm{SGS}}}^k$ is the subgrid turbulent Prandtl number for the kth phase.

Constant coefficient subgrid scale model

As suggested by Smagorinsky (1963), since the smallest turbulence eddies are almost isotropic, the Boussinesq hypothesis provides a good description of the unresolved eddies. Taking the length scale to be the filter width, the velocity scale can be expressed as the product of the length scale and the average strain rate of the resolved flow. This brings about the formulation of the Smagorinsky–Lilly model, which assumes that the subgrid scale eddy viscosity can be described in terms of a length and a velocity scale for the single fluid formulation according to

$$\mu_{T_{\text{SGS}}} = \rho C_{\text{SGS}}^2 \Delta^2 |\tilde{S}| \quad |\tilde{S}| = \sqrt{2\tilde{\mathbf{S}}\tilde{\mathbf{S}}} \tag{6.264}$$

where C_{SGS} is an empirical constant. For the multifluid approach, the subgrid scale eddy viscosity also takes the same form as expressed in Eqn (6.264). In other words,

$$\mu_{T_{\text{SGS}}}^k = \rho^k C_{\text{SGS}}^2 \Delta^2 \left|\tilde{S}^k\right| \quad \left|\tilde{S}^k\right| = \sqrt{2\tilde{\mathbf{S}}^k\tilde{\mathbf{S}}^k} \tag{6.265}$$

A theoretical analysis performed by Lilly (1966, 1967) of the decay rates of isotropic turbulent eddies in the inertial subrange of the energy spectrum suggested values of C_{SGS}^2 between 0.0289 and 0.0441. Based on the LES computations carried out by Deardoff (1970), $C_{\text{SGS}}^2 = 0.01$ was found to be most appropriate for internal flow calculation. The difference in values of C_{SGS}^2 is attributed to the effect of the mean flow strain or shear. This meant that the behavior of the small eddies is not as universal as has been surmised.

In regions close to a solid wall, the turbulent viscosity can be damped by using a damping function:

$$\mu_{T_{\text{SGS}}}^k = \rho^k C_{\text{SGS}}^2 (f_\mu \Delta)^2 \left|\tilde{S}^k\right| \tag{6.266}$$

The damping function f_μ can be prescribed according to Van Driest (1956):

$$f_\mu = 1 - \exp\left[-0.04y^+\right] \tag{6.267}$$

or formulated by Piomelli et al. (1987):

$$f_\mu = \sqrt{1 - \exp\left[-0.000064(y^+)^3\right]} \tag{6.268}$$

or be of the form proposed by Fulgosi et al. (2003):

$$f_\mu = 1 - \exp\left[-0.00013y^+ - 0.00036\left(y^+\right)^2 - 1.08 \times 10^{-5}\left(y^+\right)^3\right] \tag{6.269}$$

It should be noted that the Smagorinsky–Lilly model has been designed assuming that the simulated flow is turbulent, fully-developed and isotropic, and therefore such a model does not incorporate any information related to an eventual departure of the simulated flow from these assumptions. In order to obtain an automatic adaptation of the model for inhomogeneous flows, simulations of engineering flows are more likely to be based on the dynamic formulation of the model.

Dynamic subgrid scale model

The dynamic procedure proposed by Germano et al. (1991) is one possible approach providing the possibility of calculating a "Germano-optimal" eddy viscosity coefficient that adapts itself to the evolving flow. This particular procedure is based on the application of two different filters. In addition to the grid filter $G(|\mathbf{x}-\mathbf{x}'|)$, a test filter $\widehat{G(|\mathbf{x}-\mathbf{x}'|)}$ is explicitly applied. The test filter width $\hat{\Delta}$ is usually taken to be larger than the grid filter width Δ. For the multifluid approach, the component-weighted volume-averaging test filter operation can be defined by

$$\widehat{\widetilde{\phi(\mathbf{x}',t)}} = \frac{\widehat{\overline{\mathscr{F}^k\phi(\mathbf{x}',t)}}}{\widehat{\overline{\mathscr{F}^k}}} \tag{6.270}$$

and applying the grid filter and subsequently the test filter on the instantaneous momentum equation, the following equation for the kth phase is:

$$\frac{\partial\left(\widehat{\overline{\alpha^k}\rho^k\widetilde{\mathbf{U}^k}}\right)}{\partial t} + \nabla\cdot\left(\widehat{\overline{\alpha^k}}\rho^k\widehat{\widetilde{\mathbf{U}^k}}\otimes\widehat{\widetilde{\mathbf{U}^k}}\right) = -\nabla\left(\widehat{\overline{\alpha^k}\tilde{p}^k}\right) + \nabla\cdot\left(\widehat{\overline{\alpha^k}\tilde{\boldsymbol{\tau}}^k}\right)$$
$$-\nabla\cdot\left(\widehat{\overline{\alpha^k}\mathbf{T}^{k''}}\right) + \widehat{\overline{\alpha^k}}\rho^k\mathbf{g} + \widehat{\tilde{\mathbf{F}}^k} \tag{6.271}$$

where the subtest stresses are given by

$$\mathbf{T}^{k''} = \widehat{\overline{\rho^k}}\widehat{\mathbf{U}^k\otimes\mathbf{U}^k} - \widehat{\overline{\rho^k}}\widehat{\tilde{\mathbf{U}}^k}\otimes\widehat{\tilde{\mathbf{U}}^k} \tag{6.272}$$

If the test filter is now directly applied to the filtered-averaged momentum Eqn (6.260), the equation for the kth phase becomes

$$\frac{\partial\left(\widehat{\overline{\alpha^k}\rho^k\widetilde{\mathbf{U}^k}}\right)}{\partial t} + \nabla\cdot\left(\widehat{\overline{\alpha^k}}\rho^k\widehat{\widetilde{\mathbf{U}^k}}\otimes\widehat{\widetilde{\mathbf{U}^k}}\right) = -\nabla\left(\widehat{\overline{\alpha^k}\tilde{p}^k}\right) + \nabla\cdot\left(\widehat{\overline{\alpha^k}\tilde{\boldsymbol{\tau}}^k}\right)$$
$$-\nabla\cdot\left(\widehat{\overline{\alpha^k}\tilde{\boldsymbol{\tau}}^{k''}}\right) - \nabla\cdot\left(\widehat{\overline{\alpha^k}}\mathbf{L}^{k''}\right) + \widehat{\overline{\alpha^k}}\rho^k\mathbf{g} + \widehat{\tilde{\mathbf{F}}^k} \tag{6.273}$$

where $\mathbf{L}^{k''}$ is the Germano identity. The grid level stress tensor and the test-level tensor are identified by this identity:

$$\mathbf{L}^{k''} = \mathbf{T}^{k''} - \widehat{\boldsymbol{\tau}^{k''}} \tag{6.274}$$

with $\mathbf{L}^{k''}$ and $\widehat{\boldsymbol{\tau}^{k''}}$ given as

$$\mathbf{L}^{k''} = -\left(\widehat{\overline{\rho^k}\widetilde{\mathbf{U}^k}\otimes\widetilde{\mathbf{U}^k}} - \widehat{\overline{\rho^k}\widetilde{\mathbf{U}^k}}\,\widehat{\overline{\rho^k}\widetilde{\mathbf{U}^k}}\Big/\widehat{\overline{\rho^k}}\right) \tag{6.275}$$

$$\widehat{\boldsymbol{\tau}^{k''}} = \widehat{\overline{\rho^k}\mathbf{U}^k\otimes\mathbf{U}^k} - \widehat{\overline{\rho^k}\widetilde{\mathbf{U}^k}}\,\widehat{\overline{\rho^k}\widetilde{\mathbf{U}^k}}\Big/\widehat{\overline{\rho^k}} \tag{6.276}$$

On the basis set out above, the filtered and subtest stresses can thus be represented as

$$\boldsymbol{\tau}^{k''} - \frac{1}{3}\boldsymbol{\tau}^{k''}\delta = -2C_d\Delta^2\left|\widetilde{S^k}\right|\widetilde{\mathbf{S}^k} = C_d\beta_{ij}^k \tag{6.277}$$

$$\mathbf{T}^{k''} - \frac{1}{3}\mathbf{T}^{k''}\delta = -2C_d\hat{\Delta}^2\left|\widehat{\widetilde{S^k}}\right|\widehat{\widetilde{\mathbf{S}^k}} = C_d\alpha_{ij}^k \tag{6.278}$$

where the test filtered strain rate tensor $\widehat{\widetilde{\mathbf{S}^k}}$ is given by

$$\widehat{\widetilde{\mathbf{S}^k}} = \frac{1}{2}\left(\nabla\widehat{\widetilde{\mathbf{U}^k}} + \left(\nabla\widehat{\widetilde{\mathbf{U}^k}}\right)^T\right) - \frac{1}{3}\nabla\cdot\widehat{\widetilde{\mathbf{U}^k}}\delta$$

and the contraction of the strain rate tensor at the test-level is defined as

$$\left|\widehat{\widetilde{S^k}}\right| = \sqrt{2\widehat{\widetilde{\mathbf{S}^k}}\,\widehat{\widetilde{\mathbf{S}^k}}}$$

In Eqns (6.277) and (6.278), C_d represents a coefficient to be determined, which is associated with the respective model constant C_{SGS}^2 in the previous section, and $\hat{\Delta}$ is a test filter associated with the cut-off length scale $\hat{\Delta} > \Delta$, usually taken as $\hat{\Delta} = 2\Delta$. Substituting the filtered and subtest stresses into Eqn (6.274) gives

$$\mathbf{L}^{k''} - \frac{1}{3}\mathbf{L}^{k''}\delta = L_{ij}^{k''} - \frac{1}{3}L_{ll}^{k''}\delta_{ij} = C_d\alpha_{ij}^k - \widehat{C_d\beta_{ij}^k} \tag{6.279}$$

From the above equation, Lilly (1992) suggested a least-squares approach to evaluate the local values of C_d. By assuming that C_d is the same for both filtering operations, the error

$$e_{ij} = \left(L_{ij}^{k''} - \frac{1}{3}L_{ll}^{k''}\delta_{ij}\right) - C_d\alpha_{ij}^k - \widehat{C_d\beta_{ij}^k} \tag{6.280}$$

is minimized by requiring $\partial e_{ij}e_{ij}/\partial C_d = 0$, which yields

$$C_d = \frac{\left(L_{ij}^{k''} - \frac{1}{3}L_{ll}^{k''}\delta_{ij}\right)M_{ij}^{k}}{M_{mn}^{k}M_{mn}^{k}} \tag{6.281}$$

with $M_{ij}^k = \alpha_{ij}^k - \widehat{\beta_{ij}^k}$. The numerator in Eqn (6.281) can attain both positive and negative values. This indicates that the model allows the possibility of accounting for the backscatter of the turbulence energy, which is the energy transferred from the *small eddies* to the *large eddies*. Such occurrences are prevalent in real flows although the long time average energy transport is from the *large eddies* to the *small eddies*.

A negative viscosity has a tendency of causing severe numerical instability and the denominator may become zero, which would make the constant C_d indeterminate. Ghosal et al. (1995) removed the mathematical inconsistency by generalizing the least square method into a constrained variational problem consisting of the minimization of the integral of the error over the entire domain, with the additional constraint that the coefficient be nonnegative. This led to a rigorous problem of solving the Fredholm integral equation of the second kind, which requires the integral to be iteratively solved using under-relaxation to improve convergence. The cost is comparable to that for the Poisson equation for the pressure and can be rather expensive. Piomelli and Liu (1995) developed, however, a simpler constrained model where Eqn (6.279) is recast in the form

$$L_{ij}^{k''} - \frac{1}{3}L_{ll}^{k''}\delta_{ij} = C_d\alpha_{ij}^k - \widehat{C_d^*\beta_{ij}^k} \tag{6.282}$$

where an estimate of the coefficient denoted by C_d^* is assumed to be known. Equation (6.281) can be locally minimized by the following contraction:

$$C_d = \frac{\left(L_{ij}^{k''} - \frac{1}{3}L_{ll}^{k''}\delta_{ij} - \widehat{C_d^*\beta_{ij}^k}\right)\alpha_{ij}^k}{\alpha_{mn}^k\alpha_{mn}^k} \tag{6.283}$$

It is noted that the denominator in the above expression is positive definite. Normally, the coefficient C_d^* can be obtained by either the *zeroth-order* approximation by taking the value at the previous time-step $C_d^* = C_d^{n-1}$ or evaluated using a *first-order* approximation formulated in the form $C_d^* = C_d^{n-1} + \frac{t^n - t^{n-1}}{t^{n-1} - t^{n-2}}(C_d^{n-1} - C_d^{n-2})$.

Comments

1. The dynamic eddy-viscosity model has a tendency to yield a negative viscosity and thus may lead to ill-posed equations—the denominator may become zero, which would make

the constant C_d indeterminate—and unstable simulations. This problem can be solved by some averaging and/or clipping the negative values to Eqn (6.281), for example, in order to dampen large local fluctuations. The averaging can be performed either by plane-averaging along a homogeneous direction or local-averaging over the test filter cell. In complex flows, an average over a small time interval is employed instead. The apparently *ad hoc* averaging procedure that recovers the statistical notion of energy transfer from the resolved to the subgrid scales and removal of negative eddy velocity effectively stabilizes the dynamic model. Nevertheless, this fact still precludes the computation of a fully inhomogeneous flow.

2. A more elegant way of overcoming the problem is to adopt the subgrid scale model of Vreman (2004). This model does not involve any averaging or clipping procedures. A superior feature of this model is its capability to predict zero viscosity in regions where the flow is laminar. The subgrid scale eddy viscosity has been formulated for the single fluid formulation according to

$$\mu_{T_{\mathrm{SGS}}} = \rho C_v \Pi \tag{6.284}$$

where C_v is the model coefficient, which can be related to the Smagorinsky–Lilly constant C_{SGS} by $C_v \approx 2.5 C_{\mathrm{SGS}}^2$, and

$$\Pi = \sqrt{\frac{B_\beta}{\alpha_{ij}\alpha_{ij}}}, \quad \alpha_{ij} = \frac{\partial U_j}{\partial x_i}, \quad \beta_{ij} = \sum_{m=1}^{3} \Delta^2 \alpha_{mi}\alpha_{mj},$$
$$B_\beta = \beta_{11}\beta_{22} - \beta_{12}\beta_{12} + \beta_{11}\beta_{33} - \beta_{13}\beta_{13} + \beta_{22}\beta_{33} - \beta_{23}\beta_{23} \tag{6.285}$$

For the multifluid approach, the subgrid scale eddy viscosity can also be expressed in the same form as Eqns (6.284) and (6.285):

$$\mu_{T_{\mathrm{SGS}}}^k = \rho^k C_v \Pi^k \tag{6.286}$$

where the term Π^k is now given by

$$\Pi^k = \sqrt{\frac{B_\beta^k}{\alpha_{ij}^k \alpha_{ij}^k}}, \quad \alpha_{ij}^k = \frac{\partial U_j^k}{\partial x_i}, \quad \beta_{ij}^k = \sum_{m=1}^{3} \Delta^2 \alpha_{mi}^k \alpha_{mj}^k,$$
$$B_\beta^k = \beta_{11}^k\beta_{22}^k - \beta_{12}^k\beta_{12}^k + \beta_{11}^k\beta_{33}^k - \beta_{13}^k\beta_{13}^k + \beta_{22}^k\beta_{33}^k - \beta_{23}^k\beta_{23}^k \tag{6.287}$$

The dissipation and the exact subgrid dissipation have been shown to vanish for precisely the same class of flows as exemplified in Vreman (2004), especially in transitional and near-wall regions.

6.9 Summary

In the computational multiphase fluid dynamics framework, all solution methods for Eulerian models are required to employ some form of discretization. This entails in the first instance the construction of a suitable mesh. Depending on the complexity of the flow geometry, the problem demands a judicious application of different mesh systems in the form of structured or unstructured elements. The finite volume method is described to derive the appropriate discretized forms of the transport equations governing the conservation of mass, momentum and energy in the Eulerian framework. Numerical approximations of the terms associated with the time derivative, advection, diffusion and source within the transport equations are discussed. Solution algorithms, which require suitable matrix solvers to handle the system of algebraic equations, are described through the use of a segregated or coupled approach. The latter approach is deemed necessary for multiphase flow problems when tight coupling between the phases is required.

In the population balance framework, specific numerical approaches to solve the population balance equation are described. Useful solution methods, such as the class method and method of moments to determine the particle size distribution and ease of coupling with models associated the computational multiphase fluid dynamics framework, are presented.

In the Lagrangian framework, solution methods for the Lagrangian tracking of particles are described not only for the discrete element method but also molecular dynamics and Brownian dynamics. This is because the concept of Lagrangian tracking simply entails following the motion of individual particles through the consideration of the linear and angular momentum equations (*Newton's second law*). In principle, the Lagrangian particle methods primarily differ by the particle interaction laws and the imposition of random forcing in models to mimic collisions or interaction with particles. Appropriate solution methods to aptly capture the physical particle behavior at a certain range of time and length scales, and for specific types of discrete elements, are discussed.

Last, some pertinent aspects of turbulence modeling are introduced. For the multifluid model, the frameworks based on RANS and LES are presented. In RANS, the widely used two-equation k-ε model is considered for multiphase flow problems via a straightforward generalization of its original formulation for single phase flows. The SST for multiphase flow problems is also presented. In LES, appropriate subgrid scale models to resolve the small-scale motions are discussed while the large-scale motions are directly solved depending on a scale the underlying computational mesh will allow.

CHAPTER 7

Some Applications of Population Balance with Examples

7.1 Introduction

During the past few decades, the demand of industrial applications and the extent of model development in resolving the motion of multiphase flows in conjunction with the tracking of swarms of bubbles, drops, and particles has been growing rapidly within the scientific community. Without a doubt, such accomplishments could only have been made possible through the dedication of researchers and code developers in implementing population balance models into a number of computer codes available commercially. Through the fundamental development of the population balance equation in Chapter 3, the cornerstone of any population balance models lies in the heart of mechanistic descriptions of the interactions of bubbles, drops, and particles, as well as the judicious choice of efficient numerical solution techniques. In Chapters 4 and 5, fundamental considerations and associated theories of mechanistic models for different classifications of multiphase flows have been presented. In Chapter 6, some basic concepts of numerical solution techniques for solving the population balance equation have been described.

Through discussions from previous chapters, the mathematical formulation of the population balance framework has been presented in a manner where it is sufficiently generic and not confined to any particular multiphase flow systems. To accurately model any particular multiphase flow system, specific mechanistic models such as coalescence and the break up of fluid particles or aggregation and breakage of solid particles are, nonetheless, required to be implemented in order to achieve an adequate closure for the framework and interpretation of underlying physical phenomena that would occur within the particular multiphase system of interest.

Having understood the many theoretical important aspects associated with the computational multiphase fluid dynamics and population balance frameworks and the variety of solution methods, it is therefore fitting that this chapter draws together the different strands of knowledge gathered by aptly describing the dominant physical phenomena of typical multiphase flow systems and, more importantly, the application and conversion of theory into practice. From a practical viewpoint, it is envisaged that selected examples in the subsequent sections will assist in establishing some concrete steps toward modeling population balance problems that are frequently encountered in different classifications of multiphase flows.

7.2 Population Balance Solutions to Gas–Liquid Flow

7.2.1 Background

Gas–liquid flow can be found in many natural and industrial processes and often features complex interphase mass, momentum, and energy transfers. One typical example of a natural process is the application of water plumes in dealing with environment protection problems such as aeration of lakes, mixing of stagnant water and destratification of water reservoirs. To promote sufficient three-dimensional mixing in the system, it is essential to determine the interfacial momentum transfer and better understand currents induced by gas bubbles evolving in the surrounding liquid, thereby establishing the consequent mixing and partition of energy in the body of the liquid.

One significant industrial process is the venting of mixture vapors to liquid pools in chemical reactors. Here, bubble column reactors are employed in many biochemical and petrochemical industries. Such reactors are known as excellent systems for processes which require large interfacial area for gas–liquid mass transfer and efficient mixing for reacting species due to a host of gas–liquid reactions (oxidation, hydrogenation, halogenation, aerobic fermentation, etc.). In bubble column reactors, the size of gas bubbles is an important parameter influencing their performance. It determines the bubble rise velocity and the gas residence time, which in turn governs the gas hold up, the interfacial area, and subsequently the gas–liquid mass transfer rate. More significantly, the prevalence of bubble–bubble interaction (e.g. coalescence and break up mechanisms as discussed in Chapter 4) can profoundly influence the overall performance by altering the interfacial area that is available for mass transfer between the phases. For practical engineers or designers, the challenging task is to ensure the reactor/system is always running under optimal/efficient flow conditions, which requires a thorough understanding of the evolution of bubble size distribution and its associated interfacial mass, momentum, or even heat transfer behavior. Modeling of bubble mechanistic behavior using the population balance approach represents a crucial aspect in the rational design of bubble column reactors.

7.2.2 Modeling Interfacial Momentum Transfer for Gas–Liquid Flow

The physical characteristics of different flow regimes and associated bubble shape categories in a co-current flow of air/water within a vertical pipe are shown in Figure 7.1. At low gas volume fractions, the flow is an amalgam of individual ascending gas bubbles co-flowing with the liquid. This flow regime, referred to as the bubbly flow regime, can be further subdivided into two subregimes—bubble flow at low liquid flow rates and dispersed bubble flow at high liquid flow rates. As the gas volume fraction increases, a pattern is exhibited whereby slugs of highly aerated liquid move upward along the pipe. These so-called Taylor bubbles have characteristics of a spherical cap nose and are somewhat

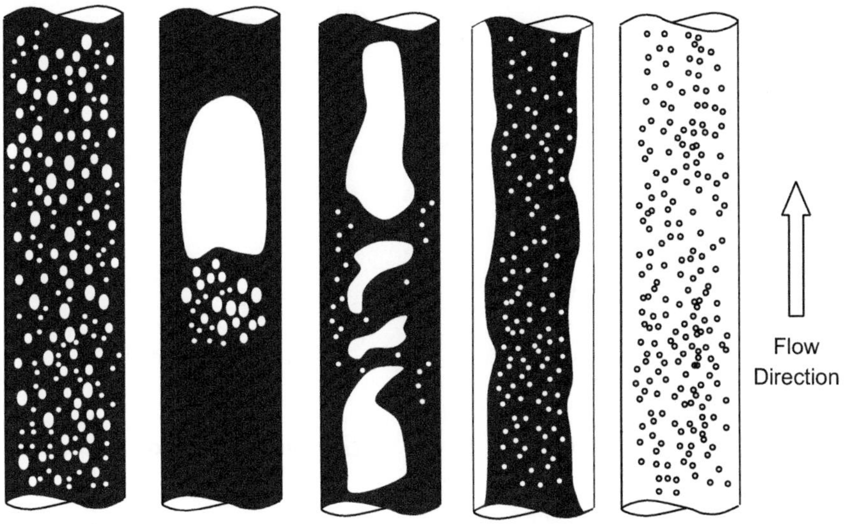

Figure 7.1
Flow regimes for air/water in a vertical pipe.

abruptly terminated at the bottom edge. The elongated gas bubbles are separated by liquid slugs which may have smaller bubbles near the skirt. Sizes of the slug units, Taylor bubbles, and liquid slugs may vary considerably. This slug pattern is often avoided in the design because it causes undesirable flow instability. Subsequently, large unsteady volumes of gas accumulate within these mixing motions and produce the flow regime known as churn-turbulent flow with increasing volume fractions. Here, liquid may be flowing up and down in an oscillatory fashion. At very high gas velocities, an annular pattern is observed whereby parts of the liquid flow along the pipe and other parts as droplets become entrained in the gas flow. In the gas–liquid interface, especially for sufficient high gas velocity, there may be large amplitude waves that break up during the flow process. The breaking of these waves is the continuous source of the deposition of droplets in the gas or vapor core. At even higher gas velocities, a dispersed pattern exists. There is now a considerable amount of liquid in the gas core.

For different flow regimes, it is clear that the topological distribution of bubbles and its interface with the liquid exhibit substantial variations which eventually dictate the resultant interfacial momentum transfer. Figure 7.2 illustrates some typical interfacial forces acting on bubbles, which can have a substantial effect on the phase distribution within the flow system. These interfacial forces create the deformable interface and dictate the shape of bubbles, as well as contacting local turbulent flow structure at microscopic level. It remains impractical to fully resolve all of these forces, which generally requires tracking all interfaces of millions of bubbles. In Chapter 2, interfacial forces in the interpenetrating media framework can be

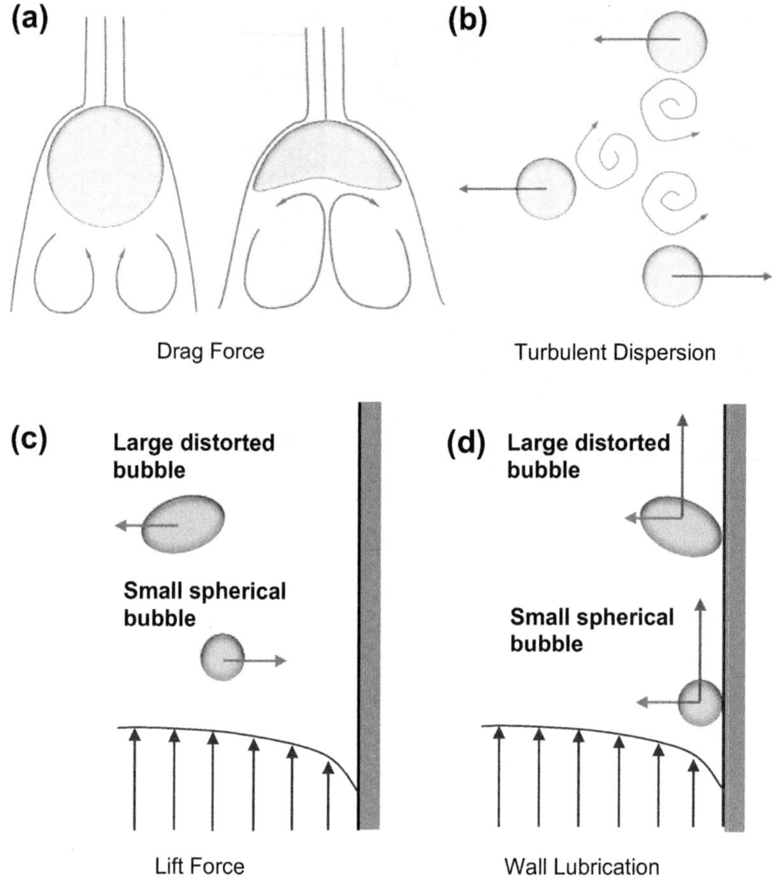

Figure 7.2
Typical interfacial forces acting on bubbles: (a) drag force, (b) turbulent dispersion force, (c) lift force, and (d) wall lubrication force.

expressed as sinks or sources in the momentum balance equations of the multifluid model. Inclusion of appropriate interfacial forces is crucial to the modeling of gas–liquid flow. The physical behavior of different interfacial forces and its formulation are discussed below.

Drag force

Unlike solid particles, gas bubble interfaces could deform as a result of the shear and form drag of the external fluid motions (Figure 7.2(a)). This can be modeled according to

$$\mathbf{F}_{\mathrm{D}}^{\mathrm{c,drag}} = -\mathbf{F}_{\mathrm{D}}^{\mathrm{p,drag}} = \frac{1}{8} C_{\mathrm{D}} a_{\mathrm{if}} \rho^{\mathrm{c}} |\mathbf{U}^{\mathrm{p}} - \mathbf{U}^{\mathrm{c}}| (\mathbf{U}^{\mathrm{p}} - \mathbf{U}^{\mathrm{c}}) \tag{7.1}$$

where C_{D} is the drag coefficient and a_{if} is the interfacial area concentration. Obviously, spherical bubbles would be subjected to different interfacial drag force from cap bubbles.

It is not surprising that the drag coefficients of a bubble should be determined in accordance with its shape and flow regime. The drag coefficients based on the correlations by Ishii and Zuber (1979) for different flow regimes are normally employed for gas–liquid flows. The function $C_D(\mathrm{Re_b})$, known as the drag curve, can be correlated for individual bubbles across several distinct regions based on the bubble's Reynolds number: $(\mathrm{Re_b} = \alpha^c \rho^c |.\mathbf{U}^p - \mathbf{U}^c|.D_s/\mu^c)$

Stokes region $(0 \le \mathrm{Re_b} < 0.2)$

$$C_D = \frac{24}{\mathrm{Re_b}} \tag{7.2}$$

Viscous region $(0 \le \mathrm{Re_b} < 1000)$

$$C_D = \frac{24}{\mathrm{Re_b}}\left(1 + 0.1\mathrm{Re_b^{0.75}}\right) \tag{7.3}$$

Turbulent region $(\mathrm{Re_b} \ge 1000)$

$$C_D = 0.44\,E \quad \text{Newton regime} \tag{7.4}$$

$$C_D = \frac{2}{3}\sqrt{\mathrm{Eo}}\,E \quad \text{Distorted regime} \tag{7.5}$$

$$C_D = \frac{8}{3}E' \quad \text{Churn–turbulent regime} \tag{7.6}$$

To account for the deformed bubble interface, the Ishii and Zuber modification for Newton and distorted regimes takes the form of the Eotvos number, which is defined by

$$\mathrm{Eo} = \frac{g(\rho^c - \rho^p)D_s^2}{\sigma} \tag{7.7}$$

where ρ^p is the density of the dispersed (gas) phase, σ is the surface tension coefficient, and D_s is the Sauter bubble diameter. Furthermore, a multiplying factor, E, in terms of the void fraction, is also introduced

$$E = \left[\frac{1 + 17.67(1 - \alpha^p)^{6/7}}{18.67(1 - \alpha^p)}\right]^2 \tag{7.8}$$

where α^p is the volume fraction of the disperse phase. For churn-turbulent flow, the multiplication factor E' takes the form

$$E' = \left(1 - \alpha^d\right)^2 \tag{7.9}$$

In the case of separated flows (annular or stratified), the interfacial drag force can also be modeled by the same expression as in Eqn (7.1). A different expression for interphase drag is nonetheless given by

$$\mathbf{F}_{\mathrm{D}}^{\mathrm{c,drag}} = -\mathbf{F}_{\mathrm{D}}^{\mathrm{p,drag}} = \frac{1}{8} f\, a_{\mathrm{if}} \rho^{\mathrm{c}} |\mathbf{U}^{\mathrm{p}} - \mathbf{U}^{\mathrm{c}}| (\mathbf{U}^{\mathrm{p}} - \mathbf{U}^{\mathrm{c}}) \tag{7.10}$$

where f is the interfacial friction factor. For stratified flows, f may be prescribed according to a constant wall friction coefficient or determined as a function of void fraction, liquid Reynolds number, and gas Reynolds number to account for wave roughness and hydrodynamic conditions. For annular flows, f for the liquid film can be described by a standard laminar correlation based on Wallis (1969) in the turbulent region. It is exclusively correlated in terms of the gas Reynolds number and average volume fraction of the liquid film along the wall.

Turbulent dispersion force

In most practical systems, turbulent flow characteristics of both gas and liquid phases also play a significant role in the phase distribution and population balance of bubbles. Under the influence of turbulent eddies, as discussed in Chapter 4, bubbles could coalesce via random collision or break up caused by turbulent shearing. On the other hand, turbulent eddies could also assist in the dispersion of bubbles, resulting in a more even phase distribution throughout the domain (Figure 7.2(b)). To consider such turbulent assisted bubble dispersion, the turbulence dispersion force taken as a function of turbulent kinetic energy in the continuous phase and the gradient of the volume fraction can be expressed in the form according to Antal et al. (1991) as

$$\mathbf{F}_{\mathrm{D}}^{\mathrm{c,dispersion}} = -\mathbf{F}_{\mathrm{D}}^{\mathrm{p,dispersion}} = C_{\mathrm{TD}} \rho^{\mathrm{c}} k^{\mathrm{c}} \nabla \alpha^{\mathrm{c}} \tag{7.11}$$

Values of constant C_{TD} ranging from 0.1 to 0.5 have been employed successfully for bubbly flows with diameters of the order of millimeters. In some situations, values up to 500 have been required (Lopez de Bertodano, 1998; Moraga et al., 2003). However, Burns et al. (2004) derived an alternative model for the turbulence dispersion force based on the consistency of Favre averaging, which is given by

$$\mathbf{F}_{\mathrm{D}}^{\mathrm{c,dispersion}} = -\mathbf{F}_{\mathrm{D}}^{\mathrm{p,dispersion}} = C_{\mathrm{TD}} C_{\mathrm{D}} \frac{\mu_{\mathrm{T}}^{\mathrm{p}}}{\rho^{\mathrm{p}} \mathrm{Sc}_{\mathrm{b}}} \left(\frac{\nabla \alpha^{\mathrm{p}}}{\alpha^{\mathrm{p}}} - \frac{\nabla \alpha^{\mathrm{c}}}{\alpha^{\mathrm{c}}} \right) \tag{7.12}$$

where C_{TD} is normally set to a value of unity, $\mu_{\mathrm{T}}^{\mathrm{p}}$ is the turbulent viscosity of the dispersed phase, and Sc_{b} is the turbulent bubble Schmidt number with an adopted value of 0.9. In the above equation, the constant C_{D} depicts the drag coefficient which essentially describes the interfacial drag force. This model therefore clearly depends on the details of the drag characteristics of the gas–liquid systems. For situations where an appropriate value of C_{TD} is not readily obtained through the turbulent dispersion force in Eqn (7.11), the Favre-averaged turbulent dispersion force formulated in Eqn (7.12) is recommended.

Lift force

For vertical gas–liquid flows, in addition to drag force that predominantly governs bubble slip velocity in the axial direction, bubbles rising in liquid are also subjected to various lateral forces which are commonly referred to as nondrag forces. The lift force is one of the many forces which has been shown to have a profound influence on the phase distribution and transition of flow regime. It is commonly believed that the lift force is induced by the horizontal velocity gradient, albeit the mechanism is yet to be fully understood. This interfacial force density is thereby correlated with the slip velocity and local vorticity of the continuous phase (curl of the velocity vector), which acts perpendicular to the direction of relative motion between two phases

$$\mathbf{F}_{\mathrm{D}}^{\mathrm{c,lift}} = -\mathbf{F}_{\mathrm{D}}^{\mathrm{p,lift}} = C_{\mathrm{L}}\alpha^{\mathrm{p}}\rho^{\mathrm{c}}(\mathbf{U}^{\mathrm{p}} - \mathbf{U}^{\mathrm{c}}) \times (\nabla \times \mathbf{U}^{\mathrm{c}}) \tag{7.13}$$

For the lift coefficient C_{L} in the above equation, Lopez de Bertodano (1992) and Takagi and Matsumoto (1998) suggested a value of $C_{\mathrm{L}} = 0.1$. Drew and Lahey (1979) proposed $C_{\mathrm{L}} = 0.5$ based on the objectivity argument for an inviscid flow around a sphere. The constant of $C_{\mathrm{L}} = 0.01$, as suggested by Wang et al. (1987), has been found to be appropriate for viscous flows. Nonetheless, based on experimental observations (Lucas et al., 2007) and numerical simulation (Bothe et al., 2006), it has been discovered that the direction of the lateral lift force is sensitive to the bubble size. Consequently, small bubbles driven by positive lift forces are separated from those opposite directed large bubbles, which migrate toward the center of the pipe. As a result, large distorted bubbles are gathered at the center of the pipe center which encourage more bubble coalescence, forming a cap or even larger Taylor bubbles, leading to flow transition from bubbly flow to slug flow. Adopting a constant lift coefficient is, thereby, inappropriate to predict the shift of bubble migration.

To consider the change of lift force subject to bubble size variation, Tomiyama (1998) developed an Eotvos number dependent correlation that allows negative coefficients to be realized in the case where the bubble size is sufficiently large, which results in a negative lateral lift force forcing large bubbles to be migrated toward the center of the flow channel. For air–water system under room temperature, the correction gives negative coefficient when the bubble size is larger than 5.8 mm. The lift coefficient can be expressed as

$$C_{\mathrm{L}} = \begin{cases} \min\left[0.288 \tan\mathrm{h}(0.121\mathrm{Re}_{\mathrm{b}}), f(\mathrm{Eo}_{\mathrm{p}})\right] & \mathrm{Eo} < 4 \\ f(\mathrm{Eo}_{\mathrm{d}}) = 0.00105\mathrm{Eo}_{\mathrm{p}}^3 - 0.0159\mathrm{Eo}_{\mathrm{p}}^2 - 0.0204\mathrm{Eo}_{\mathrm{p}} + 0.474 & 4 \le \mathrm{Eo} \le 10 \\ -0.29 & \mathrm{Eo} > 10 \end{cases} \tag{7.14}$$

where the modified Eotvos number $\mathrm{Eo_p}$ is defined by

$$\mathrm{Eo_p} = \frac{g(\rho^{\mathrm{c}} - \rho^{\mathrm{p}})D_{\mathrm{H}}^2}{\sigma} \tag{7.15}$$

in which D_{H} in the equation is the maximum bubble horizontal dimension that can be evaluated through the empirical correlation of Wellek et al. (1966).

$$D_{\mathrm{H}} = D_{\mathrm{s}}\left(1 + 0.163\mathrm{Eo}^{0.757}\right)^{1/3} \tag{7.16}$$

This particular correlation has been widely adopted, with encouraging results reported in the literature.

Wall lubrication force

Another significant non-drag lateral force in a gas–liquid flow system is the wall lubrication force (Figure 7.2(d)). Experiments have found that bubbles concentrate in the near wall region; however, they do not attach to the wall surface. This results in a low void fraction at the vicinity of the wall area. The wall lubrication force is caused by the surface tension of bubbles, which tends to push the bubble away from the wall. The force is sensitive to the bubble size, where large bubbles result in a stronger force counteracting the lift force. According to Antal et al. (1991), this force can be modeled as

$$\mathbf{F}_{\mathrm{D}}^{\mathrm{c,lubrication}} = -\mathbf{F}_{\mathrm{D}}^{\mathrm{d,lubrication}} = -\frac{\alpha^{\mathrm{p}}\rho^{\mathrm{c}}\left[(\mathbf{U}^{\mathrm{p}} - \mathbf{U}^{\mathrm{c}}) - ((\mathbf{U}^{\mathrm{p}} - \mathbf{U}^{\mathrm{c}}) \cdot \mathbf{n}_{\mathrm{w}})\mathbf{n}_{\mathrm{w}}\right]^2}{D_{\mathrm{s}}} \times \underbrace{\left(C_{\mathrm{w1}} + C_{\mathrm{w2}}\frac{D_{\mathrm{s}}}{y_{\mathrm{w}}}\right)}_{C_{\mathrm{w}}}\mathbf{n}_{\mathrm{w}} \tag{7.17}$$

where y_{w} is the distance from the wall boundary and $\mathbf{n}_{\mathrm{w}}$ is the outward vector normal to the wall. The wall lubrication constants determined through numerical experimentation for a sphere are $C_{\mathrm{w1}} = -0.01$ and $C_{\mathrm{w2}} = 0.05$. Following a recent proposal by Krepper et al. (2005), the model constants have been modified according to $C_{\mathrm{w1}} = -0.0064$ and $C_{\mathrm{w2}} = 0.016$. To avoid the emergence of attraction force, the force is set to zero for large y_{w}.

Virtual mass force

The virtual mass or added mass force arises because of the acceleration of the gas bubbles, which could become significant under certain flow conditions, particularly for systems with strong accelerating liquid flows. The resultant virtual mass force is generally taken to be proportional to the relative phase acceleration, which can be expressed as

$$\mathbf{F}_{\mathrm{D}}^{\mathrm{c,virtual\ mass}} = -\mathbf{F}_{\mathrm{D}}^{\mathrm{p,virtual\ mass}} = \alpha^{\mathrm{p}}\rho^{\mathrm{c}}C_{\mathrm{VM}}\left(\frac{D\mathbf{U}^{\mathrm{p}}}{Dt} - \frac{D\mathbf{U}^{\mathrm{c}}}{Dt}\right) \tag{7.18}$$

where D/Dt is the material derivative. The virtual mass effect is significant when the dispersed phase density is much smaller than the continuous phase density. For an inviscid flow around an isolated sphere, the constant C_{VM} is taken to be equivalent to 0.5. Nevertheless, this constant is highly dependent on shape and concentration and could be modified by further multiplying a factor E'' with C_{VM} in order to account for the effect of surrounding bubbles, which is given by Zuber (1964)

$$E'' = \frac{1 + 2\alpha^{p}}{1 - \alpha^{p}} \tag{7.19}$$

It should be noted that the virtual mass force is dependent on the fluid acceleration. The term will vanish if the acceleration approaches zero. In other words, virtual mass force becomes negligible for steady-state simulations.

7.2.3 Worked Examples

Specific approaches to resolve isothermal bubbly flows using the two-fluid formulation based on the interpenetrating media framework and a range of population balance models are discussed via relevant worked examples described below. All numerical results presented have been computed through the use of the commercial computer code of ANSYS-CFX.

Dispersed bubbly flows in vertical pipes

Two population balance approaches, grounded on the "one-group" approach based on the transport equation for the average bubble number density and MUSIG model for predicting the bubble size distribution of gas–liquid bubbly flows under isothermal conditions, are demonstrated in this worked example. Experimental data of isothermal gas–liquid bubbly flow in a vertical pipe as carried out by Liu and Bankoff (1993a, b) and Hibiki et al. (2001) are utilized to appropriately assess the relative merits of both approaches.

A schematic diagram of the experimental setup of Liu and Bankoff (1993a, b) is shown in Figure 7.3(a). The test section is a 2800 mm long, vertical acrylic pipe with an internal diameter $D = 38$ mm. Bubbles are produced and injected into the test section at the bottom. Local radial measurements are obtained at the axial location $z/D = 36.0$. Liquid velocity is attained by using hot-film anemometers, while the local void fraction and the gas velocity are measured by a two-point resistivity probe. A total of 48 flow conditions covering the range of superficial gas velocities of 0.027–0.347 m/s and superficial liquid velocities of 0.376–1.391 m/s is investigated. The bubble diameters are controlled in a narrow range of 2–4 mm during the experiments.

Similar to the setup of Liu and Bankoff (1993a), a separate isothermal air–water flow system in the experiment conducted by Hibiki et al. (2001) comprises an acrylic round pipe test

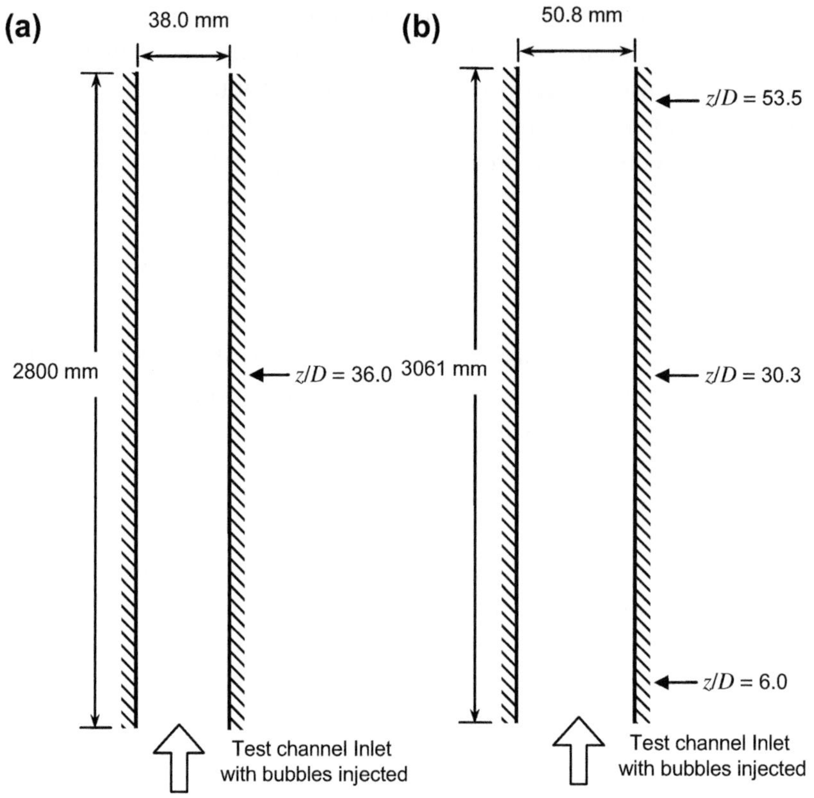

Figure 7.3
Schematic drawing of experimental test sections: (a) Liu and Bankoff (1993a) and (b) Hibiki et al. (2001).

section with an inner diameter $D = 50.8$ mm and a length of 3061 mm such as shown in Figure 7.3(b). Local flow measurements using the double sensor and hot-film anemometer probes are carried out at three axial (height) locations of $z/D = 6.0$, 30.3 and 53.5, and 15 radial locations of $r/R = 0$ to 0.95. Experiments are performed for a range of superficial liquid and gas velocities, which cover most of bubbly flow regions including finely dispersed bubbly flow and bubbly-to-slug transition flow regions.

The primary aim in this worked example is to compare two population balance approaches for simulating the bubbly flow regime. Six flow conditions are investigated (Figure 7.4). The solid lines represent the different flow regime transition boundaries based on the model by Taitel et al. (1980). All six flow conditions lie within the bubbly flow region. Details of the flow conditions are summarized in Table 7.1. For ease of discussion, experiments by Liu and Bankoff (1993a) and Hibiki et al. (2001) are hereafter referred as Exp. 1 and Exp. 2, respectively.

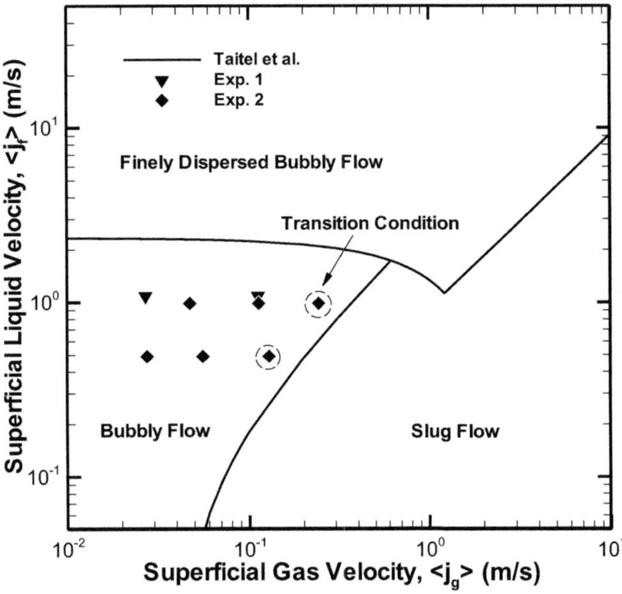

Figure 7.4
Map of flow regime and the bubbly flow conditions studied in this worked example.

Numerical features

In modeling isothermal gas—liquid flow, the two-fluid formulation model, which is based on solving two sets of equations governing the conservation of mass and momentum, is adopted. Details of the model can be found in Chapter 2. The advection term is discretized using a second-order scheme weighted between central and upwind differencing while the diffusion term is approximated according to the second-order central differencing scheme. For the time derivative term, approximation is done via a second order backward Euler time discretization scheme. In accounting for bubble induced turbulent flow, numerical investigation has revealed that the standard $k-\varepsilon$ model has a tendency of predicting an unrealistically high gas void fraction peak close to the wall (Frank et al., 2004; Cheung et al., 2007). The shear stress transport (SST) model by Menter (1993) has been shown to provide more realistic prediction of gas volume fraction or void fraction close to the wall. The SST model is thereby employed herein. For all flow conditions, reliable convergence is achieved within 2500 iterations when the root mean square (rms) pressure residual drops below 1.0×10^{-7}. A fixed physical time scale of 0.002 s is employed for all steady-state simulations.

As aforementioned, interfacial momentum exchange in isothermal bubbly flow exhibits a dominant effect on the two-phase flow behavior. The interfacial forces considered are due to drag, lift, wall lubrication, and turbulence dispersion forces. First, the drag force resulting from shear and form drag is modeled according to Ishii and Zuber (1979) in which the drag coefficient C_D is evaluated based upon the correlation of several distinct Reynolds number

Table 7.1: Bubbly flow conditions and inlet boundary conditions.

Superficial Liquid Velocity, $\langle j_f \rangle$(m/s)	Superficial Gas Velocity, $\langle j_g \rangle$(m/s)	
Liu and Bankoff (1993a) Experiment		
1.087	0.0270	0.1120
$[\alpha_g\|_{z/D=0.0}(\%)]$	[2.5]	[10.0]
$[D_s\|_{z/D=0.0}(mm)]$	[3.0]	[3.0]
Hibiki et al. (2001) Experiment		
0.491	0.0275	0.0556
$[\alpha_g\|_{z/D=0.0}(\%)]$	[5.0]	[10.0]
$[D_s\|_{z/D=0.0}(mm)]$	[2.5]	[2.5]
0.986	0.0473	0.1130
$[\alpha_g\|_{z/D=0.0}(\%)]$	[5.0]	[10.0]
$[D_s\|_{z/D=0.0}(mm)]$	[2.5]	[2.5]

regions. Second, the rise of bubbles in a liquid that are subjected to a lateral lift force is correlated to the relative velocity and the local liquid vorticity. For the lift coefficient, C_L, the Eotvos number dependent correlation proposed by Tomiyama (1998) is adopted. Third, the wall lubrication force, which is due to surface tension is adopted to prevent bubbles from attaching on the solid walls, thereby resulting in a low gas void fraction in the vicinity of the wall area. Modeled according to Antal et al. (1991), the model constants of $C_{w1} = -0.0064$ and $C_{w2} = 0.016$ are utilized. Fourth, the turbulent dispersion force expression in terms of Farve-averaged variables proposed by Burns et al. (2004) is employed. By default, the turbulent dispersion coefficient $C_{TD} = 1$ and the turbulent Schmidt number $\sigma_{t,g} = 0.9$ are adopted.

For the single transport equation of the average number density, three forms of coalescence and break up mechanisms by Wu et al. (1998), Hibiki and Ishii (2002) and Yao and Morel (2004) are employed while for the MUSIG model, the coalescence and break up mechanisms by Prince and Blanch (1990) and Luo and Svendsen (1996) are adopted. The transport equation for the average bubble number density transport with appropriate sink or source terms (Table 7.2) describing the coalescence and break up rates of bubbles is implemented within the ANSYS-CFX code. The in-built MUSIG model is applied for the computer simulations. In the present study, bubbles ranging from 0 to 10 mm diameter are equally divided into ten size groups (Table 7.3). Based on these two population balance approaches, the bubble Sauter mean diameter is obtained and used to determine the appropriate interfacial forces in the balance momentum equations.

Numerical simulation

Numerical simulations are performed on a 60° radial sector of the pipe with symmetry boundary conditions at both vertical sides. At the inlet of the test section, as the diameters of

Table 7.2: Coalescence and break up rates for the one-group approach.

$$S_{N_1} = \underbrace{\Phi_1^{RC}}_{\text{Coalescence}} + \underbrace{\Phi_1^{TI}}_{\text{Break-up}}$$

Wu et al. (1998)	$\Phi_1^{RC} = -C_{RC1}\dfrac{(\alpha^P)^2\varepsilon^{1/3}}{D_s^{11/3}\alpha_{\max}^{1/3}\left(\alpha_{\max}^{1/3}-(\alpha^P)^{1/3}\right)}\left[1-\exp\left(-C_{RC2}\dfrac{\alpha_{\max}^{1/3}(\alpha^P)^{1/3}}{\alpha_{\max}^{1/3}-(\alpha^P)^{1/3}}\right)\right]$ $\Phi_1^{TI} = C_{TI}\dfrac{\alpha^P(\varepsilon^c)^{1/3}}{D_s^{11/3}}\left(1-\dfrac{We_{cr}}{We}\right)\exp\left(-\dfrac{We_{cr}}{We}\right)$ $C_{RC1}=0.021,\ C_{RC2}=3.0,\ C_{TI}=0.0945,\ We_{cr}=2.3,\ \alpha_{\max}=0.8$
Hibiki and Ishii (2002)	$\Phi_1^{RC} = -C_{RC1}\dfrac{(\alpha^P)^2(\varepsilon^c)^{1/3}}{D_s^{11/3}(\alpha_{\max}-\alpha^P)}\exp\left(-C_{RC2}\dfrac{(\rho^l)^{1/2}(\varepsilon^c)^{1/3}D_s^{5/6}}{\sigma^{1/2}}\right)$ $\Phi_1^{TI} = C_{TI1}\dfrac{(\alpha^P)^2(1-\alpha^P)\varepsilon^{1/3}}{D_s^{11/3}(\alpha_{\max}-\alpha^P)}\exp\left(-C_{TI2}\dfrac{\sigma}{\rho^c(\varepsilon^c)^{2/3}D_s^{5/3}}\right)$ $C_{RC1}=0.03,\ C_{RC2}=1.29,\ C_{TI1}=0.03,\ C_{TI2}=1.37,\ \alpha_{\max}=0.8$
Yao and Morel (2004)	$\Phi_1^{RC} = -C_{RC1}\dfrac{(\alpha^P)^2(\varepsilon^c)^{1/3}}{D_s^{11/3}}\dfrac{\exp(-C_{RC2}\sqrt{We/We_{cr}})}{(\alpha_{\max}^{1/3}-\alpha^P)/\alpha_{\max}^{1/3}+C_{RC3}\alpha_1^g\sqrt{We/We_{cr}}}$ $\Phi_1^{TI} = C_{TI1}\dfrac{\alpha^P(1-\alpha^P)(\varepsilon^c)^{1/3}}{D_s^{11/3}}\dfrac{\exp(-We_{cr}/We)}{1+C_{TI2}(1-\alpha^P)\sqrt{We/We_{cr}}}$ $C_{RC1}=2.86,\ C_{RC2}=1.017,\ C_{RC3}=1.922,\ C_{TI1}=1.6,\ C_{TI2}=0.42,$ $We_{cr}=1.42,\ \alpha_{\max}=0.52$

Table 7.3: Diameters of each discrete bubble class for MUSIG model.

Class No.	Central Class Diameter, $d_{p,i}$ (mm)
1	0.5
2	1.5
3	2.5
4	3.5
5	4.5
6	5.5
7	6.5
8	7.5
9	8.5
10	9.5

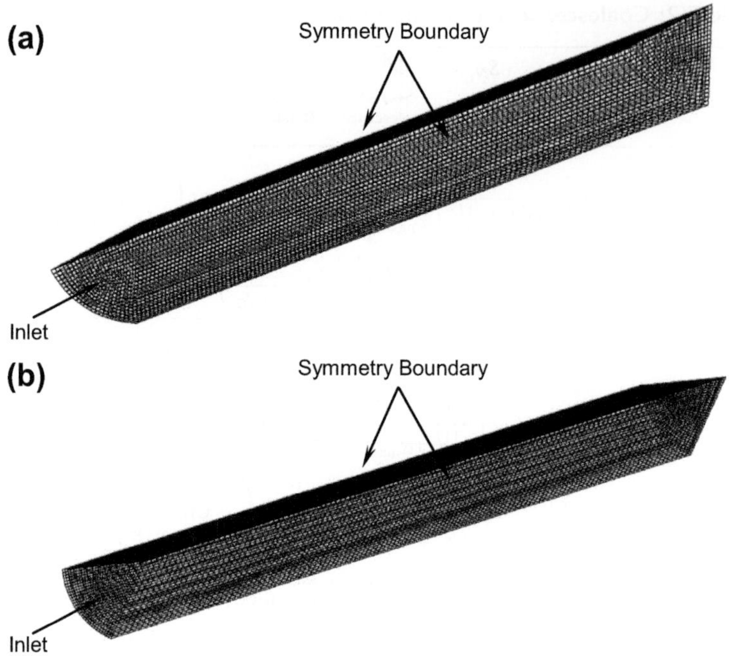

Figure 7.5
Mesh distribution of computational models: (a) Liu and Bankoff (1993a) and (b) Hibiki et al. (2001).

the injected bubbles are unknown, uniformly distributed superficial liquid and gas velocities, void fraction, and bubble size are specified in accordance with the flow condition described. Details of the boundary conditions can be referred to in Table 7.1. At the pipe outlet, a relative averaged static pressure of zero is specified. A three-dimensional mesh containing hexagonal elements is generated over the entire pipe domain. Mesh distribution of the computational models for the two test sections is depicted in Figure 7.5.

Six mesh structures, corresponding to coarse, medium, and fine with three different mesh levels, are tested covering the range of 4000–69,120 elements for Exp. 1 and 4000–108,100 elements for Exp. 2 (Table 7.4). Comparing the predicted results between the medium and fine mesh, small discrepancies are observed. The maximum differences between these two mesh levels in the two experimental flow conditions are less than 5%. It can, therefore, be concluded that the fine mesh level is sufficient for obtaining grid independent solutions. Hereafter, predicted results are all obtained based upon the fine mesh.

Numerical results

Based on the experiments performed by Hibiki et al. (2001), insignificant development of the bubble Sauter mean diameter has been observed along the axial direction. From a

Table 7.4: Details of numerical meshes adopted for the grid sensitivity study.

	Liu and Bankoff (1993a)		Hibiki et al. (2001)	
	$L \times W \times H$	Total	$L \times W \times H$	Total
Coarse	10 × 10 × 40	4000	10 × 10 × 40	4000
Medium	20 × 20 × 40	16,000	26 × 26 × 80	54,080
Fine	24 × 24 × 120	69,120	30 × 30 × 120	108,000

phenomenological viewpoint, this implies that the coalescence and break up rates of bubbles attain near equilibrium condition. For maintaining a balance between these terms, it is imperative that the coalescence rate is reduced by a factor of 1/10 in the transport equation of the average bubble number density. Similarly, coalescence and break up calibration factors with values equal to 0.05 and 1.0 are introduced into the coalescence and break up rates for the MUSIG model. One plausible explanation for this discrepancy could be attributed to the lack of resolution of the SST model in predicting the turbulence energy dissipation under the two-phase flow condition. It should be emphasized that the reduction and calibration factors are introduced herein by the mere means used for engineering estimation, which may be case sensitive and subject to flow conditions. Although adjustment of the reduction and calibration factors could invariably obtain "better" results, it would lose the predictive nature of the models and the common ground for comparison. Therefore, values of these factors are fixed for all the cases and flow conditions that are presented next.

Experimental data of Liu and Bankoff (1993a, b)

Figure 7.6 illustrates the void fraction distributions obtained from the three coalescence and break up mechanisms employed in the transport equation for the average bubble number density and the MUSIG model comparing with the measured data at the dimensionless axial position $z/D = 36.0$. From a physical viewpoint, the phase distribution patterns along the radial direction of the bubble column correspond to four basic types of distributions: wall peak, intermediate peak, core peak, and transition, such as categorized by Serizawa and Kataoka (1988).

In the bubbly flow regime, the maximum void fraction located close to the wall demonstrates the flow phase distributions typically known as the wall peak behavior, which is mainly due to the positive lift force pushing the small bubbles toward the pipe wall. As depicted in Figure 7.6, a well-developed wall peaking behavior is recorded in the experiment and successfully captured by both models. In the case of low gas superficial velocity (i.e. $[\alpha_g] = 2.5\%$, Figure 7.6(a)), all the coalescence and break up bubble mechanistic models of the average bubble number density approach underestimate the void fraction at the core of the pipe. In contrast, the MUSIG model provides a closer prediction with the experiment. However, void fractions at the core of the high gas superficial velocity case as shown in

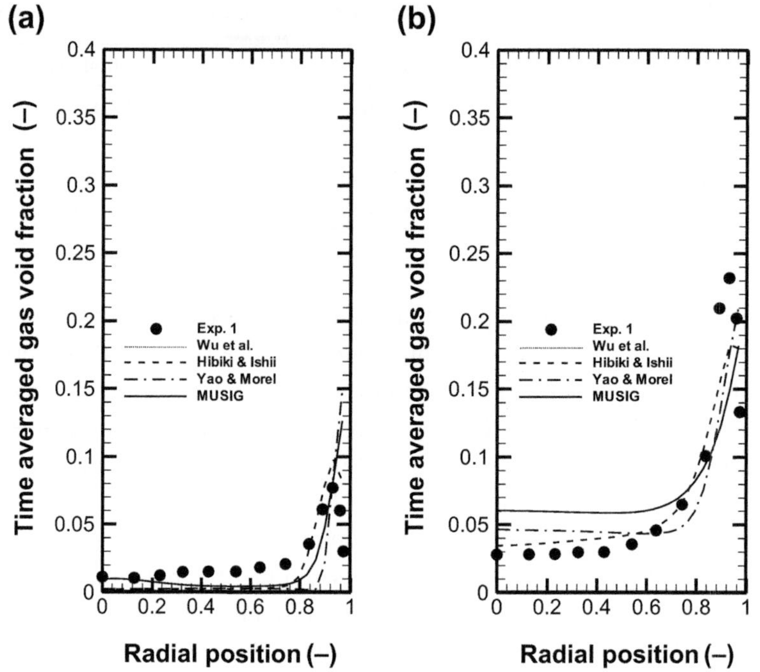

Figure 7.6

Predicted radial void fraction distribution at $z/D = 36.0$ and experimental data of Liu and Bankoff (1993b): (a) Low gas superficial velocity and (b) High gas superficial velocity.

Figure 7.6(b) are slightly over-predicted by all models. One possible reason for the over-prediction of the void fraction distribution could be due to the uncertainties associated with the application of the turbulence model, which is unable to adequately predict the appropriate values of turbulent energy dissipation and which subsequently affect the bubbles' coalescence or break up rate.

The measured and predicted radial profiles of the liquid velocity are presented in Figure 7.7. In contrast to a single-phase flow, the introduction of bubbles into the liquid flow has the tendency to enhance or reduce the liquid flow turbulence intensity as indicated by Serizawa and Kataoka (1990). In the case of enhanced turbulence, such as depicted in Figure 7.7(b), the liquid velocity profile at the core is flattened by the additional turbulence while having a relatively steep decrease almost mimicking a step change close to the pipe wall. Since the recorded liquid velocity at the wall is not zero, the numerical results expose some uncertainties of the experiment (Politano et al., 2003). Nevertheless, the predicted velocity profiles, particularly the sharp decrease of the decreasing velocities close to the wall, are successfully captured by all models and compared reasonably well with measurements. The MUSIG model appears to yield marginally better agreement than the other models. This could be attributed to the higher resolution of the particle (or bubble) size distribution via the

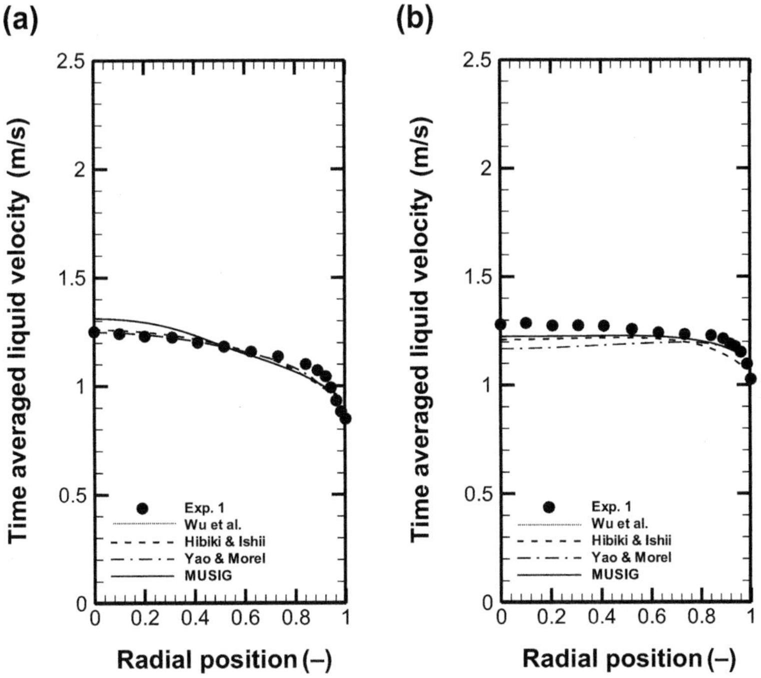

Figure 7.7
Predicted radial liquid velocity profile at $z/D = 36.0$ and experimental data of Liu and Bankoff (1993a): (a) Low gas superficial velocity and (b) High gas superficial velocity.

MUSIG model. By introducing multiple size groups to discretize the range of bubble sizes that could possibly exist within the flow instead of a single average variable parameter determined through the average bubble number density approach, the bubble Sauter diameter is seen to be better resolved, which leads to enhanced prediction of the liquid velocities.

Experimental data of Hibiki et al. (2001)

Figure 7.8 compares the gas void fraction profiles obtained from the average bubble number density and MUSIG models with the measured data in four different bubbly flow conditions. For the low void fraction cases, i.e. liquid superficial velocity $\langle j_f \rangle = 0.491$ m/s, wall peaking profiles are well established at the first measuring station of $z/D = 6.0$ due to the considerably low liquid velocities and gas velocities. However, the radial void fraction profile subsequently evolves along the axial direction, becoming a well-developed void fraction wall peak at the location of $z/D = 53.5$ for liquid superficial velocity $\langle j_f \rangle = 0.986$ m/s. The phenomenological evolution of the wall peaking behaviors is properly captured by both models.

Local radial gas and liquid velocity distributions at the measuring station of $z/D = 53.5$, close to the outlet of the pipe, are illustrated in Figure 7.9. For the cases of liquid superficial velocity $\langle j_f \rangle = 0.491$ m/s, except for the simulation result exemplified in Figure 7.9(f), the

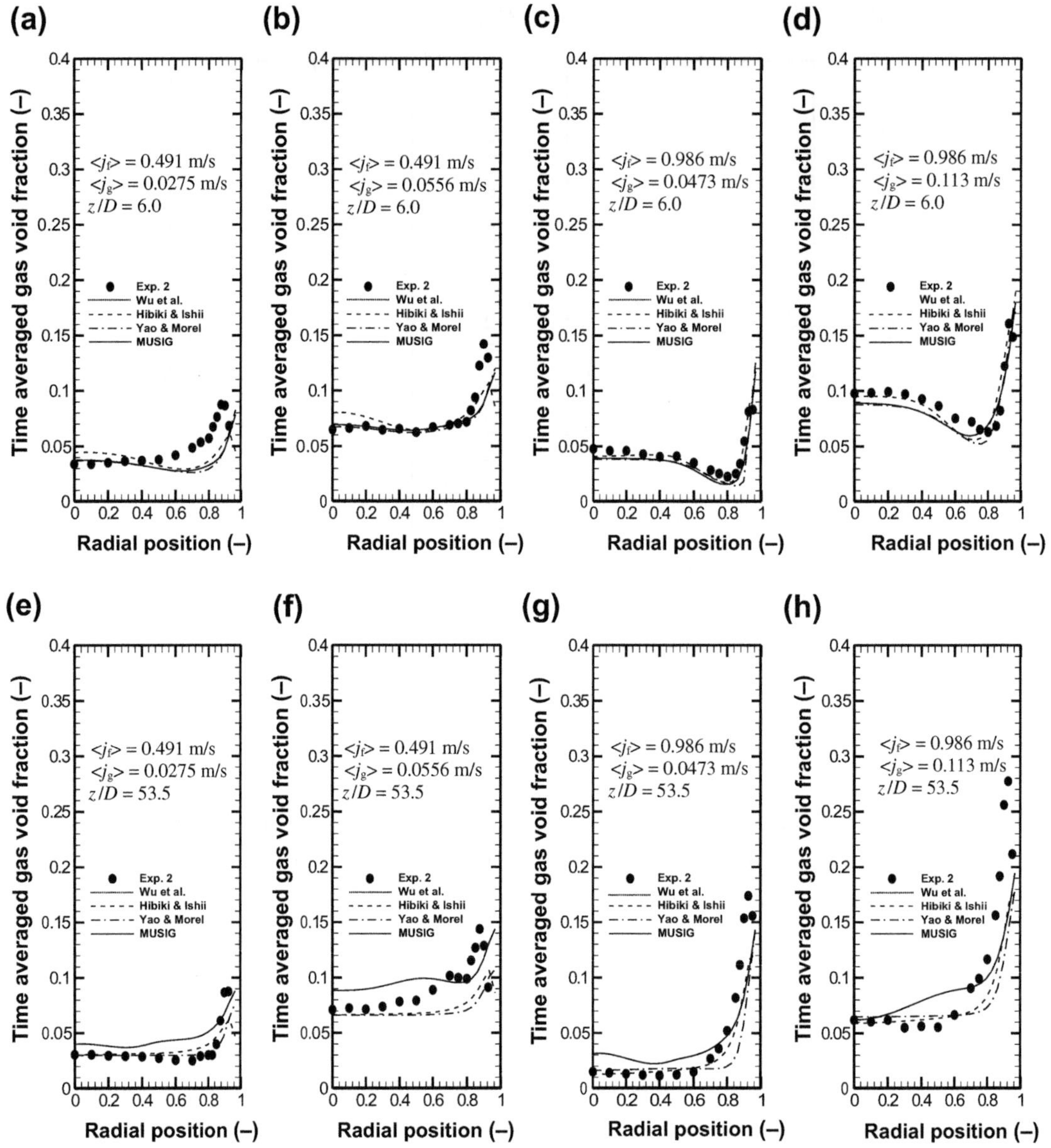

Figure 7.8

Predicted radial void fraction distribution and experimental data of Hibiki et al. (2001): (a)–(d) $z/D = 6.0$ and (e)–(h) $z/D = 53.5$.

predictions for all the average bubble number density models of the gas and liquid velocity compare favorably with the experimental data. Generally, liquid velocities at the core of the pipe are underpredicted at the location of $z/D = 53.5$ (Figure 7.9(f)). Similar observations are also found for the predictions of the gas or liquid superficial velocity $\langle j_f \rangle = 0.986$ m/s

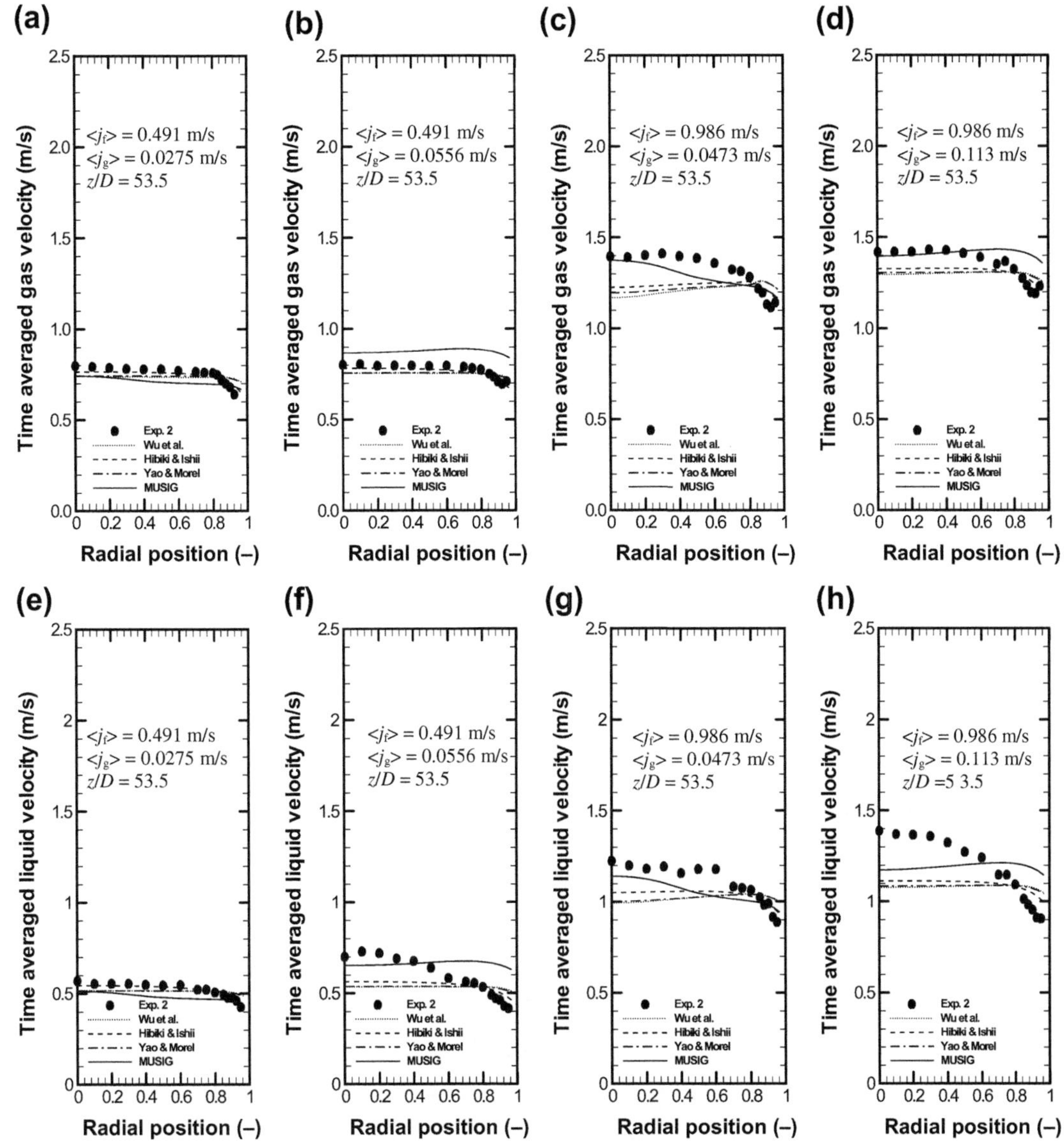

Figure 7.9

Predicted radial gas and liquid velocity profile and experimental data of Hibiki et al. (2001) at $z/D = 53.5$: (a)–(d) for gas velocity and (e)–(h) for liquid velocity.

such as those shown in Figure 7.9(c), (d), (g), and (h). Nevertheless, predictions of the MUSIG model as depicted in Figure 7.9(a–d) are found to be noticeably better than those of the average bubble number density models. Although the liquid velocities at the core are still underpredicted for liquid superficial velocity $\langle j_f \rangle = 0.986$ m/s, the MUSIG model, in general, still yields better agreement due to the higher resolution of the bubble size distribution, which would indirectly enhance the liquid velocity predictions by the provision of a more detailed description of the interfacial forces within the interfacial momentum transfer between the air and water.

Figure 7.10 shows the predicted and measured mean Sauter diameter distributions at two measuring stations, corresponding to those of the void fraction profiles in Figure 7.9. As

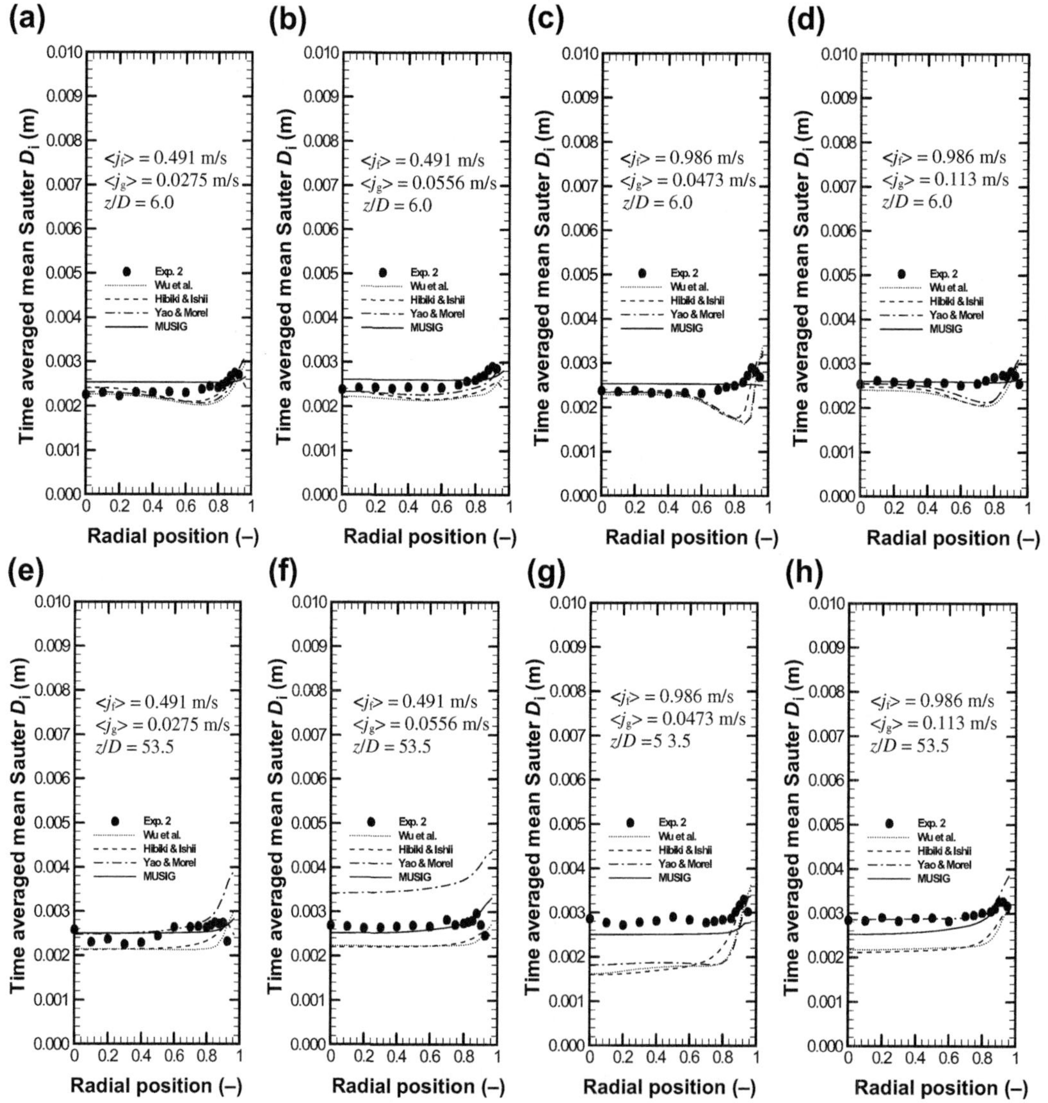

Figure 7.10

Predicted bubble Sauter diameter distribution and experimental data of Hibiki et al. (2001): (a)–(d) $z/D = 6.0$ and (e)–(h) $z/D = 53.5$.

measured by Hibiki et al. (2001), the Sauter bubble diameter profiles are almost uniform along the radial direction with some increase in size in the vicinity of the wall. The slightly larger bubbles that are formed near the wall may be due to the tendency of small bubbles migrating toward the wall creating a higher concentration of bubbles, thereby increasing the likelihood of bubble coalescence. Generally speaking, predictions from all models agree reasonably well with the measurements. For all flow cases and locations, the MUSIG model shows remarkable agreement with the measurements and is superior in determining the bubble size distribution than the average bubble number density (ABND) models. Compared to the single average variable parameter of the ABND models, the higher resolution through the use of multiple size groups captures the dynamic changes of the evolving distribution of different bubble sizes. Since the bubble Sauter diameter is generally closely coupled with the interfacial momentum forces (i.e. drag and lift forces), better predictions of the bubble Sauter diameter could significantly improve the numerical results. Unfortunately, as extra transport equations are required in the numerical calculations, additional computational effort is required at the expense in solving these equations. Computational efficiency and accuracy are issues of continuing debate. During calculations, the MUSIG model requires around twice the computational effort compared to that for the ABND models.

Based on the assumption of spherical bubbles, the local interfacial area concentration (IAC) profiles can be related to the local void fraction and bubble Sauter diameter according to $a_{if} = 6\alpha^{p}/D_s$. The measured and predicted local interfacial area concentration profiles for the two respective axial locations are depicted in Figure 7.11. The IAC radial profiles roughly follow the same trend of the void fraction distribution as stipulated in Figure 7.8. Similar to the comparison for the void fraction distribution, predictions of all models at the two measuring stations are in satisfactory agreement with measurements. However, the peak values of IAC close to the wall are better predicted by the MUSIG model. The more accurate MUSIG model could have benefited from the accurate prediction of the Sauter bubble diameter and void fraction values. In Figure 7.11(f) and Figure 7.11(h), the IAC values are overpredicted at the core region as is clearly reflected by both models. Nevertheless, the predictions of the MUSIG model generally appear to yield marginally better agreement than the transport equation for the *e* average bubble number density.

Conclusion

In this worked example, the one-group population balance approach based upon the transport equation for the average bubble number density and the MUSIG model in conjunction with the two-fluid model are assessed for their feasibility in handling gas—liquid bubbly flow under isothermal conditions. Three forms of the average bubble number density transport equation incorporating the three respective coalescence and break up mechanisms of Wu et al. (1998), Hibiki and Ishii (2002) and Yao and Morel (2004) along with the coalescence and break up

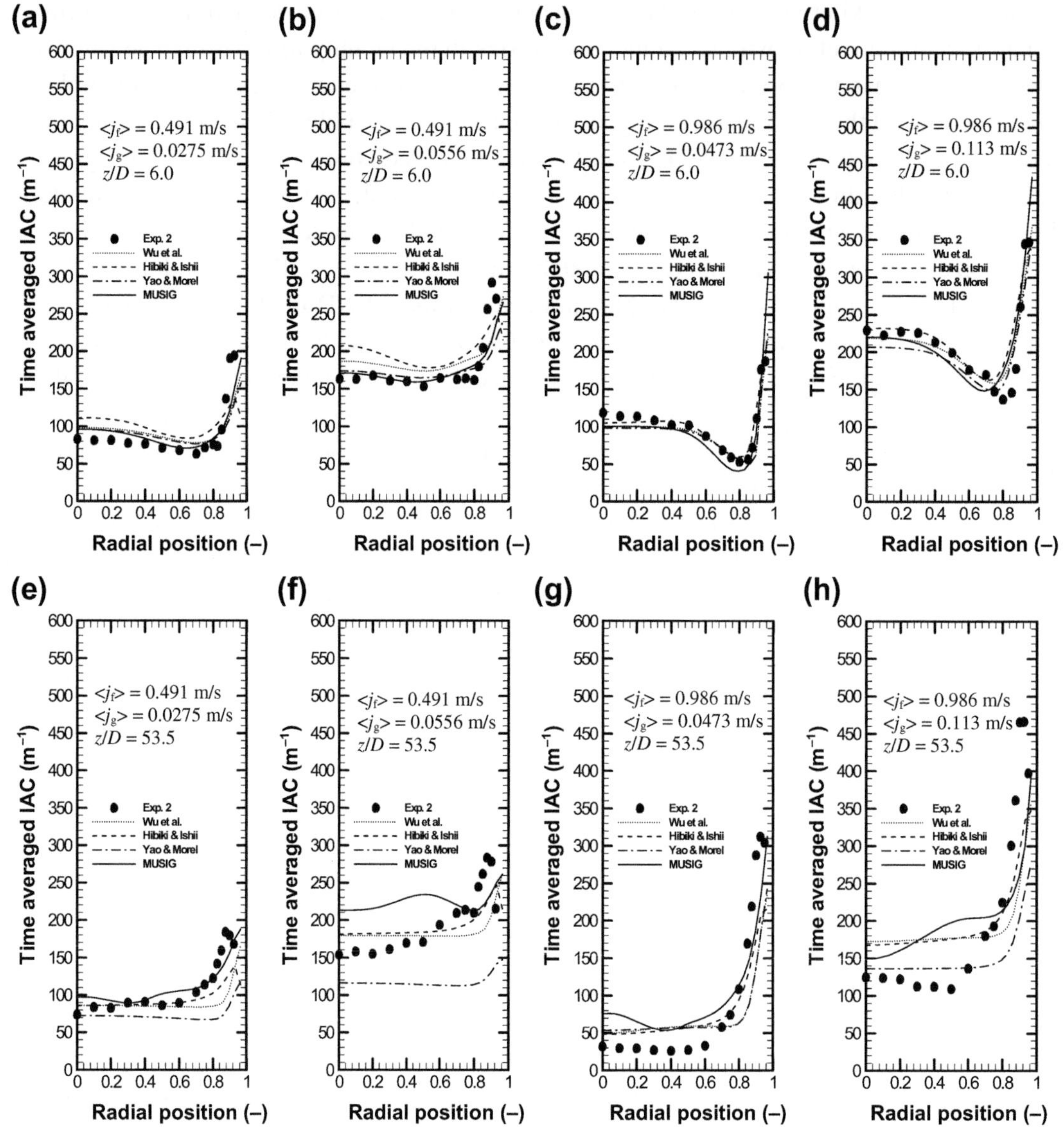

Figure 7.11

Predicted interfacial area concentration (IAC) distribution and experimental data of Hibiki et al. (2001): (a)–(d) $z/D = 6.0$ and (e)–(h) $z/D = 53.5$.

mechanisms of Prince and Blanch (1990) and Luo and Svendsen (1996) for the MUSIG model are compared against the experimental data of Liu and Bankoff (1993a, b) and Hibiki et al. (2001). Interfacial momentum transfers that embrace various interfacial force including drag, lift, wall lubrication, and turbulent dispersion force are also accounted for. In general, both population balance approaches yield close agreement with measurements for the void fraction, interfacial area concentration, Sauter bubble diameter, and gas and liquid velocities. Predictions of Sauter bubble diameter through the MUSIG model, nonetheless, attain remarkable agreement. This is attributed to its superiority in resolving the bubble size distribution, as compared to the single average variable parameter through the transport equation for the average bubble number density. As a result, predictions for the gas and liquid velocity of the MUSIG model are, in general, greatly enhanced and notably better. Numerical results clearly show that the dynamical changes of the bubbles with different sizes require higher resolution and can be achieved using the multiple size groups approach. As expected, the trade-off in adopting such an approach is at the expense of additional computational burden in solving the extra transport equations for each bubble class. Computations using the MUSIG model are twice as slow when the average bubble number density is applied under the same computational resources. Nonetheless, predictions made using the transport equation for average bubble number density are found to yield satisfactory agreement with measurements, though they appear marginally inferior in some degree to the MUSIG model results. Transport equations for the average bubble number density can thus be considered as a viable option, especially for industrial practitioners, who often demand a rapid design tool in simulating bubbly flows with reasonable accuracy. For the case of acquiring highly accurate Sauter bubble diameter distribution, the MUSIG model serves as the best alternative in handling gas–liquid flow.

Transition cap–bubbly flows in vertical pipe

Coalescence and break up mechanisms adopted in the previous section have been principally based on the assumption of interaction between spherical bubbles. Nevertheless, cap bubbles which are precursors to the formation of slug units in the slug flow regime, as well as the accumulation of large unsteady gas volumes within these mixing regions which produce the churn-turbulent flow regime with increasing volume fraction, become ever more prevalent at high gas velocity conditions. In this worked example, the modeling framework that includes the classification of bubbles of different sizes and shapes into different groups entails the consideration of additional transport equations to aptly describe the transport phenomena of these distinct groups of bubbles such as those proposed in Ishii et al. (2002) and Hibiki and Ishii (2009). A schematic diagram of the TOPFLOW experimental setup of Prasser et al. (2007) is shown in Figure 7.12. In this test facility, a large size vertical cylindrical pipe with a height of 9000 mm and an inner diameter of 195.3 mm was adopted. Water was circulated from the bottom to the top with a constant temperature of 30°C, maintained by a heat exchange installed in the water reservoir. A variable gas injection system was constructed equipped with gas injection units at 18 different axial positions from $Z/D = 1.1-39.9$. Three levels of air chambers were installed at each injection unit. The upper and the lower chambers have 72 annular distributed orifices of 1 mm diameter for small bubble injection, while the central chamber has 32 annularly distributed orifices of 4 mm diameter for large bubble injection. A fixed wire-mesh sensor

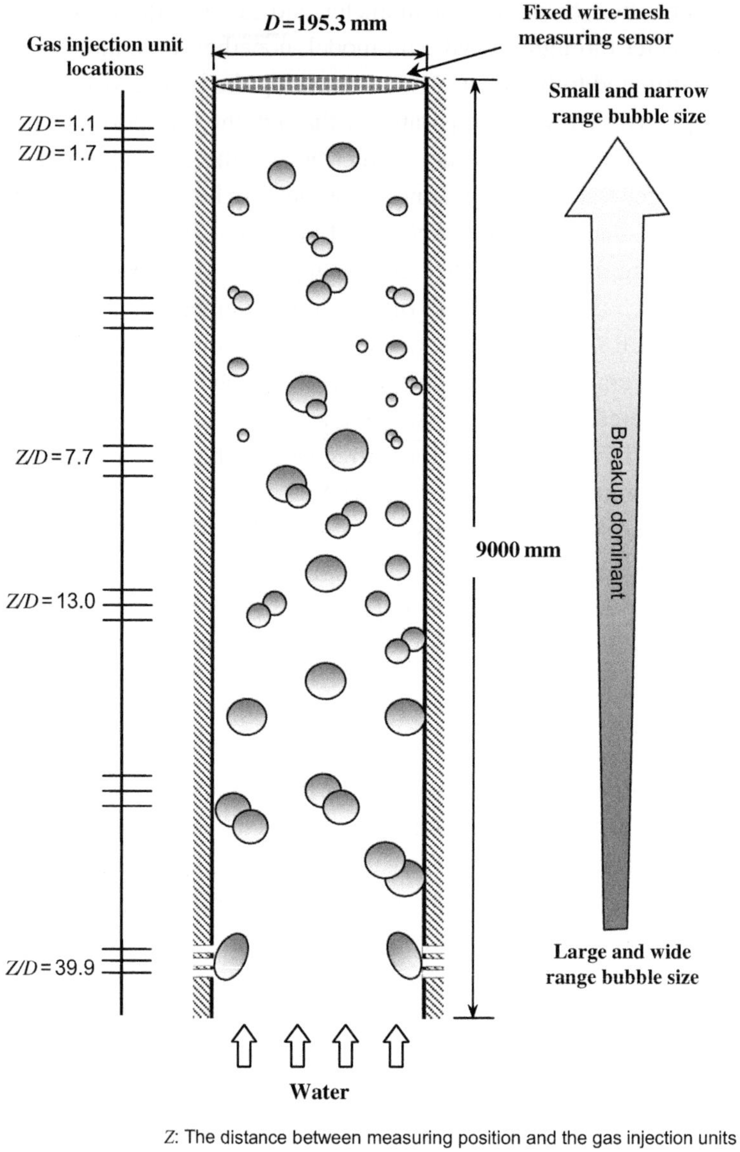

Figure 7.12
Schematic drawing of the test section of TOPFLOW experiment.

was installed at the top of the pipe where data relating to the gas volume fraction, as well as bubble size distribution, was registered.

In the TOPFLOW experiment, with the gas injection orifices located on the circumference of the pipe, highly concentrated bubbles were formed within the wall proximity. This closely packed swarm of bubbles was then immediately merged with others, forming larger bubbles.

It can be inferred that this rapid coalescence of bubbles exhibited a bimodal bubble size distribution with a wide spectrum of bubble sizes ranging between 0 and 60 mm such as measured by the sensor. These bimodal distributed bubbles from the injection unit gradually collapsed to the single-peaked distribution at the top of the test section. With such a dynamical change in bubble size, there is no doubt that bubbles with various shapes (i.e. spherical and cap) were introduced into the system and sequentially broke up into smaller bubble sizes due to the dominance of the break up mechanism. The main aim of this example is to demonstrate the formulation of a two-group model (i.e. spherical and cap bubbles) based on two transport equations for the averaged bubble number density in capturing the intragroup and intergroup interactions. The two-group formulation necessitates the treatment of bubbles into two distinct groups, namely Group 1 bubbles consisting of spherical bubbles and Group 2 bubbles consisting of cap bubbles. The Group 1 bubbles exist in the range of minimum bubble size to the maximum distorted bubble diameter, as suggested by Ishii and Zuber (1979):

$$D_{\mathrm{d,max}} = 4\sqrt{\frac{\sigma}{g\Delta\rho}} \tag{7.20}$$

whereas Group 2 bubbles exist in the range of the above limit to some maximum stable size limit:

$$D_{\max} = 40\sqrt{\frac{\sigma}{g\Delta\rho}} \tag{7.21}$$

The two-group model is assessed for the specific test cases 41, 63, 85, and 96 of the TOPFLOW experiment. Table 7.5 tabulates the flow conditions for the four selected test cases. One should note that these test cases are specifically chosen because of their distinctive flow regimes. Test cases 41 and 63 should be dominated by only Group 1 bubbles because the gas–liquid flow is principally bubbly flow. With increasing gas superficial velocities, test cases 85 and 96 denote possible transitional flow characteristics from spherical to cap bubbles—cap flow. Here, Group 2 bubbles should become more prevalent and significant as the existence of Group 1 bubbles gradually diminishes.

Table 7.5: Flow conditions for the four selected cases in this worked example.

	TOPFLOW Experiment			
	Case 41	Case 63	Case 85	Case 96
$[\langle j_l\rangle\vert_{Z/D=0}]$ (m/s)	1.017	1.017	1.017	1.017
$\left[\langle j_g\rangle\Big\vert_{Z/D=0}\right]$ (m/s)	0.0096	0.0235	0.0574	0.0898
$[\alpha_g\vert_{Z/D=0}]$ (%)	[0.94]	[2.26]	[5.34]	[8.11]

Numerical features

The spherical and cap bubbles, pressure, and temperature can be assumed to be approximately the same. However, the two bubble types cannot be considered to have a common velocity. Therefore, the original two-fluid model needs to be extended to the three-fluid model, which solves three sets of governing equations for all (liquid, spherical bubble, and cap bubble) three fluids simultaneously. One should note that such an extension is rather straightforward with the same formulation as the two-fluid model but with an additional set of mass and momentum equation for cap bubbles. Details of the two-fluid model can be found in the work of Cheung et al. (2012). Similarly to the previous worked example, the advection term is discretized using a second-order scheme weighted between central and upwind differencing, while the diffusion time is approximated according to the second-order central differencing scheme. For the time derivative term, it is approximated via a second-order backward Euler time discretization scheme. In handling bubble induced turbulent flow, the SST model is adopted.

To be consistent with the three-fluid model, spherical and cap bubbles are individually subjected to different interfacial forces. For spherical bubbles, the drag coefficient $C_{D,1}$ has been correlated for several distinct Reynolds number regions for individual bubbles according to Ishii and Zuber (1979). Nevertheless, for cap bubbles, the drag coefficient $C_{D,2}$ can be approximated to be 8/3 (Ishii and Chawla, 1979; Tomiyama, 1998). For the lift coefficient, the Eotvos number dependent correlation proposed by Tomiyama (1998) is adopted. For Group 2 bubbles, as they exist in the range above maximum distorted bubble diameter shown in Eqn (7.20), the lift coefficient becomes negative according to the correlation. For the wall lubrication force, the model from Antal et al. (1991) is adopted with the model constants $C_{w1} = 0.0064$ and $C_{w2} = 0.016$. The Farve-averaged turbulent dispersion model by Burns et al. (2004) is employed. By default, the turbulent dispersion coefficient $C_{TD} = 1$ and the turbulent Schmidt number $\sigma_{t,g} = 0.9$ is employed.

The two-group average bubble number density model, as discussed in Chapter 3, is adopted. The source and sink terms due to coalescence and break up of bubbles are specified with constitutive relations according to Hibiki and Ishii (2000) (Table 7.6). The two-group average bubble number density transport equations together with the source and sink terms were implemented through the CFX Command Language. For the purpose of computational efficiency, similar to the previous worked example, the flow was assumed to be axisymmetric so that the numerical simulations were performed on a 60° radial sector of the pipe with symmetry boundary conditions at both vertical sides. To represent the wall injection method, 12 equally spaced point sources of the gas phase were placed at the circumference of the 60° radial sector. The gas injection rate at each point source was assumed to be identical. Based on the grid sensitivity test performed, grid independent solutions have revealed that computational meshes which consisted of 48,000 elements did not appreciably change even

Table 7.6: Coalescence and break up rates for two-group approach.

Group 1 Bubbles
$S_{N_1} + R_{N_{12}} = \underbrace{\Phi_1^{RC} + \Phi_1^{TI}}_{\text{intragroup}} + \underbrace{\Phi_{12}^{WE} + \Phi_{12}^{TI}}_{\text{intergroup}}$
$\Phi_1^{RC} = -C_{RC1}\dfrac{(\alpha_1^P)^2\varepsilon^{1/3}}{D_{b,1}^{11/3}(\alpha_{RC,\max} - (\alpha_1^P + \alpha_2^P))} \times \exp\left(-K_{RC1}\dfrac{D_{b,1}^{5/6}(\rho^c)^{1/2}(\varepsilon^c)^{1/3}}{\sigma^{1/2}}\right)$ $\Phi_1^{TI} = C_{TI1}\dfrac{\alpha_1^P(1-\alpha_1^P)(\varepsilon^c)^{1/3}}{D_{b,1}^{11/3}(\alpha_{TI,\max} - (\alpha_1^P + \alpha_2^P))}\exp\left(-K_{TI1}\dfrac{\sigma}{\rho^c D_{b,1}^{5/3}(\varepsilon^c)^{2/3}}\right)$ $\Phi_{12}^{WE} = -C_{WE12}\dfrac{(\alpha_1^P)^2}{D_{b,1}^3 D_{b,2}}(v_{axial,2} - v') \times \exp\left[-K_{WE12}\dfrac{(\rho^c)^{1/2}(\varepsilon^c)^{1/3}}{\sigma^3}\left(\dfrac{D_{b,1}D_{b,2}}{D_{b,1}+D_{b,2}}\right)^{5/6}\right]$ $\Phi_{12}^{TI} = C_{TI12}\dfrac{\alpha_2^P(1-(\alpha_1^P+\alpha_2^P))(\varepsilon^c)^{1/3}}{D_{b,2}^{11/3}(\alpha_{TI,\max} - (\alpha_1^P + \alpha_2^P))} \times \exp\left(-K_{TI12}\dfrac{\sigma\{(D_{b,2}^3 - D_{b,1}^3)^{2/3} + (D_{b,1}^2 - D_{b,2}^2)\}}{\rho^l D_{b,2}^{11/3}(\varepsilon^c)^{2/3}}\right)$ $C_{RC1} = 0.0318$, $K_{RC1} = 0.441$, $C_{TI1} = 0.00508$, $K_{TI1} = 2.34$, $C_{WE12} = 0.066$, $K_{WE12} = 1.34$, $C_{TI12} = 50.35$, $K_{TI12} = 3.08$
Group 2 Bubbles
$S_{N_2} = \underbrace{\Phi_2^{WE} + \Phi_2^{TI}}_{\text{intragroup}}$
$\Phi_2^{WE} = -C_{WE2}\dfrac{(\alpha_2^P)^2}{D_{b,2}^4}(v_{axial,2} - v') \times \exp\left\{-K_{WE2}\dfrac{D_{b,2}^{5/6}(\rho^c)^{1/2}(\varepsilon^c)^{1/3}}{\sigma^{1/2}}\right\}$ $\Phi_2^{TI} = C_{TI2}\dfrac{\alpha_2^P(1-(\alpha_1^P+\alpha_2^P))(\varepsilon^c)^{1/3}}{D_{b,2}^{11/3}(\alpha_{TI,\max} - (\alpha_1^P + \alpha_2^P))}\exp\left(-K_{TI2}\dfrac{\sigma}{\rho^c D_{b,2}^{5/3}(\varepsilon^c)^{2/3}}\right)$ $C_{WE2} = 1.5$, $K_{WE2} = 0.754$, $C_{TI2} = 0.0000286$, $K_{TI2} = 0.8$

though finer computational meshes were subsequently tested. Compared to the finer mesh, the predicted cross-sectional averaged volume fractions were found only within differences of 2%. For all flow conditions, a reliable convergence criterion based on the rms residual of 1.0×10 was adopted for the termination of numerical calculations.

Numerical results

Figure 7.13 depicts the comparison of axial profiles for the predicted area-averaged void fraction against the experimental data of test cases 41, 63, 85, and 96. The steady-state,

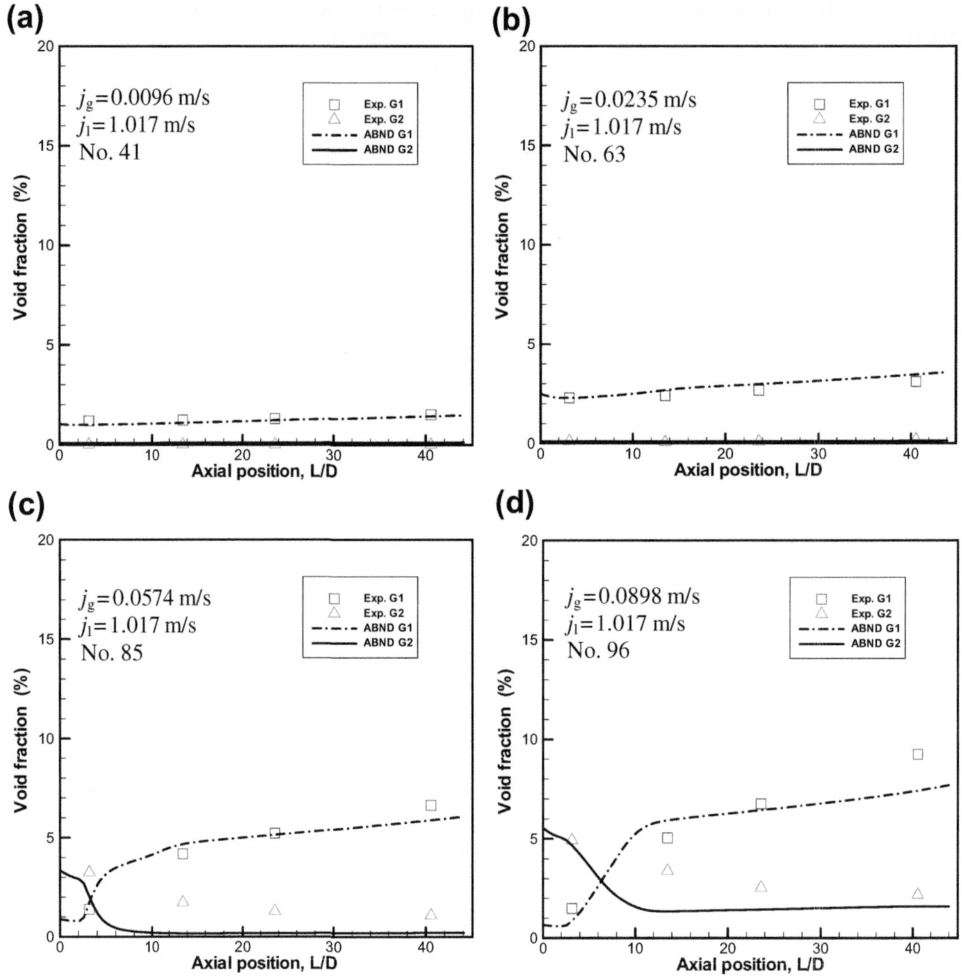

Figure 7.13
Evolution of measured and predicted void fraction profiles along axial direction with increasing superficial velocities j_g and j_l for test cases: (a) 41, (b) 63, (c) 85, and (d) 96. (For a color version of this figure, the reader is referred to the online version of this book.)

one-dimensional equation governing the conservation of mass of the gas phase in vertical flow is given by

$$\frac{\mathrm{d}}{\mathrm{d}y}(\langle\rho^{g}\rangle\langle\alpha^{g}\rangle\langle v^{g}\rangle) = 0 \tag{7.22}$$

where $\langle\,\rangle$ denotes area-averaged quantities. Equation (7.22) can be alternatively expressed as

$$\frac{1}{\langle\rho^{g}\rangle}\frac{\mathrm{d}}{\mathrm{d}y}(\langle\rho^{g}\rangle) + \frac{1}{\langle\alpha^{g}\rangle}\frac{\mathrm{d}}{\mathrm{d}y}(\langle\alpha^{g}\rangle) + \frac{1}{\langle v^{g}\rangle}\frac{\mathrm{d}}{\mathrm{d}y}(\langle v^{g}\rangle) = 0 \tag{7.23}$$

Using the ideal gas law, and because the flow is isothermal, Eqn (7.23) becomes

$$\frac{1}{\langle\alpha^{g}\rangle}\frac{\mathrm{d}}{\mathrm{d}y}(\langle\alpha^{g}\rangle) = -\frac{1}{\langle P\rangle}\frac{\mathrm{d}}{\mathrm{d}y}(\langle P\rangle) - \frac{1}{\langle v^{g}\rangle}\frac{\mathrm{d}}{\mathrm{d}y}(\langle v^{g}\rangle) \tag{7.24}$$

It can be seen from the above equation that the first term on the right-hand side is always positive because the pressure gradient is negative, and it increases with the increasing total mass flow rate. The second term on the right-hand side can, however, take positive and negative values but as the velocity of the gas phase increases, its effect decreases. In total, the area-averaged void fraction is expected to increase along the vertical pipe with the increase of the mass flow rate and the velocity of the gas phase. This trend can be seen for Group 1 bubbles for all of the test cases in Figure 7.13. For test cases 85 and 96, the prevalence of cap bubbles can be seen to be evident near the inlet of the pipe such as indicated by the higher void fraction of Group 2 bubbles. Nevertheless, the decreasing trend of Group 2 bubbles illustrates the significant mass transfer occurring from Group 2 bubbles to Group 1 bubbles, which is effected through the intergroup mechanisms of bubble interaction governed by the break up of cap bubbles due to the impact of turbulent eddies as the flow develops downstream along the vertical large pipe.

The axial profiles of area-averaged IAC of test cases 41, 63, 85, and 96 are compared in Figure 7.14. In general, because of the dominant contribution of Group 1 bubbles, the total IAC is nearly proportional to the void fraction of Group 1 bubbles. However, the presence of cap bubbles coflowing with spherical bubbles becomes significant, particularly for test case 96 at the downstream of the flow, resulting in a rather significant axial void fraction and IAC distributions of Group 2 bubbles such as those already exemplified for the same test case in Figures 7.13 and 7.14. Both axial profiles of both predicted area-averaged void fraction and IAC show reasonable agreement with the measured data.

Figure 7.15 illustrates the axial profiles of the volume equivalent bubble diameter of test cases 41, 63, 85, and 96. For all test cases, the axial profiles of the volume equivalent bubble diameter of Group 1 bubbles (i.e. dashed line) were predicted very well against the axial evolution of measured sizes. This indicated that the mechanisms governed by coalescence due to random collision driven by liquid turbulence and the break up due to the impact of turbulent eddies are sufficient to capture the appropriate bubble interaction behaviors for spherical bubbles occurring within the two-phase flow. Nevertheless, the inability of the numerical model to predict the axial profiles of the volume equivalent bubble diameter of Group 2 bubbles (i.e. solid line), especially for test cases 85 and 96, demonstrates the need for further insights and the development of appropriate mechanisms to better capture the prevailing bubble interaction behaviors governing Group 2 bubbles.

The radial migration of bubbles plays an important role in the evolution of the local flow structure, which is mainly determined by bubble coalescence and break up. Figure 7.16 shows

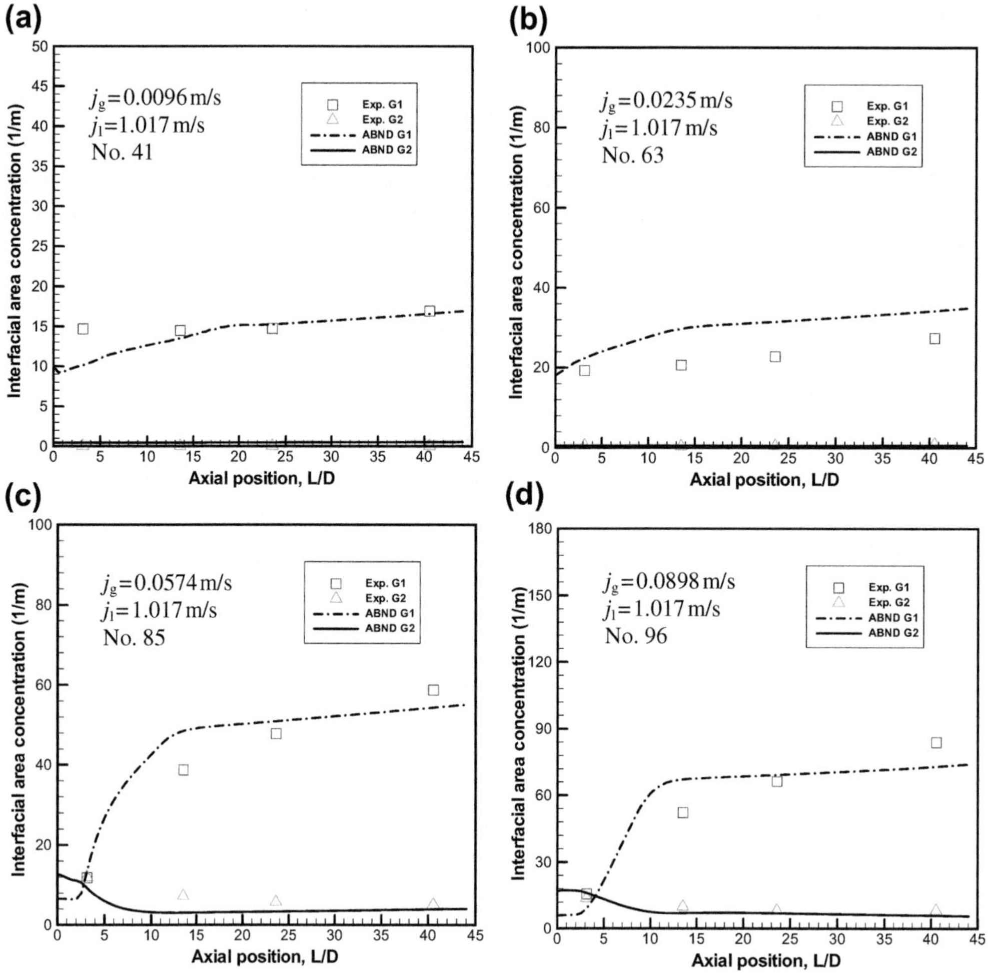

Figure 7.14

Evolution of measured and predicted IAC profiles along axial direction with increasing superficial velocities j_g and j_l for test cases: (a) 41, (b) 63, (c) 85, and (d) 96. (For a color version of this figure, the reader is referred to the online version of this book.)

the radial distribution of time-averaged local void fraction profiles at $L/D = 3.1$ (close to the gas injection units) and 39.9 (longest distance from the injection) for test cases 63 and 96. For test case 63, the wall void peaking characteristic for Group 1 bubbles is clearly observed near the inlet. Owing to the radial separation of small and large bubbles, the transition from wall void peaking to core void peaking occurs near the outlet, as demonstrated in Figure 7.16. At higher gas superficial velocity, the wall void peaking is nonetheless found near the inlet for Group 2 bubbles in test case 96. This is probably due to the presence of large bubbles being introduced immediately after the injection location. Further downstream, sufficiently large Group 1 and Group 2 bubbles move toward the center of the pipe because of the lift

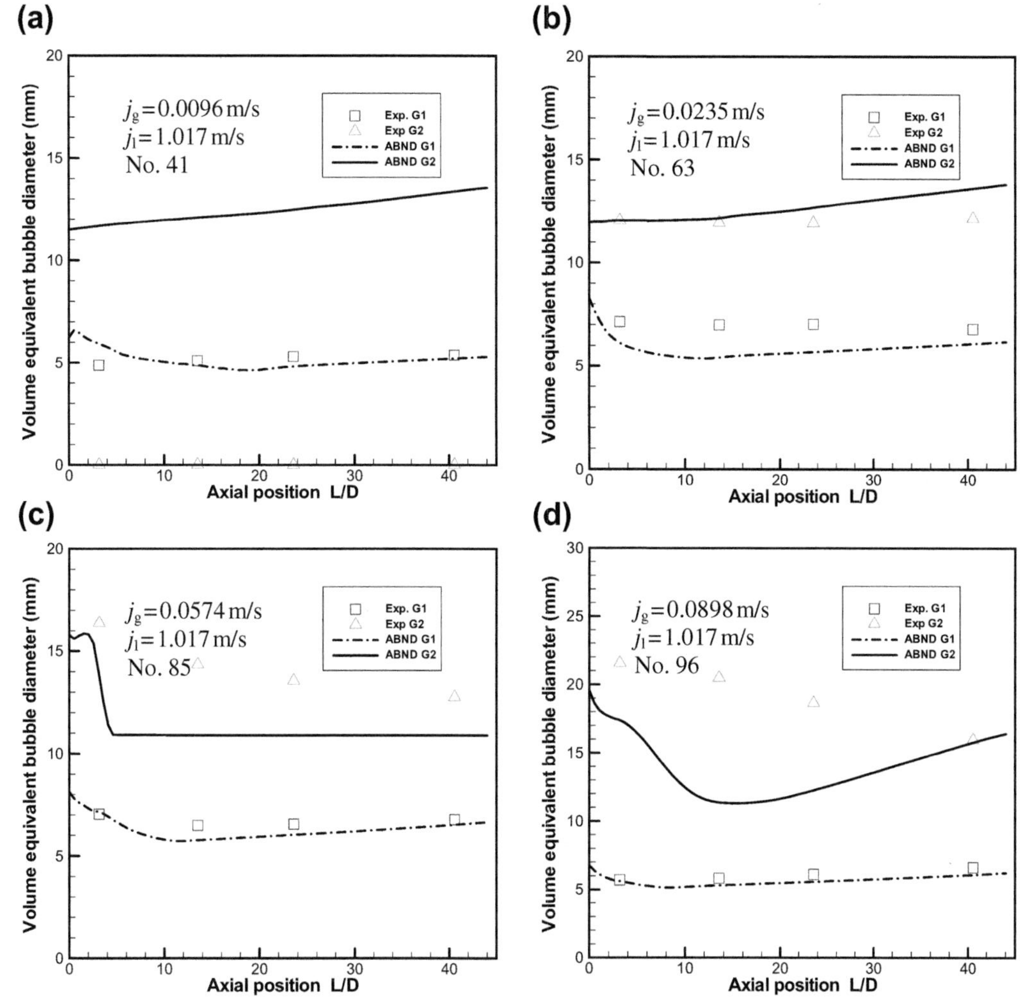

Figure 7.15

Evolution of measured and predicted volume equivalent bubble diameter profiles along axial direction with increasing superficial velocities j_g and j_l for test cases: (a) 41, (b) 63, (c) 85, and (d) 96. (For a color version of this figure, the reader is referred to the online version of this book.)

force acting on the bubbles. Also, because of the low turbulent kinetic energy at the center, these bubbles have less probability of break up and thereby further coalesce to form larger bubbles due to random collisions and wake entrainment, resulting in core void peaking near the outlet. This mechanism represents the key for the transition from bubbly-to-cap flow.

Figure 7.17 illustrates the radial distribution of IAC profiles at $L/D = 3.1$ and 39.9 for the same test cases shown in Figure 7.16. If the radial profiles of the volume equivalent bubble diameters are uniform, the radial profiles of IAC would follow the similar characteristic distribution of the void fraction. This has been particularly evident when the flow has been

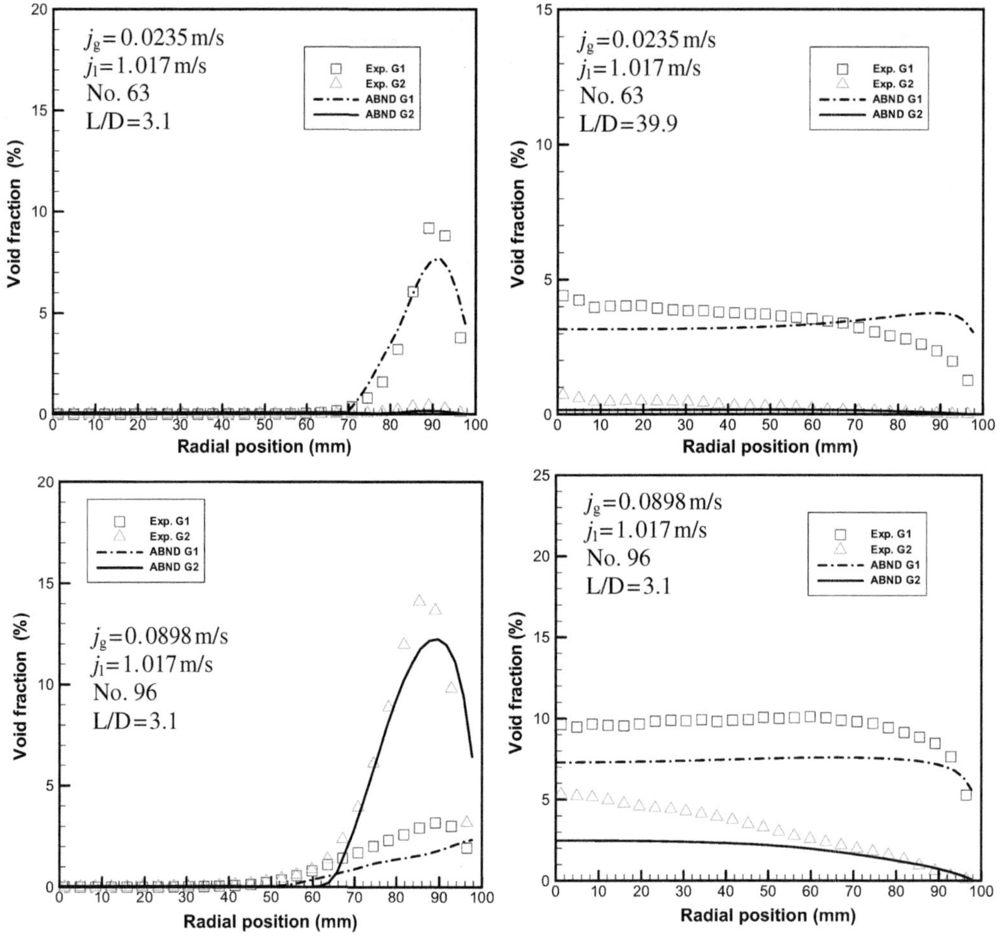

Figure 7.16

Measured and predicted radial void fraction profiles at two different axial locations ($L/D = 3.1$ and 39.9) for test cases: (upper) 63 and (lower) 96. (For a color version of this figure, the reader is referred to the online version of this book.)

sufficiently developed near the outlet of the pipe such as depicted by the radial distribution of volume equivalent bubble diameters at $L/D = 39.9$ in Figure 7.18. For the two-phase flow in a vertical large pipe with the inner diameter of 200 mm, except at $L/D = 39.9$ where the radial distribution of volume equivalent bubble diameters shows significant variation due to chaotic turbulent flow being experienced near the injection location, the current volume equivalent bubble diameter profiles are in accordance with the observed uniform profiles in experiments by Shen et al. (2005) with some decreases in size near the wall of the pipe. Because of the existence of dominant large bubbles within the flow, secondary flows that induce large bubbles not only provide the bubbles with the opportunity to migrate into the center region of the pipe but also cause the bubbles to coalesce and to form bigger bubbles or

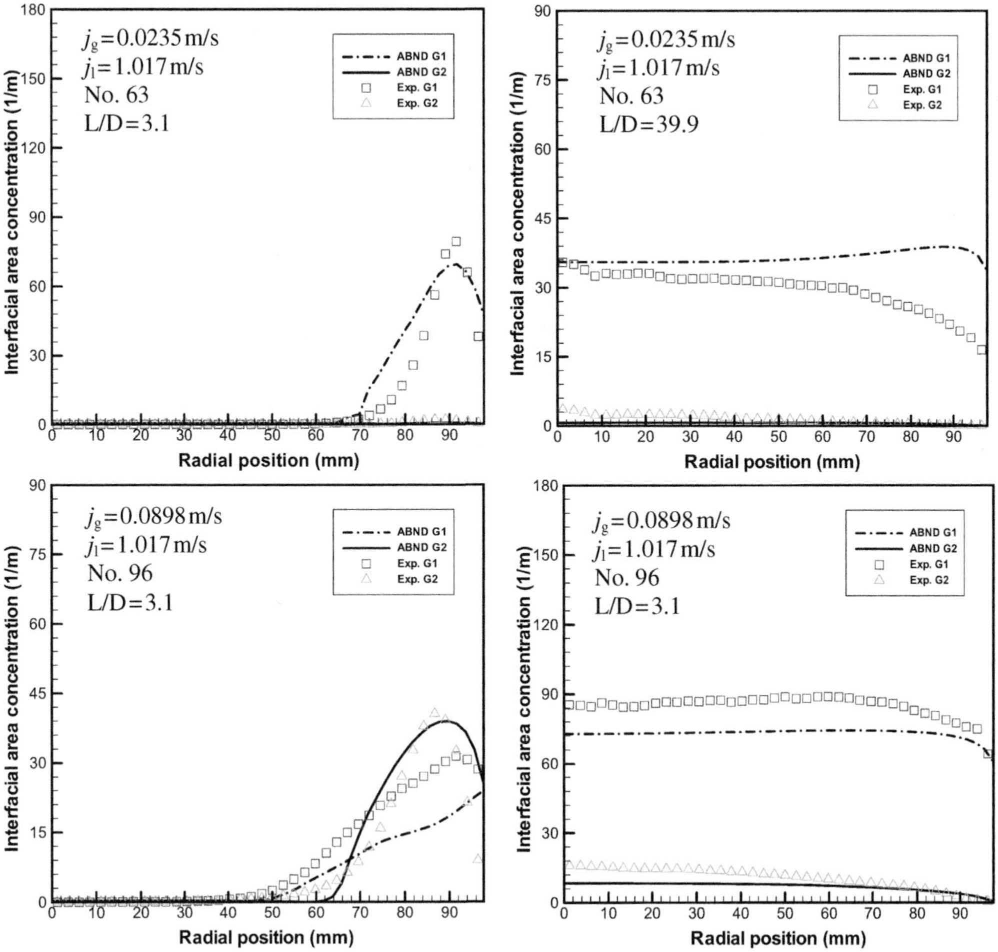

Figure 7.17

Measured and predicted radial volume equivalent bubble diameter profiles at two different axial locations ($L/D = 3.1$ and 39.9): (upper) 63 and (lower) 96. (For a color version of this figure, the reader is referred to the online version of this book.)

to break up into smaller bubbles around the large bubbles, resulting in the apparent decrease in the volume equivalent bubble diameter with the flow development. Local radial profiles of predicted area-averaged void fraction, IAC, and volume equivalent bubble diameter, show reasonable agreement with the measured data, especially near the outlet ($L/D = 3.1$).

Figure 7.19 depicts the radial distribution of gas velocity profiles for Group 1 and Group 2 bubbles at $L/D = 3.1$ and 39.9 for test cases 63 and 96. The developing two-phase flow has been captured rather well by the current numerical model especially when comparing the predicted gas velocity profiles against the measured gas velocities, not only at the location near the outlet of the pipe but also at the location near the inlet of the pipe, in spite of the

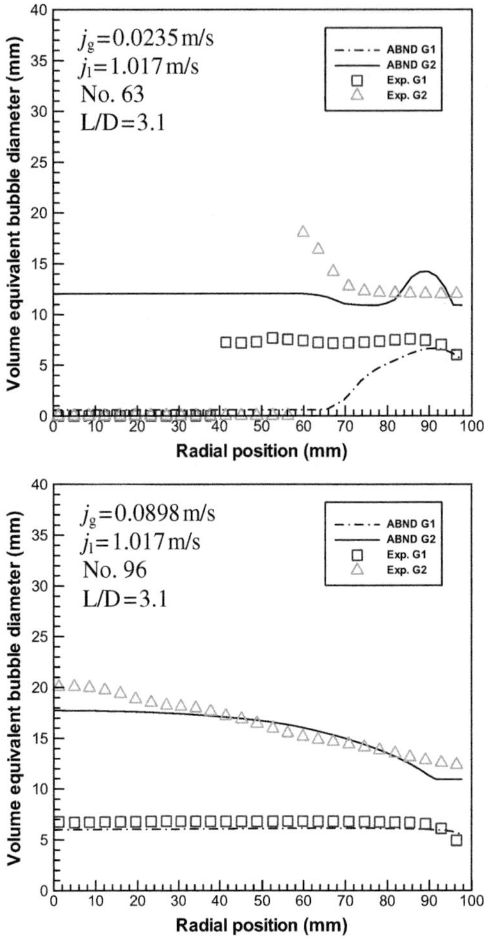

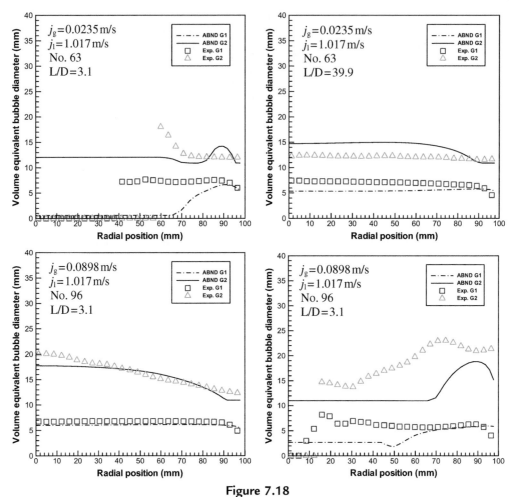

Figure 7.18

Measured and predicted radial volume equivalent bubble diameter profiles at two different axial locations ($L/D = 3.1$ and 39.9) for test cases: (upper) 63 and (lower) 96. (For a color version of this figure, the reader is referred to the online version of this book.)

rather chaotic turbulent flow being experienced shortly after the injection of bubbles into the bulk two-phase flow.

Conclusion

In relation to the development of the three-fluid model and two-group transport equations for the average bubble number density, this worked example shows the formulation considering the two groups of bubbles, such as spherical bubbles being Group 1 and cap bubbles being Group 2. Based on the proposal of Hibiki and Ishii (2000), possible interaction mechanisms for Group 1 and Group 2 bubbles have been utilized to close the equations via the consideration of coalescence due to random collisions driven by

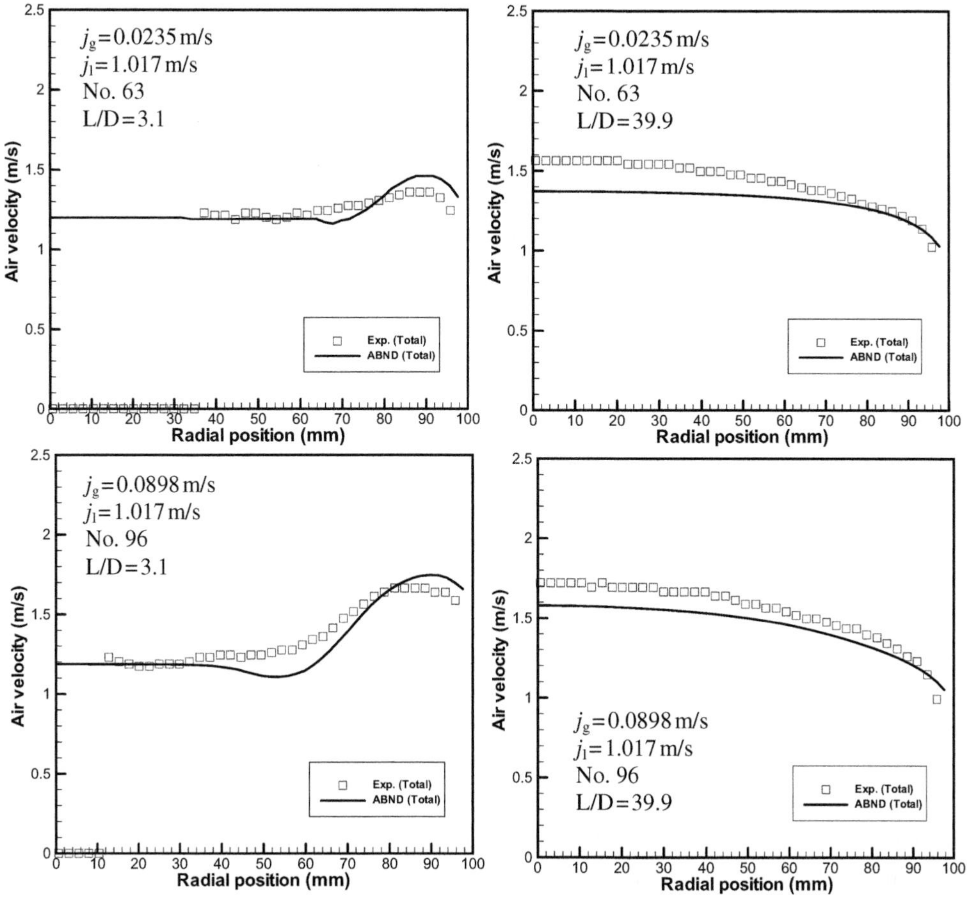

Figure 7.19

Measured and predicted radial gas velocity profiles at two different axial locations ($L/D = 3.1$ and 39.9) for test cases: (upper) 63 and (lower) 96. (For a color version of this figure, the reader is referred to the online version of this book.)

turbulence, coalescence due to wake entrainment, and break up upon the impact of turbulent eddies.

Preliminary assessment on the bubble interaction mechanisms of coalescence and break up has been performed by considering random collision, wake entrainment, and turbulent impact. The predicted results showed that the significant coalescence and break up between Group 1 and Group 2 bubbles could be adequately captured via the intergroup mechanisms which yielded satisfactory mass transfer between these two groups of bubbles. Nevertheless, further insights and development of appropriate intragroup mechanisms to better capture the prevailing bubble interaction behaviors governing Group 2 bubbles are still required.

Overall, the initial utilization of the three-fluid model and two-group average bubble number density equations have shown to be promising in the prediction of interfacial transport in the gas–liquid flow. Nevertheless, additional insights into the appropriate bubble interaction behaviors and development of intragroup mechanistic models for Group 2 bubbles need to be performed in order for the current numerical model to successfully capture flow transitions of slug or churn-turbulent bubbles as observed in a number of test cases discussed in Section 7.1. Analogously, this also needs to be performed for the intergroup mechanistic models between Group 1 and Group 2 bubbles.

7.3 Population Balance Solutions to Liquid–Liquid Flow

7.3.1 Background

Population balance problems in liquid–liquid systems share similar characteristics and behavior to flows in bubble columns. Two phase liquid–liquid contacting is one of the widely adopted multiphase processes which can be found in many industrial applications including hydrometallurgical, pharmaceutical, and food industries for extracting valuable substances via mass transfer across the two-liquid interface. From a physical viewpoint, drop size distribution and its associated heat and mass transfer throughout the extraction process are significant parameters affecting the final product quality. The stirred tank is a widely employed reactor to facilitate the above mass transfer process. Figure 7.20 shows a typical stirred tank which often consists of an agitator (i.e. propeller or turbine) for continuous mixing. Turbulence introduced by the agitator is aimed at controlling the mixing of two reactants and encourages the break up of dispersed liquid drops to achieve a larger contact

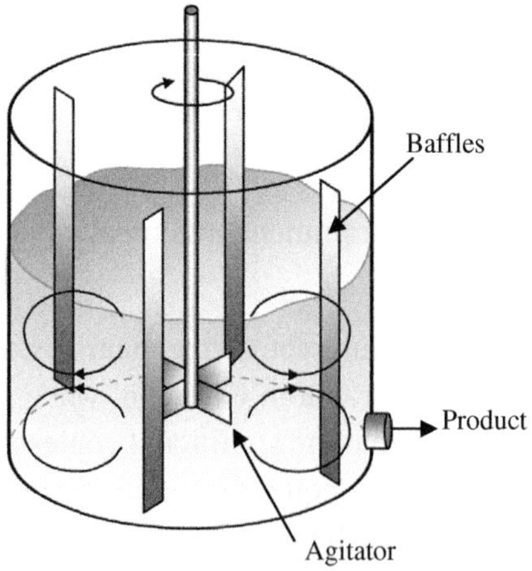

Figure 7.20
Schematic of flow structure in a typical stirred tank with baffles. (For a color version of this figure, the reader is referred to the online version of this book.)

area. The presence of battles destroys the circular flow structure created by the agitator, thereby prohibiting vortex formation. Baffles also promote flows in the axial direction, resulting in a better mixing rate (Alcamo et al., 2005). On the other hand, baffles also make the turbulence become heterogeneous, leading to a highly nonuniform drop size distribution within the tank. As the drop size distribution is affected by turbulence induced droplet coalescence and break up, it is challenging to model the population balance of droplets in a stirred tank.

7.3.2 Multiblock Model for Heterogeneous Turbulent Flow Structure in a Stirred Tank

In this section, a multiblock model proposed by Alopaeous et al. (1999) is presented to solve the population balance of droplets in a stirred tank. The multiblock model is a useful numerical tool to qualitatively investigate the evolution of droplet size distribution under the influence of various flow conditions in a stirred tank. Figure 7.21(a) depicts the typical turbulent flow structure and characteristic region that can be experienced in a stirred tank. A secondary turbulent structure creates flow circulation regions near the free surface and the bottom of the tank. Furthermore, turbulent flow exhibits a highly localized characteristic and distinguished feature within the tank caused by the agitator and other geometrical

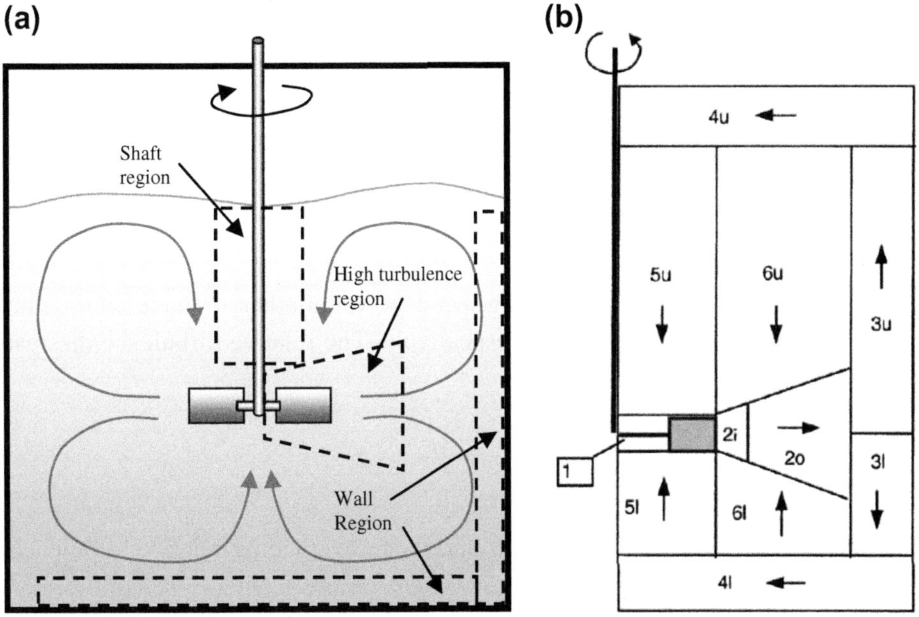

Figure 7.21

Turbulent flow structure and characteristic regions in a stirred tank (a) and subregions chosen for the multiblock model (b). (For a color version of this figure, the reader is referred to the online version of this book.) *After Alopaeus et al. (1999).*

arrangements. The agitator turbulence is several orders of magnitude greater than in other regions. Turbulence is also substantially higher in some regions due to the influence of walls or baffles at the bottom of the tank. Similarly, the rotating motion of the shaft also induces high turbulence in the vicinity. As shown in Figure 7.21(a), the agitator also creates two main secondary circulation regions in both the upper and lower parts of the tank. With such flow structure and localized turbulence intensity, Alopaeus et al. (1999) proposed the utilization of a multiblock model that consists of 11 subregions within the stirred tank. Population balance of droplets is solved explicitly by assessing and calibrating coalescence and break up parameters. Figure 7.21(b) shows the size and locale of subregions chosen for the multiblock model. As the flow is assumed to be symmetric around the impeller axis, the multiblock model is taken to be only one-half of the size of the tank. Local turbulent quantity (i.e. turbulence dissipation rate) and flow field within subregions are evaluated based on the flow model proposed by Bourne and Yu (1994). To simplify the calculation, each subregion is assumed to be mixed perfectly and the turbulence within to be homogeneous. All energy input from the agitator is assumed to be fully dissipated by the turbulence dissipation mechanism.

Based on the above assumption, the turbulent energy dissipation rate of each region is given by

$$\varepsilon_i^{\mathrm{c}} = \frac{\int_{r_0}^{r_1}\int_{z_0}^{z_1} 2\pi\varepsilon^{\mathrm{c}}(z,r)\mathrm{d}z\mathrm{d}r}{V_i} \tag{7.25}$$

where r and z are the coordinates in the radial direction and the axial direction, respectively, and V_i represents the individual volume of each subregion. Since all energy is dissipated into turbulence, the average turbulent energy dissipation rate can be evaluated in accordance with

$$\varepsilon_{\mathrm{avg}}^{\mathrm{c}} = \frac{N_{\mathrm{P}} D_i^5 N^3}{V} \tag{7.26}$$

In Eqn (7.26), N, N_{p}, D_{i}, and V are the rotating speed, power number, diameter of the impeller, and the whole volume of the stirred tank, respectively. The relative turbulence dissipation rate of each subregion is then defined as

$$\varphi_i = \frac{\varepsilon_i^{c}}{\varepsilon_{\mathrm{avg}}^{\mathrm{c}}} \tag{7.27}$$

Table 7.7 tabulates the volume of subregions and its corresponding relative turbulence dissipation rate adopted in the simulation.

To ensure energy conservation, scaling of relative turbulence dissipation was made such that

$$\sum V_i \varphi_i \tag{7.28}$$

Table 7.7: Relative dissipations and volumes of subregions.

Subregion	V_i	φ_i
1	0.0073	34
2l	0.0096	12
2o	0.0755	4.6
3u	0.2052	0.56
3l	0.0846	0.56
4u	0.0950	0.073
4l	0.1000	0.073
5u	0.0584	1.1
5l	0.0219	1.1
6u	0.2603	0.092
6l	0.0825	0.092

On the other hand, internal flow patterns representing the convection terms in the population balance equation are needed for droplet transport between subregions. A dimensionless flow value (i.e. pumping number between subregions) is defined as

$$Q_{i,j}^{*} = \frac{Q}{ND_i^3} \tag{7.29}$$

where Q represents the flow rate between subregions which is evaluated from experimentation, computational fluid dynamics (CFD) simulation, or even the empirical equation of Bourne and Yu (1994):

$$Q_{i,j}^{*} = 2.33\left(\frac{r}{D_i^3}\right) - 0.379 \tag{7.30}$$

Table 7.8 summarizes the flow direction and dimensionless flow values between subregions. In comparison to the flow field obtained from the numerical simulation with over 55,000 elemental volumes, the flow values and turbulent dissipation rates are found to be comparable. Given that the flow values and turbulent dissipation rates are consistent, the exact locations of the subregions are insignificant.

The merit of using the above multiblock model is the ability to assess the performance of the population balance model in a fast and robust way without hinderance by a large amount of mesh normally used in the CFD simulation. In this worked example, implementation of the multblock model is exemplified. It also serves to demonstrate the feasibility of using such a model for preliminary numerical population balance modeling.

Figure 7.22 shows a typical nonideal stirred tank studied in this example. As depicted, the stirred tank is 3.0 m in height with a 3.0 m inner diameter. The diameter of the impeller is 1.0 m with the specific impeller power number of 5.0. Synthesis consists of two immiscible

Table 7.8: Dimensionless flow values and the direction between subregions.

Flow Direction	Q^*
1 → 2i	0.7860
2i → 2o	1.1006
2o → 3u	1.3641
2o → 31	1.0492
3u → 4u	1.3641
31 → 41	1.0492
4u → 5u	0.4443
4u → 6u	0.9198
41 → 51	0.3417
41 → 61	0.7075
5u → 1	0.4443
51 → 1	0.3417
6u → 2i	0.1778
6u → 2o	0.7420
61 → 2i	0.1368
61 → 2o	0.5707

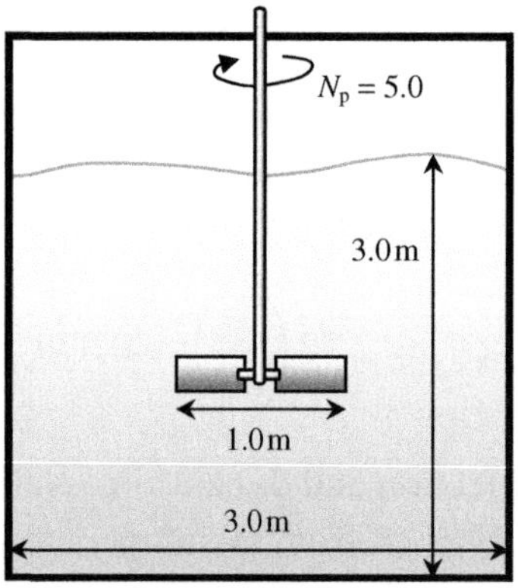

Figure 7.22
Schematic of the nonideal stirred tank (i.e. semibatch reactor) studied in this example. (For color version of this figure, the reader is referred to the online version of this book.)

liquid phases stirred continuously within the tank. Volume fraction of the dispersed phase of synthesis inside is assumed to be 0.4. Two systems corresponding to the beginning and the end of the synthesis process are modeled. Physical properties of the two systems are tabulated in Table 7.9. From the table, it should be noted that the surface tension becomes particularly low at the end of the synthesis process (i.e. Case 2). Along with the considerably high dispersed phase volume fraction (i.e. 0.4), flow structures and turbulence within the system are likely to be heterogeneous.

Table 7.9: Physical properties of the synthesis for the two studied systems.

	System 1	System 2
Viscosity (dispersed), μ_d	0.00067 Pa/s	0.0205 Pa/s
Viscosity (continuous), μ_c	0.00126 Pa/s	0.002 Pa/s
Density (dispersed), ρ_d	805 kg/m^3	923 kg/m^3
Density (continuous), ρ_c	1064 kg/m^3	1193 kg/m^3
Surface tension, σ	0.024 N/m	0.002 N/m

7.3.3 Worked Example

The prediction of the evolution of droplet size distribution in liquid–liquid flow under the influence of various flow conditions in a stirred tank is described below in order to demonstrate the application of the multiblock model proposed by Alopaeous et al. (1999).

Numerical features

In modeling nonideal stirred tank, turbulent dissipation and flow values are assumed to be homogeneous in each subregion of the multiblock model. The population balance equation for the dispersed phase is discretized into a number of classes such as the MUSIG model. The advection term in the population balance equation can be expressed as

$$N_{i,j,\text{in}} = \sum_{k=1}^{\text{nb}} \frac{Q_{k,j} N_{i,k}}{V_j} \tag{7.31}$$

$$N_{i,j,\text{out}} = \sum_{k=1}^{\text{nb}} \frac{Q_{j,k} N_{i,j}}{V_j} \tag{7.32}$$

where $N_{i,j}$ is the flow of number density for drop size class i per unit volume of block j. The subscripts in and out in Eqns (7.31) and (7.32) denote the convection speed in and out ofblock j. Q and $V$are the flow values between and the volume of the subregion, respectively, which were discussed previously. For continuous flow operation, additional terms are required to consider the convection in and out of the tank. The relative velocity between the continuous and dispersed phases can be evaluated from the modified Stokes law concerning which the influence of centrifugal forces is considered. This is based on the assumption that droplets in the system are sufficiently small, rigid, and spherical in shape. According to the modified Stokes law, the terminal velocity is given by

$$v_t = \frac{\tilde{g} a \Delta\rho}{18\mu} \tag{7.33}$$

where $\tilde{g}$ is the total acceleration due to gravity and centripetal acceleration, a is the drop diameter, and μ is the viscosity. The apparent dispersion viscosity and density for the system are calculated in accordance with

$$\mu_{\text{disp}} = \frac{\mu^{\text{c}}}{1-\alpha^{\text{p}}}\left(1+1.5\alpha\frac{\mu^{\text{p}}}{\mu^{\text{c}}+\mu^{\text{p}}}\right) \tag{7.34}$$

$$\rho_{\text{disp}} = \alpha^{\text{p}}\rho^{\text{p}} + \alpha^{\text{c}}\rho^{\text{c}} \tag{7.35}$$

All of the above values are used for evaluating properties of the whole system, including power input, impeller Reynolds number, and Kolmogroff length scale (Vermeulen et al., 1955). For the coalescence and break up processes of drops, coalescence and the break up mechanisms of Coulaloglou and Tavlarides (1977) and Bapat and Tavlarides (1985) are adopted. Beta distribution is adopted for the break up daughter distribution. For qualitative investigation, empirical constants of mechanisms are adopted according to the value proposed by Hsia and Tavlarides (1980). Two steady-state cases with different impeller speeds (i.e. 0.5 and 1.0 s^{-1}) are studied for both systems. Two transient cases with the impeller speed increasing from 0.5 to 1.0 s^{-1} and decreasing from 1.0 to 0.5 s^{-1} are also investigated to reveal the time-dependent droplet size evolution.

Numerical results

The population balance equation is discretized in accordance with the use of the class method in which the optimum number of classes is always case dependent and unknown. The effect of the number of classes is first investigated using the multiblock model. Figure 7.23 shows the relative errors for the predictions of three simulation cases with different numbers of drop classes. Since there is no analytical solution to the population balance equation, relative errors are evaluated by comparing the predictions with the results obtained from a maximum reasonable number of drop classes or a solution which is found to be independent of the number of classes. It can be seen that the relative error appears to be monotonic decreasing with the increment of the number of classes. This is due to the fact that the consideration of more drop classes results in better resolution of the droplet size distribution function and minimizes the numerical error caused by discretization. In all cases for both systems, 15 drop classes appear to be the optimum choice of a compromise between accuracy and computational requirement. Relative errors are found to be less than a few percent in all cases. For higher impeller speed, relative errors are slightly higher, which could be attributed to the higher turbulent dissipation causing a more rigorous droplet break up rate within the system. Comparing both systems with different syntheses, the differences in the physical properties of the two systems are found to be insignificant vis-a-vis the predictions and their attendant relative errors. In general, the steady-state solution requires fewer classes than the transient simulation.

After studying the influence of the number of classes, time-dependent droplet size changes within the tank are investigated. Two transient cases, where the emulsion in System 2 is initially stabilized by an impeller which is rotating at two given speeds ($N = 0.5$ or 1.0 s^{-1}), are studied. From the start of the simulations, impeller speed is either increased

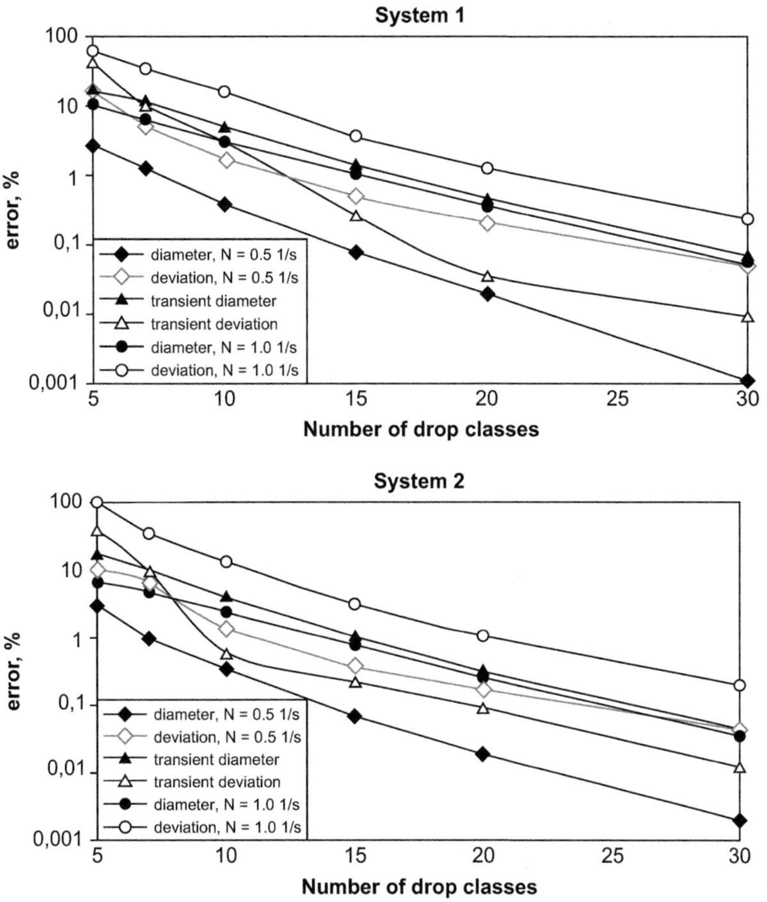

Figure 7.23

Relative errors for the three simulation cases with different numbers of drop classes. *After Alopaeous et al. (1999).*

from 0.5 to 1.0 s^{-1} or decreased from 1.0 to 0.5 s^{-1}. Fifteen drop classes are adopted for all simulations. In all simulations, the flow structure is assumed to be fully established immediately after the step change of impeller speed. Figure 7.24 shows the time-dependent changes of Sauter drop diameters in selected subregions for the two transient cases. In general, it can be observed that droplets break into smaller droplets when the impeller speed increases. Droplets are merged together forming larger droplets if the impeller speed decreases. Also shown is that the Sauter drop diameters in all three subregions exhibit a similar trend throughout the two transient processes. In both cases, droplets in subregion 1 are smaller than those in other regions, while droplets in subregion 6u are larger than in others. This is consistent with the relative turbulent dissipation rate in each subregion. With the highest turbulent dissipation in subregion 1, droplets tend to break up into those with much smaller sizes. Furthermore, one should also notice that the transient change is much slower

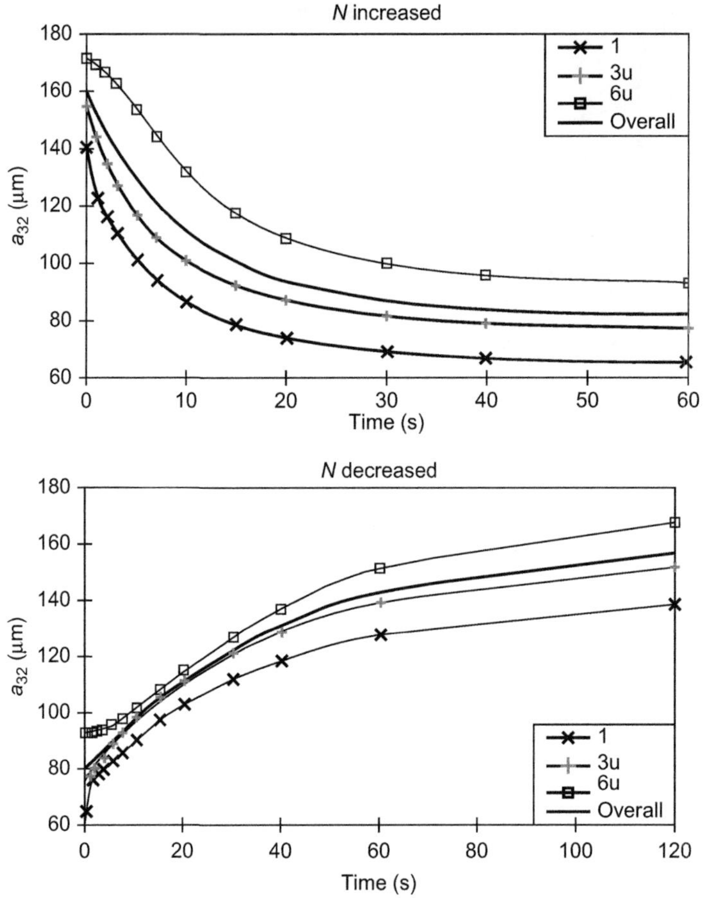

Figure 7.24

Time-dependent changes of Sauter drop diameter (a_{32}) in selected subregions in the multiblock model. *After Alopaeous et al. (1999).*

when the impeller speed is decreasing. According to the predictions, the system takes around 2 min to reach steady state with decreasing impeller speed. Meanwhile, the system becomes steady after 60 s with increasing speed.

A closer examination on the effect of turbulent dissipation can be carried out by comparing the steady-state droplet size distributions in three different subregions under two impeller speeds, as shown in Figure 7.25. For higher impeller speed, droplet size distributions in three subregions become relatively narrow, skewing toward a smaller drop size. In contrast, droplet size distributions cover a wider range of droplet size in lower impeller speed. With higher speed, local turbulent dissipation within each subregion becomes higher, leading to a more vigorous droplet break up process. The resultant droplet size distributions are, therefore, skewing toward smaller droplet size. Similarly, with lower turbulent dissipation rate, the

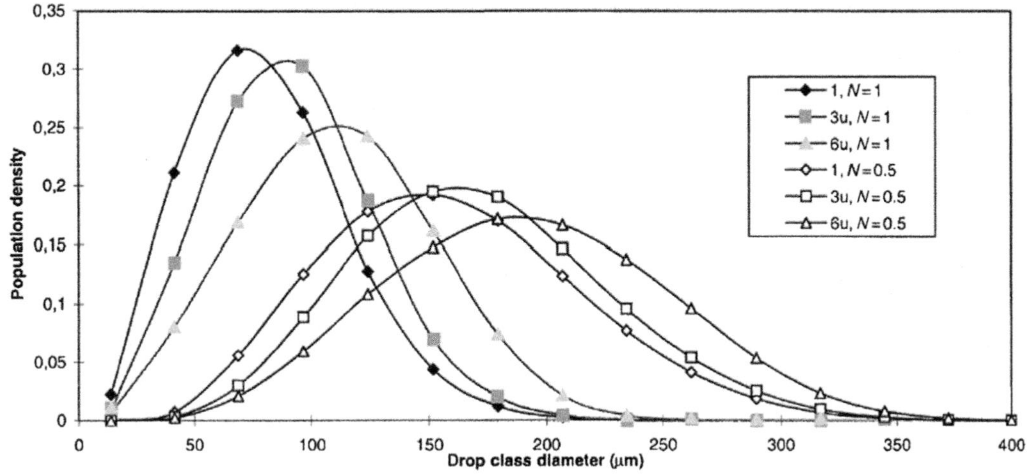

Figure 7.25

Steady-state droplet size distributions in three different subregions under two impeller speeds. *After Alopaeous et al. (1999).*

break up rates of droplets are relatively minor. In general, droplet size distributions in three subregions exhibit a very similar trend under two impeller speeds. With higher speed, the difference among subregions becomes larger due to the greater difference of absolute turbulent dissipation.

In order to assess the performance of the multiblock model, the same transient cases are simulated using a single block model (i.e. only one region to represent the whole tank). Turbulent dissipation is assumed to be homogeneous throughout the stirred tank. Figure 7.26 shows the relative difference of the predicted Sauter drop diameters between the multiblock and single block models. Although both models have the same system average turbulent dissipation, discretization using multiblock somehow predicts a slower transient response in comparison with the single block model. Such differences becomes more amplified in simulating the case with increasing impeller speed. This could be attributed to the localized turbulent dissipation rates adopted in the multiblock model. Droplet sizes thereby become inhomogeneous within the tank in transient cases resulting in significant differences from the predictions of the single block model. With the same average turbulent dissipation, when a steady state has been established in both cases, the differences between two models become insignificant.

Conclusion

In this worked example, a multiblock model for solving the population balance of droplets in a nonideal stirred tank has been introduced. Since turbulence becomes heterogeneous within the stirred tank, significant error might be introduced using a homogeneous single block model. The multiblock model simplifies the numerical procedures by scaling the flow values and local turbulent dissipation rate using reasonable assumptions. More importantly, the

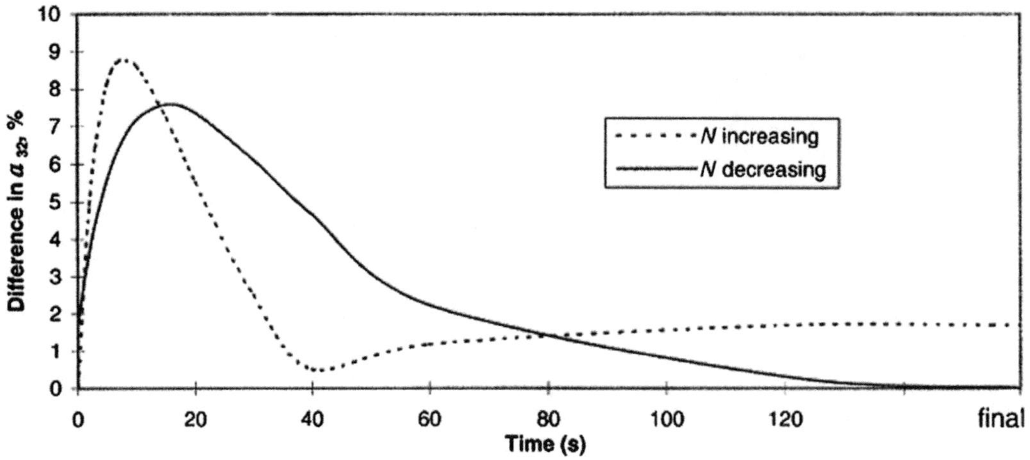

Figure 7.26
Relative differences of the predicted Sauter drop diameters (a_{32}) between the multiblock and the single block models. *After Alopaeous et al. (1999).*

multiblock model represents flow and turbulent structure in a more realistic manner. Droplet coalescence and break up mechanisms by Coulaloglou and Talvarides (1977) and Bapat and Talvarides (1985) are adopted to predict the evolution of droplet size distribution. Empirical constants of mechanisms are adopted according to the values proposed by Hsia and Tavlarides (1980). In general, the multiblock model yields reasonable predictions which align with qualitative observations. With higher energy input from the impeller, droplets in the system tend to break up into smaller droplets, resulting in a narrower drop size distribution. By comparing the droplet size in three subregions, the significance of localized turbulence and its effect on the drop size distribution have been exemplified. Although flow values and turbulence are estimated, the multiblock model presented here is sufficiently generic to describe the inhomogeneity in a stirred tank. With such a simplified numerical procedure, the model could be utilized to carry out parametric studies for the population balance of droplets. As demonstrated, it can be adopted for assessing the optimum number of drop classes for discretization. In a similar way, the model could also be used to calibrate empirical constants in coalescence and break up mechanisms or even test and compare performance of various coalescence and break up mechanisms (Alopaeous et al. 2002).

7.4 Population Balance Solutions to Gas–Particle Flow

7.4.1 Background

Fluidized bed reactors are widely used for their superiority in providing heat and mass transfer between the gas phase or liquid phase and solid particles, as well as the efficient mixing of reacting species. These reactors can be found in numerous unit operations in the chemical, petroleum, pharmaceutical, agricultural, food, and biochemical industries. Depending on the flow rate, the properties of the particles, and the type of fluid, several different states of fluidization are possible. Figure 7.27 illustrates a situation where fluid is passed upward through a bed of fine particles at different flow rates. The first situation of the fluidized bed reactor is the fixed bed. Here, the fluid merely flows through the empty crevices between the particles, and the bed remains normally stationary. When the gas or liquid flow increases, which subsequently results in the expansion of volume of the bed, the bed attains a state of minimum fluidization, and the particles begin to exhibit a fluid-like behavior. With further increase of gas or liquid flow, the fluidized bed reactor of a liquid–particle system behaves differently from a gas–particle system. The former expands smoothly and stays relatively homogeneous with no large variations of the particle concentration, while, in the latter, the excess gas starts to form channels and bubbles within the bed. These bubbles that are formed near the gas entry points at the bottom can rise up through the bed and finally burst when they reach the bed surface. They change in size and shape on the way up and collide with other bubbles or break up into smaller bubbles. At much higher flow rates, the nature of the particle bed in a gas–particle flow system changes to a state known as turbulent fluidization. Here, the bed height surface becomes less defined and the bubble shapes become more distorted, resulting in very complex turbulent-like motion of particle strands and clusters.

In many practical cases, solid particles exhibit a particle size distribution that changes continuously during the operating conditions. Consider the fluidized bed of polyolefin reactors, such as those described in Fan et al. (2004). At the point above the gas distributor, small catalyst particles are introduced, and, when they are exposed to the gas flow containing monomer, polymerization occurs. During the early stage, the catalyst particles undergo fragmentation into a large number of small particles which are rapidly encapsulated by newly

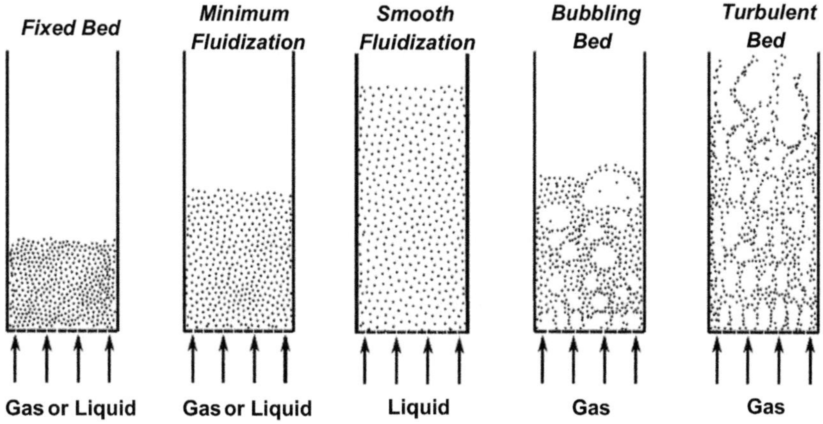

Figure 7.27
Schematic illustration of different states of fluidization.

formed polymer and thereby grow continuously. Segregation occurs due to the differences in polymer particle sizes. Fully grown polymer particles thereafter migrate to the bottom, where they are removed by the reactor. Smaller prepolymerized particles and fresh catalyst particles have a tendency to migrate to the top of the reactor and continue to react with the monomers. On one hand, polymer particles under certain undesirable operating conditions can stick to one another and form large agglomerates which can possibly undergo sintering and cause defluidization, especially when the reactor operates at a temperature close to the softening temperature of the polymer. On the other hand, particles can become brittle if the bed is too cold and may fracture, forming small fragments that elutriate with the gas flow. In order to rigorously account for particle-related phenomena, the population balance equation must be solved along with the balance equations for the gas phase.

7.4.2 Modeling Gas–Particle Flow via Direct Quadrature Method of Moment Multifluid Model

In this worked example, the fluidized bed is assumed to be isothermal with no chemical reactions. The particle size distribution is, thus, taken to be affected only by the aggregation and breakage in order to demonstrate the important link between population balance equations and the continuity and momentum balance equations.

The multifluid model is applied where the gas is considered as the continuous phase and the particles as secondary or disperse phases. Each particle is characterized by a specific diameter, density, and other properties. The primary and dispersed phases are characterized by volume fractions and, by definition, are subjected to algebraic constraint satisfying: $\sum_{k=1}^{N} \alpha^k = 1$. For the momentum balance equation for the gas, a simple Newtonian closure is employed for the stress tensor, while for the particles—especially for viscous or rapidly shearing regime, kinetic theory is used for the stress tensor. More details on the constitutive relations for the respective stress tensors can be found in Fan et al. (2004). For particle–fluid interaction, the drag force and body force due to gravity are accounted for. The drag force can be expressed as

$$\mathbf{F}_{\mathrm{D}}^{k,\mathrm{drag}} \equiv \sum_{l=1}^{N} B_{kl}\left(\mathbf{U}^l - \mathbf{U}^k\right) \tag{7.36}$$

If $\alpha^{\mathrm{c}} = \alpha^1$, the interphase drag term B_{kl} can be expressed as

$$B_{kl} = \begin{cases} \frac{3}{4} C_{\mathrm{D}} \dfrac{\alpha^{\mathrm{c}} \rho^{\mathrm{c}} (1-\alpha^{\mathrm{c}}) \left|\mathbf{U}^l - \mathbf{U}^k\right|}{d_{\mathrm{p},k}} (\alpha^{\mathrm{c}})^{-2.65} & \text{if } \alpha^{\mathrm{c}} > 0.8 \\[2ex] 150 \dfrac{(1-\alpha^{\mathrm{c}})^2 \mu^{\mathrm{c}}}{\alpha^{\mathrm{c}} d_{\mathrm{p},k}^2} + 1.75 \dfrac{(1-\alpha^{\mathrm{c}}) \rho^{\mathrm{p}} \left|\mathbf{U}^l - \mathbf{U}^k\right|}{d_{\mathrm{p},k}} & \text{if } \alpha^{\mathrm{c}} \le 0.8 \end{cases} \tag{7.37}$$

and the drag coefficient C_D is given by

$$C_D = \begin{cases} 0.44 & \text{if } Re_{p,k} > 1000 \\ \dfrac{24}{Re_{p,k}}\left(1 + 0.15 Re_{p,k}^{0.687}\right) & \text{if } Re_{p,k} \le 1000 \end{cases} \tag{7.38}$$

$$Re_{p,k} = \frac{\alpha^c \rho^c |\mathbf{U}^l - \mathbf{U}^k| d_{p,k}}{\mu^c}$$

For particle–particle interaction, an additional drag force, which can be expressed by the enduring contact force in the plastic regime, is incorporated into the momentum balance equations for the particles. The force can be expressed by

$$\mathbf{F}_D^{l,\text{drag}} \equiv \sum_{m=1}^{N} (B_{lm} + B')(\mathbf{U}^m - \mathbf{U}^l) \tag{7.39}$$

A simplified version of B_{lm} as described by Syamlal et al. (1993) is

$$B_{lm} = \frac{3(1 + e_{lm})\left(\dfrac{\pi}{2} + \dfrac{C_{flm}\pi^2}{8}\right)\alpha_l^p \rho_l^p \alpha_m^p \rho_m^p (d_{p,l} + d_{p,m})^2 g_{0_{lm}} |\mathbf{U}^l - \mathbf{U}^k|}{2\pi\left(\rho_l^p d_{p,l}^3 + \rho_m^p d_{p,m}^3\right)}$$

$$B' = \begin{cases} 0 & \text{if } \alpha^c > \alpha^{c^*} \\ 2.0 \times 10^8 \left(\alpha^c - \alpha^{c^*}\right) & \text{if } \alpha^c \le \alpha^{c^*} \end{cases} \tag{7.40}$$

where e_{lm} and C_{flm} are the coefficient of restitution and coefficient of friction, between lth and mth particles, respectively. The radial distribution function $g_{0_{lm}}$ is that derived by Lebowitz (1964) for a mixture of hard spheres

$$g_{0_{lm}} = \frac{1}{\alpha^c} + \frac{3 d_{p,l} d_{p,m}}{(\alpha^c)^2 (d_{p,l} + d_{p,m})} \sum_{\lambda=1}^{N} \frac{\alpha_\lambda^p}{d_{p,\lambda}} \tag{7.41}$$

The direct quadrature method of moment is adopted to describe the evolution of the particle size distribution for the polydisperse gas-particle flow. The transport equations for the weights (N_l) and weighted abscissae ($N_l L_l$) remain the same except the unknowns α in Eqn (6.172) are now evaluated based on particle length (L). For this case, the moment transform of the term S_{m_k} becomes

$$\begin{aligned} S_{m_k} = {} & \frac{1}{2}\sum_l \sum_m N_l N_m (L_l + L_m)^{k/3} a(L_l, L_m) - \sum_i \sum_j L_l^k N_l N_m a(L_l, L_m) \\ & + \sum_i \sum_j L_l^k N_m b(L_m) P(L_m, L_l) - \sum_i \sum_j L_l^k N_l b(L_m) \end{aligned} \tag{7.42}$$

In Eqn (7.41), the aggregation frequency can be expressed according to kinetic theory by the aggregation kernel as

$$a(L_l, L_m) = K_a(L_l + L_m)^2 \left(\frac{1}{L_l^3} + \frac{1}{L_m^3} \right)^{1/2} \tag{7.43}$$

where $K_a = \Psi_a g_{0_{lm}} (3\theta^p/\rho^p)^{1/2}$ with Ψ_a being the success factor for aggregation and the average temperature of particle θ^p that is determined by

$$\theta^p = \frac{\alpha_l^p \rho_l^p \theta_l^p + \alpha_m^p \rho_m^p \theta_m^p}{\alpha_l^p \rho_l^p + \alpha_m^p \rho_m^p} \left(\rho_l^p \frac{\pi}{6} d_{p,l}^3 + \rho_m^p \frac{\pi}{6} d_{p,m}^3 \right) \tag{7.44}$$

Similarly, the breakage kernel can be written as

$$b(L_l) = K_b(L_l + L_m)^2 \left(\frac{1}{L_l^3} + \frac{1}{L_m^3} \right)^{1/2} \tag{7.45}$$

where $K_b = \Psi_b \sum_{m=1}^{N} N_m g_{0_{lm}} (3\theta^p/\rho^p)^{1/2}$ with Ψ_b being the success factor for breakage. In Fan et al. (2004), Ψ_a and Ψ_b are assumed to be constants. Two fragment distribution functions are investigated, namely, the symmetric and erosion fragmentation (see Fragment Distribution Function of Particles in Section 5.3.1, p.146).

7.4.3 Worked Example

The evolution of particle size distribution in a laminar fluidized bed for gas–particle flow based on the investigation carried out by Fan et al. (2004) is described below. All numerical results presented have been computed through the use of multifluid code MFIX (see Fragment Distribution Function of Particles in Syamlal et al., 1993).

Numerical features

A second-order spatial discretization method is adopted. The linkage between velocity and pressure is achieved via the SIMPLE scheme. Due to the strong coupling between the phases through the drag forces, the PEA algorithm is used to treat the interphase coupling between gas and particles. An automatic time-step adjustment with an average time step of 3×10^{-4} s is adopted to speed up the calculation of the simulation.

The schematic diagram of the two-dimensional fluidized bed geometry is illustrated in Figure 7.28. The width is 10.1 cm and the height is 50.0 cm. The total number of grid cells are 15×50 with a cell width along the width of 0.67 cm and a cell height along the height of 1.0 cm. The initial static bed height is 15.9 cm. Gas at a velocity of 20 cm/s is injected from

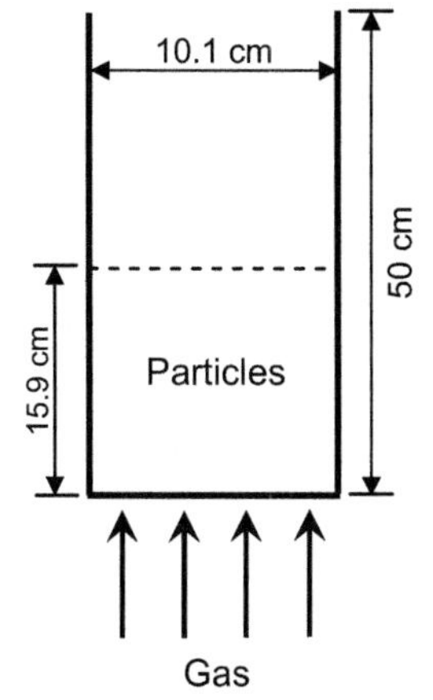

Figure 7.28
Schematic diagram of fluidized bed geometry.

the bottom. The density and viscosity of air at room temperature is used. Nonetheless, the physical properties of the particles are: density, $\rho^P = 2530$ kg/m^3, coefficient of restitution, $e = 0.8$, and the pack bed void fraction, $\alpha^{c*} = 0.38$.

Numerical results

The effect of the number of nodes $N = 2$, 3, and 4 are compared. Initial conditions of particle diameters and volume fractions for the same initial particle size distribution are given in Table 7.10.

In order to demonstrate the effectiveness of the direct quadrature method of moment in predicting the particle size distribution, constant values of the aggregation and breakage kernels are first used to assess the solution method. In Case 1, the aggregation kernel is set equal to 1×10^{-5} m^3/s^{-1} and the breakage kernel to 0.1 s^{-1}. In Case 2, the aggregation kernel is set equal

Table 7.10: Initial values of particle diameters and solid phase volume fractions for $N = 2$, 3, and 4 for the same particle size distribution ($m_0 = 32{,}050.825$ cm^{-3}, $m_1 = 670.285$ cm^{-2}, $m_2 = 15.245$ cm^{-1}, $m_3 = 0.385$, $m_4 = 1.09 \times 10^{-2}$ cm, $m_5 = 3.43 \times 10$ cm^2, $m_6 = 1.18 \times 10^{-5}$ cm^3, and $m_7 = 4.28 \times 10^{-7}$ cm^4).

	N	$l = 1$	$l = 2$	$l = 3$	$l = 4$
Particle diameter, $d_{p,l}$ (µm)	2	183	356		
	3	174	263	409	
	4	171	225	316	420
Solid volume fraction, α_l^p	2	0.274	0.356		
	3	0.196	0.229	0.205	
	4	0.157	0.157	0.157	0.157

After Fan et al. (2004).

to 1×10^{-5} m^3/s^{-1} and the breakage kernel to 1.0 s^{-1}. Hence, Case 1 is aggregation dominant while Case 2 is breakage dominant. Figures 7.29 and 7.30 show the particle size distribution at the middle of the fluidized bed for Cases 1 and 2 at different times. For the aggregation dominated case, it can be seen that smaller particles aggregate, larger particles are produced, and the volume fraction of smaller particles decreases with time. A broad distribution of particle sizes exists at 15 s in the bed. For the breakage dominated case, particles begin to become smaller due to breakage, resulting in smaller particles being produced due to excessive breakage. The particle size distributions appear to be rather different at different times.

Corresponding to the above two cases, Figures 7.31 and 7.32 depict the volume-average mean particle size d_{32} through the realization of aggregation and breakage kernels stipulated in Eqns (7.42) and (7.44) for aggregation dominated, denoted as Case 3, and breakage dominated, denoted as Case 4. In Case 4, the aggregation-success and breakage-success factors are set to values of 0.001 and 0.0001. In case 5, the aggregation-success and breakage-success factors have the same value of 0.001. As can be seen from Figure 7.30, d_{32} increases with time. Initially, particles begin to aggregate. Due to their increase in size, they migrate toward the bottom of the bed. These particles near the bottom continue to aggregate until defluidization occurs. Generally, d_{32} values predicted through $N=2$, 3, and 4 are similar except that some discrepancies with regard to the defluidization dynamics are observed. In Case 5, the volume-average mean particle sizes d_{32} predicted by the direct quadrature method of moment for $N=2$, 3, and 4 are very similar. Nevertheless, different values of N may produce different results, particularly when segregation becomes significant.

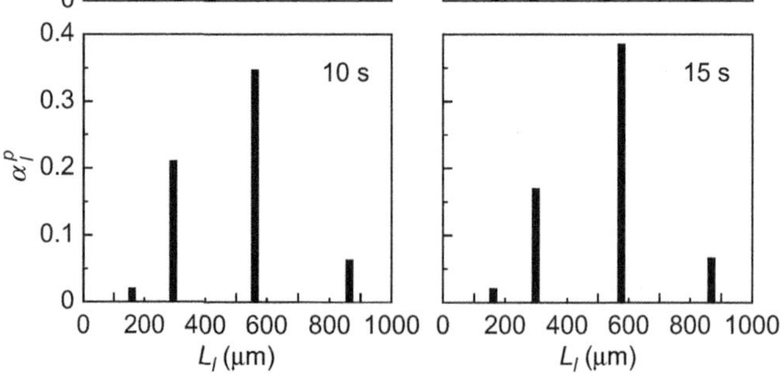

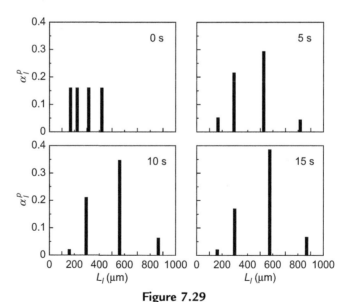

Figure 7.29
Particle size distribution at the middle of the fluidized bed at 0, 5, 10, and 15 s for Case 1. *After Fan et al. (2004).*

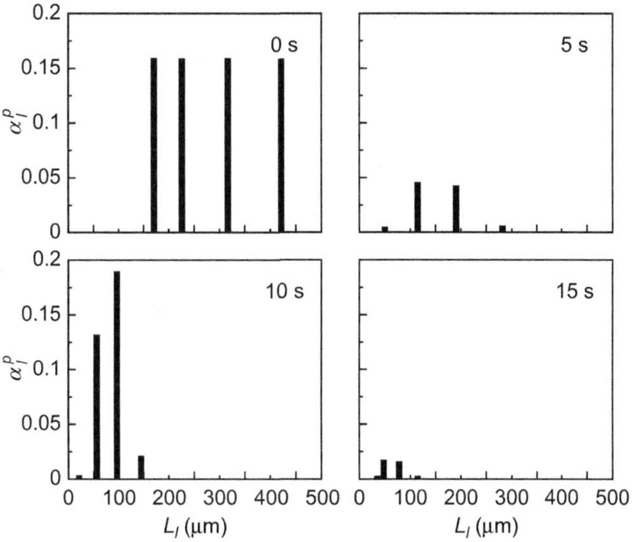

Figure 7.30
Particle size distribution at the middle of the fluidized bed at 0, 5, 10, and 15 s for Case 2. *After Fan et al. (2004).*

All the above simulations have adopted symmetric fragmentation. Figure 7.33 illustrates a comparison between symmetric fragmentation and erosion for the volume-average mean particle size using $N=3$. The decrease of d_{32} with time for symmetric fragmentation and erosion appears to be rather similar. Nevertheless, a closer investigation of the volume-average abscissas, particle volume fractions, and weights (see Figure 7.34) reveals that

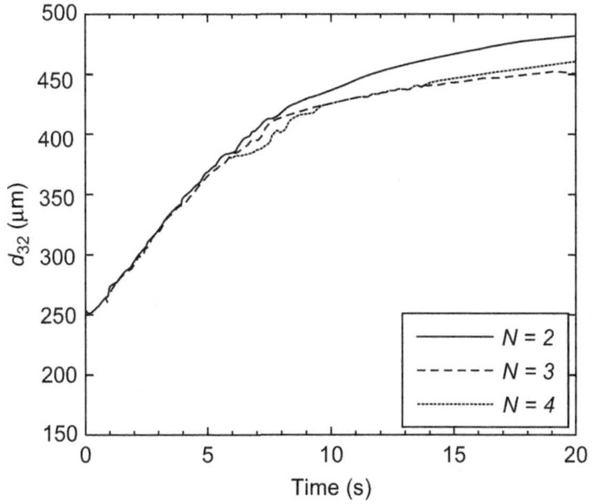

Figure 7.31
Volume-average mean particle diameter for Case 3. *After Fan et al. (2004).*

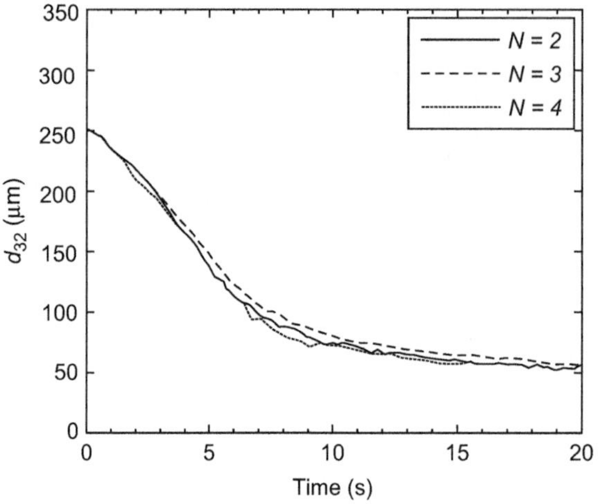

Figure 7.32
Volume-average mean particle diameter for Case 4. *After Fan et al. (2004).*

smaller fragments generated via the erosion process become smaller at a faster rate when compared to symmetric fragmentation.

Conclusion

This worked example demonstrates the effectiveness of the direct quadrature method of moment multifluid model in predicting the evolution of the particle size distribution due to aggregation and breakage of particles based on constant and kinetic-theory kernels in fluidized bed reactors. Both types of kernel are able to describe the phenomena of particle growth, segregation, and

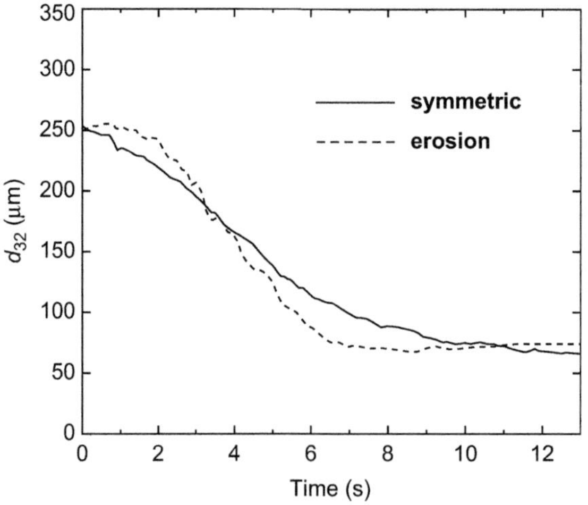

Figure 7.33
Effect of symmetric fragmentation and erosion on the volume-average mean particle diameter. *After Fan et al. (2004).*

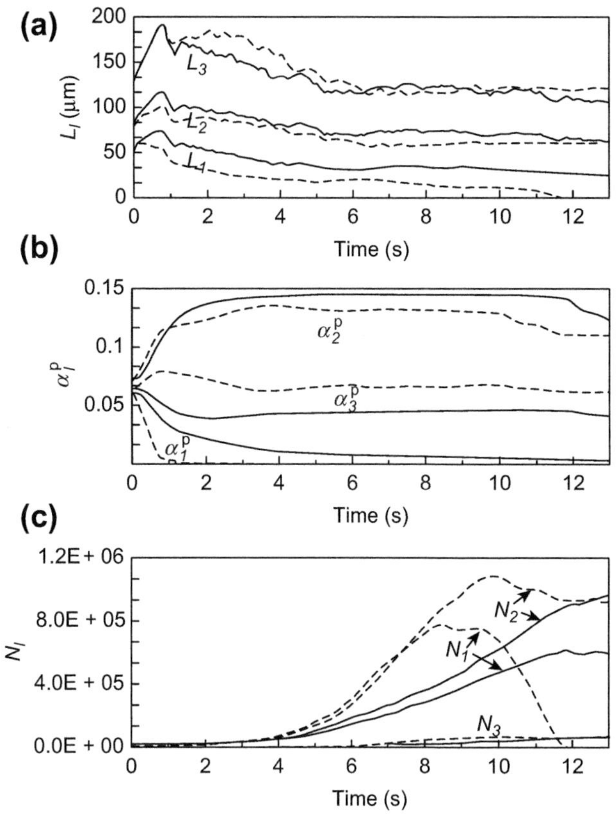

Figure 7.34

Comparison between symmetric fragmentation (solid lines) and erosion (dashed lines): (a) abcissae, (b) solid volume fraction, and (c) weights. *After Fan et al. (2004).*

elutriation due to aggregation and breakage. The performance of the model using two, three, and four nodes is assessed. It can be shown that three nodes appear to be appropriate for representing the particle size distribution. Erosion has been found to be a less effective breakage mechanism in the presence of aggregation. This is because particles reduce their volume by a factor of two with symmetric fragmentation, while erosion can result in the formation of a fragment whose volume can be much smaller than the volume of the other fragment.

7.5 Population Balance Solutions to Liquid–Particle Flow

7.5.1 Background

Population balance of solid particles suspended in a continuous liquid and size evolution due to aggregation and breakage processes are discussed in this section. The presence of aggregation and breakage of solid particles in liquid flow can be found in many technologically important systems, especially in areas of biology, crystallization, and polymer science. As summarized in Chapter 5, aggregation processes can be categorized as perikinetic coagulation due to Brownian motion, shear-induced orthokinetic coagulation, and differential sedimentation aggregation. The shear-induced orthokinetic coagulation which becomes significant for particle size range 1−40 μm is one of the most significant processes in a practical multiphase system. Discussions of the worked exampled are centered on the fundamental work by Wang et al. (2005) where shear-induced aggregation and breakage processes in laminar Taylor-Couette flow has been investigated. The Taylor-Couette flow is a classical flow phenomenon which occurs in between the gap of two concentric cylinders of which the inner cylinder is rotating at a specific angular speed while the outer cylinder is fixed or rotating in the opposite direction. Figure 7.35 illustrates three typical flow regimes of laminar Taylor-Couette flow.

In the case where the outer cylinder is fixed, centrifugal flow is driven by the rotating inner cylinder. As the angular velocity increases, flow within the gap undergoes a progression of flow instability from laminar Couette flow to laminar vortex flow, leading to a wavy vortex flow and, subsequently, turbulent vortex flow. The flow transition between regimes is related to the azimuthal Reynolds number, which is given by

$$\mathrm{Re_a} = \frac{\rho \omega r_i d}{\mu} \tag{7.46}$$

where ω is the angular velocity of the inner cylinder and d is the annular gap between the inner and the outer cylinder (i.e. $r_o - r_i$). Based on the definition given in Eqn (7.45), a critical azimuthal Reynolds number $\mathrm{Re_c}$, depending on the specific geometry of the flow device,

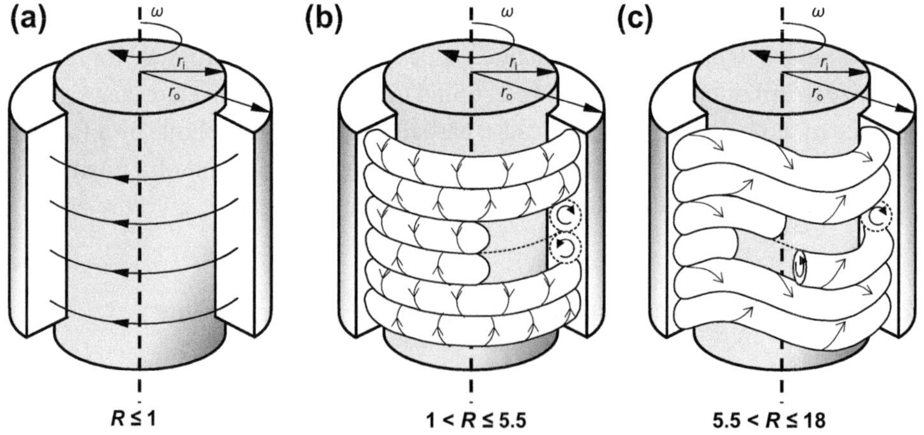

Figure 7.35
Three typical flow regimes of laminar Taylor-Couette flow: (a) laminar Couette flow, (b) laminar vortex flow, and (c) wavy vortex flow.

can be realized for the onset of Taylor instability. As shown in Figure 7.35(a) a, the range of azimuthal Reynolds number ratio (i.e. $R = \mathrm{Re_a}/\mathrm{Re_c}$) can be used to characterize the different flow regimes. In laminar Couette flow, the flow is characterized by a pure azimuthal flow without any axial velocity component. When the ratio is higher than 1 (i.e. $R > 1$), the Couette flow becomes unstable yielding a steady-state secondary flow pattern that appears in axisymmetric vortical structures (known as Taylor vortices) (Figure 7.35(b)). As the outer cylinder is fixed, the shear is significantly higher in the vicinity of the inner cylinder in comparison to other regions (especially near the outer cylinder). Further increasing the angular speed of the inner cylinder, the vortices are subjected to greater instability, forming a wavy vortex flow (Figure 7.35(c)). At sufficient high speed (typically $R > 18$), the vortices transform into turbulent vortex flows.

7.5.2 Modeling Liquid—Particle Flow via Quadrature Method of Moment

In retrospect, the quadrature method of moment is based on the same fundamental concepts that have been used in the formulation of the direct quadrature method of moment. The main difference is that the weights and abscissae for the quadrature method of moment are now found by solving the nonlinear system of algebraic equations that result from the application of the quadrature approximation for the moments of the size distribution based on particle length as the internal coordinate:

$$m_k = \sum_{i=1}^{N} N_i L_i^k \tag{7.47}$$

where N is the number of nodes. While the direct quadrature method of moment solves the weights and weighted abscissae directly from transport equations, as described in Section 6.6.3, the quadrature method of moment solves the transport equations for the moments (Eqns (6.152) and (6.153)) and recovers the weights and abscissae via the use of the product-difference algorithm from the low-order moments (Gordon, 1968). A brief description of the algorithm is given below.

The first step of the product-difference algorithm is to construct a matrix **Q** with components ($Q_{i,j}$) beginning from the moments. The first and second columns of **Q** are

$$Q_{i,1} = \delta_{i1} \tag{7.48}$$

$$Q_{i,2} = (-1)^{i-1} m_{i-1} \tag{7.49}$$

for $i \in 1, \ldots, 2N+1$. Since the final weights can be corrected by multiplying by the true moment m_0, the calculations can proceed by assuming a normalized distribution (i.e. $m_0 = 1$). Therefore, the remaining components can be found from the product-difference algorithm according to

$$Q_{i,j} = Q_{i,j-1}Q_{i+1,j-2} - Q_{i,j-2}Q_{i+1,j-1} \tag{7.50}$$

for $j \in 3, \ldots, 2N + 1$ and $i \in 1, \ldots, 2N + 2 - j$. For $N = 2$, the matrix **Q** becomes

$$\mathbf{Q} = \begin{bmatrix} 1 & 1 & m_1 & m_2 - m_1^2 & m_3 m_1 - m_2^3 \\ 0 & -m_1 & -m_2 & -m_3 + m_2 m_1 & 0 \\ 0 & m_2 & m_3 & 0 & 0 \\ 0 & -m_3 & 0 & 0 & 0 \\ 0 & 0 & 0 & 0 & 0 \end{bmatrix} \tag{7.51}$$

An N-dimensional vector α can be defined as

$$\alpha_i = \begin{cases} 0 \\ \dfrac{Q_{1,i+1}}{Q_{1,i}Q_{1,i-1}} & \text{for } i \in 2, \ldots, 2N \end{cases} \tag{7.52a}$$

With this vector, such as described in Eqns (7.52a and 7.52b), a symmetric tridiagonal matrix can be obtained from the sums and products of α_i. In other words,

$$\begin{cases} a_i = \alpha_{2i} + \alpha_{2i-1} & \text{for } i \in 1, \ldots, 2N - 1 \\ b_i = -\sqrt{\alpha_{2i+1}\alpha_{2i-1}} & \text{for } i \in 1, \ldots, 2N - 2 \end{cases} \tag{7.52b}$$

where a_i and b_i are the diagonal and codiagonal components of the Jacobi matrix given by

$$\mathbf{J}_N = \begin{bmatrix} a_1 & b_1 & & & & & \\ b_1 & a_2 & b_2 & & & & \\ & \cdots & \cdots & \cdots & & & \\ & & b_{i-1} & a_i & b_{i+1} & & \\ & & & \cdots & \cdots & \cdots & \\ & & & & b_{N-2} & a_{N-1} & b_{N-1} \\ & & & & & b_{N-1} & a_N \end{bmatrix} \tag{7.53}$$

In order to determine the weights (N_i) and abscissae (L_i), the eigenvalues (λ) and eigenvectors (v) of the above matrix **J** need to be calculated. Since **J** is symmetric and diagonal, the abscissae are directly the eigenvalues, i.e. $L_i = \lambda_i$, while the weights can be found from the first components of the respective eigenvectors: $w_j = m_0 v_{j1}^2$. By using the actual values of m_0, the nonnormalized weights are subsequently obtained.

Appropriate aggregation and breakage kernels are adopted. As discussed in Chapter 5, the aggregation due to shear-induced collision in laminar flow is handled by the kernel derived by Smoluchowski (1917). The collision efficiency can be expressed in terms of the flow number (Eqn 5.14), which incorporates considerations of shear-induced and van der Waals forces for both permeable and impermeable particles. For the breakage kernel, both exponential and power-law kernels are employed. Fragment distribution function for symmetric fragmentation is considered.

7.5.3 Worked Example

In this worked example, numerical studies by Wang et al. (2005) which focus on the laminar Taylor vortex regimes are discussed below. All numerical results presented have been computed through the use of commercially available computer code FLUENT 6.0.

Numerical Features

CFD techniques are used to predict the velocity and shear distribution in a Taylor-Couette device. Predictions from the CFD simulations are validated against particle image velocimetry (PIV) measurements. A schematic illustration of the Taylor-Couette device is shown in Figure 7.36. The flow device consists of two concentric cylinders of 432 mm length. The inner cylinder with a radius of 34.9 mm is constructed of polished stainless steel. For optical measurements, the outer cylinder is made of transparent precision glass Pyrex tubing with an inner radius of 48.6 mm. The annular gap between the two cylinders is 13.7 mm. For this device, the critical azimuthal Reynolds number, Re_C, is 82.8. Surfactant-free polystyrene latex spheres were used as suspended particles with diameters of 9.6 μm ± 7.4%. The isoelectric point for these particles is at a pH of around 3.8. The density of solution is carefully adjusted to 1.055 g/cm^3 in order to avoid sedimentation and radial migration effects. The viscosity of the solution is 1.097×10^{-3} kg/m s.

For ease of numerical computations, the Taylor-Couette flow is considered to be axisymmetric. A two-dimensional computational domain which spans the meridional section is adopted for

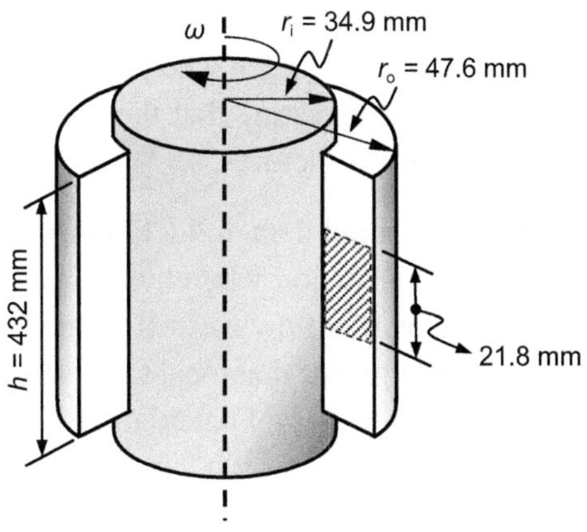

Figure 7.36
Schematic illustration of the Taylor-Couette device and of the computational domain used in the simulation (gray shaded section).

simulation. Two specific flow conditions are considered. The azimuthal Reynolds number ratios of the two conditions are $R = \mathrm{Re_a}/\mathrm{Re_c} = 4$ (Laminar vortex flow) and $R = \mathrm{Re_a}/\mathrm{Re_c} = 12$ (Laminar wavy vortex flow). Based on the PIV measurements, the critical azimuthal Reynolds number is $\mathrm{Re_c} = 84.5$. For mesh resolution, grid sensitivity studies show that a minimum of 30 radial nodes is required to obtain a grid independent solution. Due to a steep velocity gradient near the inner cylinder region, five layers of mesh are required within 1 mm from the surface of the inner cylinder. In the axial direction, the end effect is assumed to be negligible. It is, therefore, sufficient only to simulate two vortex pairs for validation. The length of computational domain is thereby set as 2.18 cm, which corresponds to the axial wave length obtained from experiment. A minimum of 120 nodes and 180 nodes are placed in the axial direction for the cases $R = 4$ and $R = 12$, respectively. Periodic boundary conditions are applied at both axial ends and all walls are treated as nonslip boundary.

The transport equations for all six moments ($2N$ where $N = 3$) are implemented in FLUENT 6.0 employing user defined function. A steady-state flow field is first obtained from the CFD simulations. The convergence criteria are set to be 10^{-6} for all variables. There are three different shear flow conditions (i.e. $R = 3.48$, 6.96 and 10.44) with initial particle number densities, $m_0 = 4.02 \times 10^{11}$, 4.378×10^{11} and 4.02×10^{11}, respectively. All variables are assumed to be uniform as the initial field in all simulations.

Numerical results

Figure 7.37 shows the comparison between PIV measurement and CFD prediction velocity field of vortices for the cases $R = 4$ and $R = 12$. For case $R = 4$, to reduce measurement noise, the PIV results are representation of a time-averaged velocity field based on 20 sequential field measurements. Similarly, 200 sequential field measurements fields are used for time averaging in the case $R = 12$. In general, high velocity flows are generated near the inner cylinder regions in both cases. Significant velocity reductions are also observed at the top due to the non-slip condition of the fixed outer cylinder. The predicted velocity fields are in good agreement with the PIV measurements, suggesting that the mesh resolution and CFD model are capable of capturing the general flow structures.

A closer comparison between PIV measured data and CFD results is shown in Figures 7.38 and 7.39, which further supports the conclusion whereby CFD calculations can accurately represent the time-dependent laminar vortex flow as well as the time-averaged velocity for laminar wavy vortex flow. Figure 7.39 suggests that local time average shear rates can be used to determine the aggregation and breakage kernel calculations.

The shear history of an inert particle is exemplified in Figure 7.40. Since the shear is significantly larger in the region near the inner cylinder than in the outer cylinder or in the middle of the annular gap, particles in the liquid flow tend to experience a range of different magnitudes of shear rates as they circulate in the vortices. The histogram of shear rates in

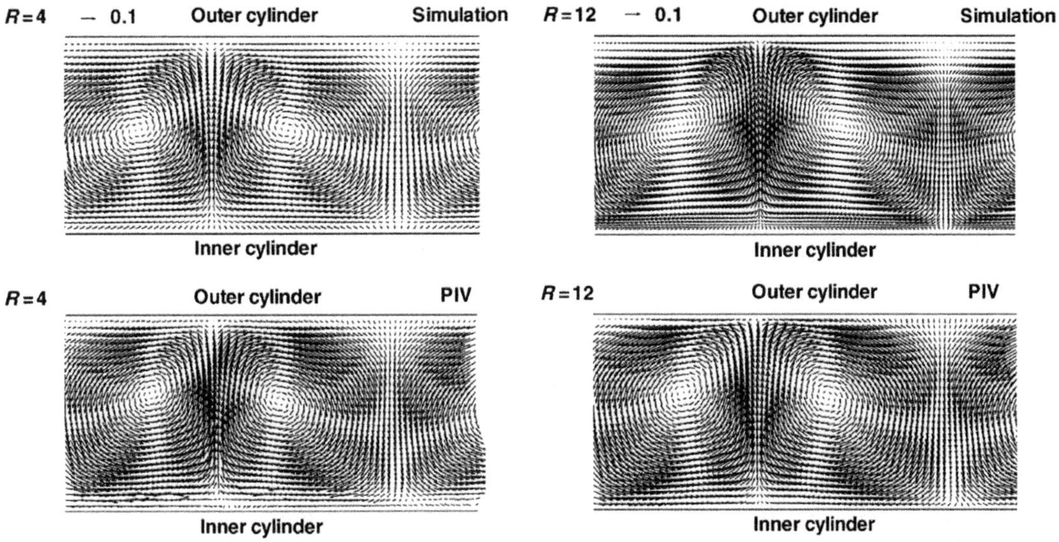

Figure 7.37

Comparison of PIV measurement (bottom) and CFD prediction (top) velocity field for cases $R = 4$ and $R = 12$. *After Wang et al. (2005).*

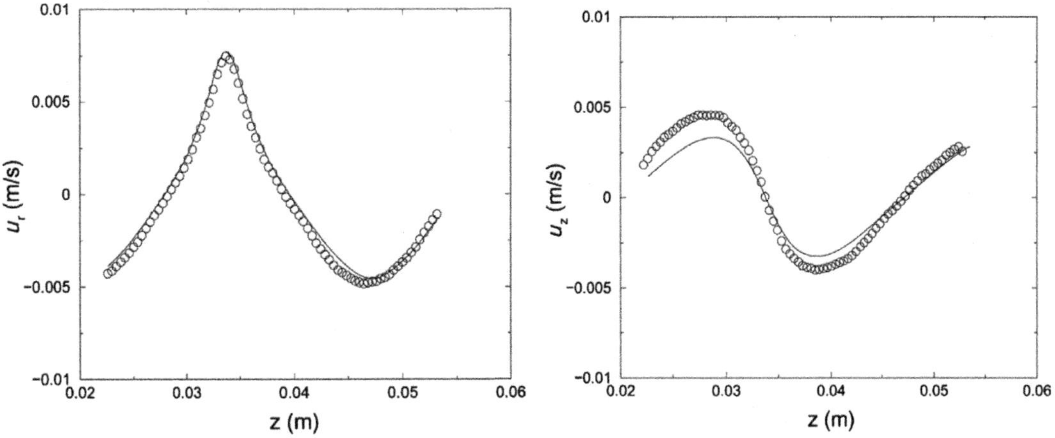

Figure 7.38

Comparison of measured and predicted radial, u_r, and axial, u_z, velocity profiles at $r = 39.4$ mm for $R = 4$. *After Wang et al. (2005).*

Figure 7.41 further shows that the shear rate distribution is rather wide, which can be taken to be similar to a log-normal distribution rather than a Gaussian distribution. Because of the nonlinearity exhibited by the shear rate, it is paramount that the local shear instead of the spatial average shear is employed in the calculation of the aggregation and breakage rates.

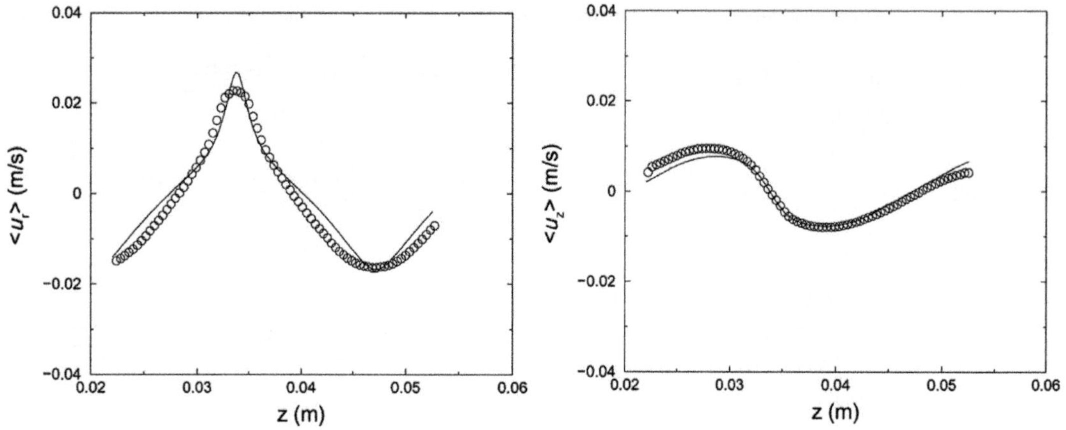

Figure 7.39
Comparison of measured and predicted mean radial, $<u_r>$, and mean axial, $<u_z>$, velocity profiles at $r=39.4$ mm for $R=12$. *After Wang et al. (2005).*

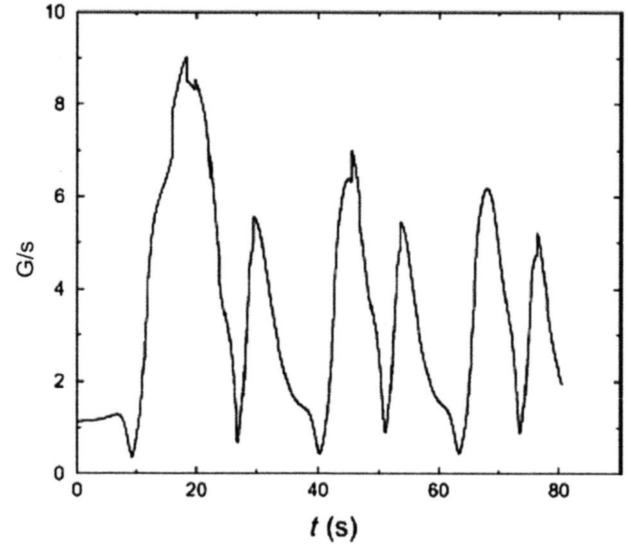

Figure 7.40
Shear history of inert particle for $R=3.48$. *After Wang et al. (2005).*

Figure 7.42 depicts the evolution of the mean particle size d_{43} ($=m_4/m_3$) for three Reynolds number ratios: $R=3.48$, 6.96, and 10.44. Symmetric fragment distribution is considered. The first stage of aggregation is due to the formation of doublets, which lasts up to 60 min. Larger aggregates are subsequently formed during exponential growth before reaching a plateau at a later time. It can be seen that the final mean aggregate size is dependent on the azimuthal Reynolds number, and, hence, the shear rate can be seen in the figure. As the spatial

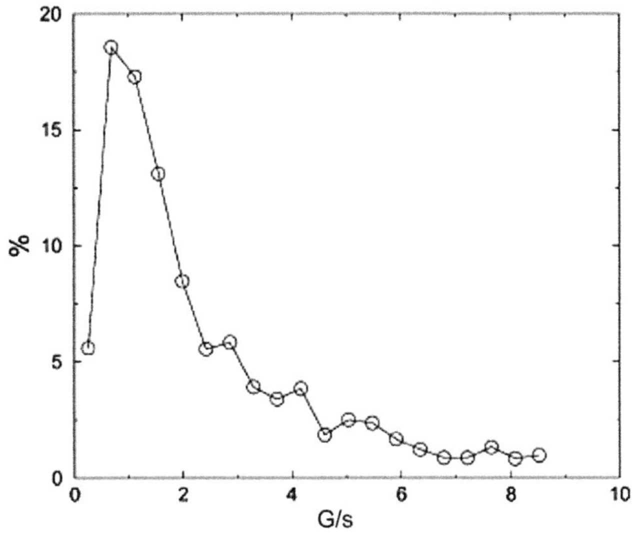

Figure 7.41
Histogram of shear rates for $R = 3.48$. *After Wang et al. (2005).*

average shear rate increases, the steady-state value of d_{43} decreases. This indicates that shear-induced breakage of aggregates is important, since higher shear rates produce more frequent collisions which lead to a higher chance of large aggregates being formed. In contrast, during the first two stages of aggregation, larger mean shear rates result in larger

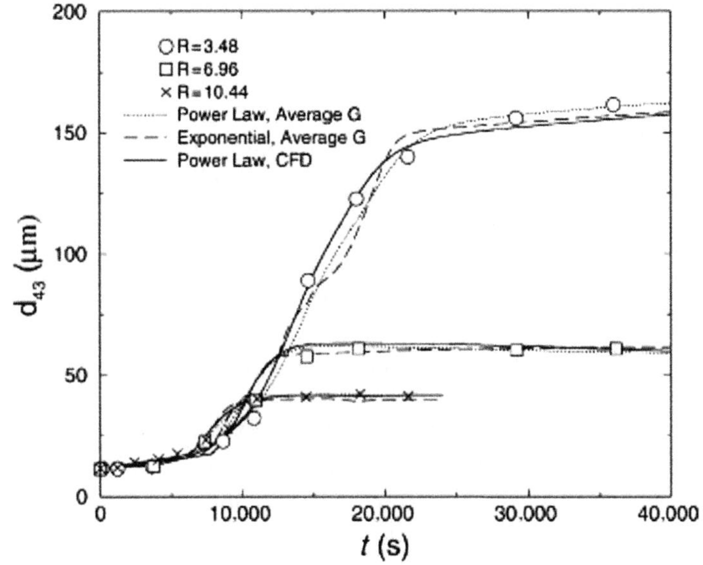

Figure 7.42
Evolution of mean particle size d_{43} with time at different shears ($R = 3.48$, 6.96, and 10.44) and at particle volume fraction $2.3 \pm 0.1 \times 10^{-4}$. *After Wang et al. (2005).*

mean particle size growth rates, which in turn suggests that aggregates have not fully grown to a stage where they are large enough to be vulnerable to shear-induced breakage.

Conclusion

This worked example demonstrates the use of the quadrature method of moment to predict the evolution of the particle size distribution due to aggregation and breakage of particles in a Taylor-Couette flow. In order to obtain an accurate shear field for the simulation of aggregation and breakage, the CFD model is validated through the velocity field measurements obtained via PIV. It can be seen that the mean particle size grows before reaching a steady-state value. With increasing shear rates, the steady-state value of the mean particle size decreases, which suggests that breakage of aggregates is the primary reason for the cessation of particle growth.

7.6 Summary

The examples in this chapter have been purposefully selected to demonstrate the application of population balance modeling in conjunction with the use of multifluid model based on interpenetrating media framework to resolve different classifications of multiphase flows.

Population balance solutions to gas–liquid flow are demonstrated through the consideration of two worked examples. First, the one-group approach based on a single transport equation for the average bubble number density and the MUSIG model are considered for dispersed bubbly flows in vertical pipes. For the former, three forms of coalescence and break up mechanisms of Wu et al. (1998), Hibiki and Ishii (2002), and Yao and Morel (2004) are employed, while for the latter, the coalescence and break up mechanisms of Prince and Blanch (1990) and Luo and Svendsen (1996) are adopted. The relative merits of both approaches are assessed. Second, the two-group approach based on two transport equations for the average bubble number density is proposed for transition cap-bubbly flows in vertical pipe. Coalescence and break up mechanisms based on the constitutive relations according to Hibiki and Ishii (2000) are applied to achieve closure for the transport equations. Initial utilization of the three-fluid model and two-group average bubble density equations have shown to be promising in the prediction of interfacial transport in transition cap-bubbly flows.

Population balance solutions to liquid–liquid flow are exemplified through a stirred tank worked example. In this particular case, the multiblock model suggested by Alopaeous et al. (1999) is described in order feasibly to predict the evolution of droplet size distribution. For the coalescence and break up processes of drops, the coalescence and break up mechanisms of Coulaloglou and Tavlarides (1977) and Bapat and Tavlarides (1985) are adopted. The multiblock model simplifies the numerical procedures by scaling the flow value and local turbulent dissipation rate with some reasonable assumptions in order that the droplet size distribution is predicted more accurately by the applied drop coalescence and break up mechanisms.

Population balance solutions to gas–particle flow are realized through the worked example of a fluidized bed reactor. By considering the particles which exhibit a fluid-like behavior, the multifluid model is applied. The particle size distribution is determined through the consideration of the direct quadrature method of moment. Realistic kernels based on kinetic theory by Fan et al. (2004) have been shown to describe the phenomena of particle growth, segregation, and elutriation due to aggregation and breakage of particles in a fluidized bed reactor.

Population balance solutions to liquid–particle flow in a Taylor-Couette device are provided via the consideration of the quadrature method of moment to predict the evolution of particle size distribution. The worked example given focuses on the importance of obtaining an accurate shear field for the determination of the local shear-induced aggregation and breakage.

The window of opportunity for the consideration of population balance and multifluid models are plentiful across different classifications of multiphase flows. These combined models represent the critical technology for current multiphase flow investigation. Because of the many successful applications that have been achieved thus far, some emerging areas on the use of the population balance approach are further expounded in the next chapter.

CHAPTER 8

Future of the Population Balance Approach

8.1 Introduction

The generic concept of population balances has existed for well over the last four decades. It was not until very recently that population balances have been considered as a modeling tool for the prediction of particle size distributions in multiphase flows. Identification of population balance models that are essential to consolidate their applicability with regard to particular complex processes of particles has been considerably enhanced in view of the increasing fidelity of particle size measurement techniques.

Particles that may undergo the processes of coalescence and break up in gas–liquid or liquid–liquid flows, or aggregation and breakage in gas–particle flows or liquid–particle flows, such as those that have been thoroughly elaborated upon in this book, can significantly dictate the behavior of the multiphase system under consideration. For a practical multiphase system on the industrial scale, solutions of population balances of particles must invariably be sought for particles undergoing displacements in turbulent flow patterns. The mutual coupling that needs to be realized between the discrete particle phase and continuous fluid phase makes the computational problem challenging because the solution over the entire flow domain requires progressive calculations to be carried out in space and time between the two phases. Although there remains uncertainty regarding a number of aspects for the formulation of the population balance models together with that of the numerical solutions attained through computational multiphase fluid dynamics, the combination of these two frameworks, which still require further development, has proven to be rather effective in resolving complicated flow problems. Based on the promising features of population balance analysis of multiphase flows, some emerging areas on the use of the population balance approach are discussed in this chapter.

8.2 Emerging Areas on the Use of the Population Balance Approach

8.2.1 Natural and Biological Systems

In natural systems, the ability to forecast the production rate of fish in a controlled environment is important for commercial fishery operations. Prediction of size distributions and related growth rates as the fish population ages is certainly of considerable interest. A population balance approach to modeling fishery dynamics has been developed by

Thompson and Cauley (1979) in order to provide important information as to size/age history, growth rates, and fish size distribution predictions. This information can allow the formulation of harvesting policies, the design of optimal operating conditions, or the evaluation of other economic aspects of fish farming.

There are many similarities between the flow of gas bubbles, liquid drops, or solid particles in multiphase flows which allow population balance models to be applicable to fish systems. Treating fish as discrete particles, a population balance equation can be solved to account for their growth, birth (nucleation), and disappearance by "death" mechanisms. In the simplified model adopted by Thompson and Cauley (1979), which only includes the growth process, the model has been found to be capable of predicting the length and growth rate dependence on age, along with the additional advantage of predicting fish size distribution characteristics such as average size, standard deviation, and coefficient of variation. Figure 8.1 shows the distributions predicted by the population balance model compared to the histograms of experimental data of Brown (1957) for the growth of brown trout. Based on the initial distribution at 5 weeks, which is assumed to be a normal distribution in order to establish the initial condition necessary to solve the population balance equation, the theoretical distribution is predicted at 11 weeks. In general, it is reasonable to assume that smaller fish will suffer a greater rate of mortality than the larger fish. The lack of agreement between the model prediction and experimental data could be attributed to the absence of mortality of the fish sample in the model.

Although it is recognized that the simplified population balance model does not include the effects of life cycle changes, such as birth, mortality, possible changes in shape or attainment

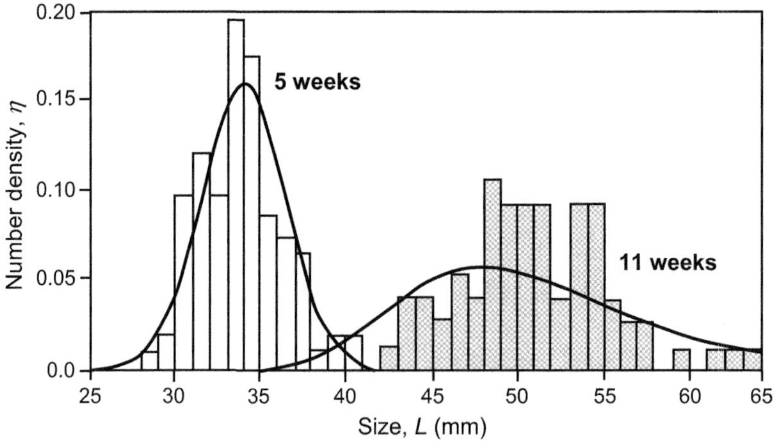

Figure 8.1

Comparison of a size distribution prediction to experimental data of Brown (1951) for growth of brown trout. The number density, η, is normalized to unity. *Source: After Thompson and Cauley (1979).*

of sexual maturity, the explicit dependence of environmental or seasonal changes or physiological events (such as achievement of sexual maturity) on fish growth rates. For example, water temperature and pH, light, food availability, or population crowding, can be further included by evaluating the dependence of these growth parameters on these environmental conditions. This will result in a more comprehensive population balance model to better characterize the actual dynamics of fish systems.

The development of a population balance model for biological systems is also of major interest not only in the manufacture of commodity chemicals, pharmaceuticals, food products, and agricultural products via the use of bioreactors but, more importantly, for the analysis of a range of biomedical problems such as cancer, a disease characterized by an imbalance between cell growth and apoptosis (Fredrickson, 2003). In particular, a cell population balance model describing the cell cycle dynamics of myeloma (malignant tumor of the bone marrow) cell cultivation has been developed by Liu et al. (2007). To describe the cell cycle dynamics of myeloma, a three-stage population balance model is solved for the G1 phase, the S phase, and the G2M phase. The G1 phase accounts for the synthesis of the metabolites required for DNA replication, the S phase accounts for DNA synthesis, the G2 phase accounts for the gap between DNA synthesis and cell division that serves to ensure proper replication of the DNA, and the M-phase characterizes cell division. In Liu et al. (2007), the G2 phase and M phase are combined in their analysis. Figure 8.2 illustrates the cell cycle of the myeloma cell line. Cell volume accumulates throughout the whole cell cycle. It is postulated that cells in the G1 phase are not ready to enter into the S phase unless they attain a certain volume. Similarly, a critical cell volume is also required for cell division in the G2M phase. The S to G2M transition is assumed to be dependent on the level of DNA content. Cell death is assumed to occur ubiquitously.

With regard to all the population balance equations of the G1 phase, S phase, and G2M phase, the transient development of each respective cell number density function of the three phases and the advection along the internal coordinate, such as the cell volume of the state space, appears on the left-hand side of the equations. For the *first* equation for the G1 phase, the source and sink terms on the right-hand side are represented by the accumulation, growth,

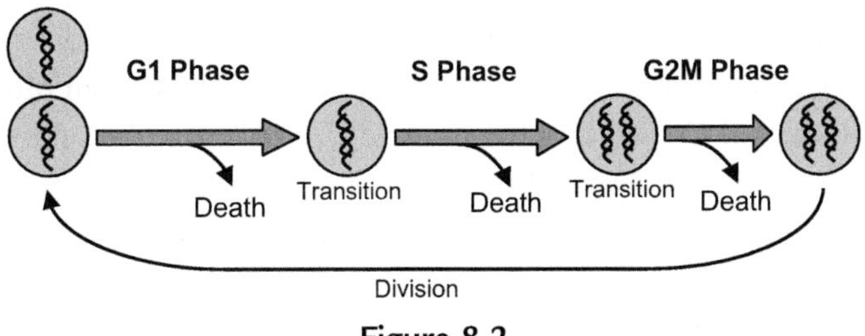

Figure 8.2
Schematic diagram of cell cycle of myeloma cell line. *Source: After Liu et al. (2007).*

source, transition, death, and dilution in sequence. The growth term accounts for the loss of cells due to the fact that they grow into bigger cells. The source term represents the rate of birth of cells originating from the division of all bigger cells in the G2M phase. The transition term accounts for the loss of cells due to transition to the next phase. The death term expresses the loss of cells due to cell death. The dilution term represents the dilution of cell number density caused by feeding. For the *second* equation for the S phase, an additional term denoting the loss of cells stemming from DNA synthesis is presented by the advection along the internal coordinate of the relative DNA content on the left-hand side of the equation. The sink terms on the right-hand side are represented by the death and dilution. For the *third* equation for the G2M phase, the source term on the right-hand side denotes the source cells, which are derived from the S phase. The sink terms on the right-hand side are represented by the division term that expresses the loss of cells due to cell division, death, and dilution. More details on the mathematical formulation of the various terms in the population balance equations can be found in Liu et al. (2007).

This three-stage population balance model, which has been formulated in terms of cell volume and relative DNA content by Liu et al. (2007), has demonstrated the capability of describing myeloma cell population dynamics through the prediction of the evolution of the cell volume and relative DNA content distributions. The model predictions of viable and dead cell densities have been found to be in reasonable agreement with the measured data. The model can be further improved to account for sophisticated intrinsic physiological behaviors and to incorporate more complex kinetics during the cell cycle of the myeloma cell line.

8.2.2 Bulk Attrition

Population balance models are increasingly being utilized to describe particle attrition in the processing of particles which can arise due to mechanical action, phase change within particles, chemical reaction within particles, or rapid temperature transients. Particle attrition as a result of mechanical damage can be found in the deformation of bulk particle material, such as in moving beds, during flow in solids storage hoppers, such as in granular flow, belt conveying, and power mixing systems. In most instances, the process of attrition must be avoided or minimized for such systems (Ouchiyama et al., 2005). Fluidized bed jet attritors which employ sonic gas jets can provide substantial particle attrition in order to control the particle size in the fluid bed. Controlling the particle size of coke is of enormous importance in such systems since large particles will result in slugging and poor circulation while too many fine particles with a size less than 70 μm result in agglomeration and poor fluidization (McMillan et al., 2007). In both pressurized and oxygen-enriched fluidized bed combustion, the process of attrition can increase the reaction rates by removing the sulfated layers on limestone particles. Limestone is widely used in these systems not only as a sulfur dioxide adsorbent but also as a bed material to transport heat from the fuel particles to boiler tubes. Depending on the environment being experienced by the particles in these systems, the primary fragmentation of these particles, in some circumstances, may lead to a significant loss of nonreacted fine particles. The knowledge of particle size distribution is therefore an important factor since it can dramatically affect the fluidization behavior, heat transfer, and pollutant formation (Saastamoinen et al., 2010).

In the absence of phase change, chemical reaction or temperature transients, two mechanisms of attrition have been considered—particles undergoing abrasion and fracture (or fragmentation). The mechanisms of abrasion and fracture are schematically illustrated in Figure 8.3. In abrasion, an elemental process of degradation of a particle is considered to yield one daughter particle of almost similar size to the parent particle along with a large number of extremely very fine small daughter particles. The occurrence of abrasion usually results in the daughter particles having smooth edges. Nevertheless, when breakage arises due to fracture, daughter particles of intermediate sizes can be found during the attrition process. In addition, the daughter particles generally have an appearance of sharp jagged edges. The key aspect in

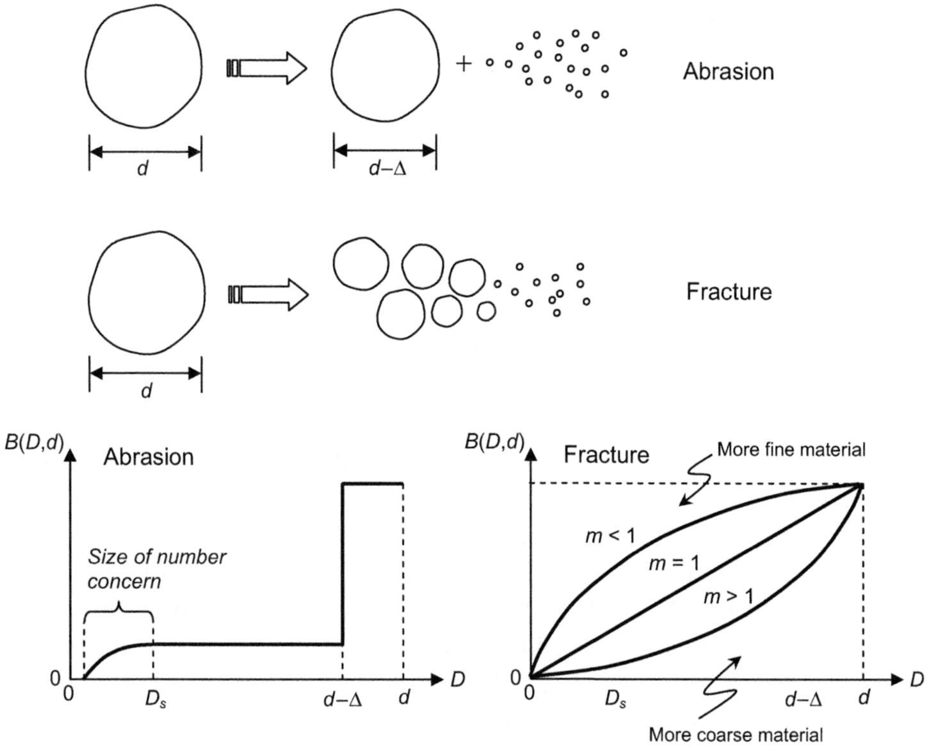

Figure 8.3
Mechanisms of abrasion and fracture and associated breakage functions $B(D, d)$. *Source: After Ouchiyama et al. (2005).*

modeling such mechanisms is the formulation of appropriate source terms in the population balance equation which can be realized via proper characterization of breakage functions associated with the daughter size distributions for abrasion and fracture. Ouchiyama et al. (2005) proposed breakage functions describing the respective attrition processes which are depicted in the same figure (Figure 8.3). It can be seen that the process of particle abrasion leads to only a small amount of change in the size of the particle. The breakage index m as shown in the figure for attrition due to fracture, which needs to be appropriately determined, can be taken as a property of the particles which may be dependent upon the characteristics such as the brittleness and local microstructure of the material.

8.2.3 Crystallization

The application of population balance models to describe discrete growing elements has been instrumental in the modeling of crystallization. Population balance provides the dynamic evolution of crystal size distribution in all stages of nucleation and growth. This approach is particularly well developed in the analysis of crystallization systems because it can take into account all possible scenarios such as secondary nucleation, aggregation, and breakage (Gadewar and Doherty, 2004). Not only that, but it has been shown to be a powerful tool for studying the effects of various operating conditions such as impurity, solvents, and cooling rates. It is also a process modeling tool for product development, process design and optimization, and for control because it deals with the population of crystals and their interactions with processing conditions (Christofides et al., 2007).

Crystal morphology prediction has become an increasingly important consideration because products that are obtained through the crystallization process often have specific crystal shape and size distributions which can have an impact on the downstream processing, final product properties, and end-use performance. Population balance modeling assumes a monosize dimension through the definition of the particles, as the diameter of a sphere having the same volume of particles obviously misses important information about the evolutionary behavior for the morphology of crystals as a population. Motivated by this loss of morphological information, the development of a morphological population balance model has been realized. A multidimensional morphological population balance model (based on polyhedral solid shapes) was proposed by Ma and his colleagues, which has the capability to predict complicated crystal shapes (Ma et al., 2008). In essence, the model simulates the size-related dimension evolution of crystals for each identified independent crystal face. After identifying the initial facet structures of the crystals and determining the appropriate kinetic parameters for the crystal growth rates of each face, the corresponding population balance equation to predict the location variations for each of the independent crystal face allows the temporal reconstruction of crystal shapes. Figure 8.4 depicts the crystal shape evolution with time for potash aluminum with the crystal exhibiting a pure octahedral diamond-like morphology;

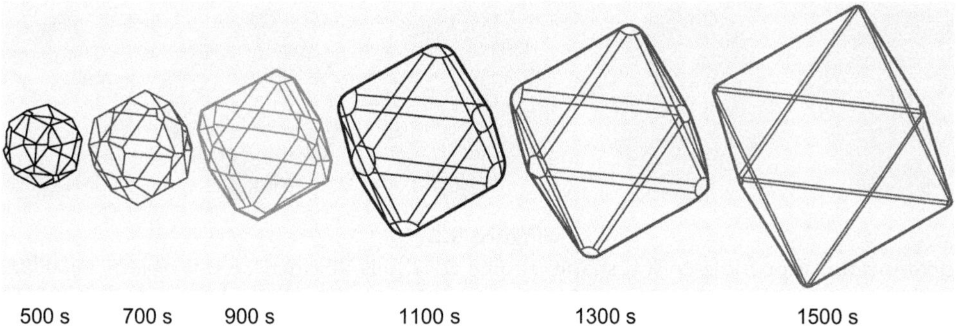

Figure 8.4

Crystal shape evolution during the crystallization process of potash aluminum. *Source: After Ma et al. (2008).* (For a color version of this figure, the reader is referred to the online version of this book.)

such shape has been confirmed via observations by Amara et al. (2004). Similarly, Ma and his colleagues extended their model to predict the crystal growth behavior of protein crystal such as the enzyme Hen Egg White (HEW) lysozyme, which can be seen in Figure 8.5. The crystal shape of the HEW lysozyme through time can be clearly observed to evolve from needle-like crystals to plate-like crystals during crystallization (Liu et al., 2010).

Another factor that can possibly affect shape evolution is attrition, particularly when small fragments of larger crystal get chipped off through damage. If the crystal has distinct and sharp corners, the attrition rate is significantly larger in comparison to the same mass of a spherical crystal. Through repeated impacts of the initially sharp cornered crystal, the crystal shape can be extensively modified to the point where the attrition rates can be affected. This, in turn, changes the material properties of the remaining crystal, which do not necessarily stay the same throughout the attrition process. Further repeated impacts on the crystal may lead to fatigue and eventually result in breakage. Briesen (2009) developed a multidimensional morphological population balance model to predict the removal, or rounding off, of the sharp

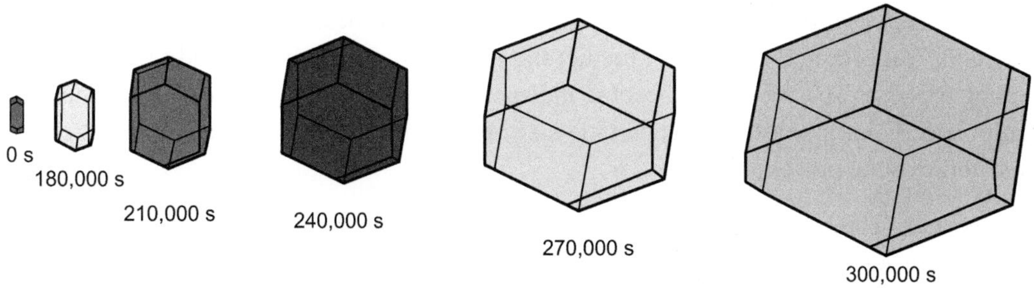

Figure 8.5

Crystal shape evolution during the crystallization process of HEW lysozyme. *Source: After Liu et al. (2010).* (For a color version of this figure, the reader is referred to the online version of this book.)

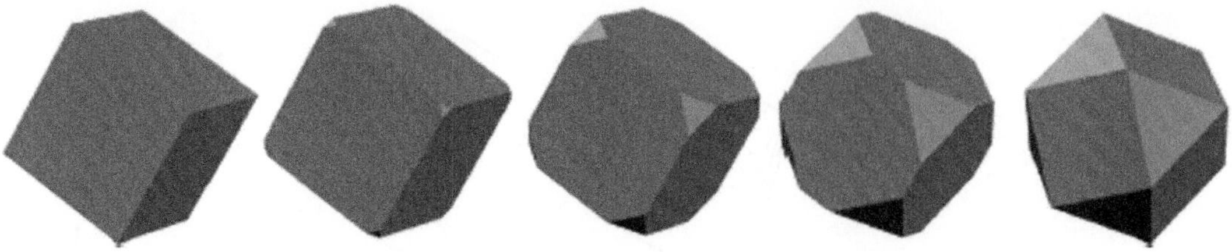

Figure 8.6
Characterization of possible crystal shape leading to a fully abraded crystal. *Source: After Briesen (2009).* (For a color version of this figure, the reader is referred to the online version of this book.)

corners of a crystal. Beginning from a perfect cubic shape, the model incorporates the characterization of possible crystal shapes leading to a fully abraded crystal such as that illustrated in Figure 8.6.

8.2.4 Synthesis of Nanoparticles

For flame-generated nanoparticle production systems, processes of particle transport, filtration, and dispersion all require the ability to predict aggregate formation in various fluid flows and its effect on flow. The synthesis process of nanosized particles, such as commercially manufactured carbon blacks, fumed silica, and pigmentary titania, can encapsulate a wide range of complex particle and fluid dynamics. Commonly produced through a flame process in a high-temperature environment, the coagulation, surface growth, and sintering of these particles are strongly coupled with the solid-phase chemistry and gas-phase chemistry, local temperature, and fluid flow at different length scales ranging from laminar flow to turbulent flow. On the one hand, the competition between the gas-phase and surface reactions can normally dictate the dominant trends of particle formation and growth. On the other hand, the effect of turbulence significantly influences the mixing rates as well as the reaction rates. Both of these aspects can have a dramatic impact on the final product due to various processing conditions. In the quest to optimally design flame reactors and better understand the basic phenomena regarding the evolution of these particles, numerical simulations, especially the ability to predict the particle size distribution and resolve the coupling of these processes, have become indispensable in order to gain deeper insights and to tailor specific properties such as particle size and shape, and crystal phase and, in return, better improve the process itself.

Tsantilis et al. (2002) employed the population balance model to investigate the significance of various particle formation pathways during the flame synthesis of nanosized titania particles by titanium tetraisopropoxide (TTIP) oxidation in a premixed methane–oxygen flame. The pathways are: (1) pure gas-phase thermal decomposition, (2) gas-phase hydrolysis and (3) gas-phase and surface thermal decomposition of TTIP. Dynamics of the growth of nanosized titania particles are investigated via an efficient moving sectional model accounting for simultaneous gas-phase and surface reactions followed by coagulation and sintering, therefore covering the whole spectrum of particle evolution until the final collection point, typically governed by coagulation and small sintering rates. By adopting the sectional method which involves discretizing the particle state space, the population balance equation for the discrete sections of the titania number concentration is solved. In this equation, the forcing terms are the coagulation due to Brownian collision spanning the free-molecular, transition and continuous regimes and suitable reactions rates for the gas-phase chemical reaction (nucleation), surface growth, and sintering. It should be noted that the sectional method tends to resolve the full distribution; it is thus more expensive than the moment method. The current model can be extended by studying other processes involving gas and surface growth reactions such as soot formation and synthesis of nanosized particles of controlled size, where monitoring the very early stages of particle formation and growth is of great importance.

Akroyd et al. (2011) investigated the modeling of detailed fully coupled nanosized particle formation in a turbulent reacting flow. They proposed a projected fields (PF) method which approximates the joint composition probability density function (PDF) transport equation that includes the Interaction by Exchange with the Mean (IEM) micromixing model (Villermaux and Devillon, 1972). It is worthwhile to differentiate between the *model* and the *numerical method* in the PF method. The PDF transport equation is considered to be the model, while the weighted field approximation and projection are taken to be the *numerical method*. For this reason, the reference to a PF method arises. In order to aptly describe the evolution of nanosized particles in a turbulent reacting flow, the PF method is combined with the population balance model based on the method of moments with interpolative closure (MOMIC) and detailed titania chemistry. The forcing terms in MOMIC are the inception, (nucleation) due to Brownian collision spanning the free-molecular, transition and continuous regimes and surface growth. Considering the example of the flame process via the chloride process for the industrial synthesis of titania, simulations have shown that inception occurs in a mixing zone near the reactor inlets. Most of the mass of the nanosized particles arises from surface growth downstream of the mixing zone with a narrower size distribution occurring in the regions of higher surface growth. The methodology has provided a reasonable description of the velocity field and gas-phase composition PDF with reasonable computational effort.

8.3 Summary

New areas of emerging applications that have been described in this chapter have certainly made population balance analysis a significant component of the modeling repertoire. This chapter focuses on the various attributes of population balance models that include the prediction of size distributions and related growth rates of fish population in natural systems and the study of cell cycle dynamics of cancerous tumors in biological systems, consideration of particle attrition as a result of abrasion and fracture due to mechanical action, morphology prediction relating to the dynamic evolution process of crystals in all stages of nucleation and growth, and the synthesis of nanoparticles in particle production processes. It should be noted from that the latest trends and the development of population balance models as presented have not been aimed to be exhaustive or comprehensive but rather to direct the attention of the reader to the bigger picture of modeling commensurate with the potential of applications across a wide range of different challenging and complex problems. As the future continues to unravel, it is anticipated that computational multiphase fluid dynamics and population balance modeling will remain at the forefront of intensifying research and development as long as the fundamentals of the vast majority of fluid flows and particle processes remain unresolved.

References

Chapter 1

Agrawal, K., Loezos, P.N., Syamlal, M., Sundaresan, S., 2001. The role of mesoscale structures in rapid gas—solid flows. J. Fluid Mech. 445, 151—185.
Fredrickson, A.G., Ramkrishna, D., Tsuchiya, H.M., 1967. Statistics and dynamics of prokaryotic cell populations. Math. Biosci. 1, 327—374.
Friedlander, S.K., 2000. Smoke, Dust and Haze: Fundamentals of Aerosol Dynamics, second ed. Oxford University Press, Oxford.
Fujita, M., Yamaguchi, Y., 2007. Multiscale simulation method for self-organization of nanoparticles in dense suspension. J. Comp. Phys. 223, 108—120.
Hidy, G.M., Brock, J.R., 1970. The Dynamics of Aerocolloidal Systems. Pergamon, Oxford.
Hulburt, H.M., Katz, S., 1964. Some problems in particle technology: a statistical mechanical formulation. Chem. Eng. Sci. 19, 55—574.
Lu, J., Tryggvason, G., 2008. Effect of bubble deformability in turbulent bubbly upflow in a vertical channel. Phys. Fluids 20-040701.
Marshall, J.S., 2009. Discrete-element modeling of particulate aerosol flows. J. Comp. Phys. 228, 1541—1561.
Pandis, S.N., Seinfeld, J.H., 1998. Atmospheric Chemistry and Physics: From Air Pollution to Climate Change. John Wiley & Sons, New York.
Quan, S.P., Lou, J., Schmidt, D.P., 2009. Modeling merging and breakup in the moving mesh interface tracking method for multiphase flow simulations. J. Comp. Phys. 228, 2660—2675.
Ramkrishna, D., Borwanker, J.D., 1973. A puristic analysis of population balance. Chem. Eng. Sci. 28, 1423—1435.
Ramkrishna, D., 1979. Statistical models of cell populations. Adv. Biochem. Eng. 11, 1—47.
Ramkrishna, D., 1985. The status of population balances. Rev. Chem. Eng. 3, 49—95.
Ramkrishna, D., 2000. Population Balances. Theory and Applications to Particulate Systems in Engineering. Academic Press, San Diego.
Randolph, A.D., Larson, M.A., 1964. A population balance for countable entities. Can. J. Chem. Eng. 42, 280—281.
Serizawa, A., 2003. Some remarks on mechanisms of phase distribution in an adiabatic bubbly pipe flow. Multiphase Sci. Tech. 15, 79—88.

Chapter 2

Adeniji-Fashola, A., Chen, C.P., 1990. Modeling of confined turbulent fluid-particle flows using Eulerian and Lagrangian schemes. Int. J. Heat Mass Transf. 33, 691—701.
Agee, L.J., Banerjee, S., Duffey, R.B., Hughes, E.D., 1978. Some Aspects of Two-Fluid Models and Their Numerical Solutions, Second OECD/NEA Specialists Meeting on Transient Two-Phase Flow. CEA, France.
Bagchi, P., Balachandar, S., 2002. Effect on free rotation on the motion of a solid sphere in linear shear flow at moderate Re. J. Fluid Mech. 473, 379—388.

Banerjee, S., Chan, A.M.C., 1980. Separated flow model I. Analysis of the averaged and local instantaneous formulations. Int. J. Multiphase Flow 6, 1–24.

Besnard, D.C., Harlow, F.H., 1988. Turbulence in multiphase flow. Int. J. Multiphase Flow 14, 679–699.

Brackbill, J.U., Johnson, U.N., Kashiwa, B.A., Vander-Heyden, W.B., 1997. Multiphase Flows and Particle methods, CFD Annual Conference of Computational Fluid Dynamics, Victoria, Canada.

Burry, D., Bergeles, G., 1993. Dispersion of particles in anisotropic turbulent flows. Int. J. Heat Mass Transf. 4, 651–664.

Chen, P.P., Crowe, C.T., 1984. On the Monte-Carlo method for modeling particle dispersion in turbulence. ASME FED 10, 37–42.

Clift, R., Grace, J.R., Weber, M.E., 1978. Bubbles, Drops and Particles. Dover Publications, New York.

Crowe, C.T., Sharma, M.P., Stock, D.E., 1998. Particle-source-in cell (PSI-Cell) model for gas-droplet flows. J. Fluid Eng. 99, 325–332.

Delhaye, J.M., Achard J.L., 1976. On the averaging operators introduced in two-phase flow modeling, Proceedings of CSNI Specialists Meeting on Transient Two-phase Flow, Toronto, Canada.

Drew, D.A., Passman, S.L., 1999. Theory of Multicomponent Fluids. Springer-Verlag, Berlin.

Drew, D.A., 1983. Mathematical modeling of two-phase flow. Ann. Rev. Fluid Mech. 15, 261–291.

Elghobashi, S.E., 1994. On predicting particle-laden turbulent flows. Appl. Sci. Res. 52, 309–329.

Faeth, G.M., 1983. Recent advances in modeling particle transport properties and dispersion in turbulent flow. In: Proceedings of ASME-JSME Thermal Engineering Conference, vol. 2, pp. 517–534.

Fede, P., Simonin, O., 2006. Numerical study of the subgrid fluid turbulence effects on the statistics of heavy colliding particles. Phys. Fluids 18, 045103.

Gosman, A.D., Iosnnides, E., 1983. Aspects of computer simulation of liquid-fueled combustors. J. Energy 7, 482–490.

Gouesbet, G., Berlemont, A., 1999. Eulerian and Lagrangian approaches for predicting the behaviour of discrete particles in turbulent flows. Prog. Energy Combust. Sci. 25, 133–159.

Hoomans, B.P.B., Kuipers, J.A.M., Briels, W.J., Swaaij, W.P., 1996. Discrete particle simulation of bubble and slug formation in a two-dimensional gas-fluidized bed: a hard-sphere approach. Chem. Eng. Sci. 51, 99–118.

Ishii, M., Hibiki, T., 2006. Thermo-Fluid Dynamics of Two-Phase Flow. Springer-Verlag, Berlin.

Joseph, D.D., Lundgren, T.S., Jackson, R., Saville, D.A., 1990. Ensemble averaged and mixture theory equations for incompressible fluid-particle suspensions. Int. J. Multiphase Flow 16, 35–42.

Kallio, G.A., Stock, D.E., 1986. Turbulent particle dispersion: a comparison between Lagrangian and Eulerian modeling approaches. ASME FED 35, 23–34.

Kashima, B.A., Rauenzahn, R.M., 1994. A multimaterial formalism, ASME Fluids Engineering Summer Meeting, New York, USA.

Kolev, N.I., 2005. Multiphase Flow Dynamics 1: Fundamentals, second ed. Springer-Verlag, Berlin.

Lhuillier, D., 1996. The macroscopic modelling of multi-phase mixtures. In: Schaflinger, U. (Ed.), Flow of Particles in Suspensions. Springer-Verlag, Wien, New York.

Li, A., Ahmadi, G., 1992. Dispersion and deposition of spherical particles from point sources in a turbulent channel flow. Aerosol Sci. Technol. 16, 209–226.

Milojevic, D., 1990. Lagrangian stochastic-deterministic (LSD) predictions of particle dispersion in turbulence. Part. Part. Syst. Charact. 7, 181–190.

Mostafa, A.A., Mongia, H.C., 1987. On the modeling of turbulent evaporating sprays: Eulerian versus Lagrangian approach. Int. J. Heat Mass Transf. 30, 2583–2593.

Mostafa, A.A., Mongia, H.C., 1988. On the interaction of particles and turbulent fluid flow. Int. J. Heat Mass Transf. 31, 2063–2075.

Panton, R.J., 1968. Flow properties for the continuum viewpoint of a non-equilibrium gas particle mixture. J. Fluid Mech. 31, 273–304.

Rivero, M., Magnaudet, J., Fabre, J., 1991. Quelques Résultats Nouveaux Concernant les Forces Hydrodynamiques sur une Sphère Solide ou une Bulle Sphérique. C. R. Acad. Sci. Paris Sér. II 314, 1499–1506.

Saffman, P.G., 1965. The lift on a small sphere in a slow shear flow. J. Fluid Mech. 22, 385–400.
Shirolkar, J.S., Coimbra, C.F.M., McQuay, M.Q., 1996. Fundamental aspects of modeling turbulent particle dispersion in dilute flows. Prog. Energy Combust. Sci. 22, 363–399.
Shuen, J.S., Chen, L.D., Faeth, G.M., 1983. Evaluation of a stochastic model of particle dispersion in a turbulent round jet. Chem. Eng. J. 29, 167–170.
Vernier, P., Delhaye, J.M., 1968. General two-phase flow equation applied to the thermohydrodynamics of boiling nuclear reactors. Acta Tech. Belg. Energie Primaire 4, 3–43.
Yadigaroglu, G., Lahey Jr., R.T., 1976. On the various forms of the conservation equations in two-phase flow. Int. J. Multiphase Flow 2, 477–494.

Chapter 3

Cheung, S.C.P., Yeoh, G.H., Tu, J.Y., 2007a. On the modeling of population balance in isothermal vertical bubbly flows – average bubble number density approach. Chem. Eng. Proc. 46, 742–756.
Cheung, S.C.P., Yeoh, G.H., Tu, J.Y., 2007b. On the numerical study of isothermal bubbly flow using two population balance approaches. Chem. Eng. Sci. 31, 164–1072.
Cheung, S.C.P., Yeoh, G.H., Qi, F.S., Tu, J.Y., 2012. Classification of bubbles in vertical gas-liquid flow: Part 2 – a model evaluation. Int. J. Multiphase Flow 39, 135–147.
Friedlander, S.K., 2000. Smoke, Dust and Haze: Fundamentals of Aerosol Dynamics, second ed. Oxford University Press, Oxford.
Jakobsen, H., 2008. Chemical reactor Modelling. Chapter 9, 807–865. Springer Berlin Heidelberg.
Kocamustafaogullari, G., Ishii, M., 1995. Foundation of the interfacial area transport equations and its closure relations. Int. J. Heat Mass Transf. 38, 481–493.
Luo, H., Svendsen, H., 1996. Theoretical model for drop and bubble break-up in turbulent dispersions. AIChE J. 42, 1225–1233.
Luo, H., 1993. Coalescence, break-up and liquid recirculation in bubble column reactors, PhD dissertation, Norwegian Institute of Technology, Norway.
Prince, M.J., Blanch, H.W., 1990. Bubble coalescence and break-up in air-sparged bubble column. AIChE J. 36, 1485–1499.
Qi, F.S., Yeoh, G.H., Cheung, S.C.P., Tu, J.Y., Krepper, E., Lucas, D., 2012. Classification of bubbles in vertical gas–liquid flow: Part 1 – an analysis of experimental data. Int. J. Multiphase Flow 39, 121–134.
Ramkrishna, D., 2000. Population Balances. Theory and Applications to Particulate Systems in Engineering. Academic Press, San Diego.
Tsouris, C., Tavlarides, L.L., 1994. Breakage and coalescence models for drops in turbulent dispersions. AIChE J. 40, 395–406.

Chapter 4

Angelidou, C., Psimopoulos, M., Jameson, G.J., 1979. Size distribution functions of dispersions. Chem. Eng. Sci. 34, 671–676.
Azbel, D., Athanasios, I.L., 1983. A mechanism of liquid entrainment. In: Cheremisinoff, N. (Ed.), Handbook of Fluids in Motion. Ann Arbor Science Publishers, Ann Arbor, U.S.A., p. 473.
Batchelor, G.K., 1956. The Theory of Homogeneous Turbulence. University Press, Cambridge.
Bilicki, Z., Kestin, J., 1987. Transition criteria for two-phase flow patterns in vertical upward flow. Int. J. Multiphase Flow 13, 283–294.
Chatzi, E.G., Gavrielides, A.D., Kiparissides, C., 1989. Generalized model for prediction of the steady-state drop size distributions in batch stirred vessels. Ind. Eng. Chem. Res. 28, 1704–1711.
Chesters, A.K., 1975. The applicability of dynamic-similarity criteria to isothermal, liquid–gas, two-phase flows without mass transfer. Int. J. Multiphase Flow 2, 191–121.

Chesters, A.K., Hoffman, G., 1982. Bubble coalescence in pure liquids. Appl. Sci. Res. 38, 353–361.

Chesters, A.K., 1991. The modeling of coalescence processes in fluid–liquid dispersion: a review of current understanding. Trans. Chem. Eng. 69A, 259–270.

Clift, R., Grace, J.R., Weber, M.E., 1978. Bubbles, Drops and Particles. Dover Publications, New York.

Colella, D., Vinci, D., Bagatin, R., Masi, M., Bakr, E.A., 1999. A study on coalescence and breakage mechanisms in three different bubble columns. Chem. Eng. Sci. 54, 4767–4777.

Colin, C., Riou, X., Fabre, J., 2004. Turbulence and Shear-induced Coalescence in Gas–Liquid Pipe flows, Fifth Int. Conf. Multiphase Flow, ICMF'04, Yokohama, Japan.

Coulaloglou, C.A., Tavlarides, L.L., 1977. Description of interaction processes in agitated liquid–liquid dispersions. Chem. Eng. Sci. 32, 1289–1297.

Coulaloglou, C.A., 1975. Dispersed phase interactions in an agitated flow vessel, PhD dissertation, Illinois Institute of Technology, Chicago.

Davis, R.H., Schonberg, J.A., Rallison, J.M., 1989. The lubrication force between two viscous drops. Phys. Fluids A 1, 77–81.

de Nevers, N., Wu, J.L., 1971. Bubble coalescence in viscous fluids. AIChE J. 17, 182–186.

Doubliez, L., 1991. The drainage and rupture of a non-foaming liquid film formed upon bubble impact with a free surface. Int. J. Multiphase Flow 17, 783–803.

Duineveld, P.C., 1994. Bouncing and coalescence of two bubbles in water, PhD dissertation, University of Twente, Netherlands.

Fan, L.S., Tsuchiya, K., 1990. Bubble Wake Dynamics in Liquids and Liquid–Solid Suspensions. Butterworth-Heinemann, Stoneham, MA.

Friedlander, S.K., 1977. Smoke, Dust and Haze: Fundamentals of Aerosol Dynamics. Wiley, New York.

Fu, X.Y., Ishii, M., 2002. Two-group interfacial area transport in vertical air–water flow. I. Mechanistic model. Nucl. Eng. Des. 219, 143–168.

Grace, H.P., 1982. Dispersion phenomena in high viscosity immiscible fluid systems and application of static mixers as dispersion devices in such systems. Chem. Eng. Comm. 14, 225–277.

Harmathy, T.Z., 1960. Velocity of large drops and bubbles in media of infinite or restricted extent. AIChE J. 6, 281–288.

Hesketh, R.P., Fraser Russell, T.W., Etchells, A.W., 1987. Bubble size in horizontal pipelines. AIChE J. 33, 663–667.

Hibiki, T., Ishii, M., 2000a. One-group interfacial area transport of bubbly flows in vertical round tubes. Int. J. Heat Mass Transf. 43, 2711–2726.

Hibiki, T., Ishii, M., 2000b. Two-group interfacial area transport equations at bubbly-to-slug flow transition. Nucl. Eng. Des. 202, 39–76.

Hibiki, T., Ishii, M., 2009. Interfacial area transport equations for gas-liquid flow. J. Comp. Multiphase Flows 1, Transf. 44, 1–22.

Howarth, W.J., 1964. Coalescence of drops in a turbulent flow field. Chem. Eng. Sci. 19, 33–38.

Howarth, W.J., 1967. Measurement of coalescence frequency in an agitated tank. AIChE J. 13, 1007–1013.

Ishii, M., Kim, S., Uhle, J., 2002. Interfacial area transport: model development and benchmark experiments. Int. J. Heat Mass Transf. 45, 3111–3123.

Jeelani, S.A.K., Hartland, S., 1991. Effect of approach velocity on binary and interfacial coalescence. Chem. Eng. Res. Des. 69, 271–281.

Kalkach-Navarro, S., Drew, D.A., Lahey Jr., R.T., 1994. Analysis of the bubbly/slug flow regime transition. Nucl. Eng. Des. 151, 15–39.

Kamp, A.M., Chesters, A.K., 2001. Bubble coalescence in turbulent flows: a mechanistic model for turbulence-induced coalescence applied to microgravity bubbly pipe flow. Int. J. Multiphase Flow 27, 1363–1396.

Kennard, E.H., 1938. Kinetic Theory of Gases. McGraw-Hill, New York.

Kocamustafaogullari, G., Ishii, M., 1995. Foundation of the interfacial area transport equation and its closure relations. Int. J. Heat Mass Transf. 38, 481–493.

Kolmogorov, A.N., 1941. The local structure of turbulence in incompressible viscous fluids at very large Reynolds numbers. Dokl. Akad. Nauk. SSSR 30, 299–303.

Kuboi, R., Komasawa, I., Otake, T., 1972. Collision and coalescence of dispersed drops in turbulent liquid flow. J. Chem. Eng. Jpn. 5, 423–424.

Lafi, A.Y., Reyes, Jr., J.N. 1994. General Particle Transport Equation, Report OSU-NE-9409, Department of Nuclear Engineering, Oregon State University, Oregon.

Lasheras, J.C., Eastwood, C., Martinez-Bazan, C., Montanes, J.L., 2002. A review of statistical models for the break-up of an immiscible fluid immersed into a fully developed turbulent flow. Int. J. Multiphase Flow 28, 247–278.

Lee, C.-H., Erickson, L.E., Glasgow, L.A., 1987. Bubble breakup and coalescence in turbulent gas–liquid dispersions. Chem. Eng. Comm. 59, 65–84.

Lehr, F., Mewes, D., 1999. A transport equation for the interfacial area density applied to bubble columns. Chem. Eng. Sci. 56, 1159–1166.

Lehr, F., Mewes, D., 2001. A transport equation for the interfacial area density applied to bubble columns. Chem. Eng. Sci. 56, 1159–1166.

Lehr, F., Mewes, D., Millies, M., 2002. Bubble-size distribution and flow fields in bubble columns. AIChE J. 42, 1225–1233.

Levich, V.G., 1962. Physicochemical Hydrodynamics. Prentice Hall, Englewood Cliffs.

Liao, Y., Lucas, D., 2009. A literature review of theoretical models for drop and bubble breakup in turbulent dispersions. Chem. Eng. Sci. 64, 3389–3406.

Liao, Y., Lucas, D., 2010. A literature review on mechanisms and models for the coalescence process of fluid particles. Chem. Eng. Sci. 65, 2851–2864.

Lo, S., Zhang, D.S., 2009. Modelling of break-up and coalescence in bubbly two-phase flows. J. Comp. Multiphase Flows 1, 22–38.

Luo, H., Svendsen, H., 1996. Theoretical model for drop and bubble break-up in turbulent dispersions. AIChE J. 42, 1225–1233.

Luo, H., 1993. Coalescence, break-up and liquid recirculation in bubble column reactors, PhD dissertation, Norwegian Institute of Technology, Norway.

Martinez-Bazan, C., Montanes, J.L., Lasheras, J.C., 1999a. On the breakup of an air bubble injected into fully developed turbulent flow. Part 1. Breakup frequency. J. Fluid Mech. 401, 157–182.

Martinez-Bazan, C., Montanes, J.L., Lasheras, J.C., 1999b. On the breakup of an air bubble injected into fully developed turbulent flow. Part 2. Size PDF of the resulting daughter bubbles. J. Fluid Mech. 401, 183–207.

Miyahara, T., Tsuchiya, K., Fan, L.S., 1991. Effect of turbulent wake on bubble-bubble interactions in a gas–liquid–solid fluidized bed. Chem. Eng. Sci. 46, 2368–2373.

Nambiar, D.K.R., Kumar, R., Das, T.R., Gandhi, K.S., 1992. A new model for the breakage frequency of drops in turbulent stirred dispersions. Chem. Eng. Sci. Vol. 47, 2989–3002.

Narsimhan, G., Gupta, J.P., Ramkrishna, D., 1979. A model for transitional breakage probability of droplets in agitated lean liquid–liquid dispersions. Chem. Eng. Sci. 34, 257–265.

Oolman, T.O., Blanch, H.W., 1986. Bubble coalescence in stagnant liquids. Chem. Eng. Comm. 43, 237–261.

Park, J.Y., Blair, L.M., 1975. The effect of coalescence on drop size distribution in an agitated liquid–liquid dispersion. Chem. Eng. Sci. 30, 1057–1064.

Prince, M.J., Blanch, H.W., 1990. Bubble coalescence and break-up in air-sparged bubble column. AIChE J. 36, 1485–1499.

Richardson, J.F., Zaki, W.M., 1954. Sedimentation and fluidization, part I. Trans. Chem. Eng. 32, 35–53.

Risso, F., Fabre, J., 1998. Oscillations and breakup of a bubble immersed in a turbulent field. J. Fluid Mech. 372, 323–355.

Ross, S.L., 1971. Measurements and models of the dispersed phase mixing process, PhD dissertation, The University of Michigan, Ann Arbor.

Ruckenstein, E., Jain, R.K., 1974. Spontaneous rupture of thin liquid films. J. Chem. Soc. Faraday Trans. 2 (Vol. 70), 132–147.

Sagert, N.H., Quinn, M.J., 1976. The coalescence of H_2S and CO_2 bubbles in water. Can. J. Chem. Eng. 54, 392–398.

Schlichting, H., 1979. Boundary Layer Theory, seventh ed. McGraw-Hill, New York.

Shinnar, R., Church, J.M., 1960. Predicting particle size in agitated dispersions. Ind. Eng. Chem. 52, 253–256.

Simon, M., 2004. Koaleszenz von Tropfen und Tropfenschwärmen, PhD dissertation, die Teschinschen Universtität Kaiserslautern.

Sovova, H., 1981. Breakage and coalescence of drops in a batch stirred vessel—II. Comparison of model and experiments. Chem. Eng. Sci. 36, 1567–1573.

Stewart, C.W., 1995. Bubble interaction in low-viscosity liquids. Int. J. Multiphase Flow 21, 1037–1010.

Sun, X., Kim, S., Ishii, M., Beus, S.G., 2004. Modeling of bubble coalescence and disintegration in confined upward two-phase flow. Nucl. Eng. Des. 230, 3–26.

Tennekes, H., Lumley, J.L., 1972. A First Course in Turbulence. MIT Press, Cambridge, MA.

Tsouris, C., Tavlarides, L.L., 1994. Breakage and coalescence models for drops in turbulent dispersions. AIChE J. 40, 395–406.

Wang, T.F., Wang, J.F., Jin, Y., 2003. A novel theoretical breakup Kernel function for bubbles/droplets in a turbulent flow. Chem. Eng. Sci. 58, 4629–4637.

Wang, L., Marchisio, D.L., Virgil, R.D., Fox, R.O., 2005a. CFD simulation of aggregation and breakage processes in laminar Taylor-Coutte flow. J. Colloid Interface Sci. 282, 380–396.

Wang, T.F., Wang, J.F., Jin, Y., 2005b. Theoretical prediction of flow regime transition in bubble columns by the population balance model. Chem. Eng. Sci. 60, 6199–6209.

Wieringa, J.A., Dieren, F., van, Janssen, J.J., Agterof, W.G.M., 1996. Droplet breakup mechanics during emulsification in colloid mills at high dispersed volume fractions. Trans. Inst. Chem. Eng. 74-A, 554–562.

Williams, M.M.R., Loyalka, S.K., 1991. Aerosol Science Theory and Practice: With Special Applications to the Nuclear Industry. Pergamon Press, Oxford.

Wu, Q., Kim, S., Ishii, M., Beus, S.G., 1998. One-group interfacial area transport in vertical bubbly flow. Int. J. Heat Mass Transf. 41, 1103–1112.

Zhao, H., Ge, W., 2007. A Theoretical bubble breakup model for slurry beds or three-phase fluidized beds under high pressure. Chem. Eng. Sci. 62, 109–115.

Chapter 5

Abrahamson, J., 1975. Collision rates of small particles in a vigorously turbulent fluid. Chem. Eng. Sci. 30, 1371–1379.

Aoki, K.M., Akiyama, T., 1995. Simulation studies of pressure and density wave propagations in vertically vibrated beds of granules. Phys. Rev. E 52, 3288–3291.

Austin, L., Shoji, K., Bhatia, V., Jindal, V., Savage, K., Klimpel, R., 1976. Some results on the description of size reduction as a rate process in various mills. Ind. Eng. Chem. Process Des. Dev. 15, 187–196.

Cheng, J., Yang, C., Mao, Z.S., Zhao, C., 2009. CFD modeling of nucleation, growth, aggregation and breakage in continuous precipitation of barium sulfate in a stirred tank. Ind. Eng. Chem. Res. 48, 6992–7003.

Cundall, P.A., Strack, O.D.L., 1979. A discrete numerical model for granular assemblies. Geotechnique 29, 47–65.

Davis, R.H., Serayssol, J.M., Hinch, E.J., 1986. The elastohydrodynamic collision of 2 spheres. J. Fluid Mech. 163, 479–497.

Derjaguin, B.V., Muller, V.M., Toporov, Y.P., 1975. Effect of contact deformations on the adhesion of particles. J. Colloid Interface Sci. 53, 314–326.

Diemer, R.B., Olson, J.H., 2002. A moment methodology for coagulation and breakage problems: Part 3—generalized daughter distribution function. Chem. Eng. Sci. 57, 4187–4198.

Firth, B.A., Hunter, R.J., 1976. Flow properties of coagulated colloidal suspensions I. Energy dissipation in the flow units. J. Colloid Interface Sci. 57, 248–256.

Flesch, J.C., Spicer, P.T., Pratsinis, S.E., 1999. Laminar and turbulent shear-induced flocculation of fractal aggregates. AIChE J. 45, 1114–1124.

Greenwood, J.A., 1997. Adhesion of elastic spheres. Proc. R. Soc. Lond. A 453, 1277–1297.

Hertz, H., 1882. Ueber die Berührung fester elastischer Koerper. J. Reine Angewandte Mathematik 92, 156–171.
Hill, P.J., Ng, K.M., 1995. New discretization procedure for the breakage equation. AIChE J. 41, 1204–1216.
Hsia, M.A., Tavlarides, L.L., 1983. Simulation analysis of drop breakage, coalescence and micromixing in liquid–liquid stirred tanks. Chem. Eng. J. 26, 189–199.
Johnson, K.L., Kendall, K., Roberts, A.D., 1971. Surface energy and the contact of elastic solids. Proc. R. Soc. Lond. A 324, 301–313.
Kargulewicz, M., Iordanoff, I., Marrero, V., Tichy, J., 2012. Modeling of magnetorheological fluids by the discrete element method. J. Tribology Trans. ASME 134, 031706.
Kostoglou, M., Dovas, S., Karabelas, A.J., 1997. On the steady-state size distribution of dispersions in breakage processes. Chem. Eng. Sci. 52, 1285–1299.
Kruggel-Emden, H., Simsek, E., Rickelt, S., Wirtz, S., Scherer, V., 2007. Review and extension of normal force models for the discrete element method. Powder Tech. 130, 157–173.
Kruggel-Emden, H., Wirtz, S., Scherer, V., 2008. A study on tangential force laws applicable to the discrete element method (DEM) for materials with viscoelastic or plastic behaviour. Chem. Eng. Sci. 63, 1523–1541.
Kusters, K.A., Wijers, J.G., Thoenes, D., 1997. Aggregation kinetics of small particles in agitated vessels. Chem. Eng. Sci. 52, 107–121.
Kuwabara, G., Kono, K., 1987. Restitution coefficient in collision between two spheres. Jpn. J. Appl. Phys. 26, 1230–1233.
Langston, P.A., Tüzün, U., Heyes, D.M., 1994. Continuous potential discrete particle simulations of stress and velocity fields in hoppers: transition from fluid to granular flow. Chem. Eng. Sci. 49, 1259–1275.
Lee, J., Herrmann, H.J., 1993. Angle of repose and angle of marginal stability–molecular-dynamics of granular particles. J. Phys. A, Math. Gen. 26, 373–383.
Li, S.Q., Marshall, J.S., Liu, G.Q., Yao, Q., 2011. Adhesive particulate flow: the discrete-element method and its application in energy and environmental engineering. Prog. Energy Combust. Sci. 37, 633–668.
Lian, G., Thornton, C., Adams, M.J., 1993. A theoretical study of the liquid bridge forces between two rigid spherical bodies. J. Colloid Interface Sci. 161, 138–147.
Liu, G.Q., Li, S.Q., Yao, Q., 2009. On the applicability of different adhesion models in adhesive particulate flows. Front. Energy Power Eng. China 4, 280–286.
Loyalka, S.K., 1992. Thermophoretic force on a single particle–I. Numerical solution of the linearized Boltzmann equation. J. Aerosol Sci. 23, 291–300.
Luo, H., Svendsen, H., 1996. Theoretical model for drop and bubble break-up in turbulent dispersions. AIChE J. 42, 1225–1233.
Marchisio, D.L., Vigil, D.R., Fox, R.O., 2003. Quadrature Method of Moments for Aggregation-Breakage Processes. J. Colloid Interface Science Vol. 258, 322–324.
Markatou, P., Wang, H., Frenklach, M., 1993. A computational study of sooting limits in laminar premixed flames of ethane, ethylene and acetylene. Combust. Flame 93, 467–482.
Marshall, J.S., 2009. Discrete-element modeling of particulate aerosol flows. J. Comp. Phys. 228, 1541–1561.
Matthewson, M.J., 1988. Adhesion of spheres by thin liquid films. Philos. Mag. A 57, 207–216.
Maugis, D., 1987. Adherence of elastomers: fracture mechanics aspects. J. Adhes. Sci. Tech. 1, 105–134.
McCoy, B.J., Wang, M., 1994. Continuous-mixture fragmentation kinetics: particle size reduction and molecular cracking. Chem. Eng. Sci. 49, 3773–3785.
Mindlin, R.D., Deresiewicz, H., 1953. Elastic spheres in contact under varying oblique forces. J. Appl. Mech. Trans. ASME 20, 327–344.
Nicolson, M.M., 1949. Interaction between floating particles. Proc. Camb. Philol. Soc. 45, 288–295.
Peng, S.J., Williams, R.A., 1994. Direct measurement of floc breakage in flowing suspension. J. Colloid Interface Sci. 166, 321–332.
Pitois, O., Moucheront, P., Chateau, X., 2000. Liquid bridge between two moving spheres: an experimental study of viscosity effects. J. Colloid Interface Sci. 231, 26–31.
Ramkrishna, D., 2000. Population Balances. Theory and Applications to Particulate Systems in Engineering. Academic Press, San Diego.

Sadd, M.H., Tai, Q.M., Shukla, X., 1993. A contact law effects on wave-propagation in particulate materials using distinct element modeling. Int. J. Non-linear Mech. 28, 251–265.

Saffman, P.G., Turner, J.S., 1956. On the collision of drops in turbulent clouds. J. Fluid Mech. 1, 16–30.

Schafer, J., Dippel, S., Wolf, D.E., 1996. Force schemes in simulations of granular materials. J. Physique 6, 5–20.

Serayssol, J.M., Davis, R.H., 1986. The influence of surface interactions on the elastohydrodynamic collision of two spheres. J. Colloid Interface Sci. 114, 54–66.

Serra, T., Casamitjana, X., 1998. Effect of the shear and volume fraction on the aggregation and breakup of particles. AIChE J. 44, 1724–1730.

Smoluchowski, M.V., 1917. Versuch einer mathematischen Theorie der Koagulationskinetik kolloider Lösungen. Z. Phys. Chem. 92, 129–168.

Tabor, D., 1977. Surface forces and surface interactions. J. Colloid Interface Sci. 58, 2–13.

Talbot, L., Cheng, R.K., Schefer, R.W., Willis, D.R., 1980. Thermophoresis of particles in a heated boundary layer. J. Fluid Mech. Vol. 101, 737–758.

Thornton, C., 1997. Coefficient of restitution for collinear collisions of elastic perfectly plastic spheres. J. Appl. Mech. Trans. ASME 64, 383–386.

Tomas, J., 2003. Mechanics of nanoparticle adhesion—a continuum approach. In: Mittal, K.L. (Ed.), Particles on Surfaces 8: Detection Adhesion and Removal. VSP, Utrecht, pp. 183–229.

Tsuji, Y., Tanaka, T., Ishida, T., 1992. Lagrangian numerical simulation of plug flow of cohesionless particles in a horizontal pipe. Powder Tech. 71, 239–250.

Tsuji, Y., Kawaguchi, T., Tanaka, T., 1993. Discrete particle simulation of two dimensional fluidized bed. Powder Tech. 77, 79–87.

Tyndall, J., 1870. On dust and disease. Proc. R. Inst. 6, 1–14.

Verlet, L., 1967. Computer "experiments" on classical fluids, I. Thermodynamical properties of Lennard—Jones molecules. Phys. Rev. 159, 98–103.

Vigil, R.D., Ziff, R.M., 1989. On the stability of coagulation-fragmentation population balances. J. Colloid Interface Sci. 133, 257–264.

Vu-Quoc, L., Zhang, X., 1999. An accurate and efficient tangential force—displacement model for elastic frictional contact in particle-flow simulations. Mech. Math. 31, 235–269.

Waldmann, L., Schmitt, K.H., 1966. Thermophoresis and diffusiophoresis of aerosols. In: Davies, C.N. (Ed.), Aerosol Science. Academic Press, New York, pp. 137–162.

Walton, O.R., Braun, R.L., 1986. Viscosity, granular temperature and stress calculations for shearing assemblies of inelastic, frictional disks. J. Rheology 30, 949–980.

Wang, L., Marchisio, D.L., Virgil, R.D., Fox, R.O., 2005. CFD simulation of aggregation and breakage processes in laminar Taylor-Coutte flow. J. Colloid Interface Sci. 282, 380–396.

Yamamoto, K., Ishihara, Y., 1988. Thermophoresis of a spherical particle in a rarefied gas of a transition regime. Phys. Fluids 31, 3618–3624.

Chapter 6

Aldeniji-Fashola, A., Chen, C.P., 1990. Modeling of confined turbulent fluid-particle flows using Eulerian and Lagrangian schemes. Int. J. Heat Mass Transf. 33, 691–701.

Baldyga, J., Bourne, J., 1999. Turbulent Mixing and Chemical Reactions. Wiley, New York.

Barth, T.J., Jespersen, D.C., 1989. The Design and Application of Upwind Schemes on Unstructured Meshes. AIAA Paper, 89–366.

Barrett, J.C., Webb, N.A., 1998. A comparison of some approximate methods for solving the aerosol general dynamics equation. J. Aerosol Sci. 29, 31–39.

Boussinesq, J., 1903. Théorie Analytique de la Chaleur, vol. 2. Gauther-Villars, Paris, pp. 154–176.

Bove, S., 2005. Computational fluid dynamics of gas—liquid flows including bubble population balance, PhD thesis, Aalborg University Esbjerg, Denmark.

Cheung, S.C.P., Yeoh, G.H., Tu, J.Y., 2007a. On the modeling of population balance in isothermal vertical bubbly flows – average bubble number density approach. Chem. Eng. Proc. 46, 742–756.

Cheung, S.C.P., Yeoh, G.H., Tu, J.Y., 2007b. On the numerical study of isothermal bubbly flow using two population balance approaches. Chem. Eng. Sci. 62, 4659–4674.

Darwish, M., Moukalled, F., 2003. TVD Schemes for Unstructured Grids. International Journal of Heat and Mass Transfer Vol. 46, 599–611.

Deardoff, J.W., 1970. A numerical study of three-dimensional turbulent channel flow at large Reynolds numbers. J. Fluid Mech. 41, 4 53–480.

Debry, E., Sportisse, B., Jourdain, B., 2003. A stochastic approach to numerical simulation of the general equation for aerosis. J. Comp. Phys. 184, 649–669.

Domilovskii, E.R., Lushnikov, A.A., Piskunov, V.N., 1979. A Monte Carlo simulation of coagulation processes. Izvestkya Akademi Nauk SSSR, Fizika Atmosfery I Okeana 15, 194–201.

Dorao, C.A., Jakobsen, H.A., 2006a. Numerical calculation of the moments of the population balance equation. J. Comp. Appl. Math. 196, 619–633.

Dorao, C., Jakobsen, H., 2006b. A least squares method for the solution of population balance problems. Comp. Chem. Eng. 30, 535–547.

Dorao, C., Jakobsen, H., 2007a. Least-squares spectral method for solving advective population balance problems. J. Comp. Appl. Math. 201, 247–257.

Dorao, C., Jakobsen, H., 2007b. Time-space-property least squares spectral method for population balance problems. Chem. Eng. Sci. 62, 1323–1333.

Dorao, C.A., Lucas, D., Jakobsen, H.A., 2008. Prediction of the evolution of the dispersed phase in bubbly flow problems. Appl. Math. Model. 32, 1813–1833.

Fletcher, C.A.J., 1991. Computational Techniques for Fluid Dynamics, vols I and II. Springer-Verlag, Berlin.

Frenklach, M., 1985. Dynamics of discrete distribution for Smoluchowski coagulation model. J. Colloid Interface Sci. 108, 237–242.

Frenklach, M., 2002. Method of moments with interpolative closure. Chem. Eng. Sci. 57, 2229–2239.

Fulgosi, M., Lakehal, D., Banerjee, S., De Angelis, V., 2003. Direct numerical simulation of turbulence in a sheared air–water flow with a deformable interface. J. Fluid Mech. 482, 319–345.

Germano, M.U., Piomelli, U., Moin, P., Cabot, W.H., 1991. A dynamic subgrid-scale eddy viscosity model. Phys. Fluids A 3, 1760–1765.

Ghosal, S., Lund, T.S., Moin, P., Akselvoll, K., 1995. A dynamic localization model for large-eddy simulation of turbulent flows. J. Fluid Mech. 286, 229–255.

Gordon, R.G., 1968. Error bounds in equilibrium statistical mechanics. J. Math. Phys. 9, 655–672.

Grosch, R., Briesen, H., Marquardt, W., Wulkow, M., 2007. Generalization and numerical investigation of QMOM. AIChE J. 53, 207–227.

Heinz, S., 2003. Statistical Mechanics of Turbulent Flows. Springer, Berlin.

Hulburt, H.M., Katz, S., 1964. Some problems in particle technology: a statistical mechanical formulation. Chem. Eng. Sci. 19, 55–574.

Jayatilleke, C.L.V., 1969. The influence of prandtl number and surface roughness on the resistance of the laminar sublayer to momentum and heat transfer. Prog. Heat Mass Transf. 1, 193–321.

John, V., Angelov, I., Oncul, A.A., Thevenin, D., 2007. Techniques for the reconstruction from a finite number of its moments. Chem. Eng. Sci. 62, 2890–2904.

Kader, B., 1993. Temperature and concentration profiles in fully turbulent boundary layers. Int. J. Heat Mass Transf. 24, 1541–1544.

Karema, H., Lo, S.M., 1999. Efficiency on interphase coupling algorithms in fluidized bed conditions. Comp. Fluids 28, 323–360.

Kim, S.E., Choudhury, D., 1995. A near-wall treatment using wall functions sensitized to pressure gradient. ASME FED 217 (Separated and Complex Flows, ASME).

Krepper, E., Lucas, D., Prasser, H., 2005. On the modeling of bubbly flow in vertical pipes. Nucl. Eng. Des. 235, 597–611.

Krepper, E., Frank, T., Lucas, D., Prasser, H.M., Zwart, P.J., 2007. Inhomogeneous MUSIG model – a population balance approach for polydispersed bubbly flows, Proceedings of the Sixth International Conference on Multiphase Flow, Leipzig, Germany.

Lahey Jr., R.T., Drew, D.A., 2001. The analysis of two-phase flow and heat transfer using multidimensional, four field, two-fluidxe "fluid" model. Nucl. Eng. Des. 204, 29–44.

Lambin, P., Gaspard, J.P., 1982. Continued-fraction technique for tight-binding systems. A generalized-moments method. Phys. Rev. B 26, 4356–4368.

Launder, B.E., Spalding, D.B., 1974. The numerical computation of turbulent flows. Comp. Meth. Appl. Mech. Eng. 3, 269–289.

Launder, B.E., 1989. Second-moment closures: present and future? Int. J. Heat Fluid Flow 10, 282–300.

Leonard, B.P., 1979. A stable and accurate convective modeling procedure based on quadratic upstream interpolation. Comp. Meth. Appl. Mech. Eng. 19, 59–98.

Li, S.Q., Marshall, J.S., Liu, G.Q., Yao, Q., 2011. Adhesive particulate flow: the discrete-element method and its application in energy and environmental engineering. Prog. Energy Combust. Sci. 37, 633–668.

Liffman, K., 1992. A direct simulation Monte Carlo method for cluster coagulation. J. Comp. Phys. 100, 116–127.

Lilly, D.K., 1966. The representation of small-scale turbulence in numerical simulation experiments. NCAR Manuscript No. 281.

Lilly, D.K., 1967. The representation of small-scale turbulence in numerical simulation experiments. In: Proc. IBM Sci. Comput.Syrup. on Environmental Science. Thomas J. Watson Research Center, pp. 195–210. Yorktown Heights, N.Y., 14–16 November 1966. IBM Form 320-1951.6. IBM Form 320-1951.

Lilly, D.K., 1992. A proposed modification of the Germano subgrid-scale closure method. Phys. Fluids 4, 633–635.

Lo, S.M., 1989. Mathematical Basis of a Multiphase flow Model, United Kingdom Atomic Energy Authority, Computational fluid Dynamics Section, Report AERE R 13432.

Lo, S.M., 1990. Multiphase Flow Model in the Harwell-FLOW3D Computer Code, AEA Industrial Technology, Report AEA-InTech-0062.

Lo, S.M., 1996. Application of Population Balance to CFD Modeling of Bubbly Flow via the MUSIG Model, AEAT-1096, AEA Technology.

Lopez de Bertodano, M., Lahey Jr., R.T., Jones, O.C., 1994a. Development of a k-ε model for bubbly two-phase flow. J. Fluid Eng. 116, 128–134.

Lopez de Bertodano, M., Lahey Jr., R.T., Jones, O.C., 1994b. Phase distribution in bubbly two-phase flow in vertical ducts. Int. J. Multiphase Flow 20, 805–818.

Maisels, A., Kruis, F.E., Fissan, H., 2004. Direct simulation Monte Carlo for simulation nucleation, coagulation and surface growth in dispersed systems. Chem. Eng. Sci. 59, 2231–2239.

Marchisio, D.L., Fox, R.O., 2005. Solution of population balance equations using the direct quadrature method of moments. J. Aerosol Sci. 36, 43–73.

Marchisio, D.L., Vigil, D.R., Fox, R.O., 2003a. Quadrature method of moments for aggregation-breakage processes. J. Colloid Interface Sci. 258, 322–324.

Marchisio, D.L., Pikturna, J.T., Fox, R.O., Vigil, R.D., 2003b. Quadrature method of moments for population-balance equations. AIChE J. 49, 1266–1276.

Marchisio, D.L., Vigil, D.R., Fox, R.O., 2003c. Implementation of the quadrature method of moments in CFD codes for aggregation-breakage problems. Chem. Eng. Sci. 58, 3337–3351.

McCoy, B.J., Madras, G., 2003. Analytic solution for a population balance equation with aggregation and fragmentation. Chem. Eng. Sci. 58, 3049–3051.

McGraw, E., Wright, D.L., 2003. Chemically resolved aerosol dynamics for internal mixtures by the quadrature method of moments. J. Aerosol Sci. 34, 189–209.

McGraw, R., 1997. Description of aerosol dynamics by the quadrature method of moments. Aerosol Sci. Technol. 27, 255–265.

Menter, F.R., 1993. Zonal Two Equation k-ω Turbulence Models for Aerodynamics flows, AIAA paper 93-2906.

Menter, F.R., 1996. A comparison of some recent eddy-viscosity turbulence models. J. Fluids Eng. 118, 514–519.

Mueller, M.E., Blanquart, G., Pitsch, H., 2009a. A joint volume-surface model of soot aggregation with the method of moment. Proc. Combust. Inst. 32, 785–792.

Mueller, M.E., Blanquart, G., Pitsch, H., 2009b. Hybrid method of moments for modeling soot formation and growth. Combust. Flame 156, 1143–1155.

Olmos, E., Gentric, C., Vial, Ch., Wild, G., Midoux, N., 2001. Numerical simulation of multiphase flow in bubble column reactors. Influence of bubble coalescence and break-up. Chem. Eng. Sci. 56, 6359–6365.

Patankar, S.V., Spalding, D.B., 1972. A calculation procedure for heat, mass and momentum transfer in three-dimensional parabolic flows. Int. J. Heat Mass Transf. 15, 1787–1806.

Patankar, S.V., 1980. Numerical Heat Transfer and Fluid Flow. Hemisphere Publishing Corporation, Taylor & Francis Group, New York.

Piomelli, U., Liu, J., 1995. Large-eddy simulation of rotating channel flows using a localized dynamics model. Phys. Fluids 7, 839–848.

Piomelli, U., Ferziger, J.H., Moin, P., 1987. Models for Large Eddy Simulation of Turbulent Channel Flows Including Transpiration, Technical Report, Rep. TF-32, Dept. Mech. Eng., Stanford University.

Pochorecki, R., Moniuk, W., Bielski, P., Zdrojkwoski, A., 2001. Modeling of the coalescence/redispersion processes in bubble columns. Chem. Eng. Sci. 56, 6157–6616.

Rai, M.M., Moin, P., 1991. Direct simulation of turbulent flow using finite-difference schemes. J. Comp. Phys. 96, 15–53.

Rhie, C.M., Chow, W.L., 1983. A numerical study of the turbulent flow past an isolated airfoil with trailing edge separation. AIAA J. 21, 1525–1532.

Rodi, W., 1993. Turbulence Models and Their Application in Hydraulics. Balkema, Rotterdam.

Saad, Y., Schultz, M., 1985. Conjugate gradient-like algorithms for solving nonsymmetric linear systems. SIAM J. 44, 417–424.

Sato, Y., Sadatomi, M., Sekoguchi, K., 1981. Momentum and heat transfer in two-phase bubbly flow—I. Int. J. Multiphase Flow 7, 167–178.

Scott, W.T., 1968. Analytic studies of cloud droplet coalescence. J. Atm. Sci. 25, 54–65.

Shi, J.M., Zwart, P.J., Frank, T., Rohde, U., Prasser, H.M., 2004. Development of a Multiple velocity Multiple size Group Model for Poly-dispersed Multiphase flows, Annual Report of Institute of Safety Research, Forschungszentrum Rossendorf, Germany.

Smagorinsky, J., 1963. General circulation experiment with the primitive equations: part I. The basic experiment. Mon. Weather Rev. 91, 99–164.

Spalding, D.B., 1972. A novel finite-difference formulation for differential expressions involving both first and second derivatives. Int. J. Num. Meth. Eng. 4, 551–559.

Tennekes, H., Lumley, J.L., 1976. A First Course in Turbulence. MIT Press, Cambridge, MA.

Tersoff, J., 1989. Modeling solid-state chemistry: interatomic potentials for multicomponent systems. Phys. Rev. B 39, 5566–5568.

Thomas, J.L., Diskin, B., Brandt, A., 2003. Textbook multigrid efficiency for fluid simulations. Ann. Rev. Fluid Mech. 35, 317–340.

Timmermann, G., 2000. A cascadic multigrid algorithm for semilinear elliptic problems. Numerische Mathematik 86, 717–731.

Tokoro, C., Okaya, K., Sadaki, J., 2005. A Fast Algorithm for the Discrete Element Method by Contact Force Prediction,. Kona 23, 182–193.

Tu, J.Y., Fletcher, C.A.J., 1995. Numerical computation of turbulent gas–particle flow in a 90° Bend. AIChE J. 41, 2187–2197.

Van der Host, H.A., 1992. BICGSTAB: a fast and smoothly converging variant of the Bi-CG for the solution of nonsymmetric linear systems. SIAM J. Sci. Stat. 13, 631–644.

Van Driest, E.R., 1956. On turbulent flow near a wall. J. Aerosol Sci. 23, 1007–1011.

van Gunsteren, W.F., Berendsen, H.J.C., 1982. Algorithms for brownian dynamics. Mol. Phys. Vol. 45, 637–647.

Verlet, L., 1967. Computer "experiments" on classical fluids, I. Thermodynamical properties of Lennard–Jones molecules. Phys. Rev. 159, 98–103.

Versteeg, H.K., Malalasekera, W., 1995. An Introduction to Computational Fluid Dynamics – The Finite Volume Method. Prentice Hall, Pearson Education Ltd., England.

Vreman, A.W., 2004. An eddy-viscosity subgrid-scale model for turbulent shear flow: algebraic theory and applications. Phys. Fluids 16, 3670–3681.

Wesseling, P., 1995. Introduction to Multi-Grid Methods, CR – 195045 ICASE 95–11, NASA.

Wilcox, D.C., 1998. Turbulence Modeling for CFD. DCW Industries, Inc.

Yeoh, G.H., Tu, J.Y., 2005. Thermal-hydrodynamic modelling of bubbly flows with heat and mass transfer. AIChE J. 51, 8–27.

Yeoh, G.H., Tu, J.Y., 2009. Computational Techniques for Multiphase Flows: Basics and Applications. Butterworth-Heinemann, Elsevier, UK.

Zhu, Z.J., Dorao, C.A., Lucas, D., Jakobsen, H.A., 2009. On the coupled solution of a combined population balance model using the least-squares spectral element method. Ind. Eng. Chem. Res. 48, 7994–8006.

Chapter 7

Alcamo, R., Micale, G., Grisafi, F., Brucato, A., Ciofalo, M., 2005. Large-eddy simulation flow in an unbaffled stirred tank driven by a Rushton turbine. Chem. Eng. Sci. 60, 2303–2316.

Alopaeous, V., Koskinen, J., Keskinen, K.I., 1999. Simulation of the population balances for liquid–liquid systems in a nonideal stirred tank. Part 1–description and qualitative validation of the model. Chem. Eng. Sci. 54, 5887–5899.

Alopaeous, V., Koskinen, J., Keskinen, K.I., Majander, J., 2002. Simulation of the population balances for liquid–liquid systems in a nonideal stirred tank. Part 2–parameter fitting and the use of the multiblock model for dense dispersions. Chem. Eng. Sci. 57, 1815–1825.

Antal, S.P., Lahey Jr., R.T., Flaherty, J.E., 1991. Analysis of phase distribution and turbulence in dispersed particle/liquid flows. Chem. Eng. Comm. 174, 85–113.

Bapat, P.M., Tavlarides, L.L., 1985. Mass transfer in a liquid–liquid CFSTR. AIChE J. 31, 659–666.

Bothe, D., Schmidtke, M., Warnecke, H.J., 2006. VOF-simulation of the lift force for single bubbles in a simple shear flow. Chem. Eng. Technol. 29, 1048–1053.

Bourne, J.R., Yu, S., 1994. Investigation of micromixing in stirred tank reactors using parallel reactions. Ind. Eng. Chem. Res. 33, 41–55.

Burns, A.D., Frank, T., Hamill, I., Shi, J., 2004. The Favre averaged drag model for turbulent dispersion in Eulerian multiphase flow, Proc. Fifth Int. Multiphase Flow, ICMF-2004, Yokohama, Japan.

Cheung, S.C.P., Yeoh, G.H., Tu, J.Y., 2007. On the modeling of population balance in isothermal vertical bubbly flows – average bubble number density approach. Chem. Eng. Proc. 46, 742–756.

Cheung, S.C.P., Yeoh, G.H., Qi, F.S., Tu, J.Y., 2012. Classification of bubbles in vertical gas-liquid flow: Part 2 – a model evaluation. Int. J. Multiphase Flow 39, 135–147.

Coulaloglou, C.A., Tavlarides, L.L., 1977. Description of interaction processes in agitated liquid–liquid dispersions. Chem. Eng. Sci. 32, 1289–1297.

Drew, D.A., Lahey Jr., R.T., 1979. Application of General Constitutive Principles to the Derivation of Multi-dimensional Two-Phase Flow Equation. Int. J. Multiphase Flow Vol. 5, 243–264.

Fan, R., Marchisio, D.L., Fox, R.O., 2004. Application of the direct quadrature method of moments to polydisperse gas–solid fluidized beds. Powder Tech. 139, 7–20.

Frank, T., Shi, J., Burns, F.A.D., 2004. Validation of Eulerian multiphase flow models for nuclear safety application, Proc. Third Symp. Two-Phase Modeling and Experimentation, (Pisa, Italy).

Gordon, R.G., 1968. Error bounds in equilibrium statistical mechanics. J. Math. Phys. 9, 655–672.

Hibiki, T., Ishii, M., 2000. One-group interfacial area transport of bubbly flows in vertical round tubes. Int. J. Heat Mass Transf. 43, 2711–2726.

Hibiki, T., Ishii, M., 2002. Development of one-group interfacial area transport equation in bubbly flow systems. Int. J. Heat Mass Transf. 45, 2351–2372.

Hibiki, T., Ishii, M., 2009. Interfacial area transport equations for gas–liquid flow. J. Comp. Multiphase Flows 1, 1–22.

Hibiki, T., Ishii, M., Xiao, Z., 2001. Axial interfacial area transport of vertical bubbly flows. Int. J. Heat Mass Transf. 44, 1869–1888.

Hsia, M.A., Tavlarides, L.L., 1980. A simulation model for homogeneous dispersion in stirred tanks. Chem. Eng. J. 20, 225–236.

Ishii, M., Chawla, T.C., 1979. Local Drag Laws in Dispersed Two-Phase Flow, Technical Report ANL-79–105. Argonne National Laboratory, Chicago IL, 1979.

Ishii, M., Zuber, N., 1979. Drag coefficient and relative velocityxe "velocity" in bubbly, droplet or particulate flows. AIChE J. 25, 843–885.

Ishii, M., Kim, S., Uhle, J., 2002. Interfacial area transport: model development and benchmark experiments. Int. J. Heat Mass Transf. 45, 3111–3123.

Krepper, E., Lucas, D., Prasser, H., 2005. On the modeling of bubbly flow in vertical pipes. Nucl. Eng. Des. 235, 597–611.

Lebowitz, J.L., 1964. Exact solution of generalized Percus–Yevick equation for a mixture of hard spheres. Phys. Rev. A 133, 895–899.

Liu, T.J., Bankoff, S.G., 1993a. Structure of air–water bubbly flow in a vertical pipe – I. Liquid mean velocity and turbulence measurements. Int. J. Heat Mass Transf. 36, 1049–1060.

Liu, T.J., Bankoff, S.G., 1993b. Structure of air–water bubbly flow in a vertical pipe – II. Void fraction, bubble velocity and bubble size distribution. J. Heat Mass Transf. 36, 1061–1072.

Lopez de Bertodano, M., 1992. Turbulent bubbly two-phase flow in a triangular duct, PhD dissertation, Rensselaer Polytechnic Institute, New York.

Lopez de Bertodano, M., 1998. Two-fluid model for two-phase turbulent jet. Nucl. Eng. Des. 179, 65–74.

Lucas, D., Krepper, E., Prasser, H.M., 2007. Use of models for lift, wall and turbulent dispersion forces acting on bubbles for poly-disperse flows. Chem. Eng. Sci. 62, 4146–4157.

Luo, H., Svendsen, H., 1996. Theoretical model for drop and bubble break-up in turbulent dispersions. AIChE J. 42, 1225–1233.

Menter, F.R., 1993. Zonal Two Equation k-ω Turbulence Models for Aerodynamics flows, AIAA paper 93-2906.

Moraga, F.J., Larreteguy, A.E., Drew, D.A., Lahey Jr., R.T., 2003. Assessment of turbulent dispersion models for bubbly flows in the low stokes number limit. Int. J. Multiphase Flow 29, 655–673.

Politano, M.S., Carrica, P.M., Converti, J., 2003. A model for turbulent polydisperse two-phase flow in vertical channels. Int. J. Multiphase Flow 29, 1153–1182.

Prasser, H.M., Beyer, M., Carl, H., Gregor, S., Lucas, D., Pietruske, H., Schutz, P., Weiss, F.P., 2007. Evolution of the structure of a gas–liquid two-phase flow in a large vertical pipe. Nucl. Eng. Des. 237, 1848–1861.

Prince, M.J., Blanch, H.W., 1990. Bubble coalescence and break-up in air-sparged bubble column. AIChE J. 36, 1485–1499.

Serizawa, A., Kataoka, I., 1988. Phase distribution in two-phase flow. In: Afgan, N.H. (Ed.), Transient Phenomena in Multiphase Flow. Hemisphere, New York, pp. 179–224.

Shen, X.Z., Mishima, K., Nakamura, H., 2005. Two-phase phase distribution in a vertical large diameter pipe. International Journal of Heat and Mass Transfer 48, 211–225.

Smoluchowski, M.V., 1917. Versuch einer mathematischen Theorie der Koagulationskinetik kolloider Lösungen. Z. Phys. Chem. 92, 129–168.

Syamlal, M., Rogers, W., O'Brien, T.J., 1993. MFIX Documentation Theory Guide. US Department of Energy, Morgantown, West Virginia.

Taitel, Y., Bornea, D., Dukler, A.E., 1980. Modeling Flow Pattern Transitions for Steady Upward Gas-liquid in Vertical Tubes. Aiche Journal 26, 345–354.

Takagi, S., Matsumoto, Y., 1998. Numerical Study on the Forces Acting on a Bubble and Particle, Proc. Third Int. Conference on Multiphase Flow, Lyon, France.

Taitel, Y., Bornea, D., Dukler, A.E., 1980. Modeling Flow Pattern Transitions for Steady Upward Gas-liquid in Vertical Tubes. Aiche Journal 26, 345–354.

Tomiyama, A., 1998. Struggle with Computational Bubble Dynamics, Proc. Third Int. Conf. Multiphase Flow, Lyon, France.

Vermeulen, T., Williams, G.M., Langlosi, G.E., 1955. Interfacial area in liquid−liquid and gas−liquid agitation. Chem. Eng. Prog. 51, 85-F−94-F.

Wallis, G.B., 1969. One-Dimensional Two-Phase Flow. McGraw-Hill, New York.

Wang, S.K., Lee, S.J., Lahey Jr., R.T., Jones, O.C., 1987. 3-D turbulence structure and phase distribution measurements in bubbly two-phase flows. Int. J. Multiphase Flow 13, 327−343.

Wang, L., Marchisio, D.L., Virgil, R.D., Fox, R.O., 2005. CFD simulation of aggregation and breakage processes in laminar Taylor-Coutte flow. J. Colloid Interface Sci. 282, 380−396.

Wellek, R.M., Agrawal, A.K., Skelland, A.H.P., 1966. Shapes of liquid drops moving in liquid media. AIChE J. 12, 854−862.

Wu, Q., Kim, S., Ishii, M., Beus, S.G., 1998. One-group interfacial area transport in vertical bubbly flow. Int. J. Heat Mass Transf. 41, 1103−1112.

Yao, W., Morel, C., 2004. Volumetric interfacial area prediction in upwards bubbly two-phase flow. Int. J. Heat Mass Transf. 47, 307−328.

Zuber, N., 1964. On the dispersed two-phase flow in laminar flow regime. Chem. Eng. Sci. 19, 897−917.

Chapter 8

Akroyd, J., Smith, A.J., Laurence, R.S., McGlashan, R., Kraft, M., 2011. A coupled CFD-population balance approach for nanoparticle synthesis in turbulent reacting flows. Chem. Eng. Res. Des. 66, 3792−3805.

Amara, N., Ratsimba, B., Wilhelm, A., Delmas, H., 2004. Growth rate of potash alum crystals: comparison of silent and ultrasonic conditions. Ultrason. Sonochem. 11, 17−21.

Briesen, H., 2009. Two-dimensional population balance modeling for shape dependent crystal attrition. Chem. Eng. Sci. 64, 661−672.

Brown, M.E., 1957. The growth of brown trout (Salmo trutta Linn.). J. Exp. Biol. 28, 4473−4491.

Christofides, P.D., Li, M., Madler, L., 2007. Control of particulate processes: recent results and future challenges. Powder Tech. 175, 1−7.

Fredrickson, A.G., 2003. Population balance equations for cell and microbial cultures and revisited. AIChE J. 49, 1050−1059.

Gadewar, S.B., Doherty, M.F., 2004. A dynamic model for evolution of crystal shape. J. Cryst. Growth 267, 239−250.

Liu, Y.H., Bi, J.X., Zeng, A.P., Yuan, J.Q., 2007. A population balance model describing the cell cycle dynamics of myeloma cell cultivation. Biotechnol. Prog. 23, 1198−1209.

Liu, J.J., Ma, C.Y., Hu, Y.D., Wang, X.Z., 2010. Modelling protein crystallisation using morphological population balance models. Chem. Eng. Res. Des. 88, 437−446.

Ma, C.Y., Wang, X.Z., Roberts, K.J., 2008. Morphological population balance for modeling crystal growth in face directions. AIChE J. 54, 209−222.

McMillan, J., Briens, C., Berutti, F., Chan, E., 2007. Particle attrition mechanism with a sonic gas jet injected into a fluidized bed. Chem. Eng. Sci. 62, 3809−3820.

Ouchiyama, N., Rough, S.L., Bridgwater, J., 2005. A population balance approach, to describing bulk attrition. Chem. Eng. Sci. 60, 1429−1440.

Saastamoinen, J.J., Shimizu, T., Tourunen, A., 2010. Effect of attrition on particle size distribution and SO_2 capture in fluidized bed combustion under high CO_2 partial pressure conditions. Chem. Eng. Sci. 65, 550−555.

Thompson, R.W., Cauley, D.A., 1979. A population balance model for fish population dynamics. J. Theor. Biol. 81, 289−307.

Tsantilis, S., Kammler, H.K., Pratsinis, S.E., 2002. Population balance modeling of flame synthesis of titania nanoparticles. Chem. Eng. Sci. 57, 2139−2156.

Villermaux, J., Devillon, J.C., 1972. Représentation de la coalescence et de la redispersion des domaines de ségrégation dans unfluide par un modèle d'interaction phénoménologique. Second Int. Symp. Chem. React. Eng, Elsevier, New York, pp. 1−13.

Index

Note: Page numbers with "f" denote figures; "t" tables; "b" boxes.

A

Abrasion, mechanism of, 333–334, 333f
Adams–Bashforth method, 194, 242
Added mass force. *See* Virtual mass force
Advection term, 60, 183–184
  basic approximation of, 186–191
  for multifluid model, 177
Aerocolloidal system, 12–13
Agglomeration, 5, 12–13, 173–175, 332–333
Aggregation due to interparticle collision, 139–144
Aggregation kernel, 312
  due to Brownian motion induced collisions, 140–141
  due to differential sedimentation induced collisions, 142–142
  due to shear-induced collisions, 141–142
Aggregation/coalescence matrix, 222–223
Aggregation/coalescence processes, 82–84
Agitator turbulence, 299–300
Applications of population balance, 263–328
  gas–liquid flow. *See* Gas–liquid flow, population balance solutions to
  gas–particle flow. *See* Gas–particle flow, population balance solutions to
  liquid–liquid flow. *See* Liquid–liquid flow, population balance solutions to
  liquid–particle flow. *See* Liquid–particle flow, population balance solutions to
Aspect ratio, 175–176
Attractive force, 143–143, 163–164
Attrition, 332–336

B

Baffles, typical stirred tank with, 298–299, 298f
Basset history force, 45
180° bend geometry, 173–175, 173f, 174f, 175f
β-function, 224–225
Biconjugate Gradient Stabilized, 216–217
Biological systems, 329–332
Birth process, 70–71, 73
Boltzmann theory of gases, 12–13, 75–76
Boundary conditions
  Dirichlet, 63, 89
  for multiphase flow, 62–67
  Neumann, 63–66, 89
Breakage due to hydrodynamic stresses, 145–148
  breakage kernel of particles, 145–146
  fragment distribution function of particles, 146–148
Breakage functions, mechanism of, 333f

Breakage kernel, 312
Breakage/break-up processes, 80–82
Break-up of fluid particles, mechanisms and kernals of, 114–133
daughter particle size distribution, 128–133
due to interfacial stability, 128
due to shear-off, 128
due to turbulent shearing, 115–127
break-up efficiency, criterions for, 125–127
hitting eddy greater than a critical value, 119–123
hitting eddy greater than a smallest daughter particle interfacial force, 124–125
kinetic energy greater than a critical value, 116–117
surface volume fluctuation greater than a critical value, 118–119
due to viscous shear force, 127–128
Brownian dynamics, 234–235, 237–240
Brownian motion, 5, 240, 317–318
aggregation kernel due to, 140–141
Bubbly flow conditions and inlet boundary conditions, 274t
Bubbly flow regime, 264–265
Bulk attrition, 264–265, 332–334
Buoyancy-induced collision, collision frequency due to, 104

C

Capillary number, 127, 162–163
Cartesian coordinates, 172–173, 178
Cartesian gradients, 189–190
Cartesian grids, 170–171
Cell cycle dynamics of myeloma, 330–331
Central differencing scheme, 187
CFX Command Language, 288–289
Churn-turbulent flow, 264–265
Class method, 219–223
Clausius–Mossotti function, 165–165, 165–166
Coalescence, 3, 5, 7–11, 10f, 82–84, 329
and break-up rates of bubble, 274, 275t, 288–289, 289t
efficiency of, 93–94
due to critical approach velocity model, 113–114
due to energy model, 112–113
due to film drainage model, 105–112, 105f, 106f, 107f
of fluid particles, 92–114
buoyancy-induced collision, collision frequency due to, 104
capture in turbulent eddy, collision frequency due to, 104
for multigroup bubbles, 133–136, 134f
for one-group bubbles, 133–136
random collision, collision frequency due to, 94–98, 95f
turbulent fluctuation, collision frequency due to, 94–98, 95f
for two-group bubbles, 133–136, 135f
velocity gradient-induced collisions, collision frequency due to, 103–104
wake entrainment, collision frequency due to, 98–103, 99f, 101f, 102f
Collision efficiency of particles, 142–144
Collision frequency, 93
due to buoyancy-induced collision, 104
due to capture in turbulent eddy, 104
due to random collision, 94–98, 95f

due to turbulent fluctuation, 94–98, 95f
due to velocity gradient-induced collisions, 103–104
due to wake entrainment, 98–103, 99f, 101f, 102f
Computational fluid dynamics (CFD), 301, 321
Computational multiphase fluid dynamics, 17–68
boundary conditions, 62–67
differential form of transport equations, 59–62
discrete element framework, Lagrangian description on, 43–59
equations of motion, 43–44
fluid–particle interaction, 44–48
particle–particle interaction, 49–59
Eulerian formulation based on interpenetrating media framework, 17–42
effective conservation equations, 37–42
energy conservation, 27–34, 28f, 30f
interfacial exchange terms, physical description of, 34–37, 35f
mass conservation, 19–22, 19f, 21f
momentum conservation, 23–27, 24f
generic form of transport equations, 59–62
integral form of transport equations, 59–62
and population balance approach, coupling between, 87–89
Conservation of energy, 27–34, 28f, 30f, 245
effectiveness of, 37–42
interfacial characteristics of, 34–37, 35f
Conservation of mass, 19–22, 19f, 21f, 245
defined, 19
effectiveness of, 37–42
interfacial characteristics of, 34–37, 35f
Conservation of momentum, 23–27, 24f, 245
effectiveness of, 37–42
interfacial characteristics of, 34–37, 35f
Conservative potentials, 240–241
Constant coefficient subgrid scale model, 255–256
Continuous phase vector, 72
Continuous potentials, normal force due to, 151–152
Control volume integration, 177
Coordinate system, 70–71, 71f
Coupled equation system, 217–218
Critical approach velocity model, coalescence efficiency due to, 94, 113–114
Crossing trajectory effect, 53
Crystallization, 334–336
Cunningham correction to Stokes' drag law, 47
Cunningham slip correction factor, 141–141

D

Damping function, 256
Daughter particle size distribution, 128–133
Death process, 70–71, 73
Deformable particles, coalition efficiency of
with fully mobile interfaces, 109–110
with immobile interfaces, 107–108
with partially mobile interfaces, 108–109
Dense gas–particle flows, 1–2
Density function, 70–71
Differential form, of multiphase flow transport equations, 59–62
Differential sedimentation induced collisions, aggregation kernel due to, 142–142

Diffusion term, basic approximation of, 184–186
Dilute gas–particle flows, 1–2
Dimensionless flow values, 301
Direct numerical simulation (DNS), 58–59
Dirichlet boundary condition, 63, 89
Discrete element framework, Lagrangian description on, 43–59
  equations of motion, 43–44
  fluid–particle interaction, 44–48
  particle–particle interaction, 49–59
Discrete element method, 5, 234–235, 240–243. *See also* Soft-sphere model
Discretized equations, algebraic form of, 194–199
Dispersed bubbly flows in vertical pipes, 271–277
  numerical features, 273–274
  numerical results, 276–277
  numerical simulate, 274–276
Dispersed phase flows, 1
  complexity of, 2–4
Dissipation force, 152–152
DMT theory, 160–162
Donor-cell concept, 187
Drag curve, 266–267
Drag force, 266–268, 266f, 310
Droplet size distributions, 306–307
Dust fouling, 137–138
Dynamic subgrid scale model, 257–260

E

Eddy lifetime, 53–56
Eddy viscosity, 244
Electric field vector, 165–166
Energy conservation. *See* Conservation of energy
Energy model, coalescence efficiency due to, 94, 112–113
Enhanced wall treatment, 251–252
Ensemble-averaging, 18, 88–89
Enthalpy equation, 60
Eotvos number, 267, 269–270, 273–274
Equations of motion, 43–44
Eulerian fluid length scale, 57–58
Eulerian formulation of multiphase flows, 17–42
  effective conservation equations, 37–42
  energy conservation, 27–34, 28f, 30f
  interfacial exchange terms, physical description of, 34–37, 35f
  mass conservation, 19–22, 19f, 21f
  momentum conservation, 23–27, 24f
Eulerian models, solution methods for, 170–172
Eulerian reference frame, 169
Eulerian–Eulerian approach. *See* Discrete element framework, Lagrangian description on
Eulerian–Lagrangian approach, 43
Explicit Adams–Bashford method, 192
Explicit method, 192

F

False numerical diffusion, 187–188
Farve-averaged variables, 88–89, 273–274
Field–particle interaction, 164–167
Film drainage model, coalescence efficiency due to, 93–94, 105–112, 105f, 106f, 107f
  compressing force, 110–112
  contact time, 110–112

deformable particles
with fully mobile interfaces, 109–110
with immobile interfaces, 107–108
with partially mobile interfaces, 108–109
nondeformable rigid particles, 106–107
Finite difference method, 170–171
Finite element method, 171
Finite volume method, 169, 171, 177–184
First-order accurate explicit method, 241–242
Fluid force, 5–7
Fluidization, different states of, 309–310, 309f
Fluidized bed geometry, 312–313, 313f
Fluidized bed reactors, 308–309
Fluid–particle interaction, 44–48
Force
due to continuous potentials, 151–152
due to hysteretic, 154–156
due to linear viscoelastic, 153–153
due to nonlinear viscoelastic, 153–154
sliding, twisting and rolling resistance, 158–159
tangential, 156–158
Fourier's law of heat condition, 64–65
Fourth-order Runge Kutta method, 194
Four-way coupling problem, 48, 48f
Fracture, mechanism of, 333f
Fragment distribution function of particles, 146–148
Fredholm's integral equation, 259
Frenkiel function, 56–57
Front tacking method, 7
Full Approximation Scheme multigrid, 214
Fully explicit method, 171
Fully implicit method, 171, 192
Future of population balance approach, 329–338
bulk attrition, 332–334
crystallization, 334–336
emerging areas on the use of, 329–337
natural and biological systems, 329–332

G
G1 phase, 331–332
γ-function, 224–225
Gas–liquid flow, population balance solutions to, 264–298
background, 264
dispersed bubbly flows in vertical pipes, 271–277
numerical features, 273–274
numerical results, 276–277
numerical simulate, 274–276
experimental data
of Hibiki et al., 272f, 279–285
of Liu and Bankoff, 272f, 277–279
modeling interfacial momentum transfer, 264–271
drag force, 266–268, 266f
lift force, 266f, 269–270
turbulent dispersion force, 266f, 268–269
virtual mass force, 270–271
wall lubrication force, 266f, 270
transition cap-bubbly flows in vertical pipe, 285–298
numerical features, 288–289
numerical results, 289–295
Gas–liquid flows, 1
examples of, 3t
Gas–particle flow, population balance solutions to, 308–317
background, 308–310
via direct quadrature method of moment multifluid model, 310–312

Gas–particle flow, population balance solutions to (*Continued*)
worked example, 312–317
numerical features, 312–313
numerical results, 313–316
Gas–particle flows, 1–2
examples of, 3t
Gas–particle liquid–particle flows, mechanistic models for, 137–168
aggregation due to interparticle collision, 139–144
aggregation kernel
due to Brownian motion induced collisions, 140–141
due to differential sedimentation induced collisions, 142–142
due to shear-induced collisions, 141–142
collision efficiency of particles, 142–144
soft-sphere model, 150–167
particle–particle interaction due to adhesion, 159–167
particle–particle interaction without adhesion, 151–159
solid particle aggregation, mechanisms and kernel models of, 138–144
solid particle breakage, mechanisms and kernel models of, 144–150
breakage kernel of particles, 145–146
fragment distribution function of particles, 146–148
main breakage mechanism, 148–150
Gauss rule, 22, 26
Gauss' divergence theorem, 62, 73
Gaussian function, 224–225
Gaussian quadrature closure, 228
Gauss–Siedel method, 212
Generalized Minimal Residual (GMRES) method, 214
Generic form, of multiphase flow transport equations, 59–62
Granular flows, 1–2
Grid gene, 172
Group 2 bubbles, 291

H

Half-normal function, 225
Hard-sphere model, 49–52, 50f, 241
Hen Egg White (HEW) lysozyme, 334–335, 335f
Heterogeneous turbulent flow structure in a stirred tank
worked example, 303–308
numerical features, 303–304
numerical results, 304–307
Hybrid differencing scheme, 188
Hybrid mesh, 173–175
for 180° bend geometry, 174f
Hydrodynamic stresses, breakage due to, 145–148
breakage kernel of particles, 145–146
fragment distribution function of particles, 146–148
Hydrogen-abstraction-carbon-addition mechanism, 148–150
Hysteretic, normal force due to, 154–156

I

Integral form, of multiphase flow transport equations, 59–62
Interfacial area concentration (IAC) profiles, 283
Interfacial attractive, 163–164
Interfacial momentum exchange in isothermal bubbly flow, 273–274

Interfacial stability,
break-up due to, 128
Interparticle collision,
aggregation due to,
139–144
Interphase mass, 13
Interphase slip algorithm
(IPSA), 171–172,
203–207
Interphase slip algorithm-
coupled (IPSA-C),
171–172, 207–210

J
Jacobi method, 212
Jacobian Matrix
Transformation,
228–229
JKR theory, 160–160,
161–162

K
Kernel
aggregation, 312
breakage, 312
Knudsen number,
140–140, 141–141
Kolmogorov length scale,
58–59
Kolmogorov scale of
turbulence, 47,
95–96, 244
Krylov subspace, 214–216
K–ε model, 273

L
Lagrange interpolation,
226–228
Lagrangian models,
solution methods for,
234–243
Brownian dynamics,
238–240
discrete element method,
240–243
molecular dynamics,
235–238
Lagrangian reference
frame, 169–170
Langevin equation, 238
Large eddies, 258–259
Large Eddy Simulation
(LES), 58–59, 170,
253–260
constant coefficient
subgrid scale model,
255–256
dynamic subgrid scale
model, 257–260
Leap-frog algorithm, 237,
242
Least-squares method, 233
Leibnitz rule, 22, 26
Lennard-Jones potential,
235–236, 240–241
Lift force, 266f, 269–270
Limestone, 332–333
Linear viscoelastic, normal
force due to, 153–153
Liquid bridging, 162–163
Liquid–liquid flow,
population balance
solutions to, 298–308
background, 298–299
multiblock model,
299–302
worked example, 303–308
numerical features,
303–304
numerical results,
304–307
Liquid–liquid flows, 1
examples of, 3t
Liquid–particle flow,
population balance
solutions to, 317–326
background, 317–319
via quadrature method of
moment, 319–321
worked example,
321–326
numerical features,
321–322
numerical results,
322–326
Liquid–particle flows, 2
examples of, 3t
Log-law, 251–252
Log-normal function, 225

M
Mach number, 41
Magnus force, 46
Markov-chain model,
56–57
Markovian approximation,
56–57
Mass conservation. *See*
Conservation of mass
Mass ratio, 147–147
Mass residual, 200
Mass-loading ratio, 47
Matrix solvers, 210–217
Mesh distribution of the
computational
models, 274–276,
276f
Mesh systems, 172–177
Method of moments, 226
with interpolative closure,
234

Method of moments with interpolative closure (MOMIC), 337
Molecular dynamics, 234–238
Momentum conservation. *See* Conservation of momentum
Monte Carlo simulation, 53–54, 218–219
Multiblock model for heterogeneous turbulent flow structure in a stirred tank, 299–302
Multifluid model, quadrature method of. *See* Quadrature method of moment
Multigrid methods, 212, 213f, 214, 214f
Multigroup multibubble class model, 133–134, 134f
Multiphase flows, turbulence modeling for. *See* Turbulence modeling, for multiphase flows
Multiphase flows
- application of, 1–2
- classification of, 1–2
- complexity of, 2–4
- examples of, 3t
- multiscale characteristics of, 5–11, 5f
- population balance modeling for, need of, 12–13

Multiphase system, 329
Multiple-size-group (MUSIG) model, 219–220, 223, 274, 275t, 276–279, 282–283
Myeloma cell line, cell cycle of, 331f

N

Nanoparticle, synthesis of, 336–337
Natural systems, 329–332
Navier–Stokes equation, 7
N-dimensional vector, 320
Near-wall modeling, 250–253
Net generation of particles, 84–85
Neumann boundary condition, 63–66, 89
Newton's Law of viscosity, 39–40
Newton's second law of motion, 23, 43
Nondeformable rigid particles, coalition efficiency of, 106–107
Nondrag forces, 269
Nonequilibrium wall functions, 251–252
Nonideal stirred tank, 302f
Nonlinear viscoelastic, normal force due to, 153–154
No-slip condition, 63–64
Numerical discretization, 177–199
- advection term, basic approximation of, 186–191
- algebraic form of, 194–199
- diffusion term, basic approximation of, 184–186
- finite volume method, 177–184
- time-advancing solutions, basic approximation of, 191–194

Numerical quadrature, 228–233
Numerical solvers, 199–218
- coupled equation system, 217–218
- matrix solvers, 210–217
- segregated approach
  - interphase slip algorithm (IPSA) for, 203–207
  - interphase slip algorithm-coupled (IPSA-C) for, 203, 207–210
  - iterative calculations for, 199–203

O

One-group approach, 87
One-way coupling problem, 48, 48f

P

Partial Elimination Algorithm, 171
Partially implicit method, 171, 199

Partially-latching-linear-spring model, 154–154
Particle adhesion effects, 160f–160f
Particle attrition, 332–333
Particle integral time scale, 53
Particle Lagrangian time scale, 53
Particle phase space, 70
Particle relaxation time, 53
Particle size distribution, 329, 332–333, 336
Particle state continuum, 72
  infinitesimal control volume of, 75f
Particle state vector, 71–72
  rate of change of, 72
Particle transport and adhesion
  problem of, 164–165
  under the action of magnetic fields, 166–167
Particle–particle interactions, 49, 58–59
  due to adhesion, 159–167
    field–particle interaction, 164–167
    interfacial attractive, 163–164
    liquid bridging, 162–163
    van der Waals force, 159–162
  hard-sphere model, 49–52, 50f
  modes of, 152f–152f
  in turbulent flows, 47–48
  turbulent transport of particle, 52–59, 52f
  without adhesion, 151–159
    normal force due to continuous potentials, 151–152
    normal force due to hysteretic, 154–156
    normal force due to linear viscoelastic, 153–153
    normal force due to nonlinear viscoelastic, 153–154
    sliding, twisting and rolling resistance, 158–159
    tangential force, 156–158
Particles, net generation of, 84–85
Polyaromatic hydrocarbons (PAH), 148–150
Polyhedral mesh, 173–175, 175f
Poly-olefin reactors, 309–310
Population balance approach, 69–90, 138–138
  aggregation process, 82–84
  and computational multiphase fluid dynamics, coupling between, 87–89
  basic consideration for, 73–76
  breakage/break-up process, 80–82
  coalescence process, 82–84
  defined, 69–70
  particles, net generation of, 84–85
  practical considerations for, 85–87
  transport equations, integrated forms of, 77–80
Population balance equation, solution methods for, 218–234
  class method, 219–223
  numerical quadrature, 228–233
  standard method of moments, 223–228
Population balance modeling, 329
  for multiphase flows, need of, 12–13
Power-law breakage kernel, 146–146
Power-law differencing scheme, 188
Predictor-corrector method, 242
Pressure gradient force, 46
Probability density function (PDF), 337
Projected fields (PF) method, 337

Q

Quadrature method of moment, 228–230, 229f
  modeling gas–particle flow via, 310–312
  modeling liquid–particle flow via, 319–321

R
Random collision, collision frequency due to, 94–98, 95f
Reduced gravity force, 46
Relative twisting rate, 158–158
Relative velocity, 40–41
Reynolds flux, 40–41, 244
Reynolds number, 45–46, 58–59
Reynolds rule, 22, 26
Reynolds stress model, 40–41, 244–245, 247
Reynolds theorem, 73
Reynolds-averaged equations and closure, 244–253
  near-wall treatment, 250–253
  shear stress transport (SST) model, 247–250
  two-equation k-ε model, 246–247
Reynolds-averaged Navier–Stokes (RANS) framework, 58, 170, 261
Rhie–Chow interpolation expressions, 200
Rolling resistance, 158–159
Runge–Kutta method, 236, 243

S
S phase, 331–332
S to G2M transition, 331
Saffman lift force, 45–46
Sauter bubble diameter distributions, 282–283
Sauter drop diameters, 306f, 307, 308f
Second order upwind differencing scheme, 189
Second-order approximations, 192
Second-order explicit Adams–Bashford method, 193, 242
Second-order fully implicit method, 193
Semi-implicit Crank–Nicolson method, 192
Semi-implicit simultaneous solution of nonlinearly coupled equations (SINCE), 171, 198
Separated phase flows, 1
Shear stress transport (SST) model, 247–250, 273
Shear-induced collisions, aggregation kernel due to, 141–142
Shear-off, break-up due to, 128
SIMPLE (Semi-Implicit for Method Pressure-Linkage Equations), 203
Skewness, 176–177
Sliding resistance, 158–159
  tangential force for, 156–157
Slip velocity. *See* Relative velocity
Slug pattern, 264–265
Slurry flows. *See* Liquid–particle flows
Smagorinsky–Lilly model, 255–256
Small eddies, 258–259
Soft-sphere model, 241
  particle–particle interaction due to adhesion, 159–167
    field–particle interaction, 164–167
    interfacial attractive, 163–164
    liquid bridging, 162–163
    van der Waals force, 159–162
  particle–particle interaction without adhesion, 151–159
    normal force due to continuous potentials, 151–152
    normal force due to hysteretic, 154–156
    normal force due to linear viscoelastic, 153–153
    normal force due to nonlinear viscoelastic, 153–154
    sliding, twisting and rolling resistance, 158–159
    tangential force, 156–158

Solid particle aggregation, mechanisms and kernel models of, 138–144
Solid particle breakage, mechanisms and kernel models of, 144–150
  breakage due to hydrodynamic stresses, 145–148
    breakage kernel of particles, 145–146
    fragment distribution function of particles, 146–148
  main breakage mechanism, 148–150
Solution methods, 169–170
  for Eulerian models, 170–172
  for Lagrangian models, 234–243
    Brownian dynamics, 238–240
    discrete element method, 240–243
    molecular dynamics, 235–238
  mesh systems, 172–177
  for population balance equation, 218–234
    class method, 219–223
    numerical quadrature, 228–233
    standard method of moments, 223–228
  numerical discretization, 177–199
    advection term, basic approximation of, 186–191
    algebraic form of, 194–199
    diffusion term, basic approximation of, 184–186
    finite volume method, 177–184
    time-advancing solutions, basic approximation of, 191–194
  numerical solvers, 199–218
    coupled equation system, 217–218
    interphase slip algorithm (IPSA), 171–172, 203–207
    interphase slip algorithm-coupled (IPSA-C), 171–172, 207–210
    iterative calculations for the segregated approach, 199–203
    matrix solvers, 210–217
Soot production, 148–150
Space average, 18
Spectral method, 171
Standard *k*-*ε* model, 246–247
Standard method of moments, 223–228
Stefan–Boltzmann constant, 47
Stochastic model, 53–54, 56–58
Stokes number, 44
Stokes' drag law, Cunningham correction to, 47
Structured mesh, 172
  for 180° bend geometry, 173f
Swarm velocity, 100
Symmetric fragmentation, 147–147

T

Tabor parameter, 161–162
Tangential force, 156–158
  sliding resistance, 156–157
Tangential spring stiffness, 158–158
Taylor bubbles, 3, 264–265
Taylor series expansions, 170–171
Taylor vortices, 318–319
Taylor-Couette flow, 317–318, 318f, 321–322, 321f
Terminal velocity, 303–304
Thermophoresis, phenomenon of, 167–167
Third-order Quadratic Upstream Interpolation for Convective Kinetics (QUICK) scheme, 189
Third-order Runge Kutta method, 194
Three-phase flows, 2
  examples of, 3t

Three-stage population balance model, 331–332
Time average, 18
Time-advancing solutions, basic approximation of, 191–194
Titanium tetraisopropoxide (TTIP), 336–337
TOPFLOW experimental setup, 285–286, 286f
Trajectory models, 61t
Transition cap-bubbly flows in vertical pipe, 285–298
  numerical features, 288–289
  numerical results, 289–295
Transport equations
  differential form of, 59–62
  Favre-averaged form of, 44
  general form of governing
    multifluid model, 61t
    trajectory model, 61t
  generic form of, 59–62
  integral form of, 59–62
  for multiphase flow, 59–62
    enthalpy, 60
    mass, 59
    x-momentum, 59
    y-momentum, 59
    z-momentum, 59
Transport equations
  integrated forms of, 77–80
Turbulence, 298–300
Turbulence modeling, for multiphase flows, 170, 244–260
Large Eddy Simulation (LES), 253–260
  constant coefficient subgrid scale model, 255–256
  dynamic subgrid scale model, 257–260
Reynolds-averaged equations and closure, 244–253
  near-wall treatment, 250–253
  shear stress transport (SST) model, 247–250
  two-equation k-ε model, 246–247
Turbulent dispersion force, 266f, 268–269
Turbulent dissipation, 307
Turbulent dissipation rate, 300
Turbulent fluctuation, collision frequency due to, 94–98, 95f
Turbulent fluidization, 308–309
Turbulent Prandtl number, 244–245
Turbulent shearing, break-up due to, 115–127
  break-up efficiency, criterions for, 125–127
  hitting eddy greater than a critical value, 119–123
  hitting eddy greater than a smallest daughter particle interfacial force, 124–125
  kinetic energy greater than a critical value, 116–117
  surface volume fluctuation greater than a critical value, 118–119
Turbulent transport of particle, 52–59, 52f
Turbulent viscosity, 244
Twisting resistance, 158–159
Two-equation k-ε model, 54–55, 65–66, 246–248
Two-group approach, 87
Two-way coupling problem, 48, 48f

U
Unloading stiffness, 155–156
Unstructured mesh, 172
  for 180° bend geometry, 173f
Upwind concept, 187
Upwind scheme, 187–188

V
Van der Waals forces, 5, 8f, 9f
Velocity gradient-induced collisions, 103–104
Velocity vector, 204–206
Velocity-Verlet algorithms, 237
Verlet algorithm, 236

Vertical pipes
  dispersed bubbly flows in, 271–277
    numerical features, 273–274
    numerical results, 276–277
    numerical simulate, 274–276
  flow regimes for air/water in, 264–265, 265f
  transition cap-bubbly flows in, 285–298
    numerical features, 288–289
    numerical results, 289–295
Virtual mass force, 270–271
Viscous shear force, break-up due to, 127–128
Volume-averaging, 88–89

W

Wake entrainment, collision frequency due to, 98–103, 99f, 101f, 102f
Wall lubrication force, 266f, 270
Wall peak behavior, 277–278
Warp angle, 176–177
Weber number, 115–116, 120

Printed and bound by CPI Group (UK) Ltd, Croydon, CR0 4YY
08/05/2025
01864786-0004